A Reference Guide to Endocytosis

A Reference Guide to Endocytosis

Edited by **Seth Cole**

Callisto Reference

New York

Published by Callisto Reference,
106 Park Avenue, Suite 200,
New York, NY 10016, USA
www.callistoreference.com

A Reference Guide to Endocytosis
Edited by Seth Cole

International Standard Book Number: 978-1-63239-011-0 (Hardback)

Printed in China.

Contents

Preface

The world is advancing at a fast pace like never before. Therefore, the need is to keep up with the latest developments. This book was an idea that came to fruition when the specialists in the area realized the need to coordinate together and document essential themes in the subject. That's when I was requested to be the editor. Editing this book has been an honour as it brings together diverse authors researching on different streams of the field. The book collates essential materials contributed by veterans in the area which can be utilized by students and researchers alike.

This book is a valuable resource which provides essential information on endocytosis. It is a collection of systematic theories and researches regarding the entrance of exterior substances into cells. As discussed in this book, endocytosis adjusts various procedures such as homeostasis of the cell, sign transduction, and entrance of viruses. In addition to the investigational methods analyzed in every chapter, the complete book is devoted to investigational approaches that will be useful for all the researchers dealing with this subject. For those more technically oriented, certain chapters look at the potential and utilization of endocytic pathways for therapeutic principles.

Each chapter is a sole-standing publication that reflects each author´s interpretation. Thus, the book displays a multi-facetted picture of our current understanding of application, resources and aspects of the field. I would like to thank the contributors of this book and my family for their endless support.

Editor

Section 1

Regulation of Tissue Homeostasis by the Endocytic Pathway

Chapter 1

The Role of Endocytosis in the Creation of the Cortical Division Zone in Plants

Ichirou Karahara, L. Andrew Staehelin and Yoshinobu Mineyuki

Additional information is available at the end of the chapter

1. Introduction

Control of the pattern of cell division is essential for the proper development of multi-cellular organisms. In animal cells, cytokinesis is mediated by a contractile ring in which the cleavage force is produced by an acto-myosin system. Furthermore, the future site of cell division in animal cells (the site where contraction starts in the cell cortex) is determined by the position of the aster during the later stages of mitosis. In contrast, plant cytokinesis involves the assembly of a cell plate from Golgi-derived vesicles. The division site in plants (the cell cortex where the cell plate fuses with the parental cell walls) is defined by a band of cortical microtubules (MTs) – the preprophase band (PPB) of MTs – that mark the division site during prophase. The PPB MTs subsequently disassemble when the cells enter prometaphase. However, some positional information, or positional memory, is retained in the cell cortex/plasma membrane where the PPB of MTs was located, and the cell plate edge grows towards and fuses with this predetermined division site. Thus, how MTs demarcate the future division site during PPB development, and how the division site memory is created and maintained in the PPB region until the end of cytokinesis, are important questions related to the regulation of division plane positioning in plants.

Several potential cell division plane-positioning molecules have been identified, and these have been classified into "positive memory" and "negative memory" types of molecules. However, how these molecules contribute to the creation of positional memory information has yet to be determined. Early electron microscopists reported the presence of vesicles in the forming PPB regions, and it was suggested that these vesicles might contribute to the creation of a PPB memory site (cortical division zone) either via exocytosis or endocytosis. By using high-pressure freezing to preserve the cells for electron microscopical analysis, we have been able to demonstrate that the vesicles are generated by endocytosis, and with the help of electron tomography, we have been able to quantify the distribution of vesicles in the cell cortex. The latter analysis has demonstrated that clathrin-mediated endocytosis is enhanced in the PPB region compared to the cell cortex outside the PPB or in the cell cortex of interphase cells. Thus, creation of the cortical division zone appears to involve increased rates of clathrin-mediated endocytosis in the PPB region. Based on these results, we propose that removal of membrane proteins by endocytosis at the division site plays a critical role in the formation of PPB "memory" structures. In this chapter, we will discuss in greater detail how endocytosis at the future site of cell division contributes to the regulation of the plane of cell division in plants.

2. Creation and demarcation of the cortical division zone[1] in plants

The division site is defined as the region where a new division plane is inserted into a cell at the end of cell division (Gunning, 1982). Since the division plane in animal cells is inserted centripetally from the cell cortex using a contractile ring, the cortical division site corresponds to a region where the cleavage furrow is initiated in the cell cortex. In plant cells, cell plate formation starts with the accumulation of Golgi-derived, cell plate-forming vesicles in the midplane of the phragmoplast MT array in the central region of the cell (Seguí-Simarro et al., 2004). Upon fusion of these vesicles, the cell plate starts to grow centrifugally until it reaches and then fuses with the plasma membrane at the cell division site, the PPB memory site. In the majority of plant cells, the final division plane is inserted in the plane defined by the equatorial plane (the plane where metaphase chromosomes arrange) existed in metaphase, and where the cell plate is initiated at the beginning of cell plate formation. This is not always the case. Figure 1 shows the process of cell division in a *Tradescantia* stamen hair cell where the equatorial plane developed in an oblique orientation (Fig. 1b). Subsequently, however, the cell plate was inserted transversely (Fig. 1e). When the mitotic apparatus of a *Tradescantia* stamen hair cell is displaced experimentally towards the distal end of the cell by centrifugation, the initially formed cell plate develops between the displaced daughter nuclei, but then gradually extends towards the cortical site where it would have been inserted if there had been no centrifugal treatment (Ôta, 1961). This experiment clearly demonstrated that the plant division site is determined prior to the separation of the chromosomes, and that the memory site, where the cell plate fuses to the parental cell wall, is maintained during and after the centrifugal treatment. When and how this cortical division site is established, and how it influences the positioning of the cell plate during cytokinesis remains to be elucidated.

---

[1] Normally the PPB is a few micrometer wide and thus this cortical band region is broader than the exact site where the cell plate attaches. Because of this, Van Damme et al. (2011) have proposed the term cortical division zone to distinguish the cortical division site, the exact region where the cell plate attaces to the cell cortex. Here we use this term if we need to distinguish the former PPB region and the exact attachment site of the cell plate.

2.1. Proteins involved in preprophase band (PPB) formation and maturation

The most prominent structural change in the region of the future cortical division site is the assembly of a PPB during the G2 and prophase stages of the cell cycle. The PPB is a band of MTs associated with vesicles in the cell cortex (Mineyuki, 1999). Pickett-Heaps & Northcote (1966a, b) provided the first description of the PPB, but did not provide an answer to the question whether the PPB predicts the division site or the position of the equatorial plane in metaphase. This problem was solved in a study of the PPB in onion guard mother cells. Onion guard mother cells are relatively small cells and the equatorial plane in metaphase orients obliquely, but the cell plate is inserted longitudinally (Miehe, 1899). In these cells, the PPB orients longitudinally, thereby predicting the future division site and not the orientation of the equatorial plane (Palevitz & Hepler, 1974). Misorientation of the division planes occurs in cells in which the PPBs are prevented from forming by experimental manipulation (Mineyuki et al., 1991a), as well as in mutant cells that cannot form PPBs (Traas et al., 1995).

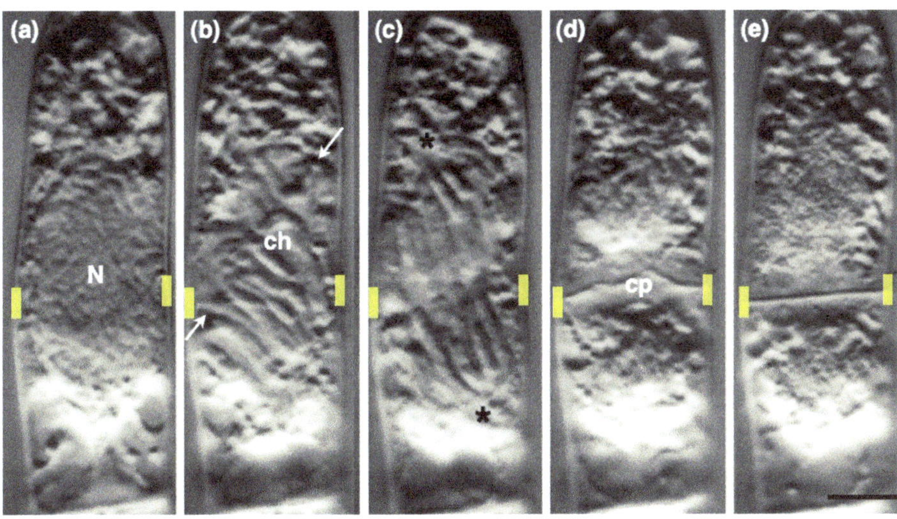

Figure 1. Cell division of a stamen hair cell of *Tradescanta virginiana*. (a) prophase, (b) metaphase, (c) anaphase, (d) just after the cell plate has reached the cell wall, (e) 18 min after (d), the cell plate becomes flatten. This cell is the same cell used in the experiment of Fig. 2 in Mineyuki & Gunning (1990). White arrows in (b) show the position of the equatorial plane. Stars (*) in (c) mark the spindle pole region. Rectangles colored yellow show the cortical division zone. N, nucleus; ch, chromosomes; cp, cell plate. Bar = 10 μm.

PPB MTs originate during the G2 phase in the form of a broad band (Fig. 2b), which narrows during prophase. The narrow MT band localizes to the region of the ultimate division site (Fig. 2c). This MT band disappears when the nucleus enters prometaphase but leaves behind positional information that aids in the subsequent orientation and function of the cell plate with the plasma membrane (Fig. 2d, e). Some MT associated proteins (MAPs) have been shown to be associated with PPBs, and studies on loss-of-function mutants have demonstrated that the MICROTUBULE ORGANIZATION 1 (MOR1) and CLIP-associated proteins (CLASP) are involved in the organization of the MTs in PPBs (Ambrose et al., 2007; Kawamura et al., 2006; Whittington et al., 2001). FASS/TONNEAU2 (TON2), a putative regulatory B" subunit of the Thr/Ser protein phosphatase 2A, and TONNEAU1 (TON1), a protein that interacts with centrin (CEN1) and is related to a human centrosomal protein, are also essential proteins for PPB formation (Camilleri et al., 2002; Traas et al., 1995). As discussed below in greater detail, actin plays a critical role in PPB formation, and the actin-depolymerising drug cytochalasin inhibits the narrowing of the MTs (Eleftheriou & Palevitz, 1992; Mineyuki & Palevitz, 1990).

Besides guiding the cell plate towards the cortical division zone, molecules associated with the PPB memory site also have the ability to induce cell plate flattening. For example, during cell division in *Tradescantia* stamen hair cells, the cell plate tends to be fluid and wrinkled (Fig. 1d) when the cell plate edges attach to the cortical division site, but flattens thereafter (Fig. 1e). The flattening process is delayed or stops when a cell plate fails to reach the correct cortical division zone (Mineyuki & Gunning, 1990).

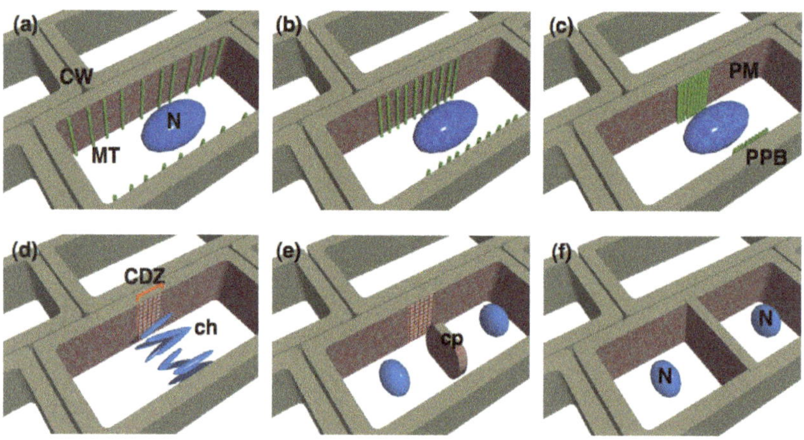

Figure 2. Schematic view of PPB development and the division plane insertion in plants. (a) interphase, (b) early PPB stage (G2~prophase), (c) late PPB stage (late prophase), (d) metaphase, (e) telophase, (f) after cell division. MT (green), microtubule; CW (light brown), cell wall; N (blue), nucleus; PM (pink), plasma membrane; CDZ (red), cortical division zone; ch, chromosome; cp, cell plate.

2.2. Candidate proteins of division site memory molecules

The PPB is considered to predict the future site of cell division in plant cells and to generate positional memory information that demarcates the cortical division zone after disappearance of the MTs. Several candidate molecules for the memory function of the cortical division zone have been described (Table 1). The first candidate molecule identified was actin. Although actin serves multiple functions during PPB development, actin filaments disappear from the PPB zone in late prophase, thereby generating an actin-depleted zone (ADZ) adjacent to the plasma membrane (Cleary et al., 1992; Liu & Palevitz, 1992). Since the ADZ remains after the disappearance of the MT band, the ADZ has been thought of as a kind of "negative memory". Another "negative memory" candidate is the kinesin-like molecule KCA1 (Vanstraelen et al., 2006), which also forms a KCA1 depleted zone (KDZ) in the same area as the ADZ. How these molecules are excluded from the cortical division zone, and how the ADZ and the KDZ are maintained during cell division remains to be determined.

Molecules such as TANGLED (TAN), a highly basic protein that can directly bind to MTs, and RanGAP1, a negative regulator of the small GTPase Ran, are accumulated in the PPB and remain there after the disappearance of the PPB MTs (Rasmussen et al., 2011; Walker et al., 2007; Xu et al., 2008). These are candidates of "positive memory" molecules. Together, the "positive" and "negative" memory molecules may be key players for guiding the cell plate to the predicted cortical division site. PHRAGMOPLAST ORIENTING KINESINs 1 and 2 (POK1, POK2), originally identified as potential partner of TAN in a yeast two-hybrid screen, are required for the correct localization of TAN and RanGAP1 to the PPB region (Müller et al., 2006), and the functional relationship between POK1/POK2 and TAN/RanGAP1 has been examined (Walker et al., 2007; Xu et al., 2008). TAN–interacting proteins DISCORDIA1 (DCD1) and ALTERNATIVE DISCORDIA1 (ADD1), maize homologs of *Arabidopsis* FASS/TON2, are two other proteins that persist at the cortical division zone after disappearance of the PPB MTs. Although DCD1/ADD1 are detectable in the cortical division zone of metaphase cells, they cannot be observed in the cortical division zone in anaphase (Wright et al., 2009).

Molecules, that appear in the cortical division zone just before the cell plate edges reach the plasma membrane, have also been identified. Adaptin-like protein, TPLATE and clathrin reappear in the cortical division zone when the cell plate edge almost reaches the cortical division site (Van Damme et al., 2006, 2011). Whether these molecules are associated with the edge region of the maturing cell plate (Seguí-Simarro et al., 2004) or with structures in the cortical division zone remains to be determined.

Based on the observation of cell plate flattening in *Tradescantia* stamen hair cells, Mineyuki and Gunning (1990) proposed that the PPB leaves behind factors involved in cell plate maturation. A MT-associated protein, AUXIN-INDUCED IN ROOT CULTURES 9 (AIR9) decorates the PPB and phragmoplast MTs and reappears at the cortical division site when the outwardly growing phragmoplast contacts the cortical division site. AIR9, then moves inward on the young cell plate to form a torus-like structure. When the cell plate is inserted outside the former PPB site no AIR9 torus is formed, suggesting that AIR9 associates with proteins that are retained in the PPB site. For this reason, AIR9 is viewed as a candidate factor involved in the regulation of cell plate maturation (Buschmann et al., 2006). A cell wall hydroxyproline-rich glycoprotein (Hall & Cannon, 2002) may also play a role in cell plate maturation.

Molecules that mark the cortical division zone after the disappearance of PPB MTs		Molecules that appear in the cortical zone at the end of cell plate insertion
Candidates of negative memories	Candidates of positive memories	
Actin (Liu & Palevitz, 1992; Cleary et al., 1992) KCA1 (Kinesin-like protein: Vanstraelen et al., 2006)	TAN (Protein having basic MT-binding domain of vertebrate APC proteins: Walker et al. 2007) RanGAP1 (RanGTPase activating protein: Xu et al., 2008) DCD1/ADD1 (Maize homologus of *Arabidosis* FASS/TON2: Wright et al., 2009)	TPLATE (Adaptin-like protein: Van Damme et al., 2006) Clathrin (Van Damme et al., 2011) AIR9 (MT associated protein: Buschmann et al., 2006) RSH (?) (Cell wall hydroxyproline-rich glycoprotein: Hall & Cannon, 2002)

Table 1. Candidates molecules for modifiers of the cortical division zone.

3. Electron tomography of high-pressure frozen cells

Electron tomography is a powerful method for visualizing and quantitatively analyzing the ultrastructural features of cells in three dimensions (Frank, 1992). In the context of PPBs, electron tomography has enabled us to obtain quantitative information on the organization of cortical and cell plate-associated MTs, and on the types and the distribution of vesicles in large volumes of cytoplasm in defined cellular domains (Austin et al., 2005; Karahara et al., 2009; Seguí-Simarro et al., 2004). This method is particularly effective when employed in conjunction with cryo-fixation, which preserves transient membrane systems much better than chemical fixation. We selected epidermal cells of onion cotyledons for the analysis of membrane structures associated with PPBs, because PPB development in this cell type is well characterized (Mineyuki et al., 1989). Most notably, in the basal region of the cotyledons, the percentage of cells undergoing mitosis is relatively high. Specimen preparation was carried out as described previously (Karahara et al., 2009). In short, a basal part of the cotyledon was cut and immediately frozen using a high pressure freezer. The high pressure-frozen samples were freeze-substituted and then embedded in Spurr's resin (Murata et al., 2002). Our electron micrographs of transverse sections of high-pressure frozen/freeze-substituted onion epidermal cells showed exceptionally-well preserved cells at the ultrastructural level (Fig. 3). Late prophase cells with a narrow PPB can be distinguished from interphase cells based on the staining pattern of the chromosomes. After the staining of 250 nm-thick tangential sections with uranyl acetate and Reynold's lead citrate, colloidal gold particles were added to both sides of the grid as fiducial markers to align the series of tilted images. These thick tangential sections of outer epidermal cell wall regions were mounted in a tilt-rotate specimen holder and observed using either a high-voltage electron microscope operating at 750 kV, or an intermediate-voltage electron microscope operating at 300 kV. The images were taken from +60° to -60° at 1.5° intervals about two orthogonal axes and collected with a digital camera attached to the electron microscopes. Tomograms were computed for each set of aligned tilts using the R-weighted back-projection algorithm. Tomograms were displayed and analyzed with Imod, the graphics component of the IMOD software package (Kremer et al., 1996).

The use of electron tomography has enabled us to identify, map and model the pits, vesicles and MTs of PPB regions in three dimension with a much higher degree of resolution than is possible with conventional ultra-thin sections obtained using an ultramicrotome (Figs. 3a, b & 4). The specimen preparation procedures employed in this study produced characteristic, high-contrast images of the triskelion complexes and lattices associated with the clathrin-coated pits, as well as of the contents of the vesicles. Although most vesicles in the cell cortex examined in the tomographic images were either dark-core, clathrin-coated vesicles (Fig. 3e) or dark-core, non-coated vesicles (Fig. 3h), we did observe some dark-core vesicles with partial coats (Fig. 3f, g). This indicates that the dark-core, non-coated vesicles could be derived from the dark-core, clathrin-coated vesicles.

To test this postulated relationship, we have also quantitatively analyzed the distance between the center of the darkly stained clathrin-coated and non-coated vesicles and the plasma membrane. If the darkly stained, non-clathrin-coated vesicles were derived from clathrin-coated, endocytic vesicles, then, on average, they should be found at a greater distance from the plasma membrane than the clathrin-coated ones. As illustrated in Fig. 5, in the cytoplasm underlying PPBs, the non-coated, darkly stained vesicles were found to be further away from the plasma membrane (74.4 ± 2.6 nm, mean ± SEM, n=168) than the clathrin-coated vesicles (52.8 ± 6.4 nm, mean ± SEM, n=29). This supports the idea that the non-coated, darkly stained vesicles were the uncoated form of the clathrin-coated vesicles on the way to endosomal compartments.

To confirm that clathrin molecules are present in the PPB, we examined the localization of clathrin in interphase and prophase cells of onion epidermal cells by immunofluorescent microscopy. The anti-clathrin heavy-chain antibodies cross-reacted with two types of intracellular structures in the onion epidermal cells, large, brightly stained objects and small, dim structures. The small, dim fluorescent structures seen in the confocal images correspond to the clathrin-coated pits and vesicles seen in the cell cortex of thin-sectioned cells (see Fig. 6 in Karahara et al., 2009). We have roughly quantified the frequency of the anti-clathrin stained, small, dim fluorescent dots observed in the PPBs of cells visualized by immunofluorescence microscopy (see Table S1 in Karahara et al., 2009). However, because the number of clathrin-containing dim fluorescent dots per square micron was smaller than the number of clathrin-coated pits and vesicles determined by electron tomography, one fluorescent dot may in some instances correspond to a cluster of several clathrin-coated pits and vesicles (Fig. 4a, circles of dashed blue lines).

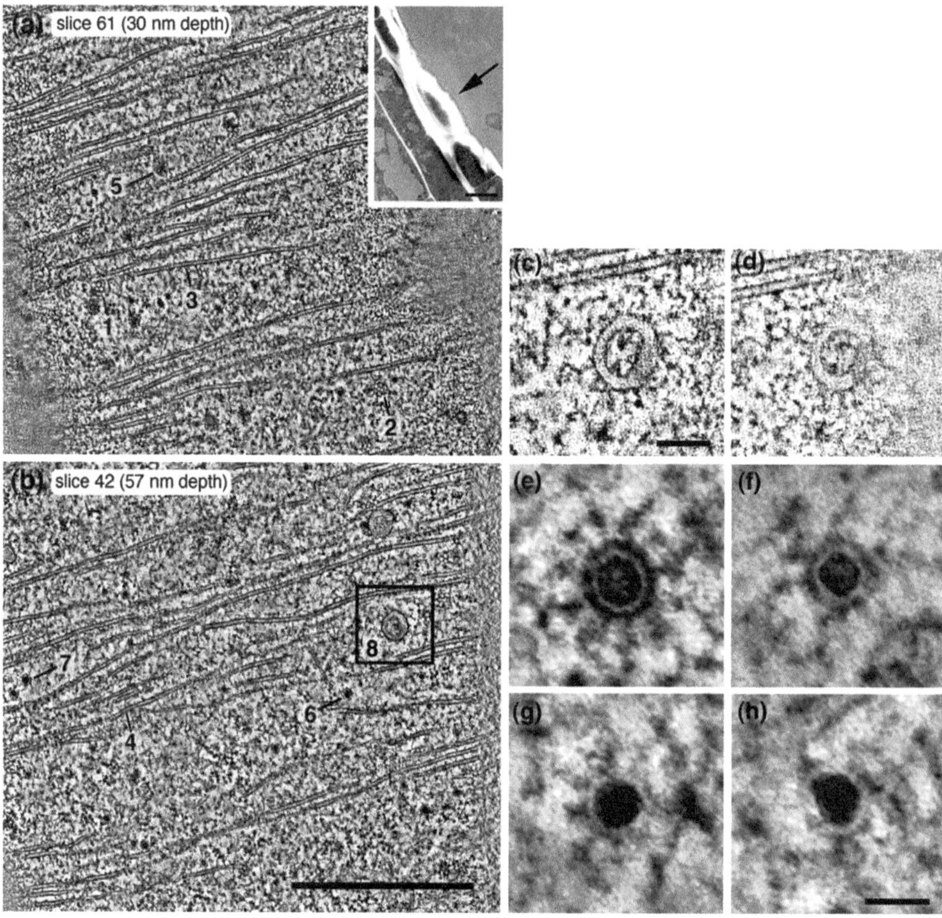

Figure 3. Tomographic images of a tangentially-sectioned PPB in a late prophase onion epidermal cell. The tomogram contained a total of 110 slices, with the higher slice numbers showing areas closer to the plasma membrane. Structures 1 and 2: Clathrin lattices associated with shallow pits. Structures 3 and 4: Two cortical MTs. Structure 5: A detached clathrin-coated vesicle. Structures 6 and 7: Partially uncoated and non-coated dark vesicles. Structure 8: Horseshoe-shaped plasma membrane infoldings. (a, b) Two images of 1.42-nm thick tomographic slices. Inset: overview electron micrograph of the 250 nm section used to make the tomogram. (c, d) Higher magnification tomographic slices images of a horseshoe-shaped plasma membrane infoldings shown in the black rectangular in (b). (c) (slice 54) and (d) (slice 60) are different sections through the same horseshoe structure framed in (b). (e-h) Gallery of 21.3-nm thick, composite tomographic slice images illustrating the morphological similarities between clathrin-coated (a), partially-coated (b, c) and non-coated (d) dense-core vesicles. Bars =(a, b) 1 μm, (c, d) 1 μm, (e-h) 50 nm and (inset Figure in (a)) 10 μm. Figure adapted from Karahara et al. (2009).

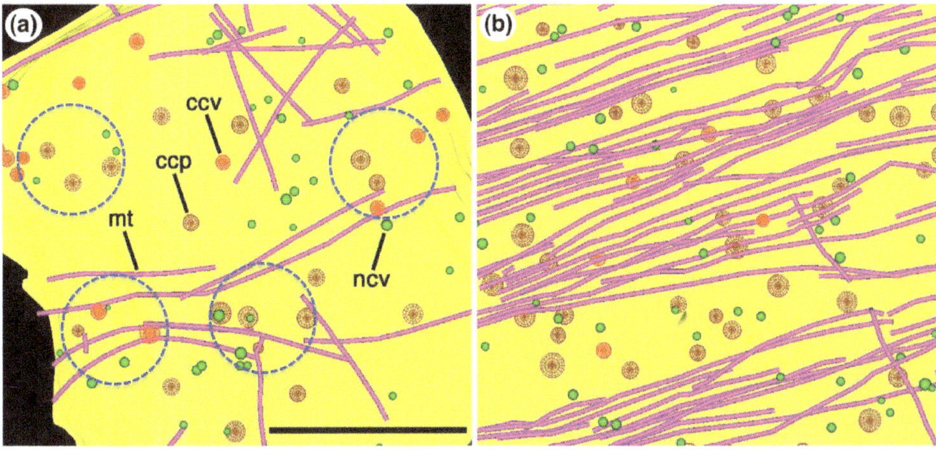

Figure 4. Tomography-based reconstructions of the cortical region at the nuclear level (i.e. PPB region) of an interphase cell (a), at the PPB region of a late prophase cell (b). ccv, clathrin coated vesicle (bright red); ccp, clathrin-coated pit (deep red); ncv, non-coated vesicle (green); mt, MT (purple); pm, plasma membrane (yellow). Clusters of several clathrin-coated pits and vesicles shown in circles of blue broken line, which may correspond to fluorescent dots seen in immunofluorescence photographs. Bar = 1 μm. Figure modified from Karahara et al. (2009).

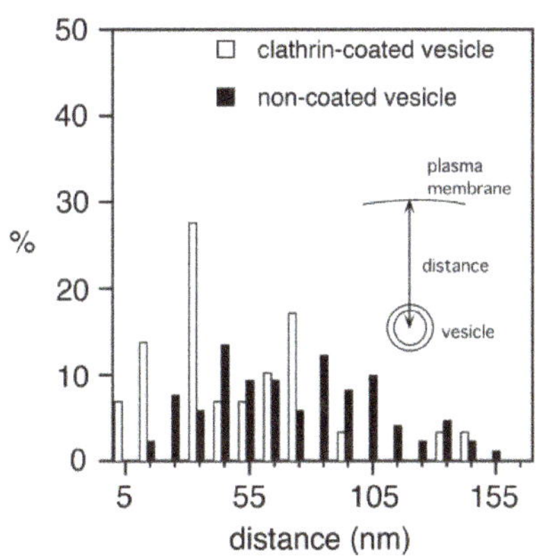

Figure 5. Histograms illustrating the distances between the center of the dark-core vesicles (clathrin-coated and non-coated) and the plasma membrane as measured in tomograms of late prophase cells. Open column, clathrin-coated vesicle; closed column, non-coated vesicle.

4. Endocytosis at the future site of cell division

Clathrin-mediated endocytosis is an attractive mechanism for locally changing the composition of the plasma membrane at the PPB site, because clathrin-coated pits are known to concentrate specific types of membrane molecules prior to budding from the plasma membrane (Bonifacino & Traub, 2003; Chen et al., 2011; Kirchhausen, 2000). The original term for endocytosis in plant cell was pinocytosis (Conner & Schmid, 2003), which included both clathrin-mediated and clathrin-independent endocytosis. The former one is considered to be the major pathway while the importance or even existence of the latter one is still being debated (Chen et al., 2011). Clathrin-mediated endosytosis in plants has been shown to involve molecules, such as adaptor proteins (Holstein, 2002; Takano et al., 2010; Van Damme et al., 2011), accessory adaptor proteins (Bar & Avni, 2009), dynamins (Bednarek & Backues, 2010), small GTPases (Naramoto et al., 2010), actin filaments (Bar et al., 2009; Lam et al., 2001), as well as post-translational protein modifications such as phosphorylation and ubiquitination (Chen et al., 2011). TPLATE, an adaptor-like protein, also appears to participate in endocytic activities associated with cell plate formation during cytokinesis (Van Damme et al., 2004, 2006, 2011).

4.1. Endocytic membrane structures in the PPB region

Quantitative analysis of the distribution of the clathrin-coated pits in the cortical region closest to the nucleus of late prophase and of interphase cells showed that the average frequency of clathrin-coated pits between the PPB (nuclear) and the non-PPB (extra nuclear) regions of the plasma membrane was reduced in the non-PPB domains (see Figure 4 in Karahara et al., 2009). Furthermore, the average frequencies of dark-core, clathrin-coated and non-coated vesicles in the cytoplasm underlying the external wall at the nuclear (PPB) and extra nuclear (non-PPB) levels of late prophase cells, and those at the nuclear level of interphase cells, demonstrated that in late prophase cells the frequency of the clathrin-coated vesicles underlying the PPB region was 3.7 fold higher than in the region outside the PPB. On the other hand, the frequency of the dark-core, clathrin-coated vesicles in the PPB region of late prophase cells was two-fold higher than in the interphase cells, and the frequency of dense-core, non-coated vesicles in late prophase cells was over three-fold higher compared to interphase cells (see Table 1 in Karahara et al., 2009). These data demonstrate that the PPB regions are sites of endocytic activity mediated by clathrin-coated vesicles.

To determine whether some of the dense core vesicles underlying the thicker, cuticle-covered outer cell wall of the epidermal cells could be secretory vesicles, we counted all of the vesicles with dark cores in the cortical cytoplasm underlying the inner and outer cell wall regions in serial thin sections of cells sectioned in the plane of their PPBs. No significant differences in the frequency of dark-core vesicles underlying inner and outer cell walls in the PPB region and in the non-PPB regions was observed. However, we did confirm the noted increase in vesicle frequency in the cytoplasm underlying the PPB, both adjacent to the thick outer and the thinner inner cell walls of the epidermal cells (see Table 2 in Karahara et al., 2009).

In plants, endocytosed vesicles have been shown to be transported via the trans Golgi network (TGN) to multivesicular bodies (MVBs), where both membrane and cargo molecules are sorted for recycling or for degradation in vacuoles (Haas et al., 2007; Kang et al., 2011; Reyes et al., 2011; Viotti et al., 2010). The onset of clathrin vesicle-mediated endocytosis from cell plates leads to a temporary increase in MVBs in apical meristem cells of *Arabidopsis thaliana* (Seguí-Simarro & Staehelin, 2006). However, in our study of onion epidermal cells we have observed few MVBs in the cortical cytoplasm (Fig. 6) and have been unable to detect any significant increase in MVBs in the PPB (Table 2). This suggests that the plasma membrane molecules endocytosed from the PPB membrane domains could be recycled back to the plasma membrane via the TGN and not transferred to the MVBs and vacuoles for degradation.

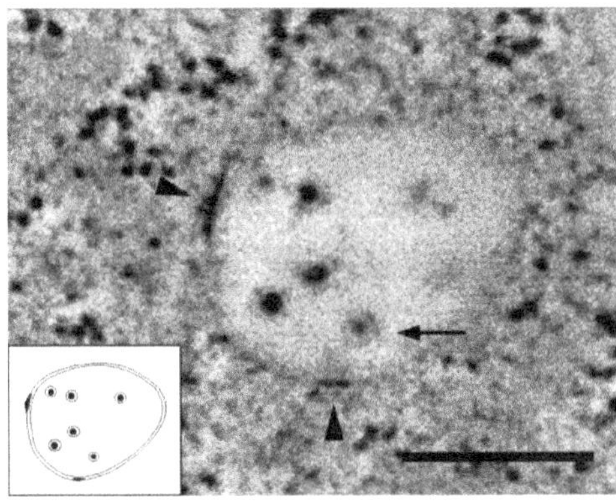

Figure 6. Electron micrograph of a thin sectioned multivesicular body (MVB) in a late prophase onion epidermal cell. The MVB contains intraluminal vesicles (arrow). Characteristic electron dense patches are seen on the surface of the MVB membrane (arrowheads). Schematic illustration depicting the MVB is shown (inset). Bar = 50 nm.

	interphase	late prophase	*P* (cell stage comparison)
nuclear level	0.16 ± 0.07	0.20 ± 0.10	0.59 (*z*=-0.53)
extra nuclear level	0.23 ± 0.06	0.11 ± 0.06	0.22 (*z*=-1.20)
P (level comparison)	1.00 (*z*=0.00)	0.77 (*z*=0.29)	

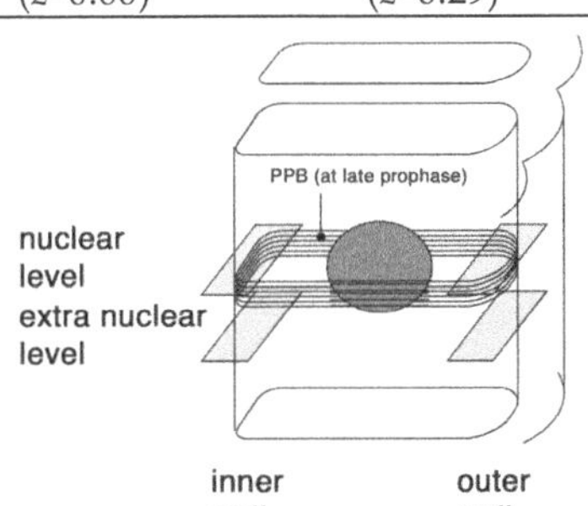

Table 2. Average frequency of MVBs observed in PPB (inner and outer cortical regions at nuclear level in late prophase cell) and in non-PPB cortical region (inner and outer cortical regions at extra nuclear level in late prophase cell and inner and outer cortical regions in interphase cell) determined from serially thin cross-sections of onion epidermal cells. The frequency of MVBs was expressed as numbers of MVBs per μm^3 (mean ± SEM, n=6). Thickness of each section was 70-90 nm. The Mann-Whitney *U*-test (two-tailed) was performed at each level. Schematic illustration depicting the plane of the sections is shown, which is adapted from Karahara et al. (2009).

Enhanced rates of endocytosis confined to PPB regions has also been observed in FM4-64 uptake studies in tobacco BY-2 cells (Dhonukshe et al., 2005). However, in our study, both the tomographic data and immunofluorescent microscopy with anti-clathrin antibodies clearly showed that the frequency of clathrin-bearing structures (clathrin-coated pits and vesicles) does not decrease abruptly at the edge of the PPB region but decreases gradually. Thus, our tomographic models demonstrate that a significant amount of the clathrin-bearing structures are also formed in the region adjacent to the PPB MTs (see Fig. 4 in Karahara et al. 2009). Based on this observation we have postulated that the formation of clathrin-coated pits is not tightly coupled to PPB MTs. Instead, the observed distribution of the MTs and of the endocytic vesicles in the PPB region can be better explained by the hypothesis that the local removal of selected molecules from the plasma membrane via endocytosis creates a membrane gradient in the PPB region that stimulates the assembly of MTs in that region. In this context, the function of the PPB MT array might be both to create a planar reference structure and an associated membrane domain in which the molecules involved in defining the division site can become organized. Therefore, the PPB region can be defined not only as a localized array of MTs but also as a localized region of clathrin-mediated endocytic activity. The fact that the PPB-associated $p34^{cdc2}$ kinase homolog (A-type cyclin-dependent kinase CDKA;1) forms a band that is narrower than the PPB (Mineyuki, 1999; Mineyuki et al., 1991b) is consistent with this idea.

4.2. Exocytic membrane structures in the PPB region

The outer tangential walls of epidermal cells are considerably thicker than the inner walls. In addition, the outer walls are covered by a cuticle. Since it is possible that there is a difference in secretory activity between the outer and the inner walls in epidermal cells, secretory activities were assessed inside and outside of the PPB region. When a secretory vesicle fuses with the plasma membrane of a plant cell, the vesicle collapses and forms a characteristic, horseshoe-shaped infolding (Staehelin & Chapman, 1987). These horseshoe-shaped membrane structures can be identified in cryofixed and freese-substituted cells and used as a diagnostic tool for assessing secretory activities (Fig. 3c and d). We have analyzed the distribution of horseshoe-shaped structures in serial thin-sectioned onion epidermal cells and have demonstrated that there was no significant difference between the frequency of horseshoe-shaped structures in the cell cortex at the nuclear level as well as at the extra-nuclear level in the late prophase cells (i.e. PPB region) and in the interphase cells (see Table 3 in Karahara et al., 2009). It has been reported that in 10% of tobacco BY-2 cells there is an increase in Golgi stacks underlying the PPB (Dixit & Cyr, 2002). To determine if onion epidermal cells also accumulate Golgi stacks in the cortical cytoplasm underlying the PPB, we have analyzed the distribution of Golgi stacks in our serial thin-sectioned cells and found that there was no significant difference between the frequency of Golgi stacks in the cell cortex at the nuclear and the extra-nuclear level in late prophase cells and interphase cells (see Table 4 in Karahara et al., 2009).

In a recent study of tobacco BY-2 cells, Toyooka et al. (2009) have described what they claimed was a new exocytic structure, and which they called secretory vesicle cluster.

However, as demonstrated in a recent electron tomography study, the so-called secretory vesicle clusters are free TGN cisternae that release their vesicles by means of cisternal fragmentation prior to the fusion of the individual secretory vesicles with the plasma membrane (Kang et al., 2011).

By quantifying the frequency of secretory structures we have demonstrated that the number of secretory events inside and outside of the PPB is essentially the same and that at this stage of the cell cycle the number of secretory events is low. The paucity of secretory structures observed in the PPB region is also consistent with the conclusion of Dixit and Cyr (2002) that Golgi secretion is not required for marking the PPB site.

4.3. Role of endocytosis in the establishment of the cortical division zone

The discovery that PPB formation involves increased rates of endocytosis at the PPB zone leads to the question as to what types of plasma membrane molecules could be selectively retrieved from this zone by means of the clathrin-coated vesicles. If molecules, that are necessary for the attachment of actin filaments or KCA1 molecules to the plasma membrane were selectively removed by endocytosis, then this could lead to the formation of actin or KCA1 depleted zones. One class of candidate proteins might be the plasma membrane-associated, actin filament-nucleating proteins called formin homology (FH) proteins (Banno & Chua, 2000; Favery et al., 2004). Several plant formins have been shown to have the ability to nucleate actin filaments, and overexpression of AtFH1 induces the formation of arrays of actin cables that project into the cytoplasm from the plasma membrane (Cheung & Wu, 2004). Thus, one possible function of the enhanced endocytic activity at forming PPBs might be the retrieval of actin-nucleating/binding proteins from these plasma membrane domains to create an actin-free zone to which the expanding cell plate is guided and where it can fuse. A similar function for the removal of KCA1 can also be envisaged. Together, our data suggest a mechanism for how endocytosis could help create a "negative memory" structure in the PPB region of preprophase cells.

5. Effects of brefeldin A on the formation of clathrin-coated membrane vesicles at the future division site

It is known that brefeldin A (BFA) interferes with the functioning of Arf proteins, which are important both for the assembly of COPI as well as for clathrin-coated vesicles that are formed both on TGN cisternae and at the plasma membrane (Nebenführ et al., 2002). To determine if BFA can also inhibit the endocytosis events associated with PPB formation, we examined the effects of BFA on the formation of clathrin-coated pits and vesicles as well as dark core vesicles in the PPB regions of epidermal cells. For the BFA treatment, we made a stock solution in methanol and diluted it in an aqueous solution of 0.1 M sucrose to achieve an effective working concentration of 100 μM BFA. The control solution contained 0.2% (v/v) methanol and 0.1 M sucrose. Onion seedlings were treated with the solution for 20 minutes before high-pressure freezing. To evaluate the responses of Golgi stacks to BFA treatment, three separate tissue regions (each including 5-6 cells) were selected, and the numbers of normal and BFA-perturbed Golgi stacks were determined.

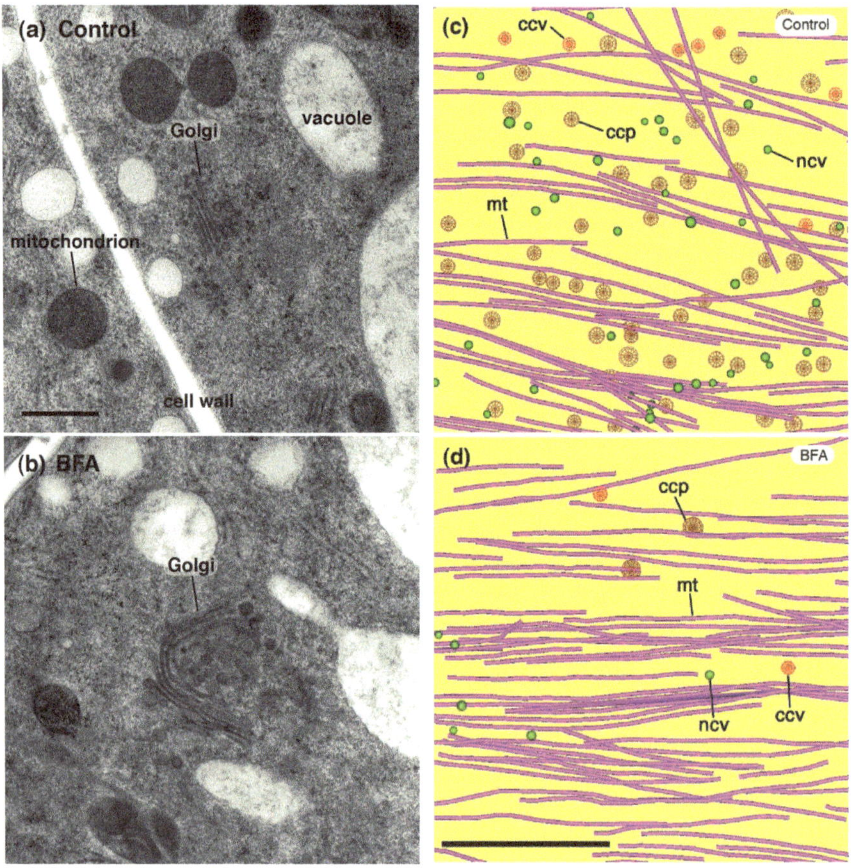

Figure 7. Effects of BFA on the Golgi architecture and on the density of clathrin-coated pits and of clathrin-coated and non-coated, dark vesicles in the PPB zone of the cytoplasm of onion epidermal cells. (a, b) Electron micrographs of thin sections showing Golgi architecture observed in a control (a) and a BFA-treated (100 μM for 20 min) (b) onion epidermal cell at the nuclear level. (c, d) Tomography-based reconstructions of the cortical region of a control (c) and a BFA-treated (d) late prophase cell. ccv, clathrin coated vesicle (bright red); ccp, clathrin-coated pit (deep red); ncv, non-coated vesicle (green); mt, MT (purple); pm, plasma membrane (yellow). Bar = (a, b) 0.7 μm and (c, d) 1 μm.

After treatment with BFA for 20 min, the onion epidermal cells contained a mixture of both normal looking and BFA-perturbed Golgi (Fig. 7a, b). In particular, the normal looking Golgi, which made up 31 ± 3 % (mean ± SEM) of the total Golgi population, displayed polar stack architecture together with one or several TGN cisternae, and resembled control Golgi (Fig. 7a). In contrast, the BFA-perturbed Golgi (69 ± 3 %) consisted of stacks that lacked a polar architecture, and whose cisternae resembled wider than normal and curved trans cisternae (Fig. 7b). Many secretory-type vesicles (84.3 ± 3.5 nm (diameter), mean ± SEM, n=73) were also seen in the vicinity of the altered stacks. In contrast to the variable appearance of the Golgi/TGN units in the BFA-treated cells, the responses of the endocytic membrane compartments to BFA were both pronounced and consistent.

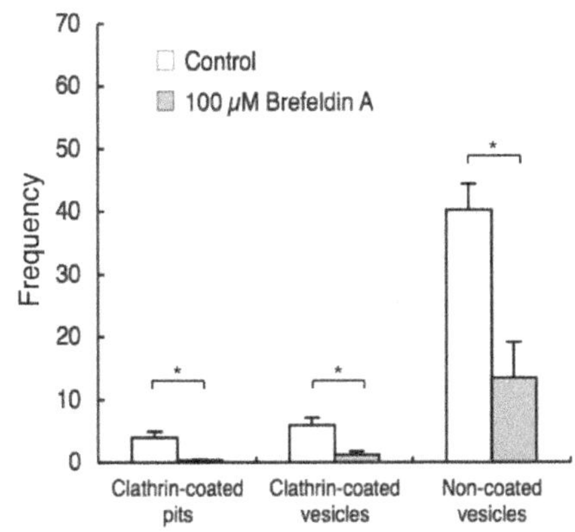

Figure 8. Average densities of clathrin-coated pits and vesicles at the nuclear level of late prophase (PPB region) of BFA-treated (100 μM for 20 min) cells. The results are based on measurements made on tomographic data sets. The frequency is expressed as number of pits per μm^2 in the case of clathrin coated pits, and numbers of vesicles per μm^3 in the case of vesicles (mean ± SEM; n=3). The Mann-Whitney *U*-test (two-tailed) was used to determine whether the difference were statistically significant compared with the control. *; P=0.0495, z=-1.964.

The number of clathrin-coated pits was decreased by ~90%, the number of clathrin-coated vesicles by ~80%, and the number of non-coated, dark-core vesicles by 67% (Fig. 8), consistent with the hypothesis that these three structures are causally related and involved in the endocytic pathway. By limiting the exposure time of the seedlings to BFA to 20 min, we have been able to differentially perturb the secretory and endocytic pathways, and thereby obtain data that are consistent with the hypothesis that the dense-core, non-coated vesicles underlying the plasma membrane are derived from clathrin-coated pits and vesicles that originate at the plasma membrane, and that they are not Golgi-derived secretory vesicles.

6. Conclusion

How the PPB marks the future site of cell division has been the subject of many studies and discussions since its discovery in 1966 (Pickett-Heaps & Northcote, 1966a, b). In a recent paper, we have quantified the distribution of clathrin-coated pits and vesicles as well as of secretory structures during PPB formation in onion epidermal cells using a combination of high-pressure freezing and electron tomography techniques. This quantitative characterization demonstrated that the rate of endocytosis is enhanced in PPB regions and suggests that the reported changes in composition of the plasma membrane of PPB regions could be brought about by the selective removal of specific plasma membrane molecules via

BFA-sensitive, clathrin-coated pits and vesicles. One possible function of the enhanced endocytic activity at forming PPBs might be the retrieval of actin-nucleating/binding proteins or KCA1 from these plasma membrane domains to create membrane zones that are depleted of such molecules. In turn, these modified plasma membrane regions could help guide the expanding cell plate to the division site and facilitate fusion of the cell plate margins to that site (Karahara et al., 2010). Thus, endocytosis appears to play an essential role in the creation of PPB "memory" structures in plants.

Author details

Ichirou Karahara
University of Toyama, Japan

L. Andrew Staehelin
University of Colorado, USA

Yoshinobu Mineyuki*
University of Hyogo, Japan

Acknowledgement

This work was supported partially by JSPS grants Nos. 21570064 and 21657011 to IK, No. 17207006 to YM, and the MEXT grant No. 17049019 to YM. We thank Mr. Takatoshi Yabuuchi at University of Hyogo for his help preparing Fig. 2.

7. References

Ambrose, J.C., Shoji, T., Kotzer, A.M., Pighin, J.A. & Wasteneys, G.O. (2007) The *Arabidopsis* CLASP gene encodes a microtubule-associated protein involved in cell expansion and division. *The Plant Cell*, 19, 2763-2775.

Austin, J.R.I., Seguí-Simarro, J.M. & Staehelin, L.A. (2005) Quantitative analysis of changes in spatial distribution and plus-end geometry of microtubules involved in plant-cell cytokinesis. *Journal of Cell Science*, 118, 3895-3903.

Banno, H. & Chua, N.H. (2000) Characterization of the Arabidopsis formin-like protein AFH1 and its interacting protein. *Plant Cell Physiology*, 41, 617–626.

Bar, M. & Avni, A. (2009) EHD2 inhibits ligand-induced endocytosis and signaling of the leucine-rich repeat receptor-like protein LeEix2. *The Plant Journal*, 59, 600-611.

Bar, M., Sharfman, M., Schuster, S. & Avni, A. (2009) The coiled-coil domain of EHD2 mediates inhibition of LeEix2 endocytosis and signaling. *PloS one*, 4, e7973.

Bednarek, S.Y. & Backues, S.K. (2010) Plant dynamin-related protein families DRP1 and DRP2 in plant development. *Biochemical Society Transactions*, 38, 797-806.

* Corresponding Author

Bonifacino, J.S. & Traub, L.M. (2003) Signals for sorting of transmembrane proteins to endosomes and lysosomes. *Annual Reviews of Biochemistry*, 72, 395-447.

Buschmann, H., Chan, J., Sanchez-Pulido, L., Andrade-Navarro, M.A., Doonan, J.H. & Lloyd, C.W. (2006) Microtubule-associated AIR9 recognizes the cortical division site at preprophase and cell-plate insertion. *Current Biology*, 16, 1938-1943.

Camilleri, C., Azimzadeh, J., Pastuglia, M., Bellini, C., Grandjean, O. & Bouchez, D. (2002) The *Arabidopsis* TONNEAU2 gene encodes a putative novel protein phosphatase 2A regulatory subunit essential for the control of the cortical cytoskeleton. *The Plant Cell*, 14, 833-845.

Chen, X., Irani, N.G. & Friml, J. (2011) Clathrin-mediated endocytosis: the gateway into plant cells. *Current Opinion in Plant Biology*, 14, 674-682.

Cheung, A.Y. & Wu, H. (2004) Overexpression of an Arabidopsis formin stimulates supernumerary actin cable formation from pollen tube cell membrane. *The Plant Cell*, 16, 257-269.

Cleary, A.L., Gunning, B.E.S., Wasteneys, G.O. & Hepler, P.K. (1992) Microtubule and F-actin dynamics at the division site in living *Tradescantia* stamen hair cells. *Journal of Cell Science*, 103, 977-988.

Conner, S.D. & Schmid, S.L. (2003) Regulated portals of entry into the cell. *Nature*, 422, 37-44.

Dhonukshe, P., Mathur, J., H lskamp, M. & Gadella, T.W.J. (2005) Microtubule plus-ends reveal essential links between intracellular polarization and localized modulation of endocytosis during division-plane establishment in plant cells. *BMC Biology*, 3, 11.

Dixit, R. & Cyr, R. (2002) Golgi secretion is not required for marking the preprophase band site in cultured tobacco cells. *The Plant Journal*, 29, 99-108.

Eleftheriou, E.P. & Palevitz, B.A. (1992) The effect of cytochalasin D on preprophase band organization in root tip cells of *Allium*. *Journal of Cell Science*, 103, 989-998.

Favery, B., Chelysheva, L.A., Lebris, M., Jammes, F., Marmagne, A., de Almeida-Engler, J., Lecomte, P., Vaury, C., Arkowitz, R.A. & Abad, P. (2004) *Arabidopsis* formin AtFH6 is a plasma membrane-associated protein upregulated in giant cells induced by parasitic nematodes. *The Plant Cell*, 16, 2529-2540.

Frank, J. (1992) *Electron tomography: Three-dimensional imaging with the transmission electron microscope*: Plenum Pub Corp.

Gunning, B.E.S. (1982) The cytokinetic apparatus: Its development and spatial regulation. In *The Cytoskeleton in Plant Growth and Development* (Lloyd, C.W. ed. London: Academic Press, pp. 229-292.

Haas, T.J., Sliwinski, M.K., Martinez, D.E., Preuss, M., Ebine, K., Ueda, T., Nielsen, E., Odorizzi, G. & Otegui, M.S. (2007) The *Arabidopsis* AAA ATPase SKD1 is involved in multivesicular endosome function and interacts with its positive regulator LYST-INTERACTING PROTEIN5. *The Plant Cell*, 19, 1295-1312.

Hall, Q. & Cannon, M.C. (2002) The cell wall hydroxyproline-rich glycoprotein RSH Is essential for normal embryo development in *Arabidopsis*. *The Plant Cell*, 14, 1161-1172.

Holstein, S.E.H. (2002) Clathrin and plant endocytosis. *Traffic*, 3, 614-620.

Kang, B.-H., Nielsen, E., Preuss, M.L., Mastronarde, D. & Staehelin, L.A. (2011) Electron tomography of RabA4b- and PI-4Kβ1-labeled *trans* Golgi network compartments in *Arabidopsis*. *Traffic*, 12, 313-329.

Karahara, I., Staehelin, L.A. & Mineyuki, Y. (2010) A role of endocytosis in plant cytokinesis. *Communicative & Integrative Biology*, 3, 1-3.

Karahara, I., Suda, J., Tahara, H., Yokota, E., Shimmen, T., Misaki, K., Yonemura, S., Staehelin, L.A. & Mineyuki, Y. (2009) The preprophase band is a localized center of clathrin-mediated endocytosis in late prophase cells of the onion cotyledon epidermis. *The Plant Journal*, 57, 819-831.

Kawamura, E., Himmelspach, R., Rashbrooke, M.C., Whittington, A.T., Gale, K.R., Collings, D.A. & Wasteneys, G.O. (2006) MICROTUBULE ORGANIZATION 1 regulates structure and function of microtubule arrays during mitosis and cytokinesis in the *Arabidopsis* root. *Plant Physiology*, 140, 102-114.

Kirchhausen, T. (2000) Clathrin. *Annual Reviews of Biochemistry*, 69, 699-727.

Kremer, J.R., Mastronarde, D.N. & McIntosh, J.R. (1996) Computer visualization of three-dimensional image data using IMOD. *Journal of Structural Biology*, 116, 71-76.

Lam, B.C.H., Sage, T.L., Bianchi, F. & Blumwald, E. (2001) Role of SH3 domain-containing proteins in clathrin-mediated vesicle trafficking in *Arabidopsis*. *The Plant Cell*, 13, 2499-2512.

Liu, B. & Palevitz, B.A. (1992) Organization of cortical microfilaments in dividing root cells. *Cell Motility and the Cytoskeleton*, 23, 252-264.

Miehe, H. (1899) Histologische und experimentelle Untersuhungen über die Anlage der Spaltöffnungen eigner Monokotylen. *Botanisches Centralblatt*, 78, 321-330, 353-359, 385-393.

Mineyuki, Y. (1999) The preprophase band of microtubules: its function as a cytokinetic apparatus in higher plants. *International Review of Cytology*, 187, 1-49.

Mineyuki, Y. & Gunning, B.E.S. (1990) A role for preprophase bands of microtubules in maturation of new cell walls, and a general proposal on the function of preprophase band sites in cell division in higher plants. *Journal of Cell Science*, 97, 527-537.

Mineyuki, Y., Marc, J. & Palevitz, B.A. (1989) Development of the preprophase band from random cytoplasmic microtubules in guard mother cells of *Allium cepa* L. *Planta*, 178, 291-296.

Mineyuki, Y., Murata, T. & Wada, M. (1991a) Experimental obliteration of the preprophase band alters the site of cell division, cell plate orientation and phragmoplast expansion in *Adiantum* protonemata. *Journal of Cell Science*, 100, 551-557.

Mineyuki, Y. & Palevitz, B.A. (1990) Relationship between preprophase band organization, F-actin and the division site in *Allium*. Fluorescence and morphometric studies on cytochalasin-treated cells. *Journal of Cell Science*, 97, 283-295.

Mineyuki, Y., Yamashita, M. & Nagahama, Y. (1991b) p34 cdc2 kinase homologue in the preprophase band. *Protoplasma*, 162, 182-186.

Müller, S., Han, S. & Smith, L.G. (2006) Two kinesins are involved in the spatial control of cytokinesis in *Arabidopsis thaliana*. *Current Biology*, 16, 888-894.

Murata, T., Karahara, I., Kozuka, T., Giddings, T.H., Staehelin, L.A. & Mineyuki, Y. (2002) Improved method for visualizing coated pits, microfilaments, and microtubules in cryofixed and freeze-substituted plant cells. *Journal of Electron Microscopy*, 51, 133-136.

Naramoto, S., Kleine-Vehn, J., Robert, S., Fujimoto, M., Dainobu, T., Paciorek, T., Ueda, T., Nakano, A., Van Montagu, M.C.E. & Fukuda, H. (2010) ADP-ribosylation factor machinery mediates endocytosis in plant cells. *Proceedings of the National Academy of Sciences USA*, 107, 21890.

Nebenführ, A., Ritzenthaler, C. & Robinson, D.G. (2002) Brefeldin A: deciphering an enigmatic inhibitor of secretion. *Plant Physiology*, 130, 1102.

Ôta, T. (1961) The role of cytoplasm in cytokinesis of plant cells. *Cytologia*, 26, 428-447.

Palevitz, B.A. & Hepler, P.K. (1974) The control of the plane of division during stomatal differentiation in *Allium*. I. Spindle reorientation. *Chromosoma*, 46, 297-326.

Pickett-Heaps, J.D. & Northcote, D.H. (1966a) Organization of microtubules and endoplasmic reticulum during mitosis and cytokinesis in wheat meristems. *Journal of Cell Science*, 1, 109-120.

Pickett-Heaps, J.D. & Northcote, D.H. (1966b) Cell division in the formation of the stomatal complex of the young leaves of wheat. *Journal of Cell Science*, 1, 121-128.

Rasmussen, C.G., Sun, B. & Smith, L.G. (2011) Tangled localization at the cortical division site of plant cells occurs by several mechanisms. *Journal of Cell Science*, 124, 270-279.

Reyes, F.C., Buono, R. & Otegui, M.S. (2011) Plant endosomal trafficking pathways. *Current Opinion in Plant Biology*, 14, 666-673.

Seguí-Simarro, J.M., Austin Ii, J.R., White, E.A. & Staehelin, L.A. (2004) Electron tomographic analysis of somatic cell plate formation in meristematic cells of Arabidopsis preserved by high-pressure freezing. *The Plant Cell*, 16, 836-856.

Seguí-Simarro, J.M. & Staehelin, L.A. (2006) Cell cycle-dependent changes in Golgi stacks, vacuoles, clathrin-coated vesicles and multivesicular bodies in meristematic cells of *Arabidopsis thaliana*: a quantitative and spatial analysis. *Planta*, 223, 223-236.

Staehelin, L.A. & Chapman, R.L. (1987) Secretion and membrane recycling in plant cells: novel intermediary structures visualized in ultrarapidly frozen sycamore and carrot suspension-culture cells. *Planta*, 171, 43-57.

Takano, J., Tanaka, M., Toyoda, A., Miwa, K., Kasai, K., Fuji, K., Onouchi, H., Naito, S. & Fujiwara, T. (2010) Polar localization and degradation of *Arabidopsis* boron transporters through distinct trafficking pathways. *Proceedings of the National Academy of Sciences USA*, 107, 5220.

Toyooka, K., Goto, Y., Asatsuma, S., Koizumi, M., Mitsui, T. & Matsuoka, K. (2009) A mobile secretory vesicle cluster involved in mass transport from the Golgi to the plant cell exterior. *The Plant Cell*, 21, 1212-1229.

Traas, J., Bellini, C., Nacry, P., Kronenberger, J., Bouchez, D. & Caboche, M. (1995) Normal differentiation patterns in plants lacking microtubular preprophase bands. *Nature*, 375, 676-677.

Van Damme, D., Bouget, F.Y., Poucke, K.V., Inzé, D. & Geelen, D. (2004) Molecular dissection of plant cytokinesis and phragmoplast structure: a survey of GFP-tagged proteins. *The Plant Journal*, 40, 386-398.

Van Damme, D., Coutuer, S., De Rycke, R., Bouget, F.Y., Inzé, D. & Geelen, D. (2006) Somatic cytokinesis and pollen maturation in *Arabidopsis* depend on TPLATE, which has domains similar to coat proteins. *The Plant Cell*, 18, 3502-3518.

Van Damme, D., Gadeyne, A., Vanstraelen, M., Inzé, D., Van Montagu, M.C.E., De Jaeger, G., Russinova, E. & Geelen, D. (2011) Adaptin-like protein TPLATE and clathrin recruitment during plant somatic cytokinesis occurs via two distinct pathways. *Proceedings of the National Academy of Sciences USA*, 108, 615-620.

Vanstraelen, M., Van Damme, D., De Rycke, R., Mylle, E., Inzé, D. & Geelen, D. (2006) Cell cycle-dependent targeting of a kinesin at the plasma membrane demarcates the division site in plant cells. *Current Biology*, 16, 308-314.

Viotti, C., Bubeck, J., Stierhof, Y.D., Krebs, M., Langhans, M., van den Berg, W., van Dongen, W., Richter, S., Geldner, N. & Takano, J. (2010) Endocytic and secretory traffic in *Arabidopsis* merge in the trans-Golgi network/early endosome, an independent and highly dynamic organelle. *The Plant Cell*, 22, 1344-1357.

Walker, K.L., Müller, S., Moss, D., Ehrhardt, D.W. & Smith, L.G. (2007) *Arabidopsis* TANGLED identifies the division plane throughout mitosis and cytokinesis. *Current Biology*, 17, 1827-1836.

Whittington, A.T., Vugrek, O., Wei, K.J., Hasenbein, N.G., Sugimoto, K., Rashbrooke, M.C. & Wasteneys, G.O. (2001) MOR1 is essential for organizing cortical microtubules in plants. *Nature*, 411, 610-613.

Wright, A.J., Gallagher, K. & Smith, L.G. (2009) *discordia1* and *alternative discordia1* function redundantly at the cortical division site to promote preprophase band formation and orient division planes in maize. *The Plant Cell*, 21, 234-247.

Xu, X.M., Zhao, Q., Rodrigo-Peiris, T., Brkljacic, J., He, C.S., Müller, S. & Meier, I. (2008) RanGAP1 is a continuous marker of the *Arabidopsis* cell division plane. *Proceedings of the National Academy of Sciences USA*, 105, 18637-18642.

Chapter 2

Roles of Cellular Redox Factors in Pathogen and Toxin Entry in the Endocytic Pathways

Jianjun Sun

Additional information is available at the end of the chapter

1. Introduction

1.1. Host-pathogen interaction: A process of co-evolution

In the mind of most human beings, microbial pathogens, including viruses, bacteria, and parasites, are the foreign invaders that cause diseases, and sometimes death. Thus, prevention and treatment of infectious diseases, by controlling and eradicating microbial pathogens, has become one of the major tasks of modern medicine. Based on current evolutionary theory, however, most of the pathogens we've seen nowadays are the species that have evolved through a close interaction with their hosts (e.g. humans and other animals). In an even broader sense, the hosts and the pathogens co-evolved through a mutual interaction. During this co-evolution, hosts and pathogens develop specific, intricate systems to either defend or invade, which in turn presents us with an intriguing picture of host-pathogen interaction. Host cells have the ability to defend themselves against the invasion of microbial pathogens. On one hand, host cells use the plasma membrane as a physical barrier that prevents pathogens from entering the cytoplasm, leaving pathogens in the harsh environment of the extracellular milieu, where pathogens are exposed to anti-microbial elements, such as antibodies, cytokines, and complement factors. On the other hand, professional phagocytic cells can engulf microbial pathogens into the phagosomes that later fuse with lysosomes, where reactive oxygen species (ROS), low pHs, and proteases can inactivate and kill the pathogens. Through the course of evolution, however, pathogens have developed effective strategies against such host defense systems. Viruses and obligate intracellular bacterial and protozoan pathogens gain access into phagocytic and non-phagocytic cells through membrane remodeling events, such as phagocytosis and macropinocytosis. In many cases, these membrane-remodeling events are controlled by cytoskeletal rearrangement mediated by pathogen-produced effector proteins (e.g. toxins) [1-3]. Within the phagosomes, pathogens have evolved a variety of mechanisms to protect themselves from damage by the hostile environment. For instance, pathogens can produce effector proteins to antagonize ROS effects [4], inhibit phagosome maturation, block phagosome-lysosome fusion, and escape from phagosomes [5]. In contrast, extracellular pathogens need to protect themselves from being engulfed by professional phagocytic cells. Such pathogens usually produce toxins to disarm the host cell defense, resulting in inhibition of phagocysis, or in cell killing [6]. Some bacterial toxins are directly translocated into the host cytoplasm through sophisticated 'molecular syringes', such as type III or type IV secretion systems [7-9], which are multi-subunit molecular machines that span the bacterial and host membranes and translocate effectors directly into host cells. Other toxins (e.g. AB toxins) are secreted by bacteria in the vicinity of the host cell and these toxins bind to specific receptors and are taken up by endocytosis [10-13]. Once internalized, bacterial toxins usually take advantage of the hostile environment in endosomes/lysosomes, and they hijack host factors, which enables translocation into the cytosol. For instance, many bacterial toxins utilize endosomal acidification (low-pH) as a trigger for conformational conversion, which activates toxins and/or facilitates release of toxins into the cytoplasm [14]. Moreover, some pathogens and toxins hijack cellular redox factors, thus allowing them enter into the host cells, which will be discussed in detail in this review.

1.2. Endocytic pathways: The portals of entry for microbial pathogens and toxins

Endocytosis is a physiological process of invagination and pinching-off pieces of the plasma membranes, and IT serves as a ubiquitous mechanism that facilitate the internationalization of various particles and molecules from the extracellular milieu into the host cytoplasm. Endocytosis plays a vital role in a diverse range of physiological processes, including maintenance of cellular homeostasis, cell polarity, and uptake of nutrients. Thus, it is not surprising that a great variety of microbial pathogens and toxins have evolved to exploit aspects of this internalization process as portals of entry into host cells. Based on the nature (e.g. size) of the extracellular substrates, endocytosis has been categorized into phagocytosis and pinocytosis. Phagocytosis is involved in engulfment of large particles (e.g. cell debris and bacterial pathogens) by professional phagocytic cells, such as macrophages, monocytes and neutrophils. Pinocytosis, on the other hand, is typically involved in uptake of small particles, such as viruses and bacterial toxins, by non-phagocytic cells. Based on the proteins involved in membrane vesiculation, endocytosis is defined by several types of mechanisms, including clathrin-mediated, caveolin-mediated, lipid raft-dependent, and macropinocytosis, etc. These different mechanisms have been described in recent excellent reviews [6,15-18].

Despite the diversity of membrane vesiculation, upon endocytosis most pathogens and toxins follow one of the two intracellular trafficking pathways (**Figure 1**). In pathway 1, pathogens and toxins travel to the early and late endosomes where some pathogens (e.g. HIV, *Chlamydia, Leishmania*) and toxins (e.g. anthrax toxin, diphtheria toxin, botulinum toxin) translocate to the cytosol; while others (e.g. *Listeria monocytogenes*) travel to the lysosomes and are then released into the cytosol. In pathway 2, instead of going through the endo-lysosomal pathway, pathogens and toxins (e.g. SV40, cholera toxin, shiga toxin, exotoxin A) traffic to the Golgi, and from the Golgi to the endoplasmic reticulum (ER), a pathway in reverse of the classical secretion pathway, called retrograde transport. An ER retention signal (e.g. KDEL) is usually required for this transport to occur [10,13,15,16,18,19].

To date, substantial evidence has suggested that cellular redox factors play an essential role in pathogen and toxin entry through endocytosis. These redox factors include protein disulfide bond isomerase (PDI), γ-interferon inducible lysosomal thiol reductase (GILT), NADPH oxidases (Nox) and some ER-chaperones. These redox factors function at various sites in the endocytic pathways that facilitate pathogen and toxin entry into the cells (**Figure 1**). These events will be discussed in detail in the later sections of this review.

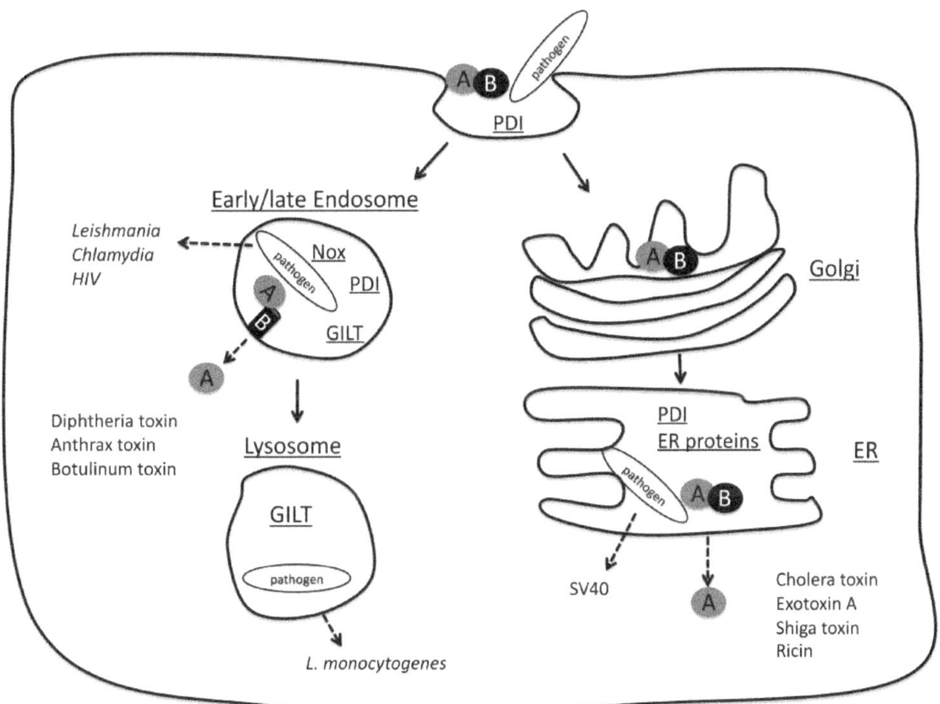

Figure 1. Interaction of cellular redox factors with microbial pathogens and toxins in the endocytic pathways. Microbial pathogens and toxins are internalized into the host cells through endocytosis. Pathway 1 (left): The pathogens and toxins travel to endosomes and/or lysosomes where they translocate to the cytosol. Pathway 2 (right): the pathogens and toxins undergo retrograde transport through Golgi to ER where they are released into the cytosol. In this cartoon, HIV, SV40, Chlamydia, Listeria monocytogenes, Leishmania, are presented as the representatives of microbial pathogens that are discussed in the review. A group of AB toxins are also shown in this cartoon, and some of these toxins are discussed in this review. The cellular redox factors, such as PDI, GILT, NADPH oxidase (Nox), are placed into the various locations of the endocytic pathways according to the current literature. Due to space limitation, the cartoon only depicts a simplified illustration. Solid arrows: intracellular trafficking; dashed arrows: translocation across the membranes.

1.3. Disulfide bond: A redox-controlled switch for pathogen and toxin entry

Disulfide bond, a covalent link between a pair of cysteine residues, plays important roles in protein structure and function. Disulfide bond has a significant impact on thermodynamics of protein folding, as it can stabilize the native conformation by disfavoring the unfolded form. In a naturally folded protein, disulfide bond maintains the protein's integrity by protecting the protein from damage by oxidants and proteolytic enzymes [20-23]. More importantly, in mature, folded proteins, some disulfide bonds can function as molecular switches that can turn "on" or "off" certain protein functions. This is usually accomplished via conformational changes induced by breaking, forming, or the isomerization of the disulfide bonds [24-26]. These diverse, reversible features of disulfide bonds can be readily manipulated by redox factors ranging from small molecular reagents (e.g. reduced/oxidized glutathione) to macromolecular redox enzymes (e.g. oxidoreductases). Moreover, these redox factors are ubiquitously present, yet un-equally distributed in the sub-cellular compartments of eukaryotic cells, which offers a distinct spatial regulation of the thiol/dithiol equilibrium [27,28]. In the course of evolution, microbial infection has become a highly regulated process. Thus, activation at the right time and at the right location is an important factor for pathogens and toxins to successfully cause an infection. Thus, a readily controlled disulfide bond "on/off" switch is evolutionarily favored. It is not surprising that pathogens and toxins use disulfide bonds as redox-controlled switches for invasion.

Increasing evidence has shown that cellular redox factors play pivotal roles in pathogen and toxin entry into the endocytic pathways, particularly through modulating the thiol-dithiol states of pathogen- and/or host-factors. For instance, cellular entry of certain bacterial toxins (e.g. diphtheria toxin [29-33], cholera toxin [34], botulinum neurotoxins [35,36], anthrax toxin [37]) are apparently dependent on the redox states (either reduced or oxidized) of the specific disulfides of either the toxin molecules, or the host receptors. At the same time, protein disulfide isomerase (PDI) [38] and other redox factors, such as gamma-interferon-inducible lysosomal thiol reductase (GILT) [39]) and NADPH oxidase [40-42], have been implicated in regulating the redox states of the disulfides. Similarly, PDI and others are also involved in the entry of numerous pathogenic bacteria (e.g. *Chlamydia* [43,44], *Listeria* [39]), viruses (e. g. HIV [45] and SV40 (46,47]) and parasites (e. g. *Leishmania* [40]) through endocytosis. This review will present the current major findings on the roles of cellular redox factors in pathogen and toxin entry with an attempt to outline the strategies and mechanisms that microbial pathogens and toxins utilize to hijack the cellular redox factors within the endocytic pathways.

2. Cellular redox factors in endocytic pathways

2.1. Redox potentials of endocytic pathways: Oxidizing or reducing?

Eukaryotic cell is organized into several distinct sub-cellular compartments, each of which maintains a distinct redox potential [28]. Relative to the extracellular milieu, which is oxidizing, it is generally believed that endocytic pathways are reducing. This notion is based on the primary function of endocytosis: i.e., the uptake and degradation of foreign and self-particles including proteins, for which a reducing potential facilitates protein unfolding. This concept has been well supported by the evidence that uptake and activation of some bacterial toxins, such as diphtheria toxin, cholera toxin and *Pseudomonas* exotoxin A, through the endocytic pathways involves reduction of disulfides. The dynamics of disulfide reduction in endocytic pathways have recently been studied with fluorescence resonance energy transfer (FRET) using a fluorescent folate conjugate, in which folate-BODIPY and Rhodamine is linked with a disulfide bond [48]. Reduction of this disulfide bond changes fluorescence from red to green, which allows real-time fluorescence imaging of the reduction in cells. Reduction was observed to occur in endosomes, with a half-life of 6 hours post-endocytosis. Using this experimental setup, reduction did not depend significantly on extracellular surface thiols or redox machinery within lysosomes or Golgi. The yielded products were sorted into different endosomes and trafficked in different directions. This excellent FRET design ensures an accurate assessment of disulfide bond reduction during normal vesicle trafficking in living cells and demonstrates that reduction occurs in the endocytic pathways. In fact, the presence of distinct redox potentials between the oxidizing extracellular space and the reducing endocytic pathways has created interest in disulfide bonds as a potential tool for drug delivery. For example, the disulfide-based bioconjugation approach has become a popular conjugation method applied in a variety of cellular drug delivery systems. Successful applications of thiol-based conjugation resulted in targeted delivery and enhanced cytosolic delivery, improved pharmacokinetics, and increased stability of the drugs [49-52].

While it is generally accepted that endocytic pathways are reducing, there is also evidence suggesting the contrary. In another independent study, a disulfide linker cleavage assay was developed whereby rhodamine red was linked to an anti-HER2 antibody through a peptide linker containing a disulfide bond [53]. Cleavage of the disulfide bond would release self-quenching of the fluorophore. In breast carcinoma SKBr3 cells, no linker cleavage was observed, as detected by fluorescence dequenching upon internalization. In contrast, the conjugate did display fluorescence dequenching when it was diverted to the lysosomal pathways, which could be an effect partly due to proteolytic degradation rather than disulfide reduction. More convincingly, the redox potentials of endocytic compartments were measured directly by expressing a redox-sensitive variant of GFP fused to various endocytic proteins. The results showed that recycling endosomes, late endosomes, and lysosomes were not reducing, but rather oxidizing and to a level comparable with conditions in the ER.

In summary, the redox potentials in the endocytic pathways appear to vary accordingly to different ligands, different cell types, and different physiological/pathological conditions. For instance, NADPH oxidase, the major enzyme that catalyzes the production of reactive oxygen species (ROS), is regulated by hormone or growth factors in normal cells, but it is constitutively activated in cancerous cells such as HeLa and hepatoma cells [54]). It is well known that in professional phagocytic cells NADPH oxidase is activated upon pathogen infection and that it produces ROS within the phagosomes, a process called oxidative burst that is described below.

2.2. Oxidative burst and NADPH oxidase in professional phagocytic cells

When professional phagocytes recognize pathogen-associated molecular patterns that are located on microbial pathogens, the phagocytes will internalize them through phagocytosis and activate a strong bacterial killing mechanism, called oxidative or respiratory burst, which is marked as an abrupt increase of superoxide formation within the phagosomes [55]. This process is mainly catalyzed by NADPH oxidase, a membrane-associated enzyme complex that is located on the phagosome membrane and generates superoxide (O_2^-) by the one-electron reduction of oxygen, using NADPH as the electron donor. Assembly and activation of NADPH oxidase requires phosphorylation of its subunits and translocation of cytosolic components to the plasma membrane [56,57]. The superoxide anion generated is enzymatically converted to hydrogen peroxide by superoxide dismutase (SOD). The generated hydrogen peroxide can serve as a precursor for hydroxyl radical (•OH) generation via a Fenton-like reaction. Hydrogen peroxide then enters cells and forms hydroxyl radical that can kill many microorganisms by reacting with different macromolecules, including proteins and DNA. Except for killing microorganisms directly, ROS can also work as secondary messengers in many signaling pathways within phagocytic cells, which promote actions of other antibacterial agents and stimulate inflammation. However, chronic inflammation induced by ROS may damage the host tissue and induce apoptosis of the phagocytic cells [56-58].

2.3. Protein disulfide isomerizes and other oxidoreductases (e.g. GILT)

2.3.1. Protein disulfide isomerase (PDI)

Enzymatic activities and sub-cellular localization:

PDI is a ubiquitous dithiol/disulfide oxidoreductase chaperone belonging to the thioredoxin oxireductase superfamily. There are around 20 PDI homologues, and the structure and function of eukaryotic PDIs have been covered in recent excellent reviews [59,60]. The prototypic PDI contains 5 domains ordered as a-b-b′-a′-c, in which two thioredoxin-like motifs (CXXC) are located in the domains a and a′, respectively. The primary function of PDI is to promote protein oxidative folding in the ER. The PDI redox-domains catalyze three redox reactions: reduction (breaking disulfide bond), oxidation (forming disulfide bond), and isomerization (exchanging disulfide bond). Independently of its redox activity, PDI also functions as a chaperone, which requires its ATPase and Ca^{2+} activities [60-62]. PDI contains a KDEL sequence at the domain c, which facilitates its retention in the ER lumen, and PDI cycles between ER and cis-Golgi through the KDEL receptor. Despite its KDEL sequence and ER retention mechanism, PDI is also involved diverse intracellular trafficking processes and is even secreted outside cell and can be found at the cell surface [59]. The cell-surface PDI is thought to localize on the plasma membrane by attachment to lipids, glycans and integral membrane proteins [59,63]. Unlike other members of thioredoxin family, PDI is not normally found in the cytosol. In addition to catalyzing protein oxidative folding in the ER, PDI has been shown to be actively involved in many other processes, such as ER-associated degradation, trafficking, calcium homeostasis, antigen presentation and host-pathogen interaction [60].

ER-located PDI:

The ER-located PDI plays an important role in host-pathogen interactions, particularly in antigen presentation and ER-mediated phagocysis of intracellular pathogens. Antigen presentation occurs through two pathways: the exogenous pathway and the endogenous pathway. In the exogenous pathway, the antigens from extracellular pathogens (e.g. fungi, bacteria, and parasites) are captured and processed in the phagosome/lysosome compartments within the long-lived antigen presenting cells (e.g. macrophages and dendritic cells) and then form complexes with MHC class II. The antigen complexed with MHC-II is then presented on the cell surface and subsequently recognized by helper CD4+ T cells. In endogenous pathway, self cell antigens and viruses synthesized within cells are degraded by the proteasome and then form complexes with MHC class I, which are presented on the cell surface and recognized by cytotoxic CD8+ T cells. The two pathways sometimes overlap and antigens can be presented by both MHC class I and class II. These include some intracellular bacterial and parasite pathogens that pass or live in the phagosomes, such as *salmonella typhimurium, Mycobacterium tuberculosis, Leishmania spp* and *Trypanosoma cruzi*. As a part of the protein folding machinery in the ER, PDI has been shown to directly regulate antigen processing of the MHC class I complex [64,65] .

Phagocysis is the main mechanism for the professional phagocytes to internalize large pathogens. A recent study has found that fusion of the ER with the plasmalemma underneath phagocytic cups is a source of membranes for the phagosome formation within macrophages. The ER-associated chaperones, including PDI, are involved in this intense membrane remodeling process [66,67]. Of particular interest to this review article is the ER-associated PDI that is involved in the translocation of the toxins (e.g. cholera toxin) and pathogens (e.g. SV40 and *Leishmania*) from the ER to the cytosol.

Secreted and cell-surface associated PDI:

While PDI enzymes are predominantly located in the ER where they act as chaperones and facilitate protein folding, they also can be secreted extracellularly and located on the cell surface. The secreted PDI and the cell surface PDI can be identified by the use of antibodies or specific ligands. Since PDI is a soluble protein, PDI is associated with the cell surface probably through electrostatic interaction with other surface-located proteins, peptides or lipids. The thioredoxin sites of PDI appear to be involved in the reducing activity of the cell exterior where protein disulfide bonds are reduced or reshuffled. Recent research has shown that the level of cell surface thiols positively correlates to the amount of cell surface PDI (68). Consistent with the fact that the thiol groups of cell surface proteins are involved in cell adhesion, PDI thioredoxin activity plays an important role in cell adhesion. In leukocyte adhesion, PDI reducing activity maintains the adhesion protein L-selectin in a particular conformation (disulfide breaking) on the cell membrane that is not accessible to proteolytic enzymes. Inhibition of PDI leads to a conformational change in L-selectin (disulfide forming), and the subsequent cleavage of L-selectin, which results in lose of cell adhesion

[22]. PDI is also involved in the integrin-mediated platelet adhesion. Integrin receptors contain "open" and "close" conformations that represent the "on" and "off" states of ligand binding, respectively. It has been suggested that PDI regulates the open and close conformations through reduction and/or reshuffling of the disulfide bonds in integrins [23,69,70]. The cell-surface PDI not only plays important roles in physiological processes, it is also involved in pathological events, particularly in the entry of pathogens and toxins into host cells as discussed in detail in this review.

PDI regulation of NADPH oxidase:

NADPH oxidase is not only the main source of ROS production during oxidative burst, it also has been described as another cell surface-associated protein with disulfide–thiol interchange activity [54,71]. NADPH oxidase proteins on the mammalian cell surface exhibit two different activities, oxidation of hydroquinones (or NADH) and protein disulfide-thiol interchange. Protein thiols on the membranes were measured by reaction with 5,5'-dithiobis-(2-nitrobenzoic acid) (DTNB; Ellman's reagent) and the results suggested that protein disulfides may be the natural electron acceptors for NADH oxidation within plasma membrane vesicles. Protein disulfides of the membranes were reduced, with a concomitant stoichiometric increase in protein thiols in the presence of NADH, while the increase in protein thiols was inhibited in parallel to the inhibition of NADH oxidation.

It is not clear how NAD(P)H oxidase catalyzes disulfide-thiol exchange. Interestingly, PDI has been shown to regulate NAD(P)H oxidase activity. In rabbit aortic smooth muscle cells PDI was found to be co-localized and co-immunoprecipitated with the oxidase subunits p22, Nox1, and Nox4. Inhibition of PDI using PDI antagonism, such as bacitracin, scrambled RNase, neutralizing antibody, or antisense oligonucleotide, resulted in inhibition of the oxidase activity, which suggests that PDI closely associates with NAD(P)H oxidase, and acts as a novel redox regulator of the oxidase [72]. Later, the PDI-mediated regulation of NADPH oxidase activity was confirmed, showing that PDI plays a role in organizing NADPH oxidase activation in a variety of physiological/pathological events [41,42,73], of which PDI association with NADPH oxidase is required for phagocytosis of Leishmania chagasi promastigotes in macrophages as is discussed later in this review.

2.3.2. Gamma-interferon (IFN-γ)-inducible lysosomal thiol reductase, GILT

Proteins internalized via the endocytic pathways are usually degraded in lysosomes, where proteolysis is facilitated by protein denaturation induced by acidic condition and by reduction of inter- and intra-molecular disulfide bonds. While high concentration of cysteines was claimed to be the physiological reducing agent in lysosomes, this small molecular reducing agent alone appears inefficient in disulfide reduction in an acidic environment, since disulfide reduction requires deprotonantion of thiols, which is not favored by acidic environments. Thus, the presence of redox enzymes within the acidic cellular compartments had been postulated for a long time [74,75]. Known redox enzymes, such as thioredoxin reductase and glutathione reductase, normally function in neutral pH environments, so they are obviously not the likely candidates. The enzymes that catalyze the reduction within acidic compartments had been elusive until recently when GILT was identified as the first thiol reductase optimally active at low pH (4.0 – 5.5) [76,77]. GILT is expressed constitutively in antigen-presenting cells, in which it is synthesized as a 35-kDa glycoprotein precursor containing a mannose-6-phosphate signal sequence and is co-localized with early endosomes. The amino- and carboxyl-terminal propeptides are cleaved in the early endosomes and the 30-kDa mature enzyme is delivered by the mannose 6-phosphate receptors through the endocytic pathways to late endosomes and lysosomes. The mature enzyme is found in MHC class II-containing compartments, where it catalyzes disulfide bond reduction to facilitate antigen processing [78]. GILT can also facilitate the transfer of disulfide-containing antigens into the cytosol, enhancing their cross-presentation by MHC class I [79]. Compared to other members of the thioredoxin family, GILT possesses seminar yet distinctive enzymatic characteristics. GILT has a similar catalytic active site (-C-X-X-C-), but does not have the common motif (-C-G-H/P-C-) that is shared by the members of the thioredoxin family. GILT shows optimal activity at pH 4.0-5.5, while the other members of the family function optimally at neutral pH. Moreover, GILT requires a reducing agent, such as DTT or cysteine (but not glutathione) to regenerate and retain its activity in vitro [76], which is consistent with the potential function of cysteine in the acidic compartments for disulfide reduction.

In addition to being constitutively expressed in antigen presenting cells, GILT is induced and up-regulated by interferon-r (IFN-γ) in other cell types via signal transducer and activator of transcription 1 [80]. GILT has been found to accumulate in macrophage phagosomes as they mature into phagolysosomes [81]. Most interestingly, GILT is a critical host factor for Listeria monocytogenes infection [39], as is discussed later in this review.

3. Roles of redox factors in entry of bacterial toxins through endocytosis

3.1. AB toxins and interchain disulfide bond

A number of proteins produced by bacterial pathogens are highly toxic to mammalian cells due to their ability to enter the cytosol and attack essential cellular metabolic and/or signal transduction pathways. These toxic proteins mostly belong to AB toxin family (82). AB toxins contain two structurally and functionally distinctive moieties: an enzymatically active A moiety that normally modifies a cellular target upon entry into the cytosol, leading to cell death or other pathological effects; and a binding/translocation B moiety that binds to cell surface receptors and translocates the A moiety into the cytosol. Commonly, an AB toxin is synthesized and secreted from the pathogen as an inactive form. This inactive precursor is activated through a proteolytic cleavage performed by either a host or a pathogen protease at a region between two cysteine residues. The cleavage results in a di-chain toxin molecule with the A moiety and the B moiety linked by a disulfide bond (**Figure 2**). AB toxin-mediated intoxication of the host cells starts with B moiety binding to the cell surface receptors, followed by receptor-mediated endocytosis. Some toxins, such as anthrax toxin, diphtheria toxins, and Clostridial neurotoxins, traffic to endosomes, where acidification triggers conformational change on B moiety that forms a protein conductive channel/pore on the endosomal membranes and translocates A moiety into the cytosol. Other toxins, including shiga toxin, cholera toxin, exotoxin A, will travel through a retrograde transport pathway to arrive at the ER. There, A moiety is released into the cytosol (**Figure 1 and 2**). In either of these two intracellular trafficking schemes, it is presumed that the interchain disulfide that links A and B moieties must be cleaved prior to translocation of A moiety into the cytosol. While the mechanism of disulfide reduction-dependent translocation is not fully understood, and may be toxin-specific, current research has provided evidence that cellular redox factors play essential roles in toxin translocation by mediating reduction of the interchain disulfide.

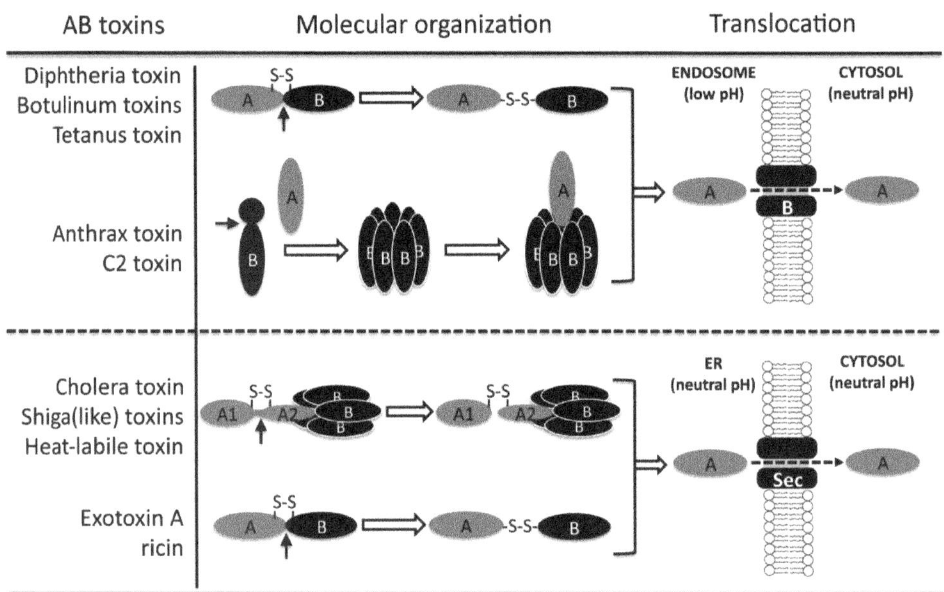

Figure 2. Molecular organization and translocation of AB toxins. Based on molecular organization (presence or absence of interchain disulfide bond) and sites of membrane translocation (endosome or ER), AB toxins are divided into four groups as indicated in this cartoon, with the representative toxins listed in each group. Group 1: the toxins (e.g. diphtheria toxin) are produced as a single-polypeptide precursor. Activation requires a proteolytic cleavage to generate a dichain molecule that is linked by an interchain disulfide bond. The toxins travel to the endosomes where the B moiety forms a pore on the endosomal membranes and translocates the A moiety into the cytosol. Group 2: the A and B moieties of the toxins (e.g. anthrax toxin and C2 toxin) are produced as separate proteins. The B moiety is activated by proteolytic cleavage and assembles into a heptameric complex that recruits the A moiety. Within the endosomes, the B moiety forms a pore on the endosomal membranes and translocates the A moiety into the cytosol. While there is no interchain disulfide bond, the disulfide bonds of anthrax toxin receptor are required for the toxin translocation, which is discussed in this review. Group 3: The proteolytic cleavage occurs in the A moiety of the toxins (e.g. Cholera toxin), resulting in two fragments, A1 and A2, that are linked by a disulfide bond. In the ER, A1 is translocated with the assistance of the ER machinery (e.g. Sec) to the cytosol, which requires the reduction of the disulfide bond. Group 4: the toxins (e.g. exotoxin A) share a similar structural organization with the toxins in Group 1, but translocation occurs in the ER, instead of the endosomes.

3.1.1. Diphtheria toxin and cell-surface PDI

Diphtheria toxin (DT) is secreted as a single polypeptide chain of 535 residues (58 kDa) from toxigenic strains of Corynebacterium diphtheriae [83]. DT is activated by a proteolytic cleavage that is catalyzed by the cellular protease furin, which results in two protomers (DT-A, 21 kDa; DT-B, 37 kDa) that are linked by a disulfide bridge (**Figure 2**). DT-A is an ADP-ribosyltransferase that ADP-ribosylates elongation factor 2 (EF-2) in the cytosol. DT-B is responsible for cell binding and translocation of DT-A to the cytosol. The receptor-binding domain at the C-terminal half of DT-B binds to the cell surface receptor (heparin-binding EGF-like growth factor) and enters the cell through the receptor-mediated, clathrin-dependent endocytosis. Within the endosomes, the acidic pH triggers a conformational change on DT-B, leading to the exposure of the hydrophobic domains and an increased tendency to interact with the membrane lipids. Thus, DT-B inserts into the membranes and forms a cation-selective channel that translocates DT-A into the cytosol, where DT-A inhibits protein synthesis by ADP-ribosylation of EF-2 [84].

An earlier study showed that membrane-impermeant sulfhydryl inhibitors (DNTB and pCMBS) markedly inhibited DT cytotoxicity, an effect that was not due to inactivation of unbound DT, inhibition of endocytosis, or impairment of endosomal acidification [32]. This indicated that the reductive cleavage of DT's interchain disulfide bond mediated by the cell surface sulfhydryls is required for the DT cytotoxicity. A later independent study of DT-mediated Vero cell intoxication showed that reduction of the single interchain disulfide bond is the rate-limiting step of the entire intoxication process, and this reduction occurred only after the toxin had passed through a low pH triggered structural change on DT-B in an early endosome [30]. Together, these studies suggested that the reductive activation of DT is catalyzed by sulfhydryls that are originally present at the cell surface and through vesiculations, become situated at the inner face of nascent endosomes. Sulfhydryl groups blocked at the cell surface will remain blocked in primary endosomes whose fluid volume still contains inhibitors. More interestingly, specific PDI inhibitors, bacitracin and anti-PDI antibodies, effectively inhibited DT-mediated cytotoxicity [33], suggesting that cell-surface PDI is involved in the translocation of DT through the reduction of the interchain disulfide bond. The PDI-catalyzed reduction appears to be specifically restricted to the site of interchain disulfide bond, but not others. In an earlier study, intramolecular disulfide bonds were generated in the DT-A domain by introducing double cysteine residues [29]. During endocytosis, the interchain disulfide bond was found to be reduced, while the engineered intramolecular disulfide bonds remained intact, which inhibited DT-A unfolding and membrane translocation. This target-specific reduction of the interchain disulfide bond implicates the existence of enzymatic specificity.

More recently, however, an assay of measuring the in vitro delivery of DT-A from the lumen of purified early endosomes to the external milieu has shown that cellular thioredoxin reductase activity plays an essential role in the cytosolic release of the DT-A, suggesting that other cellular redox enzymes, except for PDI, may also be involved in DT translocation [85].

3.1.2. Cholera toxin and ER-located PDI

Cholera toxin (CT), produced by the bacterium *Vibrio cholerae,* acts on intestinal epithelial cells in mammals to induce massive salt and water secretion, causing severe diarrhea [86,87]. CT consists of one A subunit, an ADP-ribosyltransferase that targets heterotrimeric Gs proteins, and five B subunits that bind to the cell surface receptor, ganglioside GM1. The A subunit is cleaved by a bacterial endoprotease to form two fragments, A1 and A2, that are linked by a disulfide bond (**Figure 2**). Reduction of the disulfide bond is required for translocation of A1 fragment from ER to the cytosol, where it ADP-ribosylates Gs. ADP-ribosylation of Gs results in a constitutive activation of adenylyl cyclase and an increase of cAMP levels. The A2 fragment bores through the center of the B subunit ring with a C-terminal KDEL (Lys-Asp-Glu-Leu) sequence that protrudes outwards. CT enters cells through clathrin-independent endocytic pathways, primarily through the cholesterol-rich plasma membrane domain caveolae. The ER retrieval KDEL sequence located on the A2 fragment leads the toxin through the retrograde transport pathway from Golgi network to the ER. Within the ER the A1 fragment is unfolded, released from the rest of the toxin, and translocated across the ER membrane to the cytosol [88]. It has been reported that translocation occurs through the Sec61 channel and utilizes ER-associated degradation (ERAD), which is a physiological process for retro-translocation of mis-folded proteins into the cytosol.

Although the mechanism of A1 translocation from ER into the cytosol is not clear, it is presumed that the toxin in the ER must undergo the following events: subunits must be disassembled, the disulfide bond must be reduced, and the A1 fragment must be unfolded in order to be translocated through the Sec61 channel. However, the ER is the compartment where proteins are folded, assembled and disulfide bonds are formed. This apparent paradox has raised the question of how the disassembly and unfolding of the toxin occur in such an unfavorable environment. Because of the reported role of PDI in the reduction of the interchain disulfide bond of DT, inhibitors of cell-surface PDI, such as bacitracin, DTNB, and anti-PDI antibodies, were tested [89,90]. In contrast to DT, these inhibitors had no effect on CT cytotoxicity in intact cells, suggesting that the reduction of CT does not occur at the cell surface or in the early endocytic pathways. In the presence of Triton X-100, however, these inhibitors significantly inhibited CT activity. Further study revealed that the A1 fragment is co-localized with PDI in the ER-derived membrane fraction, suggesting that PDI is the redox factor that catalyzes the CT reduction at the ER [90]. Subsequently, an excellent study showed that protein disulfide isomerase (PDI) in the ER lumen functions to disassemble and unfold the toxin once the A chain is cleaved [34]. In this reaction PDI acts as a redox-driven chaperone: that is, in the reduced state, it binds to the A chain and unfolds it, while in the oxidized state it releases the substrate. Moreover, the PDI-mediated CT translocation in the ER appears to be coordinated with a series of ER proteins, such as Ero1, Erp72, and Derlin-1 [34,91]. Together, these studies have revealed a highly coordinately operations exploited by PDI and other ER chaperone proteins in CT translocation.

3.2. Membrane translocation of Botulinum neurotoxins and anthrax toxin: two stories without cellular redox enzymes?

3.2.1. Botulinum neurotoxins: Interchain disulfide bridge remains intact throughout the translocation

Botulinum neurotoxins (BoNTs), produced by various strains of the spore-forming bacteria (e.g. *Clostridium botulinum, C. butyricum, and C. barati*), are known as the most poisonous toxins in nature [11,92]. To date, seven antigenically distinguishable BoNTs (designated from A to G) have been identified in research. The seven serotypes of BoNTs in combination with another neurotoxin of clostridia, tetanus neurotoxin (TeNT) from *C. tetani*, make up the clostridial neurotoxin family. BoNTs cause flaccid paralysis by targeting to peripheral motorneuron, while TeNT causes opposite symptoms by acting in inhibitory interneurons [82]. Despite their extreme cytotoxicity, when applied with an appropriate dose BoNTs could act as effective drugs because of their powerful neuroparalytic activity. In fact, BoNT serotype A is the first biological toxin that has received FDA approval for treatment of human diseases, such as cervical torticollis, strabismus, and dystonias.

Each BoNT is synthesized as a ~150 kDa single chain protein. This single chain precursor is subsequently cleaved into a di-chain molecule, in which the ~50 kDa light chain (LC) and the ~100 kDa heavy chain (HC) remain linked via a single disulfide bond [93,94] (**Figure 2**). The HC is composed of an N-terminal translocation domain and a C-terminal receptor-binding domain. The receptor-binding domain binds to both gangliosides and protein receptors on the cell membrane and the toxin is internalized through a receptor-mediated endocytosis [95-97]. Within the acidic endosomes, the translocation domain undergoes a conformational change to form a pore on the endosomal membranes and translocates LC to the cytosol. In the cytosol, LC, a zinc endoprotease, specifically cleaves SNARE proteins, resulting in inhibition of synaptic exocytosis [13,87,92,98].

Apparently BoNTs share similar features with DT in terms of molecular organization (e. g. The A and B moieties are linked by an interchain disulfide bond) and the mode of entry into the host cells (e.g. receptor-mediated endocytosis, low-pH-induced pore formation and translocation). Despite of these apparent similarities, however, the interchain disulfide bridge of BoNTs appears to dictate a different mechanism in translocation, relative to DT. The dynamics of the toxin translocation was elegantly examined in a single channel/single molecule assay using patch clamp recording on the cell membranes [36,99]. The disulfide bond needs to remain intact throughout LC translocation, and premature reduction of the disulfide bond even after channel formation or within the lipid bilayer arrests translocation. Consistent with this result, addition of the reducing agent TCEP before the toxin endocytosis inhibited the proteolytic activity of BoNT/B in human neuronal SHSY-5Ycells (35). It is hypothesized that the disulfide bridge between LC and HC is intact in the low pH, oxidizing environment of the endosomal lumen. Once LC is translocated across the membranes, the disulfide bridge is reduced in the neutral pH, reducing cytoplasm, which results in LC release [92]. Moreover, in an in vitro planar lipid bilayer system, BoNT can conduct the translocation of LC into the trans compartment without the presence of additional cellular factors [10] Together, these results strongly support a model that HC-LC complex embedded in the membrane is a transmembrane chaperone. The HC chaperone activity driven by a pH gradient across the endosome prevents aggregation of the LC in the acidic vesicle interior, maintains the LC in a unfolded conformation during translocation, and releases it after it refolds at the neutral-pH cytosol. In this model, the reduction of interchain disulfide bond only occurs after the LC translocation to the cytosol [36]. Although in planar lipid bilayer, reduction of the disulfide bond occurs without additional factors. Presently, it is not clear if any cytosolic redox enzymes facilitate this process in vivo.

3.2.2. Anthrax toxin: Pore formation and translocation require intact disulfide bonds of the receptors

Anthrax toxin, produced by *Bacillus anthracis,* is responsible for the major symptoms of anthrax disease [14]. Anthrax toxin is a tripartite AB toxin consisting of two A moieties, lethal factor (LF, 90 kDa) and edema factor (EF, 90 kDa), and one B moiety, protective antigen (PA, 83 kDa). Anthrax toxin-mediated intoxication of host cells starts with PA binding to the cell surface receptors. Currently, two receptors for PA have been identified: anthrax toxin receptor 1 (ANTXR1; or, tumor endothelial marker 8, TEM8) and anthrax toxin receptor 2 (ANTXR2; or, capillary morphogenesis protein 2, CMG2) [101-103]. The extracellular domains of the two receptors share over 60% of sequence homology. Both contain a conserved von willebrand factor A (VWA) domain, which binds to PA (104), and a newly defined immunoglobulin-like (Ig-like) domain [37]. Upon binding to the cell surface receptors, PA83 is cleaved by the cellular protease furin into PA63 and PA20. The PA63 self-assembles into a heptameric or an octameric complex, called prepore, to which LF and EF bind. Endocytosis of anthrax toxin is a highly regulated event, in which S-palmitoylation of the receptor cytoplasmic tail plays a role to prevent constitutive endocytosis of the toxin. The toxin-receptor complex is redistributed on the plasma membrane from the glycerophospholipidic regions to the specialized domains of lipid rafts, where receptor ubiquitination triggers endocytosis. The toxin-receptor complex is supposedly internalized into the cell through the clathrin-mediated endocytosis [12,105,106]. Within the endosomes, acidification triggers conformational change on PA and converts the prepore into a pore on the endosomal membranes, through which LF and EF are translocated to the cytosol. There, LF, a zinc-dependent protease, cleaves MAP kinase kinases, which results in lethality of the host cells; EF, an adenylate cyclase, increases cellular cAMP level, which causes edema.

Unlike many other AB toxins that are produced as a single polypeptide chain and require proteolytic cleavage to generate A and B moieties linked by an interchain disulfide bond, anthrax toxin is produced as the three separate polypeptides: PA, LF and EF (**Figure 2**). Most interestingly, anthrax toxin has no cysteine residue out of the total 2373 residues in the three proteins. This unique "cysteine-free" feature appears to exclude the possibility of exploiting redox-controlled "thiol-dithiol exchange" as a potential mechanism that regulates anthrax toxin translocation. However, our recent study showed that the disulfide bonds in the Ig-like domain of ANTXR2 were required for anthrax toxin pore formation and membrane translocation [37]. Reduction of the disulfide bonds significantly blocked anthrax toxin pore formation on the liposomal membrane and plasma membranes evidenced by the release of K+ ion, and it also blocked translocation of a model substrate across the cell membranes. More recently, purified PDI was shown to facilitate the refolding of the recombinant extracellular domain of ANTXR2 (Sun, unpublished data), indicating that the receptor disulfide bonds may be subjected to redox regulation by PDI or PDI-like oxidoreductases. The mechanism of anthrax inhibition induced by reduction of the receptor disulfide bonds is not yet fully understood.

Based on the available data, one can hypothesize that anthrax toxin translocation must require an oxidative environment or factors that favor the receptor disulfide bond formation, instead of reducing it as of DT, CT and other toxins. But how might the endosomes maintain a disulfide bond favorable environment? In the early stage of anthrax infection, macrophages are activated by the components of *Bacillus anthracis* and launch a strong oxidative burst within the phagosomes for bacterial killing immediately after phagocytosis of bacteria [4]. In addition, it is reported that anthrax is expressed and plays an essential role in several stages of infection, including the very early stage, in the newly germinated spores within macrophages [107,108]. An earlier study has also shown that anthrax lethal toxin stimulated an oxidative burst in macrophages and induced cytolysis [109]. The toxin-induced macrophage lysis was dependent on the ability of the macrophages to mount an oxidative burst and was inhibited by exogenous antioxidants. Based on the above evidence and given the fact that anthrax toxin requires a redox environment that favors receptor disulfide bond formation for translocation, an intriguing hypothesis is that the bacterium and the toxin stimulate an oxidative burst within the host cells so as to ensure anthrax toxin translocation. Consistent with this hypothesis, *B. anthracis* has evolved mechanisms defending itself against oxidative stress, including superoxide dismutases, peroxidases, and catalases, all of which suppress the damaging Fenton reaction catalyzed by reactive oxygen or nitrogen species [4,110,111]. Moreover, as mentioned above, the unusual "cysteine-free" feature of anthrax toxin might have been selected through evolution permitting the toxin to be exempt from the damaging thiol-modifications caused by oxidative stress. In summary, Bacillus anthracis and anthrax toxin may have evolved the ability to subvert oxidative burst, the host defense mechanism, for their own benefit. Instead of being damaged by oxidative burst, anthrax toxin takes advantage of its oxidizing power that maintains integrity of the receptor disulfide bonds for toxin translocation.

3.3. Thiol-activated cytolysins and GILT: Reduction of undecapeptide cysteine

Thiol-activated cytolysins are a group of pore-forming toxins that are secreted by taxonomically diverse species of gram-positive bacteria, responsible for life-threatening infections [112]. Currently, over 20 family members have been identified, including listeriolysin O (LLO) from *Listeria monocytogenes* that causes meningitis and abortion; perfringolysin O (PFO) from *Clostridium perfringens* that causes gas gangrene; and pneumolysin (PLY) from *Streptococcus pneumoniae* that causes pneumonia and meningitis. Each of these toxins is produced as a single polypeptide chain with molecular weight ranging from 50 – 80 kDa and shares high degree of sequence similarity ranging from 40 –

80%, suggesting a close structural and functional relation between them. In deed, the toxins share a common mode of action. All of the thiol-activated cytolysins are produced as water-soluble monomers and use cholesterol as cell surface receptors. Upon binding to the cholesterol, these toxins undergo cholesterol-dependent oligomerization and membrane insertion, leading to membrane damage. Thus, they are also referred to as cholesterol-dependent cytolysins, CDCs. The diameters of the ring-shaped pores can exceed 150 Å, making these toxins a widely used tool as membrane-permeabilizing agents in cell biology. Not surprisingly, all the available crystal structures of the toxins share an elongated, four-domain structure. Upon pore formation, the toxins undergo dramatic domain rearrangements that have been recently revealed by excellent fluorescence measurements and by cryo-electronic microscopy study [113-115].

Thiol-activated cytolysins share another critical common feature, that is, the toxins are activated by reducing agents and suppressed by oxidation [116,117]. The requirement of thiol-reduction for the toxin activation appears to rely on a single cysteine residue located in a highly conserved undecapeptide (also known as tryptophan-rich region) in domain 4 of the toxins [118,119]. The undecapeptide cysteine is the only cysteine present in the primary structure of the secreted toxins. Irreversible oxidation of this cysteine inhibited cytolytic activity, suggesting that this cysteine plays a central role in the cytolytic mechanism. Recently, GILT was found to be a critical host factor for *Listeria monocytogenes* infection [39]. As an intracellular pathogen, *L. monocytogenes* is internalized into the phagosomes, where it secrets LLO to form pores on the endosomal membranes that facilitate bacterial escape from the phagosomes to the cytosol. Since LLO activation requires reducing activity, the authors speculated that GILT, the only known thiol oxidoreductase present in the phagosomes, might activate LLO in vivo. The results confirmed the authors' hypothesis, showing that mice lacking GILT are resistant to *L. monocytogenes* infection; GILT activates LLO within the phagosomes by the thiol reductase activity and purified GILT activates recombinant LLO in vitro.

While the thiol(s) targeted by GILT was not directly identified, the highly conserved, undecapeptide cysteine is obviously the potential target. Besides LLO, GILT also activates streptolysin O (SLO), produced by *Streptococcus pyogenes,* as measured by the haemolysis of sheep red blood cells. But GILT failed to activate the SLO mutant that lacked the undecapeptide cysteine residue. Thus, GILT presumably targets to the characteristic cysteine residue and GILT-mediated exposure of this critical cysteine may result in a conformational change that allows the formation of the pre-pore complex and full activation.

4. Cellular entry of pathogenic bacteria, viruses, and parasites that require cellular redox factors

Available evidence has indicated that cellular redox factors are widely involved in entry of numerous microbial pathogens, ranging from bacteria, viruses and parasites. Here, we briefly review the best-characterized examples from each category.

4.1. Pathogenic bacteria: Chlamydia entry

Chlamydia trachomatis is the leading bacterial agent responsible for sexually transmitted diseases. Two biovariants of *C. trachomatis,* trachoma and lymphogranuloma venereum, cause 90 million new sexually transmitted infections per annum and 400 – 600 million cases of trachoma worldwide. As an obligatory intracellular pathogen, it requires host invasion for survival and growth. However, little is known about the molecular mechanism of Chlamydia entry into host cells. Serovar E, an adhesion molecule from *C. trachomatis,* is known to be required for invasion of genital epithelial cells, but the host factor(s) required for the pathogen entry was not known until PDI was identified as an potential mediator [43]. PDI was detected in an earlier immunoprecipitation experiment [120]., in which a biotinylated apical membrane protein receptor attached to elementary body (EB) was stripped off the surface of HE-1B cells and immunoprecipitated with anti-EB antibodies, followed by 2D SDS-PAGE and MALDI MS analysis. During EB attachment, exposure of HEC-1B cells to three different inhibitors of PDI reductive activity (DTNB, bacitracin, and anti-PDI antibodies) resulted in reduced chlamydial infection. Subsequently, a proteomic study of CHO6 cell line [121], a mutagenized cell line resistant to attachment and infection by Chlamydia, showed that CHO6 has a defect in processing of the leader sequence of PDI, which results in altered cellular distribution of PDI. PDI is abundantly localized in the ER, and surface localization is predominantly sequestered to large patches compared to the punctate pattern in the wild type CHOK1 cells. Complementation by expression of full-length PDI restored *C. trachomatis* binding and infectivity in the CHO6 mutant cell line. These data directly demonstrate that native PDI at the cell surface is required for effective chlamydial attachment and infection. Most recently, RNA interference was used to confirm that cellular PDI is essential for Chlamydial attachment to cells [44]. More precisely, the role of PDI in the process of chlamydial infection was further dissected using genetic complementation and PDI-specific inhibitors, showing that PDI has two essential and independent roles in the process of chlamydial infection. It is structurally required for chlamydial attachment, and the thiol-mediated oxido-reductive function is necessary for entry. While PDI is required for chlamydial attachment, it does not function as a receptor for the pathogen. Other host factor(s) that structurally associate with PDI may be required for chlamydial attachment.

4.2. Viral entry

As obligatory intracellular parasites, viruses can only replicate within host cells. Most viruses that infect vertebrate and insect cells exploit the endocytic pathways to enter the host cells, particularly through macropinocytosis [15,16]. Entry of enveloped virus to the host cells normally requires binding of virus to the cell surface and fusion of the viral membrane with the host cell membrane. These processes are accomplished through a coordinated interaction between viral envelope glycoproteins and host cell surface receptors, during which conformational changes of the proteins involved play an essential role in virus binding and/or membrane fusion. Increasing evidence suggests that the conformational changes are largely triggered by isomerization or reduction of the disulfide bonds catalyzed by either viral- or host- redox factors. The requirement of redox factors for viral entry is exemplified with human immunodeficiency virus (HIV), Newcastle disease virus [122-124], Sindbis virus [125] and avian leukosis virus [126], etc. For the non-enveloped virus SV40, the virus is internalized through endocytosis and retrograde transported to ER, where it makes use of the thiol-disulfide oxidoreductases, ERp57 and PDI as well as the retrotranslocation proteins, to move to the cytosol. The HIV and SV40 viruses are discussed below.

4.2.1. HIV

Infection of human immunodeficiency virus (HIV) starts with viral binding to attachment factors, such as mannose binding C-type lectin receptor and intracellular adhesion molecule on the surface of CD4+ lymphocytes. The HIV envelope glycoprotein gp120 binds to CD4 protein, the primary receptor of HIV-1, and undergoes conformational changes, which allows the virus to interact with its co-receptors, CXCR4 or CCR5. Subsequently, these interactions stimulate downstream conversion of HIV gp41 envelope subunit to a competent fusion conformation [45]. Initially, inspired by the finding that the cell-surface PDI reductive activity is required for DT entry and cytotoxicity through reduction of the interchain disulfide bond, the roles of cell-surface PDI in HIV entry into human lymphoid cells were tested with PDI inhibitors, DTNB, bacitracin and anti-PDI antibodies [127]. The result showed that HIV infection was markedly inhibited by those inhibitors, suggesting that HIV and its target cell engage in the PDI-mediated thiol-disulfide interchange and that the reduction of critical disulfides in viral envelope glycoproteins may be the initial event that triggers conformational changes required for HIV entry and cell infection. This finding revealed a novel direction in the study of HIV entry. A series of experiments have been performed to define the roles of PDI or PDI-like redox enzymes in viral entry and membrane fusion, particularly in the aspect of thiol-disulfide interchange on viral- and host-factors. PDI was first found clustered at the CD4+ lymphocyte surface in the vicinity of CD4-enriched regions and later PDI was co-precipitated with both soluble and cellular CD4 [128]. Moreover, anti-PDI antibodies and the inhibitors of its catalytic function altered HIV envelope-mediated membrane fusion, which suggests that PDI catalytic activity functions in the HIV envelope-mediated cell-cell fusion in a post-CD4 binding step [129]. It is believed that PDI-CD4 interaction at the cell surface enables PDI to reach CD4-bound viral glycoproteins. HIV gp120 is a highly disulfide-bonded molecule that attaches HIV to CD4 and co-receptor CXCR4 or CCR5, thus it becomes a potential target of cell-surface PDI. It has been shown that soluble PDI cleaved disulfide bonds in the recombinant gp120 in vitro and the gp120 bound to the CD4 on the cell surface undergoes a disulfide reduction that is prevented by the PDI inhibitors [130]. Furthermore, additional studies showed that on average two of the nine disulfides of gp120 are reduced during interaction with the lymphocyte surface after CXCR4 binding prior to fusion and that the cell surface PDI catalyzes this process. Thus, the PDI-mediated disulfide restructuring within the HIV envelope constitutes the molecular basis of the post-receptor binding conformational changes that induce fusion competence. Due to the essential role of PDI in HIV entry, PDI is regarded as a potential drug target. Most recently, a high-throughput screening of PDI-specific inhibitors identified the natural compound juniferdin as the most potent inhibitor of PDI. And derivatives of juniferdin were synthesized and used to carry out further studies, of these, compound 13 showed comparable inhibitory activity but reduced cytotoxicity, compared to juniferdin [131].

Interestingly, PDI knockdown by siRNA in U373 and HeLa cells had little effect on HIV infection as compared to the effect mediated by general thiol inhibitors [132]. This discrepancy raised the question whether the reductive activity of PDI is coupled to other redox enzymes that could enhance the redox-dependent viral membrane fusion and entry. This hypothesis has been supported by several recent studies. The extracellular portion of CD4 contains four immunoglobulin-like domains, D1 to D4. The D2 disulfide bond appeared redox-active and regulated by thioredoxin that is secreted by CD4(+) T cells. Locking the CD4 and the thioredoxin active-site dithiols in the reduced state with a hydrophilic trivalent arsenical blocked entry of HIV-1 into host cells [133]. More recently, human glutaredoxin-1 (Grx1) has been shown to efficiently catalyze gp120 and CD4 disulfide reduction in vitro [134]. Grx1 catalyzes the reduction of two disulfide bridges in gp120 in a similar manner to that of PDI. Anti-Grx1 antibodies inhibited the Grx1 activity and block HIV-1 replication in cultured peripheral blood mononuclear cells. The polyanion PRO2000, previously shown to prevent HIV entry, inhibited the Grx1- and PDI-dependent reduction of gp120 disulfides. Thus, other redox enzymes other than PDI may also be involved in HIV entry. Studies that further dissect the specific roles of PDI and other redox enzymes in HIV entry are needed to uncover the mechanism of HIV entry.

4.2.2. SV40

Simian virus 40 (SV40) is a simple, non-enveloped DNA virus that belongs to the polyoma virus family. It uses ganglioside GM1 as receptor, and enters host cells through a unique endocytic pathway, caveolae/lipid raft-mediated endocytosis [15,47,135]. After internalization, instead of trafficking to endosome/lysosome compartments, it traffics into a pH-neutral, caveolin-containing endocytic organelle, called caveosome. From there the virus moves in noncaveolar vesicles along microtubules to the ER through retrograde transport. In the ER, SV40 manages to translocate into the cytosol, and from there it enters the nucleus via nuclear pore complexes for viral replication. SV40 capsids are composed of homopentamers of the major capsid protein VP1, and VP1 is associated with one of the minor structural proteins VP2 or VP3. The virus has icosahedral symmetry and contains 72 pentamers, of which 12 are five-coordinated and the rest of 60 are six-coordinated. The pentamers are linked to each other by the interchain disulfide bonds between the residues Cys104. Isomerization of the disulfide bonds in the ER is crucial for the viral uncoating process [136-138]. Recent data has shown that SV40 makes use of the protein folding and quality control machinery in the ER for initial uncoating and membrane translocation [46]. Among all the ER-resident proteins, ERp57 and PDI more specifically regulate SV40 infection through isomerization of the disulfide bonds. Silencing of ERp57 and PDI substantially decreases SV40 infection. In addition, these ERp57 and PDI cooperate with the

ER-associated degradation (ERAD) proteins, Derlin-1 and Sel1L, facilitating a Ca^{2+}-dependent retrotranslocation from the ER to the cytosol.

4.3. Parasite entry

4.3.1. PDI, NADPH oxidase and Leishmania entry

Leishmania, a family of obligate intracellular parasites, causes leishmaniases in millions of individuals worldwide [139]. The parasites are transmitted by a variety of species of sand flies from two major genera Phlebotomus and Lutzomyia, respectively. The life cycle of Leishmania starts with a motile promastigote form in the insect host, in which they attach to the midgut wall to avoid being expelled with the blood meal. They later migrate to the digestive tract and differentiate into a non-dividing metacyclic form. The metacyclic promastimgotes are injected into the mammalian host, enter macrophages and differentiate into non-flagellated amastigotes that replicate and persist intracellularly, which provides a reservoir for transmission. Entry of promastimgotes into macrophages through phagocytosis is a critical step for leishmania infection. Inside the phagosome and/or phagolysosome vesicles, the parasites are exposed to enzymes, antimicrobial peptides, and ROS generated by NADPH oxidase activation [140]. Surprisingly, the promastimgotes are able to survive in such a stressful environment and differentiate into amastigotes, progressing to an active infectious disease. While the mechanism of pathogenesis is still elusive, current research has greatly advanced our understanding of the process. It is well known that Nox2 oxidase, a prototypic member of the NADPH oxidase family, is activated during phagocytosis of Leishmania, and uptake of the parasites is inhibited by antioxidants (e. g. catalase) [141]. This suggests that the oxidative stress induced by NADPH oxidase activation may have a favorable effect, instead of the expected anti-microbial effect, with regard to parasite infectivity. Recently, PDI has been shown to be involved in phagocytosis of Leishmania chagasi through regulation of NADPH oxidase, in which PDI was found to closely associate with the NADPH oxidase, and inhibitors of PDI (bacitracin, phenylarsine oxide, anti-PDI antibody) significantly blocked promastigote phagocytosis (40). These results correlate well with, and are supported by, the previous findings that proteomic study of macrophages showed that PDI is involved in the formation of the phagosomes during phagocytosis of some parasites, including Leishmania (66,81). And PDI is closely associated with NADPH oxidase and plays an organizer role in NADPH oxidase activation [41,72].

5. Conclusions and future perspectives

Current studies have revealed excitingly novel features of host-pathogen interaction. Microbial pathogens (bacteria, viruses, and parasites) and bacterial toxins exploit different aspects of the endocytic pathways, and hijack cellular redox factors to accomplish entry and invasion. Despite the fact that the pathogens and toxins are very diverse, the mechanism involved in the infectivity could be readily narrowed down to any of the simple redox reactions: reduction, oxidation, or isomerization of the thiol/dithiol groups on proteins from either pathogens or hosts. Therefore, these redox reactions could potentially be part of a general mechanism for pathogen and toxin invasion, and this redox-dependent entry mechanism can be an attractive target for anti-microbial and anti-toxin drug development. The specificity of regulation is supposed to heavily rely on specific interaction of the cellular redox factors with the pathogens in a timely and spatial manner in various endocytic pathways. Thus far, PDI seems to be involved in many of the pathogen and toxin entry events, which is, at least in part, due to its diverse cellular distribution, ranging from the cell surface to the ER. Further studies are needed to address this question if other members of the PDI family and other ER-residing folding machinery are involved in pathogen and toxin entry. It is exciting that GILT was identified as the first oxidoreductase active optimally in the acidic compartments. The roles of GILT in entry of pathogens or toxins other than *L. monocytogenes*/LLO warrant further investigation. Finally, some pathogens or toxins can survive through the oxidative burst within the phagosomes and even can take advantage of this host defense mechanism for invasion. Thus, more studies are needed to look into the new insights concerning the role of oxidative burst in host-pathogen interactions.

Author details

Jianjun Sun
Department of Biological Sciences, University of Texas at El Paso, El Paso, TX, USA

Acknowledgement

I apologize to those individuals whose work was not discussed in this article due to space limitations. I thank Dr. Giulio Francia and Caroline Chen for proofreading the article. The author's work is supported by grants from the NIH 5SC1GM095475, G12MD007592 and 5G12RR008124.

6. References

[1] Barbieri JT, Riese MJ, Aktories K. Bacterial toxins that modify the actin cytoskeleton. Annu Rev Cell Dev Biol. 2002;18:315–344.
[2] Aktories K, Barbieri JT. Bacterial cytotoxins: targeting eukaryotic switches. Nat Rev Micro. 2005 May 1;3(5):397–410.
[3] Ham H, Sreelatha A, Orth K. Manipulation of host membranes by bacterial effectors. Nat Rev Micro. 2011 Sep.;9(9):635–646.
[4] Shatalin K, Gusarov I, Avetissova E, Shatalina Y, McQuade LE, Lippard SJ, et al. Bacillus anthracis-derived nitric oxide is essential for pathogen virulence and survival in macrophages. Proc Natl Acad Sci USA. 2008 Jan. 22;105(3):1009–1013.
[5] van der Wel N, Hava D, Houben D, Fluitsma D, van Zon M, Pierson J, et al. M. tuberculosis and M. leprae translocate from the phagolysosome to the cytosol in myeloid cells. Cell. 2007 Jun. 29;129(7):1287–1298.
[6] Ernst JD. Bacterial inhibition of phagocytosis. Cell Microbiol. 2000 Oct.;2(5):379–386.

[7] Enninga J, Rosenshine I. Imaging the assembly, structure and activity of type III secretion systems. Cell Microbiol. 2009 Oct. 1;11(10):1462–1470.
[8] Juhas M, Crook DW, Hood DW. Type IV secretion systems: tools of bacterial horizontal gene transfer and virulence. Cell Microbiol. 2008 Dec. 1;10(12):2377–2386.
[9] Llosa M, Roy C, Dehio C. Bacterial type IV secretion systems in human disease. Mol Microbiol. 2009 Jul. 1;73(2):141–151.
[10] Lord JM, Smith DC, Roberts LM. Toxin entry: how bacterial proteins get into mammalian cells. Cell Microbiol. 1999 Sep.;1(2):85–91.
[11] Binz T, Rummel A. Cell entry strategy of clostridial neurotoxins. J. Neurochem. 2009 Jun.;109(6):1584–1595.
[12] van der Goot G, Young JAT. Receptors of anthrax toxin and cell entry. Mol Aspects Med. 2009 Dec.;30(6):406–412.
[13] Geny B, Popoff MR. Bacterial protein toxins and lipids: pore formation or toxin entry into cells. Biol. Cell. 2006 Nov.;98(11):667–678.
[14] Collier RJ, Young JAT. Anthrax toxin. Annu Rev Cell Dev Biol. 2003;19:45–70.
[15] Mercer J, Helenius A. Virus entry by macropinocytosis. Nature. 2009 May;11(5):510–520.
[16] Carrasco L. Entry of animal viruses and macromolecules into cells. FEBS Lett. 1994 Aug. 22;350(2-3):151–154.
[17] Gruenheid S, Finlay BB. Microbial pathogenesis and cytoskeletal function. Nature. 2003 Apr. 17;422(6933):775–781.
[18] Pizarro-Cerdá J, Cossart P. Bacterial adhesion and entry into host cells. Cell. 2006 Feb. 24;124(4):715–727.
[19] Falnes PO, Sandvig K. Penetration of protein toxins into cells. Curr. Opin. Cell Biol. 2000 Aug.;12(4):407–413.
[20] Mamathambika BS, Bardwell JC. Disulfide-linked protein folding pathways. Annu Rev Cell Dev Biol. 2008;24:211–235.
[21] Welker E, Wedemeyer WJ, Narayan M, Scheraga HA. Coupling of conformational folding and disulfide-bond reactions in oxidative folding of proteins. Biochemistry. 2001 Aug. 7;40(31):9059–9064.
[22] Chen VM, Hogg PJ. Allosteric disulfide bonds in thrombosis and thrombolysis. J Thromb Haemost. 2006 Dec. 1;4(12):2533–2541.
[23] Hogg PJ. Contribution of allosteric disulfide bonds to regulation of hemostasis. J Thromb Haemost. 2009 Jul. 1;7 Suppl 1:13–16.
[24] Hogg PJ. Disulfide bonds as switches for protein function. Trends Biochem Sci. 2003 Apr. 1;28(4):210–214.
[25] Schmidt B, Ho L, Hogg PJ. Allosteric disulfide bonds. Biochemistry. 2006 Jun. 20;45(24):7429–7433.
[26] Hogg PJ. Biological regulation through protein disulfide bond cleavage. Redox Rep. 2002;7(2):71–77.
[27] Halvey PJ, Watson WH, Hansen JM, Go Y-M, Samali A, Jones DP. Compartmental oxidation of thiol-disulphide redox couples during epidermal growth factor signalling. Biochem J. 2005 Mar. 1;386(Pt 2):215–219.

[28] Go Y-M, Jones DP. Redox compartmentalization in eukaryotic cells. Biochim Biophys Acta. 2008 Nov.;1780(11):1273–1290.
[29] Falnes PO, Choe S, Madshus IH, Wilson BA, Olsnes S. Inhibition of membrane translocation of diphtheria toxin A-fragment by internal disulfide bridges. J Biol Chem. 1994 Mar. 18;269(11):8402–8407.
[30] Papini E, Rappuoli R, Murgia M, Montecucco C. Cell penetration of diphtheria toxin. Reduction of the interchain disulfide bridge is the rate-limiting step of translocation in the cytosol. J Biol Chem. 1993 Jan. 25;268(3):1567–1574.
[31] Ryser HJ, Mandel R, Ghani F. Cell surface sulfhydryls are required for the cytotoxicity of diphtheria toxin but not of ricin in Chinese hamster ovary cells. J Biol Chem. 1991 Oct. 5;266(28):18439–18442.
[32] Feener EP, Shen WC, Ryser HJ. Cleavage of disulfide bonds in endocytosed macromolecules. A processing not associated with lysosomes or endosomes. J Biol Chem. 1990 Nov. 5;265(31):18780–18785.
[33] Mandel R, Ryser HJ, Ghani F, Wu M, Peak D. Inhibition of a reductive function of the plasma membrane by bacitracin and antibodies against protein disulfide-isomerase. Proc Natl Acad Sci USA. 1993 May 1;90(9):4112–4116.
[34] Tsai B, Rodighiero C, Lencer WI, Rapoport TA. Protein disulfide isomerase acts as a redox-dependent chaperone to unfold cholera toxin. Cell. 2001 Mar. 23;104(6):937–948.
[35] Shi X, Garcia GE, Neill RJ, Gordon RK. TCEP treatment reduces proteolytic activity of BoNT/B in human neuronal SHSY-5Y cells. J. Cell. Biochem. 2009 Aug. 1;107(5):1021–1030.
[36] Fischer A, Montal M. Crucial role of the disulfide bridge between botulinum neurotoxin light and heavy chains in protease translocation across membranes. J Biol Chem. 2007 Oct. 5;282(40):29604–29611.
[37] Sun J, Collier RJ. Disulfide bonds in the ectodomain of anthrax toxin receptor 2 are required for the receptor-bound protective-antigen pore to function. PLoS ONE. 2010;5(5):e10553.
[38] Stolf BS, Ioannis S, Lopes LR, Vendramin A, Goto H, Laurindo FRM, et al. Protein disulfide isomerase and host-pathogen interaction. ScientificWorldJournal. 2011;11:1749–1761.
[39] Singh R, Jamieson A, Cresswell P. GILT is a critical host factor for Listeria monocytogenes infection. Nature. 2008 Oct. 30;455(7217):1244–1247.
[40] Santos CXC, Stolf BS, Takemoto PVA, Amanso AM, Lopes LR, Souza EB, et al. Protein disulfide isomerase (PDI) associates with NADPH oxidase and is required for phagocytosis of Leishmania chagasi promastigotes by macrophages. Journal of Leukocyte Biology. 2009 Oct.;86(4):989–998.
[41] Laurindo FRM, Fernandes DC, Amanso AM, Lopes LR, Santos CXC. Novel role of protein disulfide isomerase in the regulation of NADPH oxidase activity: pathophysiological implications in vascular diseases. Antioxid Redox Signal. 2008 Jun.;10(6):1101–1113.

[42] Fernandes DC, Manoel AHO, Wosniak J, Laurindo FR. Protein disulfide isomerase overexpression in vascular smooth muscle cells induces spontaneous preemptive NADPH oxidase activation and Nox1 mRNA expression: effects of nitrosothiol exposure. Arch Biochem Biophys. 2009 Apr. 15;484(2):197–204.
[43] Dautry-Varsat A, Subtil A, Hackstadt T. Recent insights into the mechanisms of Chlamydia entry. Cell Microbiol. 2005 Dec.;7(12):1714–1722.
[44] Abromaitis S, Stephens RS. Attachment and entry of Chlamydia have distinct requirements for host protein disulfide isomerase. PLoS Pathog. 2009 Apr.;5(4):e1000357.
[45] Ryser HJ-P, Flückiger R. Progress in targeting HIV-1 entry. Drug Discov Today. 2005 Aug. 15;10(16):1085–1094.
[46] Schelhaas M, Malmström J, Pelkmans L, Haugstetter J, Ellgaard L, Grünewald K, et al. Simian Virus 40 depends on ER protein folding and quality control factors for entry into host cells. Cell. 2007 Nov. 2;131(3):516–529.
[47] Medina-Kauwe LK. "Alternative" endocytic mechanisms exploited by pathogens: new avenues for therapeutic delivery? Adv Drug Deliv Rev. 2007 Aug. 10;59(8):798–809.
[48] Yang J, Chen H, Vlahov IR, Cheng J-X, Low PS. Evaluation of disulfide reduction during receptor-mediated endocytosis by using FRET imaging. Proc Natl Acad Sci USA. 2006 Sep. 12;103(37):13872–13877.
[49] Saito G, Swanson JA, Lee K-D. Drug delivery strategy utilizing conjugation via reversible disulfide linkages: role and site of cellular reducing activities. Adv Drug Deliv Rev. 2003 Feb. 10;55(2):199–215.
[50] Bareford LM, Swaan PW. Endocytic mechanisms for targeted drug delivery. Adv Drug Deliv Rev. 2007 Aug. 10;59(8):748–758.
[51] Rajendran L, Knölker H-J, Simons K. Subcellular targeting strategies for drug design and delivery. Nat Rev Drug Discov. 2010;9(1):29–42.
[52] Jones AT, Gumbleton M, Duncan R. Understanding endocytic pathways and intracellular trafficking: a prerequisite for effective design of advanced drug delivery systems. Adv Drug Deliv Rev. 2003 Nov. 14;55(11):1353–1357.
[53] Austin CD, Wen X, Gazzard L, Nelson C, Scheller RH, Scales SJ. Oxidizing potential of endosomes and lysosomes limits intracellular cleavage of disulfide-based antibody-drug conjugates. Proc Natl Acad Sci USA. 2005 Dec. 13;102(50):17987–17992.
[54] Morré DJ, Chueh PJ, Lawler J, Morré DM. The sulfonylurea-inhibited NADH oxidase activity of HeLa cell plasma membranes has properties of a protein disulfide-thiol oxidoreductase with protein disulfide-thiol interchange activity. J. Bioenerg. Biomembr. 1998 Oct.;30(5):477–487.
[55] Halliwell B. Biochemistry of oxidative stress. Biochem Soc Trans. 2007 Nov. 1;35(Pt 5):1147–1150.
[56] Iles KE, Forman HJ. Macrophage signaling and respiratory burst. Immunol Res. 2002;26(1-3):95–105.
[57] Gwinn MR, Vallyathan V. Respiratory burst: role in signal transduction in alveolar macrophages. J Toxicol Environ Health B Crit Rev. 2006;9(1):27–39.

[58] Forman HJ, Torres M. Reactive oxygen species and cell signaling: respiratory burst in macrophage signaling. American Journal of Respiratory and Critical Care Medicine. 2002 Dec. 15;166(12 Pt 2):S4–8.
[59] Turano C, Coppari S, Altieri F, Ferraro A. Proteins of the PDI family: unpredicted non-ER locations and functions. J. Cell. Physiol. 2002 Nov.;193(2):154–163.
[60] Appenzeller-Herzog C, Ellgaard L. The human PDI family: versatility packed into a single fold. Biochim Biophys Acta. 2008 Apr. 1;1783(4):535–548.
[61] Ellgaard L, Ruddock LW. The human protein disulphide isomerase family: substrate interactions and functional properties. EMBO Rep. 2005;6(1):28–32.
[62] Kersteen EA, Raines RT. Catalysis of protein folding by protein disulfide isomerase and small-molecule mimics. Antioxid Redox Signal. 2003 Aug. 1;5(4):413–424.
[63] Terada K, Manchikalapudi P, Noiva R, Jauregui HO, Stockert RJ, Schilsky ML. Secretion, surface localization, turnover, and steady state expression of protein disulfide isomerase in rat hepatocytes. J Biol Chem. 1995 Sep. 1;270(35):20410–20416.
[64] Kim Y, Kang K, Kim I, Lee YJ, Oh C, Ryoo J, et al. Molecular mechanisms of MHC class I-antigen processing: redox considerations. Antioxid Redox Signal. 2009 Apr.;11(4):907–936.
[65] Park B, Lee S, Kim E, Cho K, Riddell SR, Cho S, et al. Redox regulation facilitates optimal peptide selection by MHC class I during antigen processing. Cell. 2006 Oct. 20;127(2):369–382.
[66] Gagnon E, Duclos S, Rondeau C, Chevet E, Cameron PH, Steele-Mortimer O, et al. Endoplasmic reticulum-mediated phagocytosis is a mechanism of entry into macrophages. Cell. 2002 Jul. 12;110(1):119–131.
[67] Tsai B, Walczak CP, Bernardi KM. Endoplasmic reticulum-dependent redox reactions control ER-associated degradation and pathogen entry. Antioxid Redox Signal. 2011 Dec. 5.
[68] Jiang XM, Fitzgerald M, Grant CM, Hogg PJ. Redox control of exofacial protein thiols/disulfides by protein disulfide isomerase. J Biol Chem. 1999 Jan. 22;274(4):2416–2423.
[69] Lahav J, Gofer-Dadosh N, Luboshitz J, Hess O, Shaklai M. Protein disulfide isomerase mediates integrin-dependent adhesion. FEBS Lett. 2000 Jun. 16;475(2):89–92.
[70] Jordan PA, Gibbins JM. Extracellular disulfide exchange and the regulation of cellular function. Antioxid Redox Signal. 2006 Feb.;8(3-4):312–324.
[71] Chueh P-J, Morré DM, Morré DJ. A site-directed mutagenesis analysis of tNOX functional domains. Biochim Biophys Acta. 2002 Jan. 31;1594(1):74–83.
[72] Janiszewski M, Lopes LR, Carmo AO, Pedro MA, Brandes RP, Santos CXC, et al. Regulation of NAD(P)H oxidase by associated protein disulfide isomerase in vascular smooth muscle cells. J Biol Chem. 2005 Dec. 9;280(49):40813–40819.
[73] de A Paes AM, Veríssimo-Filho S, Guimarães LL, Silva ACB, Takiuti JT, Santos CXC, et al. Protein disulfide isomerase redox-dependent association with p47(phox): evidence for an organizer role in leukocyte NADPH oxidase activation. Journal of Leukocyte Biology. 2011 Oct.;90(4):799–810.

[74] Gainey D, Short S, McCoy KL. Intracellular location of cysteine transport activity correlates with productive processing of antigen disulfide. J. Cell. Physiol. 1996 Aug.;168(2):248–254.
[75] Creighton TE, Zapun A, Darby NJ. Mechanisms and catalysts of disulfide bond formation in proteins. Trends Biotechnol. 1995 Jan.;13(1):18–23.
[76] Arunachalam B, Phan UT, Geuze HJ, Cresswell P. Enzymatic reduction of disulfide bonds in lysosomes: characterization of a gamma-interferon-inducible lysosomal thiol reductase (GILT). Proc Natl Acad Sci USA. 2000 Jan. 18;97(2):745–750.
[77] Phan UT, Arunachalam B, Cresswell P. Gamma-interferon-inducible lysosomal thiol reductase (GILT). Maturation, activity, and mechanism of action. J Biol Chem. 2000 Aug. 25;275(34):25907–25914.
[78] Glickman JN, Morton PA, Slot JW, Kornfeld S, Geuze HJ. The biogenesis of the MHC class II compartment in human I-cell disease B lymphoblasts. J Cell Biol. 1996 Mar.;132(5):769–785.
[79] Singh R, Cresswell P. Defective cross-presentation of viral antigens in GILT-free mice. Science. 2010 Jun. 11;328(5984):1394–1398.
[80] Luster AD, Weinshank RL, Feinman R, Ravetch JV. Molecular and biochemical characterization of a novel gamma-interferon-inducible protein. J Biol Chem. 1988 Aug. 25;263(24):12036–12043.
[81] Garin J, Diez R, Kieffer S, Dermine JF, Duclos S, Gagnon E, et al. The phagosome proteome: insight into phagosome functions. J Cell Biol. 2001 Jan. 8;152(1):165–180.
[82] Barth H, Aktories K, Popoff MR, Stiles BG. Binary bacterial toxins: biochemistry, biology, and applications of common Clostridium and Bacillus proteins. Microbiol Mol Biol Rev. 2004 Sep. 1;68(3):373–402, table of contents.
[83] Collier RJ. Diphtheria toxin: mode of action and structure. Bacteriological reviews. 1975 Mar. 1;39(1):54–85.
[84] Collier RJ. Understanding the mode of action of diphtheria toxin: a perspective on progress during the 20th century. Toxicon. 2001 Nov. 1;39(11):1793–1803.
[85] Ratts R, Zeng H, Berg EA, Blue C, McComb ME, Costello CE, et al. The cytosolic entry of diphtheria toxin catalytic domain requires a host cell cytosolic translocation factor complex. J Cell Biol. 2003 Mar. 31;160(7):1139–1150.
[86] Sears CL, Kaper JB. Enteric bacterial toxins: mechanisms of action and linkage to intestinal secretion. Microbiol. Rev. 1996 Mar.;60(1):167–215.
[87] Fujinaga Y. Transport of bacterial toxins into target cells: pathways followed by cholera toxin and botulinum progenitor toxin. J Biochem. 2006 Aug.;140(2):155–160.
[88] Lencer WI, Constable C, Moe S, Jobling MG, Webb HM, Ruston S, et al. Targeting of cholera toxin and Escherichia coli heat labile toxin in polarized epithelia: role of COOH-terminal KDEL. J Cell Biol. 1995 Nov.;131(4):951–962.
[89] Orlandi PA. Protein-disulfide isomerase-mediated reduction of the A subunit of cholera toxin in a human intestinal cell line. J Biol Chem. 1997 Feb. 14;272(7):4591–4599.

[90] Majoul I, Ferrari D, Söling HD. Reduction of protein disulfide bonds in an oxidizing environment. The disulfide bridge of cholera toxin A-subunit is reduced in the endoplasmic reticulum. FEBS Lett. 1997 Jan. 20;401(2-3):104–108.
[91] Bernardi KM, Forster ML, Lencer WI, Tsai B. Derlin-1 facilitates the retro-translocation of cholera toxin. Mol Biol Cell. 2008 Mar.;19(3):877–884.
[92] Montal M. Translocation of botulinum neurotoxin light chain protease by the heavy chain protein-conducting channel. Toxicon. 2009 Oct.;54(5):565–569.
[93] Lacy DB, Tepp W, Cohen AC, DasGupta BR, Stevens RC. Crystal structure of botulinum neurotoxin type A and implications for toxicity. Nat Struct Biol. 1998 Oct.;5(10):898–902.
[94] Lacy DB, Stevens RC. Sequence homology and structural analysis of the clostridial neurotoxins. J Mol Biol. 1999 Sep. 3;291(5):1091–1104.
[95] Dong M, Liu H, Tepp WH, Johnson EA, Janz R, Chapman ER. Glycosylated SV2A and SV2B mediate the entry of botulinum neurotoxin E into neurons. Mol Biol Cell. 2008 Dec.;19(12):5226–5237.
[96] Dong M, Tepp WH, Liu H, Johnson EA, Chapman ER. Mechanism of botulinum neurotoxin B and G entry into hippocampal neurons. J Cell Biol. 2007 Dec. 31;179(7):1511–1522.
[97] Chen C, Fu Z, Kim J-JP, Barbieri JT, Baldwin MR. Gangliosides as high affinity receptors for tetanus neurotoxin. J Biol Chem. 2009 Sep. 25;284(39):26569–26577.
[98] Yeh FL, Dong M, Yao J, Tepp WH, Lin G, Johnson EA, et al. SV2 mediates entry of tetanus neurotoxin into central neurons. PLoS Pathog. 2010;6(11):e1001207.
[99] Fischer A, Montal M. Single molecule detection of intermediates during botulinum neurotoxin translocation across membranes. Proc Natl Acad Sci USA. 2007 Jun. 19;104(25):10447–10452.
[100] Koriazova LK, Montal M. Translocation of botulinum neurotoxin light chain protease through the heavy chain channel. Nat Struct Biol. 2003 Jan.;10(1):13–18.
[101] Bradley KA, Mogridge J, Mourez M, Collier RJ, Young JA. Identification of the cellular receptor for anthrax toxin. Nature. 2001 Nov. 8;414(6860):225–229.
[102] Scobie HM, Rainey GJA, Bradley KA, Young JAT. Human capillary morphogenesis protein 2 functions as an anthrax toxin receptor. Proc Natl Acad Sci USA. 2003 Apr. 29;100(9):5170–5174.
[103] Liu S, Crown D, Miller-Randolph S, Moayeri M, Wang H, Hu H, et al. Capillary morphogenesis protein-2 is the major receptor mediating lethality of anthrax toxin in vivo. Proc Natl Acad Sci USA. 2009 Jul. 28;106(30):12424–12429.
[104] Lacy DB, Wigelsworth DJ, Scobie HM, Young JAT, Collier RJ. Crystal structure of the von Willebrand factor A domain of human capillary morphogenesis protein 2: an anthrax toxin receptor. Proc Natl Acad Sci USA. 2004 Apr. 27;101(17):6367–6372.
[105] Abrami L, Leppla SH, van der Goot FG. Receptor palmitoylation and ubiquitination regulate anthrax toxin endocytosis. J Cell Biol. 2006 Jan. 16;172(2):309–320.
[106] Abrami L, Liu S, Cosson P, Leppla SH, van der Goot FG. Anthrax toxin triggers endocytosis of its receptor via a lipid raft-mediated clathrin-dependent process. J Cell Biol. 2003 Feb. 3;160(3):321–328.

[107] Guidi-Rontani C, Weber-Levy M, Labruyère E, Mock M. Germination of Bacillus anthracis spores within alveolar macrophages. Mol Microbiol. 1999;31(1):9–17.
[108] Guidi-Rontani C, Levy M, Ohayon H, Mock M. Fate of germinated Bacillus anthracis spores in primary murine macrophages. Mol Microbiol. 2001 Nov. 1;42(4):931–938.
[109] Hanna PC, Kruskal BA, Ezekowitz RA, Bloom BR, Collier RJ. Role of macrophage oxidative burst in the action of anthrax lethal toxin. Mol Med. 1994 Nov. 1;1(1):7–18.
[110] Gusarov I, Nudler E. NO-mediated cytoprotection: instant adaptation to oxidative stress in bacteria. Proc Natl Acad Sci USA. 2005 Sep. 27;102(39):13855–13860.
[111] Cybulski RJ, Sanz P, Alem F, Stibitz S, Bull RL, O'Brien AD. Four superoxide dismutases contribute to Bacillus anthracis virulence and provide spores with redundant protection from oxidative stress. Infect Immun. 2009;77(1):274–285.
[112] Rossjohn J, Feil SC, McKinstry WJ, Tweten RK, Parker MW. Structure of a cholesterol-binding, thiol-activated cytolysin and a model of its membrane form. Cell. 1997 May 30;89(5):685–692.
[113] Ramachandran R, Tweten RK, Johnson AE. The domains of a cholesterol-dependent cytolysin undergo a major FRET-detected rearrangement during pore formation. Proc Natl Acad Sci USA. 2005 May 17;102(20):7139–7144.
[114] Johnson AE. Fluorescence approaches for determining protein conformations, interactions and mechanisms at membranes. Traffic. 2005 Dec. 1;6(12):1078–1092.
[115] Tilley SJ, Orlova EV, Gilbert RJC, Andrew PW, Saibil HR. Structural basis of pore formation by the bacterial toxin pneumolysin. Cell. 2005 Apr. 22;121(2):247–256.
[116] Geoffroy C, Gaillard JL, Alouf JE, Berche P. Purification, characterization, and toxicity of the sulfhydryl-activated hemolysin listeriolysin O from Listeria monocytogenes. Infect Immun. 1987 Jul.;55(7):1641–1646.
[117] Portnoy DA, Chakraborty T, Goebel W, Cossart P. Molecular determinants of Listeria monocytogenes pathogenesis. Infect Immun. 1992 Apr.;60(4):1263–1267.
[118] Soltani CE, Hotze EM, Johnson AE, Tweten RK. Specific protein-membrane contacts are required for prepore and pore assembly by a cholesterol-dependent cytolysin. J Biol Chem. 2007 May 25;282(21):15709–15716.
[119] Soltani CE, Hotze EM, Johnson AE, Tweten RK. Structural elements of the cholesterol-dependent cytolysins that are responsible for their cholesterol-sensitive membrane interactions. Proc Natl Acad Sci USA. 2007 Dec. 18;104(51):20226–20231.
[120] Davis CH, Raulston JE, Wyrick PB. Protein disulfide isomerase, a component of the estrogen receptor complex, is associated with Chlamydia trachomatis serovar E attached to human endometrial epithelial cells. Infect Immun. 2002 Jul.;70(7):3413–3418.
[121] Conant CG, Stephens RS. Chlamydia attachment to mammalian cells requires protein disulfide isomerase. Cell Microbiol. 2007 Jan.;9(1):222–232.
[122] Jain S, McGinnes LW, Morrison TG. Overexpression of thiol/disulfide isomerases enhances membrane fusion directed by the Newcastle disease virus fusion protein. J Virol. 2008 Dec.;82(24):12039–12048.
[123] Jain S, McGinnes LW, Morrison TG. Role of thiol/disulfide exchange in newcastle disease virus entry. J Virol. 2009 Jan.;83(1):241–249.

[124] Jain S, McGinnes LW, Morrison TG. Thiol/disulfide exchange is required for membrane fusion directed by the Newcastle disease virus fusion protein. J Virol. 2007 Mar.;81(5):2328–2339.
[125] Abell BA, Brown DT. Sindbis virus membrane fusion is mediated by reduction of glycoprotein disulfide bridges at the cell surface. J Virol. 1993 Sep.;67(9):5496–5501.
[126] Smith JG, Cunningham JM. Receptor-induced thiolate couples Env activation to retrovirus fusion and infection. PLoS Pathog. 2007 Dec. 21;3(12):e198.
[127] Ryser HJ, Levy EM, Mandel R, DiSciullo GJ. Inhibition of human immunodeficiency virus infection by agents that interfere with thiol-disulfide interchange upon virus-receptor interaction. Proc Natl Acad Sci USA. 1994 May 10;91(10):4559–4563.
[128] Fenouillet E, Barbouche R, Courageot J, Miquelis R. The catalytic activity of protein disulfide isomerase is involved in human immunodeficiency virus envelope-mediated membrane fusion after CD4 cell binding. J Infect Dis. 2001 Mar. 1;183(5):744–752.
[129] Gallina A, Hanley TM, Mandel R, Trahey M, Broder CC, Viglianti GA, et al. Inhibitors of protein-disulfide isomerase prevent cleavage of disulfide bonds in receptor-bound glycoprotein 120 and prevent HIV-1 entry. J Biol Chem. 2002 Dec. 27;277(52):50579–50588.
[130] Barbouche R, Miquelis R, Jones IM, Fenouillet E. Protein-disulfide isomerase-mediated reduction of two disulfide bonds of HIV envelope glycoprotein 120 occurs post-CXCR4 binding and is required for fusion. J Biol Chem. 2003 Jan. 31;278(5):3131–3136.
[131] Khan MMG, Simizu S, Lai NS, Kawatani M, Shimizu T, Osada H. Discovery of a small molecule PDI inhibitor that inhibits reduction of HIV-1 envelope glycoprotein gp120. ACS chemical biology. 2011 Mar. 18;6(3):245–251.
[132] Ou W, Silver J. Role of protein disulfide isomerase and other thiol-reactive proteins in HIV-1 envelope protein-mediated fusion. Virology. 2006 Jul. 5;350(2):406–417.
[133] Matthias LJ, Yam PTW, Jiang X-M, Vandegraaff N, Li P, Poumbourios P, et al. Disulfide exchange in domain 2 of CD4 is required for entry of HIV-1. Nat Immunol. 2002 Aug. 1;3(8):727–732.
[134] Auwerx J, Isacsson O, Söderlund J, Balzarini J, Johansson M, Lundberg M. Human glutaredoxin-1 catalyzes the reduction of HIV-1 gp120 and CD4 disulfides and its inhibition reduces HIV-1 replication. Int J Biochem Cell Biol. 2009 Jun. 1;41(6):1269–1275.
[135] Lin AE-J, Guttman JA. Hijacking the endocytic machinery by microbial pathogens. Protoplasma. 2010 Aug.;244(1-4):75–90.
[136] Li PP, Nakanishi A, Clark SW, Kasamatsu H. Formation of transitory intrachain and interchain disulfide bonds accompanies the folding and oligomerization of simian virus 40 Vp1 in the cytoplasm. Proc Natl Acad Sci USA. 2002 Feb. 5;99(3):1353–1358.
[137] Liddington RC, Yan Y, Moulai J, Sahli R, Benjamin TL, Harrison SC. Structure of simian virus 40 at 3.8-A resolution. Nature. 1991 Nov. 28;354(6351):278–284.
[138] Li PP, Nakanishi A, Tran MA, Salazar AM, Liddington RC, Kasamatsu H. Role of simian virus 40 Vp1 cysteines in virion infectivity. J Virol. 2000 Dec.;74(23):11388–11393.

[139] Sibley LD. Invasion and intracellular survival by protozoan parasites. Immunol Rev. 2011 Mar.;240(1):72–91.

[140] Beattie L, Kaye PM. Leishmania-host interactions: what has imaging taught us? Cell Microbiol. 2011 Nov.;13(11):1659–1667.

[141] Van Assche T, Deschacht M, da Luz RAI, Maes L, Cos P. Leishmania-macrophage interactions: insights into the redox biology. Free Radic Biol Med. 2011 Jul. 15;51(2):337–351.

Chapter 3

Key Events in Synaptic Vesicle Endocytosis

Frauke Ackermann, Joshua A. Gregory and Lennart Brodin

Additional information is available at the end of the chapter

1. Introduction

A synaptic release site is characterized by a pool of synaptic vesicles aggregated to an active zone at the presynaptic plasma membrane. When an action potential arrives, calcium channels in the active zone open to generate a steep increase in calcium concentration. Calcium binds the synaptic vesicle protein synaptotagmin which promotes its interaction with the SNARE complex and with the plasma membrane, together triggering fusion of the synaptic vesicle membrane with the plasma membrane (1). Following fusion, the synaptic vesicle membrane needs to be removed from the plasma membrane to prevent its expansion, and recycling of the vesicle components is needed to refill the pool of vesicles at the release site.

An outline of the steps in the recycling of a synaptic vesicle is depicted in Figure 1. The vesicle membrane first moves out from the active zone into the periactive zone. The mechanism behind such movement is unclear but it is critical in order to maintain the function of the active zone. Impaired clearance of vesicle components from the release site has been linked with depression of neurotransmtter release (2). After the vesicle membrane has reached the periactive zone, clathrin and accessory endocytic factors accumulate to begin the nucleation of a clathrin coat. The coat grows and invaginates until a deeply invaginated coated pit with a narrow neck has formed. The neck of the coat is then surrounded by a dynamin-containing ring or short spiral, which helps to cut off the neck. The free vesicle rapidly sheds its coat and it may be directly refilled with neurotransmitter and prepared for a new round of release. Alternatively, the primary endocytic vesicle may first fuse with an endosome, prior to undergoing a second endosomal budding step to yield a new synaptic vesicle. Although the presence of an endosomal recycling route has been well established (3,4) its precise role in vesicle cycling is not fully clear. It may potentially be used to recycle readily releaseable vesicles (5), or it may participate in refilling the reserve pool during extended periods of synaptic activity (6). The endosomal route may be used more extensively in subsets of synapses (7).

2. Clathrin-mediated endocytosis as the main synaptic vesicle recycling pathway

The model of synaptic vesicle recycling shown in Figure 1 has its origin in quick-freeze studies performed at the frog neuromuscular junction 8, and microinjection studies performed in giant synapses in lamprey and squid (9-11). In the latter type of experiments a compound - antibody, toxin or peptide - that disrupts the function of an endocytic protein (or a protein-protein interaction) is microinjected into the presynaptic cytoplasm. When the microinjected synapse is examined at rest, the synaptic structure is normal. However, repetitive stimulation uncovers defects in synaptic vesicle recycling. These include loss of synaptic vesicles, expansion of the plasma membrane, and accumulation of clathrin-coated endocytic intermediates in the periactive zone. Depending on which protein is perturbed, the structure of the accumulated intermediates may differ. For instance, if the clathrin/AP2-binding region of epsin is perturbed, enlarged coated pits with wide necks occur (Fig. 2). In contrast, if dynamin - SH3 domain interactions are perturbed, deeply invaginated coated pits with narrow necks appear (Fig. 3). If synaptojanin is perturbed, free clathrin coated vesicles accumulate as a sign of impaired uncoating (12) .

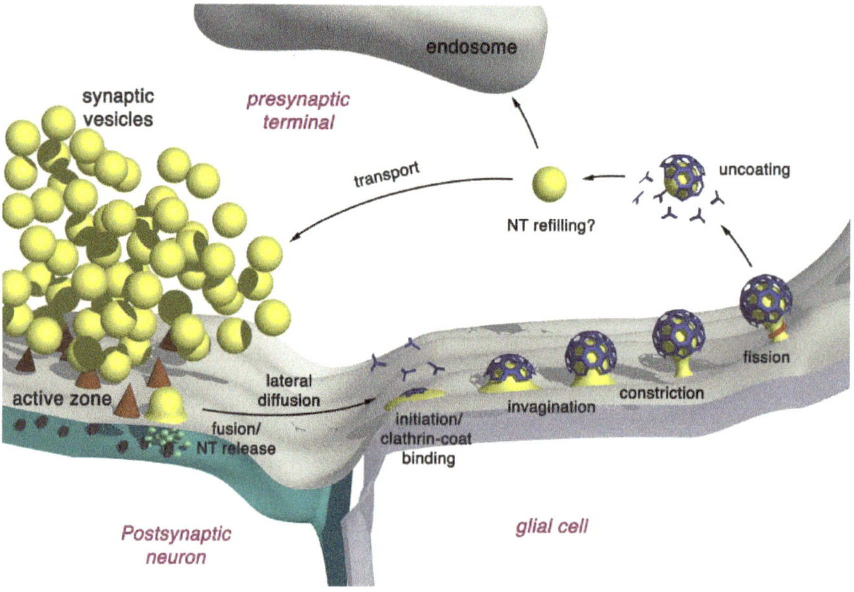

Figure 1. Model of clathrin-mediated synaptic vesicle endocytosis. Synaptic vesicles partially or completely fuse with the presynaptic membrane at the active zone and release neurotransmitter into the synaptic cleft. The membrane of the fused vesicles then diffuses laterally to the areas outside the active zone where it is retrieved by clathrin-mediated endocytosis. Clathrin-coated vesicle formation involves several morphologically distinct steps, from clathrin coat binding, invagination of the coated bud, constriction and fission of the pit 'neck' and the subsequent stripping of the clathrin coat from the newly formed vesicle. The vesicle is then either directly transported back to the cluster of synaptic vesicles or translocated to a primary endosomal compartment. During endocytosis and migration to the release site vesicles are refilled with transmitter (NT). Reproduced from ref 17.

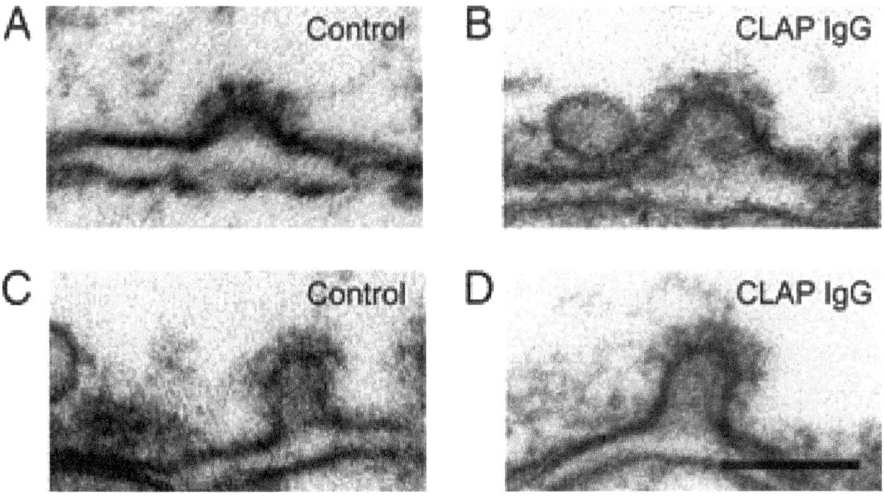

Figure 2. Microinjection of antibodies to the CLAP region of epsin increases the size of coated pits. Electronmicrographs show the periactive zone area in lamprey giant reticulospinal axons stimulated at 5 Hz following microinjection. A-B, Shallow coated pits from control (A) and CLAP antibody-injected (B). C-D, Examples of non-constricted (bucket-shaped) coated pits from control (C) and CLAP antibody-injected axons (D). Scale bars = 100 nm. Reproduced from ref 85.

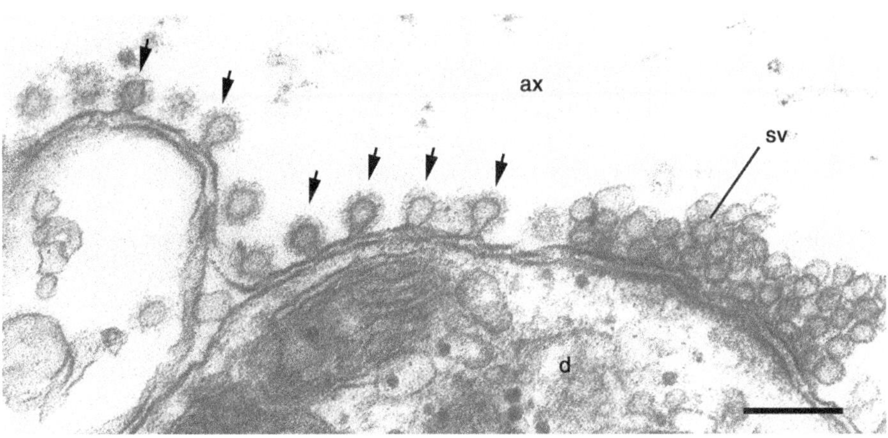

Figure 3. Microinjection of the SH3 domain of amphiphysin traps coated pits with narrow necks. The electronmicrograph shows the periactive zone area in a lamprey giant reticulospinal axon, and a synaptic release site with clustered synaptic vesicles is visible to the right. The axon was stimulated at 0.2 Hz for 30 min prior to fixation Scale bar = 200 nm. Reproduced from ref 28.

The requirement of clathrin in synaptic vesicle recycling has also been demonstrated in experiments using photoinactivation of a transgenically encoded protein (FlAsH-FALI method) (13,14). The technique is based on the use of a short tetracysteine epitope tag that covalently binds a membrane permeable dye, Lumio. When excited with fluorescent light, Lumio inactivates the tagged protein. Following tagging of the clathrin light or heavy chain, illumination followed by repetitive stimulation causes a complete loss of synaptic vesicles along with massive accumulation of plasma membrane folds in the terminals (13,14). The importance of clathrin-mediated synaptic vesicle endocytosis has also been demonstrated in real-time imaging experiments using synaptic vesicle proteins tagged with a pH-sensitive reporter (15). Granseth et al showed that brief action potential stimulation is followed by an endocytic response (due to loss of the acidic pH in the vesicles) with a time course of about 15 s. Such responses were abolished in neurons in which the expression of the clathrin heavy chain had been knocked down by RNAi (16). In addition to the studies mentioned above, a number of genetic studies performed in C. elegans, Drosophila and mice support the critical role of clathrin-mediated synaptic vesicle endocytosis (15,17). In fact, the molecular analysis once began with studies in a temperature-sensitive paralytic Drosophila mutant, *shibire* (18,19). Following the discovery that the *shibire* mutation is situated in the dynamin gene (20,21) a network of interconnected endocytic proteins could be identified (15,22).

In the present chapter we will only briefly comment on other, non-clathrin mediated mechanisms of synaptic vesicle recycling. One such mechanism of clathrin-independent membrane internalization is termed bulk endocytosis. Large membrane cisternae are internalized and subsequently converted to synaptic vesicles, but the budding mechanism involved is not well defined. In some model systems, like cerebellar granule cell synapses, bulk endocytosis has been examined in detail and it has been been found to operate under conditions of physiological stimulation (23). In many studies, however, the occurrence of bulk endocytosis in nerve terminals has been linked to non-physiological conditions, including excessive stimulation, or moderate stimulation combined with disruption of the clathrin machinery (15,24). The term kiss-and-run refers to a mode of recycling that involves a transient opening and closing of a fusion pore without loss of the vesicle´s integrity. The functional role of kiss-and-run recycling has been the matter of lively debate (25,26). Evidence in favor of the kiss-and-run phenomenon has mainly been obtained in imaging studies. Studies detailing the behavior of single pH-sensitive quantum dots trapped in individual synaptic vesicles in hippocampal boutons supports the possibility that synaptic vesicles can open transiently (27). Further studies, however, are required to determine the generality of this phenomenon and its possible implications for synaptic transmission.

3. A storage pool of endocytic proteins is associated with synaptic release sites

Early models of synaptic recycling often assumed that endocytic proteins occur in a diffusible cytoplasmic pool from which they are recruited to the plasma membrane to participate endocytosis. This appears, at least for most proteins, not to be case. In contrast, endocytic proteins have been found to be distinctly accumulated at release sites. Following the observation that an SH3 domain (of amphiphysin) bound tightly to synaptic vesicle clusters (28), it was shown by immunogold labeling that many endocytic proteins including dynamin, amphiphysin, epsin, endophilin and intersectin accumulate within the vesicle cluster (29). The cluster-associated protein pool can be mobilized by synaptic activity. Thus, after a period of repetitive stimulation the endocytic proteins partly dissociate from the synaptic vesicles and appear at the plasma membrane in the periactive zone (30-32). Such protein redistibution has also been observed in live imaging experiments. The clathrin heavy chain is found to be concentrated in the center of release sites at rest but rapidly redistibutes to the periphery upon stimulation (16). Interestingly, not only endocytic, but also other soluble presynaptic proteins, such as synapsin, NSF, rab3, and rabphilin reside in the cluster at rest and redistribute peripherally upon stimulation (33,34). In agreement, *in vitro* studies showed that these proteins bind reversibly to synaptic vesicles. It was suggested by Denker et al that the presence of a cluster of synaptic vesicles (larger than what is needed to support neurotransmitter release) provides a buffer site for proteins near synaptic release sites (34). What regulates the mobilization of proteins from the synaptic vesicle cluster? The work of Denker et al suggest that calcium is one important factor, but other factors may also be required. In a study of synapsin, Orenbuch et al found that not only calcium influx and phosphorylation of synapsin, but also exocytosis is required in order for synapsin to redistribute from synaptic vesicle clusters (Orenbuch et al J Neurochem, in press). It will be interesting to examine whether mobilization of endocytic proteins from the synaptic vesicle cluster also requires a signal associated with exocytosis.

4. Early events in synaptic vesicle endocytosis

The model of synaptic vesicle recycling depicted in Figure 1 may suggest that synaptic vesicle membrane is absent from the periactive zone until it has fused in the active zone and moved laterally. It is now becoming increasingly clear, however, that some synaptic vesicle membrane resides in the axonal plasma membrane in between periods of exo- and endocytosis. Thus, in resting hippocampal nerve terminals, extracellularly applied antibodies to the luminal domain of synaptotagmin binds the axonal surface near release sites (35). Studies employing antibodies with pH-sensitive tags have further shown that a plasma membrane-resident pool of synaptotagmin is preferentially endocytosed at the onset of a bout of endocytosis (36). These findings indicate that a subset of "readily retrievable vesicles" occur in the periactive zone and can be endocytosed rapidly upon stimulation. The protein components in this vesicular membrane pool may be sorted and packaged to facilitate rapid endocytosis (37).

With regard to the precise order of recruitment of endocytic proteins to the periactive zone information is as yet limited. This contrasts with the detailed information that has been obtained in non-neuronal cells grown on glass slides in which protein movement near the plasma membrane can be tracked by total internal reflection (TIRF) microcopy (38). These studies indicate that among the first proteins to occur at the plasma membran is the F-BAR protein FCHo1/2, followed by the scaffold proteins eps15 and intersectin. Different adaptor proteins are then recruited while clathrin shows a slow build-up terminating at scission. Dynamin is present at low levels from early stages but exhibits a peak just before scission. A similar behavior is also observed for endophilin and synaptojanin (38,39).

5. What triggers synaptic vesicle endocytosis?

The simplest answer to the question of what triggers synaptic vesicle endocytosis would be the vesicle membrane itself. It is known that (clathrin-mediated) retrieval of synaptic vesicle membrane can be temporally dissociated from action potential-induced calcium influx (40). Thus, calcium influx is not needed to trigger endocytosis. Moreover, compensatory synaptic vesicle endocytosis can occur after non-calcium-dependent triggering of exocytosis by hypertonic sucrose stimulation (41). The synaptic vesicle membrane thus appears to contain components capable of inducing its reinternalization. However, in the absence of calcium influx, the time-course of endocytosis is slower than that seen under normal conditions of calcium-triggered release. Indeed, several studies have shown that calcium can accelerate endocytosis (42,43). Several proteins have been implicated as calcium sensors for endocytosis, including calmodulin (44), calcineurin (45) and synaptotagmin (46). At present, it remains unclear whether different synapses utilize different trigger mechanisms. One of the most detailed investigations of an endocytic calcium sensor was recently performed in hippocampal neurons (41). These authors examined synaptotagmin, the trigger of fast synchronous exocytosis (47,48), which also is also implicated in endocytosis (49). Interestingly, Yao et al found that the calcium dependence of synaptotagmin in exo- and endocytosis could be uncoupled. Either the C2A or C2B domain of synaptotagmin could function as calcium sensor for endocytosis, whereas only the C2B domain effectively supported exocytosis. It was also found that retargeting of synaptotagmin to the plasma membrane abolished the calcium dependence of endocytosis but not that of exocytosis. Synaptotagmin thus appears two play distinct roles, one as a calcium sensor that triggers fast synchronous exocytosis and another as a calcium sensor that speeds up endocytosis.

6. Recycling of SNARE proteins

The role of SNARE (soluble NSF attachment protein receptors) proteins in synaptic vesicle fusion have been described in great detail (50), but the subsequent fate of the SNARE complex and its components synaptobrevin, syntaxin and SNAP25 have been less well studied. Initial studies suggested that disassembly of the SNARE proteins occurs shortly before fusion such that NSF is in a position to regulate the kinetics of neurotransmitter release (51). More recent studies, however, suggest that SNARE complex disasembly occurs much earlier, even before synaptic vesicle endocytosis. Imaging studies showed that syntaxin remains in the plasma membrane after synaptic vesicles have been endocytosed, indicating that complex dissasembly preceeds endocytosis (52). Moreover, it was shown that NSF and SNARE proteins accumulate in the periactive zone after inhibition of NSF function 53. It is quite possible that, following its disassembly, synaptobrevin participates in clathrin-mediated endocytosis. Synaptic vesicle endocytosis is impaired in synaptobrevin-deficient mice (54), and the endocytic adaptors AP180 and CALM have been found to bind synaptobrevin. Notably, these adaptors bind at a site within the SNARE domain that is only accessible after the SNARE complex has been disassembled. Together these observations indicate that SNARE complex disassembly occurs within the plasma membrane of the periactive zone prior to the onset of synaptic vesicle endocytosis, and they further suggest that synaptobrevin may facilitate clathrin-mediated endocytosis.

7. BAR domains – Membrane benders or membrane binders?

Proteins with BAR domains have attracted much interest due to the striking structural features of this domain. Two BAR domains form a dimer with a concave surface that can bind phospholipid membranes (22,55). Hence BAR domains have been implicated as inducers of curvature and as curvature-sensing modules that bind membrane domains with a given curvature. Endophilin is the BAR protein that has attracted most interest in the synaptic vesicle field, both because it is expressed in nerve terminals in organisms ranging from worms and flies to mammals, and because it interacts with dynamin and synaptojanin (56). The effect of perturbing endophilin has been tested in many studies, all of which point to an important role of the protein. Endophilin has been suggested to act at multiple steps in synaptic vesicle endocytosis. A role early in the endocytic reaction was suggested by the finding that shallow coated pits can be trapped by endophilin antibody microinjection in the lamprey giant axon. This phenomenon has also seen in Drosophila after genetic reduction of the endophilin levels (57-59). Endophilin has been detected by immunogold labeling at the rim of shallow coated pits (32). These obervations are possibly compatible with a membrane bending role of endophilin at an early stage of endocytosis, but such a function has not yet been supported by studies in mammalian models (see below). Second, a role for endophilin in recruitment of dynamin to the neck of coated pits has been proposed. Endophilin occurs at the proximal part of the neck of coated pits, and peptides competing the endophilin – dynamin interaction inhibit formation of dynamin rings as well as subsequent membrane fission (12,32 see also 60). Finally, endophilin has been linked with vesicle uncoating by its interaction with synaptojanin. In the lamprey giant axon perturbation of the endophilin – synaptojanin interaction results in accumulation of numerous free clathrin coated vesicles, in addition to deeply invaginated coated pits. In mice lacking all three endophilin genes nerve terminals were found to contain large numbers of free clathrin coated vesicles (that are nearly absent in wild-type animals) (61). Somewhat surprisingly, in the mouse model no other type of endocytic intermediate was accumulated. Moreover, in both C. elegans and Drosophila the phenotype of synaptojanin mutants closely resembled that of endophilin mutants, and endophilin was found to be required for localization of synaptojanin to nerve terminals (62,63). It is therefore likely that a principal function of endophilin in nerve terminals is to mediate recruitment of synaptojanin to the vesicle neck to support uncoating. Hence, a role of the BAR domain of endophilin as a binder rather then a bender appears most plausible.

8. New insights into dynamin function and membrane fission

Different models have been proposed to account for the role of dynamin in catalyzing endocytic membrane fission (64,65). The most recent models incorporate rich high-resolution structural information. The crystal structure of full-length dynamin has been determined by taking advantage of assembly-deficient mutants (66,67). Insight into the organization of assembled dynamin multimers has been gained by computer docking of domain crystal structures into cryo-EM images (68). These studies suggest that initial constriction of the coated pit neck, triggered by GTP binding and structural changes in the middle domain of dynamin, promotes GTP domain dimerization between tetramers in adjacent helical rungs. Assembly-stimulated GTP hydrolysis is suggested to induce a rotation that provides force, and propagation of this change could cause further constriction of the neck leading to fission. Moreover, *in vitro* studies have provided detailed insight into the dynamic behavior of dynamin at membranes. It was found that the extended dynamin spirals that form around lipid tubules in the absence of GTP (69) do not effectively promote fission. Instead assembly of short spirals followed by disassembly led to membrane fission (70). Moreover, dynamin alone can form self-limited assemblies that drive vesiculation from a lipid surface in the presence of GTP 71 (Fig. 5A).

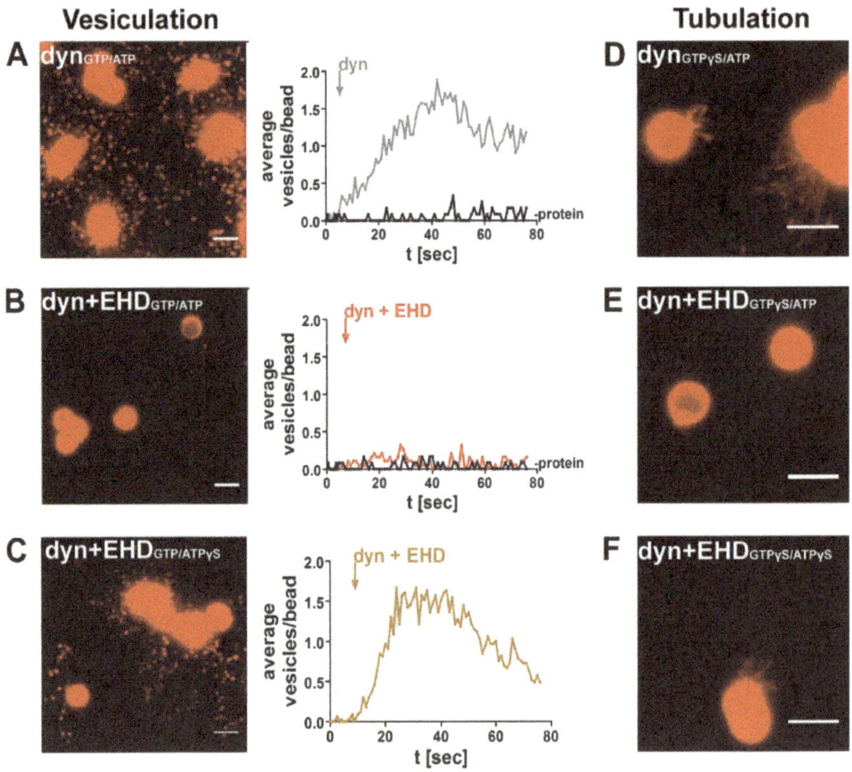

Figure 4. Dynamin-induced vesiculation and its modulation by EHD. A, Vesiculation *in vitro* from rhodamine-labeled SUPER templates induced by application of dynamin in the constant presence of GTP (1 mM GTP and ATP present in A and B). The trace with the response to dynamin (dyn) is superimposed on the trace preceeding addition of dynamin (-protein) B, Vesiculation was suppressed when dynamin was co-applied with l-EHD. C, Reduced inhibitory effect of l-EHD on dynamin-induced vesiculation in the constant presence of GTP after replacement of ATP with ATPγS (1 mM). D, Application of dynamin in the constant presence of GTPγS induced formation of narrow tubules (1 mM GTPγS and ATP present in D-F). E, Tubule formation was suppressed when dynamin was co-applied with l-EHD. F, Reduced inhibitory effect of l-EHD on dynamin-induced tubulation in the constant presence of GTPγS after replacement of ATP with ATPγS. Scale bars=5 μm. Reproduced from ref 75.

Under *in vivo* conditions the function of dynamin depends strictly on interactions with other proteins. In particular, interactions with SH3 domains are important. As indicated above, perturbation of dynamin - SH3 domain interactions blocks endocytosis at a late stage (Fig. 3). In this case, dynamin rings do not form suggesting that SH3 interactions are needed for proper recruitment or assembly of dynamin at the neck of the coated pit. Several proteins may mediate SH3 domain interactions with dynamin in nerve terminals, including amphiphysin, endophilin, intersectin, SNX9 and syndapin. It may be noted, however, that in Drosophila (and possibly in other invertebrates) neither amphiphysin nor syndapin are expressed in nerve terminals (72,73).

Recent studies performed in the lamprey giant reticulospinal synapse indicate that extrinsic proteins not only regulate the recruitment of dynamin, but they may also control the length of the dynamin spiral. Eps15 homology domain-containing proteins (EHDs) are conserved ATPases implicated in membrane remodelling, primarily in endosomal traffic. EHD1 is enriched at synaptic release sites (74), suggesting a possible involvement in the trafficking of synaptic vesicles. The role of EHD in this function has been analyzed in the lamprey giant reticulospinal synapse. EHD1/3 was detected by immunogold at endocytic structures adjacent to release sites. In antibody microinjection experiments, perturbation of EHD inhibited synaptic vesicle endocytosis and caused accumulation of clathrin-coated pits with atypical, elongated necks (Fig. 5). The necks were covered with helix-like material containing dynamin (75). To test whether EHD directly interferes with dynamin function, fluid supported bilayers were used as *in vitro* assay. EHD strongly inhibited vesicle budding induced by dynamin in the constant presence of GTP (Fig. 4A-C). EHD also inhibited dynamin-induced membrane tubulation in the presence of GTPγS (Fig. 4D-E) a phenomenon linked with dynamin helix assembly. Taken together the *in vivo* and *in vitro* results suggest that l-EHD acts to limit the formation of long, unproductive dynamin helices, thereby promoting vesicle budding (75).

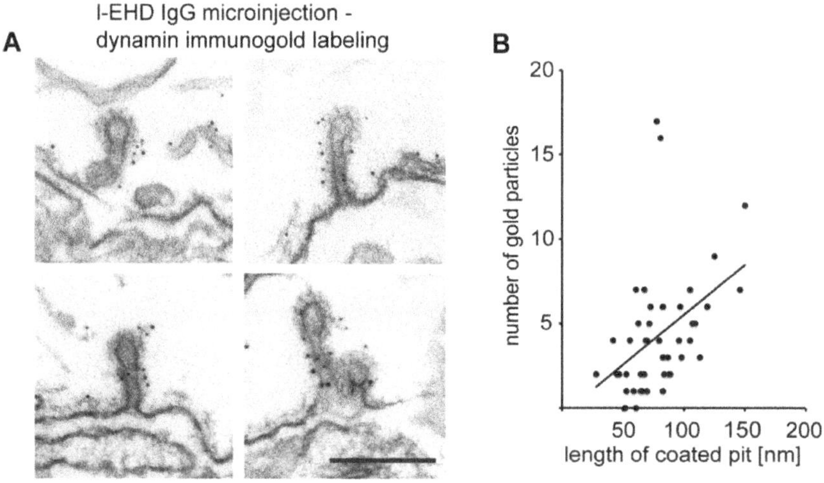

Figure 5. Immunogold localization of dynamin at endocytic pits with elongated necks trapped after perturbation of EHD. A, Examples of coated pits with long necks decorated with dynamin immunogold labeling in lamprey giant axons stimulated after microinjection of EHD antibodies. Scale bar=0.2 µm. B, Regression analysis of dynamin labeling and length of coated pits in EHD antibody-injected axons (R^2=0.43, n=45, 0.05<p<0.01, Pearson's correlation coefficient). Reproduced from ref 75.

9. Possible implications of synaptic vesicle endocytosis for disease mechanisms

Knowledge about the mechanisms of synaptic vesicle endocytosis is not only important for our understanding of synaptic information processing, but may potentially also shed light on pathogenetic mechanisms. One aspect conerns toxins and infectious agents that may hi-jack the synaptic endocytic machinery to enter neurons (76-78). Another aspect concerns the possible role of endocytosis in neurodegenerative disorders that affect synapses. For instance, Lewy body pathology, occurring in Parkinson and other disorders, may involve endocytic uptake of α-synuclein fibrils that induce intracellular fibril formation, which in turn leads to synaptic dysfunction (79). Synaptic endocytosis has also been implicated in the pathogenesis of Alzheimer's disease. For instance, part of the processing of the amyloid precursor protein (APP), into synaptotoxic amyloid beta peptides appears to occur in nerve terminals in an endocytosis-dependent manner. Thus, microdialysis studies have shown that the extracellular Aβ pool in brain is elevated by enhanced synaptic activity and lowered after inhibition of synaptic endocytosis (80-82), Accordingly, formation of Aβ in a neuronal cell line has been shown to be suppressed by knock-down of the clathrin adaptor AP180 (83). It is also interesting to note that γ-secretase, that cleaves APP, is present in nerve terminals (84).

Author details

Frauke Ackermann, Joshua A. Gregory and Lennart Brodin
Department of Neuroscience, Karolinska Institutet, Stockholm, Sweden

10. References

[1] Wojcik, S. M. & Brose, N. Regulation of membrane fusion in synaptic excitation-secretion coupling: speed and accuracy matter. *Neuron* 55, 11-24 (2007).
[2] Neher, E. What is Rate-Limiting during Sustained Synaptic Activity: Vesicle Supply or the Availability of Release Sites. *Front Synaptic Neurosci* 2, 144, doi:10.3389/fnsyn.2010.00144 (2010).
[3] Wucherpfennig, T., Wilsch-Brauninger, M. & Gonzalez-Gaitan, M. Role of Drosophila Rab5 during endosomal trafficking at the synapse and evoked neurotransmitter release. *J Cell Biol* 161, 609-624 (2003).
[4] Uytterhoeven, V., Kuenen, S., Kasprowicz, J., Miskiewicz, K. & Verstreken, P. Loss of skywalker reveals synaptic endosomes as sorting stations for synaptic vesicle proteins. *Cell* 145, 117-132 (2011).
[5] Hoopmann, P. *et al.* Endosomal sorting of readily releasable synaptic vesicles. *Proc Natl Acad Sci U S A* 107, 19055-19060 (2010).
[6] Voglmaier, S. M. & Edwards, R. H. Do different endocytic pathways make different synaptic vesicles? *Curr Opin Neurobiol* 17, 374-380 (2007).
[7] Nakatsu, F. *et al.* Defective function of GABA-containing synaptic vesicles in mice lacking the AP-3B clathrin adaptor. *J Cell Biol* 167, 293-302 (2004).

[8] Miller, T. M. & Heuser, J. E. Endocytosis of synaptic vesicle membrane at the frog neuromuscular junction. *Journal of Cell Biology* 98, 685-698 (1984).
[9] Brodin, L., Löw, P. & Shupliakov, O. Sequential steps in clathrin-mediated synaptic vesicle endocytosis. *Curr Opin Neurobiol* 10, 312-320. (2000).
[10] Brodin, L. & Shupliakov, O. Giant reticulospinal synapse in lamprey: molecular links between active and periactive zones. *Cell Tissue Res* 326, 301-310 (2006).
[11] Augustine, G. J. *et al.* Clathrin and synaptic vesicle endocytosis: studies at the squid giant synapse. *Biochem Soc Trans* 34, 68-72 (2006).
[12] Gad, H. *et al.* Fission and uncoating of synaptic clathrin-coated vesicles are perturbed by disruption of interactions with the SH3 domain of endophilin. *Neuron* 27, 301-312 (2000).
[13] Heerssen, H., Fetter, R. D. & Davis, G. W. Clathrin dependence of synaptic-vesicle formation at the Drosophila neuromuscular junction. *Curr Biol* 18, 401-409 (2008).
[14] Kasprowicz, J. *et al.* Inactivation of clathrin heavy chain inhibits synaptic recycling but allows bulk membrane uptake. *J Cell Biol* 182, 1007-1016 (2008).
[15] Dittman, J. & Ryan, T. A. Molecular circuitry of endocytosis at nerve terminals. *Annu Rev Cell Dev Biol* 25, 133-160 (2009).
[16] Granseth, B., Odermatt, B., Royle, S. J. & Lagnado, L. Clathrin-mediated endocytosis is the dominant mechanism of vesicle retrieval at hippocampal synapses. *Neuron* 51, 773-786 (2006).
[17] Shupliakov, O. & Brodin, L. Recent insights into the building and cycling of synaptic vesicles. *Exp Cell Res* 316, 1344-1350 (2010).
[18] Kelly, L. E. & Suzuki, D. T. The effects of increased temperature on electroretinograms of temperature-sensitive paralysis mutants of Drosophila melanogaster. *Proc Natl Acad Sci U S A* 71, 4906-4909 (1974).
[19] Poodry, C. A. & Edgar, L. Reversible alteration in the neuromuscular junctions of Drosophila melanogaster bearing a temperature-sensitive mutation, shibire. *J Cell Biol* 81, 520-527 (1979).
[20] Chen, M. S. *et al.* Multiple forms of dynamin are encoded by shibire, a Drosophila gene involved in endocytosis. *Nature* 351, 583-586 (1991).
[21] van der Bliek, A. M. & Meyerowitz, E. M. Dynamin-like protein encoded by the Drosophila shibire gene associated with vesicular traffic. *Nature* 351, 411-414 (1991).
[22] Frost, A., Unger, V. M. & De Camilli, P. The BAR domain superfamily: membrane-molding macromolecules. *Cell* 137, 191-196 (2009).
[23] Cousin, M. A. Activity-dependent bulk synaptic vesicle endocytosis--a fast, high capacity membrane retrieval mechanism. *Mol Neurobiol* 39, 185-189 (2009).
[24] Gad, H., Löw, P., Zotova, E., Brodin, L. & Shupliakov, O. Dissociation between Ca2+-triggered synaptic vesicle exocytosis and clathrin-mediated endocytosis at a central synapse. *Neuron* 21, 607-616 (1998).
[25] Granseth, B., Odermatt, B., Royle, S. J. & Lagnado, L. Comment on "The dynamic control of kiss-and-run and vesicular reuse probed with single nanoparticles". *Science* 325, 1499; author reply 1499 (2009).
[26] Zhang, Q., Li, Y. & Tsien, R. W. The dynamic control of kiss-and-run and vesicular reuse probed with single nanoparticles. *Science* 323, 1448-1453 (2009).

[27] Park, H., Li, Y. & Tsien, R. W. Influence of Synaptic Vesicle Position on Release Probability and Exocytotic Fusion Mode. *Science*, (2012, in press).

[28] Shupliakov, O. *et al.* Synaptic vesicle endocytosis impaired by disruption of dynamin-SH3 domain interactions. *Science* 276, 259-263 (1997).

[29] Shupliakov, O. The synaptic vesicle cluster: a source of endocytic proteins during neurotransmitter release. *Neuroscience* 158, 204-210 (2009).

[30] Evergren, E. *et al.* Amphiphysin is a component of clathrin coats formed during synaptic vesicle recycling at the lamprey giant synapse. *Traffic* 5, 514-528 (2004).

[31] Evergren, E. *et al.* Intersectin is a negative regulator of dynamin recruitment to the synaptic endocytic zone in the central synapse. *J Neurosci* 27, 379-390 (2007).

[32] Sundborger, A. *et al.* An endophilin-dynamin complex promotes budding of clathrin-coated vesicles during synaptic vesicle recycling. *J Cell Sci* 124, 133-143 (2011).

[33] Chi, P., Greengard, P. & Ryan, T. A. Synapsin dispersion and reclustering during synaptic activity. *Nat Neurosci* 4, 1187-1193 (2001).

[34] Denker, A., Kröhnert, K., Bückers, J., Neher, E. & Rizzoli, S. O. The reserve pool of synaptic vesicles acts as a buffer for proteins involved in synaptic vesicle recycling. *Proc Natl Acad Sci U S A* (in press) (2011).

[35] Fernandez-Alfonso, T., Kwan, R. & Ryan, T. A. Synaptic vesicles interchange their membrane proteins with a large surface reservoir during recycling. *Neuron* 51, 179-186 (2006).

[36] Hua, Y. *et al.* A readily retrievable pool of synaptic vesicles. *Nat Neurosci* 14, 833-839, (2011).

[37] Willig, K. I., Rizzoli, S. O., Westphal, V., Jahn, R. & Hell, S. W. STED microscopy reveals that synaptotagmin remains clustered after synaptic vesicle exocytosis. *Nature* 440, 935-939 (2006).

[38] Taylor, M. J., Perrais, D. & Merrifield, C. J. A high precision survey of the molecular dynamics of mammalian clathrin-mediated endocytosis. *PLoS Biol* 9, e1000604, doi:10.1371/journal.pbio.1000604 (2011).

[39] Henne, W. M. *et al.* FCHo proteins are nucleators of clathrin-mediated endocytosis. *Science* 328, 1281-1284 (2010).

[40] Gad, H., Low, P., Zotova, E., Brodin, L. & Shupliakov, O. Dissociation between Ca2+-triggered synaptic vesicle exocytosis and clathrin-mediated endocytosis at a central synapse. *Neuron* 21, 607-616 (1998).

[41] Yao, J., Kwon, S. E., Gaffaney, J. D., Dunning, F. M. & Chapman, E. R. Uncoupling the roles of synaptotagmin I during endo- and exocytosis of synaptic vesicles. *Nat Neurosci* 15, 243-249 (2012).

[42] Hosoi, N., Holt, M. & Sakaba, T. Calcium dependence of exo- and endocytotic coupling at a glutamatergic synapse. *Neuron* 63, 216-229 (2009).

[43] Balaji, J., Armbruster, M. & Ryan, T. A. Calcium control of endocytic capacity at a CNS synapse. *J Neurosci* 28, 6742-6749 (2008).

[44] Wu, X. S. *et al.* Ca(2+) and calmodulin initiate all forms of endocytosis during depolarization at a nerve terminal. *Nat Neurosci* 12, 1003-1010 (2009).

[45] Clayton, E. L. *et al.* The phospho-dependent dynamin-syndapin interaction triggers activity-dependent bulk endocytosis of synaptic vesicles. *J Neurosci* 29, 7706-7717 (2009).

[46] Nicholson-Tomishima, K. & Ryan, T. A. Kinetic efficiency of endocytosis at mammalian CNS synapses requires synaptotagmin I. *Proc Natl Acad Sci U S A* 101, 16648-16652,(2004).
[47] Littleton, J. T., Stern, M., Schulze, K., Perin, M. & Bellen, H. J. Mutational analysis of Drosophila synaptotagmin demonstrates its essential role in Ca(2+)-activated neurotransmitter release. *Cell* 74, 1125-1134 (1993).
[48] Geppert, M. *et al.* Synaptotagmin I: a major Ca2+ sensor for transmitter release at a central synapse. *Cell* 79, 717-727 (1994).
[49] Zhang, J. Z., Davletov, B. A., Sudhof, T. C. & Anderson, R. G. Synaptotagmin I is a high affinity receptor for clathrin AP-2: implications for membrane recycling. *Cell* 78, 751-760 (1994).
[50] Sudhof, T. C. & Rothman, J. E. Membrane fusion: grappling with SNARE and SM proteins. *Science* 323, 474-477 (2009).
[51] Schweizer, F. E. *et al.* Regulation of neurotransmitter release kinetics by NSF. *Science* 279, 1203-1206 (1998).
[52] Mitchell, S. J. & Ryan, T. A. Syntaxin-1A is excluded from recycling synaptic vesicles at nerve terminals. *J Neurosci* 24, 4884-4888 (2004).
[53] Yu, W., Kawasaki, F. & Ordway, R. W. Activity-dependent interactions of NSF and SNAP at living synapses. *Mol Cell Neurosci* 47, 19-27 (2011).
[54] Deak, F., Schoch, S., Liu, X., Sudhof, T. C. & Kavalali, E. T. Synaptobrevin is essential for fast synaptic-vesicle endocytosis. *Nat Cell Biol* 6, 1102-1108 (2004).
[55] McMahon, H. T. & Gallop, J. L. Membrane curvature and mechanisms of dynamic cell membrane remodelling. *Nature* 438, 590-596 (2005).
[56] Ringstad, N., Nemoto, Y. & De Camilli, P. The SH3p4/Sh3p8/SH3p13 protein family: binding partners for synaptojanin and dynamin via a Grb2-like Src homology 3 domain. *Proceedings of the National Academy of Sciences of the United States of America* 94, 8569-8574 (1997).
[57] Ringstad, N. *et al.* Endophilin/SH3p4 is required for the transition from early to late stages in clathrin-mediated synaptic vesicle endocytosis. *Neuron* 24, 143-154 (1999).
[58] Guichet, A. *et al.* Essential role of endophilin A in synaptic vesicle budding at the Drosophila neuromuscular junction. *EMBO J* 21, 1661-1672, doi:10.1093/emboj/21.7.1661 (2002).
[59] Andersson, F., Low, P. & Brodin, L. Selective perturbation of the BAR domain of endophilin impairs synaptic vesicle endocytosis. *Synapse* 64, 556-560 (2010).
[60] Fabian-Fine, R. *et al.* Endophilin promotes a late step in endocytosis at glial invaginations in Drosophila photoreceptor terminals. *J Neurosci* 23, 10732-10744 (2003).
[61] Milosevic, I. *et al.* Recruitment of endophilin to clathrin-coated pit necks is required for efficient vesicle uncoating after fission. *Neuron* 72, 587-601 (2011).
[62] Verstreken, P. *et al.* Synaptojanin is recruited by endophilin to promote synaptic vesicle uncoating. *Neuron* 40, 733-748 (2003).
[63] Schuske, K. R. *et al.* Endophilin is required for synaptic vesicle endocytosis by localizing synaptojanin. *Neuron* 40, 749-762 (2003).
[64] Hinshaw, J. E. Dynamin and its role in membrane fission. *Annu Rev Cell Dev Biol* 16, 483-519. (2000).

[65] Marks, B. *et al.* GTPase activity of dynamin and resulting conformation change are essential for endocytosis. *Nature* 410, 231-235. (2001).
[66] Faelber, K. *et al.* Crystal structure of nucleotide-free dynamin. *Nature* 477, 556-560 (2011).
[67] Ford, M. G., Jenni, S. & Nunnari, J. The crystal structure of dynamin. *Nature* 477, 561-566 (2011).
[68] Chappie, J. S. *et al.* A pseudoatomic model of the dynamin polymer identifies a hydrolysis-dependent powerstroke. *Cell* 147, 209-222 (2011).
[69] Danino, D. & Hinshaw, J. E. Dynamin family of mechanoenzymes. *Curr Opin Cell Biol* 13, 454-460. (2001).
[70] Bashkirov, P. V. *et al.* GTPase cycle of dynamin is coupled to membrane squeeze and release, leading to spontaneous fission. *Cell* 135, 1276-1286 (2008).
[71] Pucadyil, T. J. & Schmid, S. L. Real-time visualization of dynamin-catalyzed membrane fission and vesicle release. *Cell* 135, 1263-1275 (2008).
[72] Zhang, B. & Zelhof, A. C. Amphiphysins: raising the BAR for synaptic vesicle recycling and membrane dynamics. Bin-Amphiphysin-Rvsp. *Traffic* 3, 452-460 (2002).
[73] Kumar, V., Alla, S. R., Krishnan, K. S. & Ramaswami, M. Syndapin is dispensable for synaptic vesicle endocytosis at the Drosophila larval neuromuscular junction. *Mol Cell Neurosci* 40, 234-241 (2009).
[74] Wei, S. H. *et al.* EHD1 is a synaptic protein that modulates exocytosis through binding to snapin. *Molecular and Cellular Neuroscience* 45, 418-429 (2010).
[75] Jakobsson, J. *et al.* Regulation of synaptic vesicle budding and dynamin function by an EHD ATPase. *J Neurosci* 31, 13972-1398 (2011).
[76] Binz, T. & Rummel, A. Cell entry strategy of clostridial neurotoxins. *J Neurochem* 109, 1584-1595 (2009).
[77] Montal, M. Botulinum neurotoxin: a marvel of protein design. *Annu Rev Biochem* 79, 591-617, doi:10.1146/annurev.biochem.051908.125345 (2010).
[78] Harper, C. B. *et al.* Dynamin inhibition blocks botulinum neurotoxin type A endocytosis in neurons and delays botulism. *J Biol Chem* 286, 35966-35976 (2011).
[79] Volpicelli-Daley, L. A. *et al.* Exogenous alpha-synuclein fibrils induce Lewy body pathology leading to synaptic dysfunction and neuron death. *Neuron* 72, 57-71 (2011).
[80] Kamenetz, F. *et al.* APP processing and synaptic function. *Neuron* 37, 925-937, doi:S0896627303001247 [pii] (2003).
[81] Cirrito, J. R. *et al.* Synaptic activity regulates interstitial fluid amyloid-beta levels in vivo. *Neuron* 48, 913-922, (2005).
[82] Cirrito, J. R. *et al.* Endocytosis is required for synaptic activity-dependent release of amyloid-beta in vivo. *Neuron* 58, 42-51, (2008).
[83] Wu, F., Matsuoka, Y., Mattson, M. P. & Yao, P. J. The clathrin assembly protein AP180 regulates the generation of amyloid-beta peptide. *Biochem Biophys Res Commun* (2009).
[84] Frykman, S. *et al.* Synaptic and endosomal localization of active gamma-secretase in rat brain. *PLoS One* 5, e8948 (2010).
[85] Jakobsson, J. *et al.* Role of epsin 1 in synaptic vesicle endocytosis. *Proc Natl Acad Sci U S A* 105, 6445-6450 (2008).

Chapter 4

The Vacuole Import and Degradation Pathway Converges with the Endocytic Pathway to Transport Cargo to the Vacuole for Degradation

Abbas A. Alibhoy and Hui-Ling Chiang

Additional information is available at the end of the chapter

1. Introduction

In eukaryotes, pathways concerned with protein synthesis and those involved in protein degradation serve to maintain the levels of proteins in a cell [1]. The degradation of proteins occurs by two major pathways, the proteasomal degradation pathway and the lysosomal degradation pathway [2]. In the proteasomal degradation pathway, target proteins are ubiquitinated by a system of E1, E2 and E3 enzymes [2,3]. Thereafter, the ubiquitinated proteins are delivered to the proteasome for degradation [2,3]. In contrast, the lysosome, which contains many hydrolytic enzymes, serves as the site of degradation for a multitude of pathways. One such pathway is the macroautophagy pathway [4]. This undiscerning catabolic process, comprising of approximately 30 *ATG* genes, helps cells to endure phases of nutrient starvation and other stresses by degrading proteins and organelles in the lysosome [5-7]. In disparity, chaperone-mediated autophagy is a selective autophagy pathway that targets specific cargo proteins (having the KFERQ amino acid sequence) to the lysosome for degradation via cytosolic chaperone proteins [8-11]. Therefore, vital processes such as cell development, growth and homeostasis require autophagy and it's absence or deregulation can result in diseases such as cancer, and even neurodegeneration [12,13].

Vesicular transport facilitates the delivery of proteins to the different organelles of the cell, with the exception of transport to the nucleus, peroxisomes, endoplasmic reticulum etc [14]. These intermediate carriers of proteins range from the endosomes to the coat protein complex I (COPI) vesicles, COPII vesicles and clathrin-coated vesicles [14-16]. The anterograde transport (from ER to Golgi) of proteins is mediated by the COPII vesicles while the retrograde transport (from Golgi to ER) is mediated by the COPI vesicles [14-16]. In addition, the transport of proteins from the plasma membrane to the early endosomes, and from the Golgi to endosomes is facilitated by the clathrin-coated vesicles [14]. In yeast, the organelle that is homologous to the mammalian lysosome is the vacuole [17-18]. This organelle is essential for cellular processes such as maturation of vacuolar resident proteins, protein degradation, and for osmoregulation [18]. The transport of vacuole resident proteins into the vacuole is essential for the function of this organelle. For example, the Vps pathway transports carboxypeptidase Y (CPY) from the Golgi to the vacuole for maturation [19]. This pathway enlists the involvement of approximately 40 *VPS* genes [19]. Moreover, endocytosis is another pathway that delivers proteins from the plasma membrane and other extracellular molecules to the vacuole [18-20]. In addition, proteins can also be delivered from the cytoplasm into vacuole. For instance, the Cvt pathway delivers enzymes such as aminopeptidase I (API) and α-mannosidase from the cytoplasm to the vacuole [5,21].

Transport of proteins and organelles to vacuole can be affected by alterations in nutrient stimuli [5,21]. Upon starving *Saccharomyces cerevisiae* of nitrogen, proteins are sequestered in autophagosomes and then transported to the vacuole for degradation by the macroautophagy pathway. The target of rapamycin 1 protein (Tor1p) is a component of TORC1 that functions to regulate gene expression, ribosomal synthesis and nutrient transport [22,23]. Intriguingly, Tor1p inhibits the macroautophagy pathway. Rapamycin induces the macroautophagy pathway even in the absence of nitrogen starvation. In another instance, when yeast is replenished from growth in media containing oleic acid to that containing glucose, the peroxisomes are transported to the vacuole for degradation [24].

2. Regulation of gluconeogenic enzymes in yeast

In *Saccharomyces cerevisiae,* essential regulatory enzymes in the gluoconeogenesis pathway such as fructose-1,6-bisphosphatase (FBPase), malate dehydrogenase (MDH2), phosphoenolpyruvate carboxykinase (Pck1p) and isocitrate lyase (Icl1p) are induced when cells are grown in media depleted of glucose [25-27]. These enzymes function to synthesize glucose from non-carbohydrate carbon sources such as pvruvate or acetate. Upon supplying cells with media containing fresh glucose, these enzymes are inactivated [26-28]. This is referred to as catabolite inactivation [26-28]. FBPase is the best-studied example of catabolite inactivation [26,27]. From previous investigations, it has been determined that there may be many contributing factors; however protein degradation is the principal mechanism that inactivates FBPase.

FBPase is a suitable candidate for degradation studies for two reasons. First, expression of FBPase can be induced in response to specific stimuli [25-27]. And secondly, following glucose replenishment, FBPase is promptly degraded and exhibits a half-life of approximately 20-40 min. A key factor in targeting FBPase for degradation may be protein modification. To better illustrate this, it has been suggested that phosphorylation of FBPase may be a regulatory factor in this protein's degradation [29]. There is evidence that FBPase is phosphorylated at serine 11 and that this phosphorylation increases following glucose replenishment [30]. Protein kinase A (PKA) and the Ras2 signaling pathway mediate phosphorylation of FBPase [29-31].

3. The site of degradation of gluconeogenic enzymes

The site of degradation of FBPase is dependent on the duration of starvation. From studies conducted by the Wolf lab, it has been demonstrated that following glucose replenishment, FBPase is inactivated by ubiquitination [32-35]. There is evidence that the N-terminal proline residue is essential for the polyubiquitination of FBPase following glucose replenishment [36]. Thereafter, ubiquitinated FBPase is then delivered to the proteasome for degradation [32-36]. This is because the mutations in genes involved in the proteasome pathway such as *CIM3* results in the inhibition of FBPase degradation [32-36]. In contrast, after glucose starvation of yeast cells for 3 days, it has been determined that FBPase is phosphorylated and inactivated by PKA [29-31]. Following inactivation, FBPase is delivered to the vacuole for degradation [37-39]. For instance, the degradation of FBPase was examined using a *Δpep4Δprb1Δprc1* vacuole mutant. This mutant strain contains deletion of proteinases A, B and C. In the absence of these genes, there is retardation in the degradation of proteins that are delivered to the vacuole [38,41]. In this study, upon replenishing cells with fresh glucose following one day starvation it was observed that FBPase was degraded normally. However, following glucose replenishment after 3 days glucose starvation, FBPase degradation was inhibited. This suggests that FBPase degradation following 3 days glucose starvation is dependent on the presence of vacuolar proteinases. More recently, our lab has also demonstrated that other gluconeogenic enzymes such as MDH2, Pck1p and Icl1p also share the same degradation characteristics as FBPase. Furthermore, the re-distribution of these enzymes from the cytosol to vacuole following glucose replenishment has been validated by immunofluorescence and immunoelectron microscopy studies. At present, it is suggested that differential modification of FBPase following glucose replenishment dictates whether the protein is degraded in the vacuole or the proteasome. Such a disparate degradation behavior has been previously ascribed to the degradation of the fatty acid synthase subunit ß [42]. Depending on growth conditions, fatty acid synthase subunit ß is degraded either in the vacuole or the proteasome.

4. The vacuole import and degradation pathway

The gluconeogenic enzymes (FBPase, MDH2, Pck1p and Icl1p) are transported to the vacuole for degradation by a selective autophagy pathway [37-41]. This pathway is called the vacuole import and degradation (Vid) pathway. The genes involved in this pathway are cumulatively called *VID* genes [37-41]. For the purposes of characterizing this pathway, FBPase was selected as a marker for associated studies. By using a myriad of mutagenesis assays, our lab has identified many genes that play a role in the Vid pathway. For instance, mutants, created by subjecting cells to UV mutagenesis, have been studied for their ability to degrade FBPase. A colony blotting procedure was utilized to screen for mutants defective in FBPase degradation following glucose replenishment [40]. The results from these experiments were further validated by performing pulse-chase experiments. It was determined that while FBPase was degraded with a half-life of 20-40 min in wild-type cells, mutants degraded FBPase with a half-life ranging from 120-400 min. Moreover, all *vid* mutations were recessive as these mutants were complemented for the FBPase degradation defect upon mating with wild-type cells.

Another strategy to identify genes involved in the Vid pathway was by transposon mutagenesis. For this strategy, a transposon-lacZ/LEU2 library was transformed into wild-type cells. These mutants were then screened for FBPase degradation defects a using colony blotting procedure [43]. The identities of the mutated genes were ascertained by extracting the genomic DNA and the subsequent amplification of the nucleotide sequences adjoining the transposon insertion site via PCR. The product from the PCR was sequenced and analyzed using gene sequence alignment software from the National Center for Biotechnology Information (NCBI). Moreover, the degradation defect attributed to these mutants was confirmed by using yeast null mutants for the corresponding genes. Furthermore, the FBPase degradation phenotype was rescued upon transforming the corresponding *VID* genes into these mutants. The *vid* mutants are distinct from those affecting protein secretion (*sec*), vacuolar proteolysis (*pep*) and vacuolar protein sorting (*vps*). Upon studying the distribution of FBPase in cells of these mutants, it was inferred that the mutants can be classified into two categories. After replenishing cells with fresh glucose, some mutants depicted a more cytosolic staining of FBPase (Class A mutants) while other mutants showed FBPase to be distributed in punctate structures (Class B mutants) [40].

5. Vid vesicles: Intermediate carriers of the Vid pathway

From fractionation analysis, it was proposed that in the Vid pathway, FBPase was delivered to the vacuole for degradation via intermediate vesicles. This hypothesis was investigated by isolation and purification of FBPase-associated vesicles to near homogeneity [44]. In this investigation, wild-type cells were shifted to glucose for 30 min at 22ºC and vesicles were purified. At this temperature, there is a delay in the delivery of FBPase to the vacuole [44]. Following homogenization and subsequent centrifugation at 100,000 x g of cells, the intracellular organelles were separated by size via fractionation on a Sephacryl S-1000 column. Immunoblotting with antibodies against FBPase and organelle markers enabled in assessing the purity of the isolated FBPase-associated vesicles. FBPase was detected in two distinct peaks from the S-1000 fractionation [44]. The first peak was enriched in both the vacuole membrane marker CPY and the plasma membrane marker Pma1p [44]. Interestingly, the second FBPase peak was enriched in a number of intracellular organelle markers. These include markers for the ER (Sec62p), Golgi (Mnn1p), vacuole (CPY), mitochondria (cytochrome C), and the ER-derived COPII vesicles (Sec22p) [44]. Owing to the enrichment of the second FBPase peak with numerous intracellular organelle markers, this peak was purified by further fractionation on sucrose density equilibrium gradients. From this fractionation, it was ascertained that FBPase was present in only one peak that corresponded to a density of 1.18 – 1.22 g/ml [44]. As this density did not correspond to any of the above intracellular organelle markers, this indicated that FBPase might be contained in distinct intracellular structures. Upon examining the FBPase containing peak using electron microscopy, a uniform population of vesicles (35-50 nm in diameter) was observed [44].

An understanding of how the Vid vesicles facilitate delivery of FBPase to the vacuole is vital to understanding the degradation kinetics of the Vid pathway. As such, it was first hypothesized that if the Vid vesicles serve as intermediate carriers in the Vid pathway, then FBPase will be associated with these vesicles prior to their delivery into the vacuole. In that endeavor, studying the distribution of FBPase at 22ºC aided in examining the kinetics of FBPase association with Vid vesicles in wild-type cells [44]. In this experiment, wild-type cells were replenished with fresh glucose for various times at 22ºC. It was determined that FBPase was associated with the Vid vesicle fraction at t=30 min and was then distributed in both the Vid vesicle and vacuole fractions by 60 min [44]. Moreover, FBPase was associated with the vacuole by 90 min [44]. These results indicate that glucose induces FBPase to be distributed in Vid vesicles and that this occurs prior to delivery of this protein to the vacuole. CPY, which was used as a control in this experiment, was not affected by glucose under these same conditions [44]. In order to ascertain whether FBPase was sequestered into the lumen of the Vid vesicles, the vesicles were purified and then incubated in the presence or absence of proteinase K [44]. The underlying principle of this assay is that FBPase that is sequestered into the lumen of the Vid vesicles will be unaffected by proteinase K digestion and that FBPase that is peripherally associated with the vesicles will be digested by proteinase K. It was determined that FBPase was stable when incubated with proteinase K, which indicated that this protein was sequestered in the lumen of Vid vesicles [44]. Addition of 2% Triton X-100 to permeabilize the Vid vesicle membrane resulted in digestion of FBPase by proteinase K. Thus, a portion of FBPase is sequestered inside Vid vesicles. However, these observations do not rule out the prospect of low amounts of FBPase being associated with the vesicles peripherally [44].

6. The biogenesis and trafficking of Vid vesicles to the vacuole

Owing to the unique nature of the Vid vesicles, innumerable questions need to be answered. Questions ranging from elucidating the origin of the vesicles to characterizing the mechanism by which FBPase is sequestered are imperative for better understanding the Vid pathway. In addition, if Vid vesicles are intermediary carriers of cargo protein in the Vid pathway, the vesicles should contain proteins that are essential for the import of FBPase into the vesicles and also for transport of FBPase from the vesicles to the vacuole. In that endeavor, *VID24* was characterized as a gene involved in the degradation of FBPase in the Vid pathway. This gene was identified by chromosomal walking [45].

The *VID24* gene encodes a protein with a molecular weight of 41 kDa. Vid24p has been characterized as a peripheral protein that is distributed to the Vid vesicles [45]. Under glucose starvation conditions, Vid24p is expressed at low levels in wild-type cells. Following glucose replenishment, Vid24p is detected at increased levels from 20 to 120 min. It has been suggested that glucose induces *de novo* synthesis of Vid24p as addition of cyclohexamide with glucose was determined to inhibit induction of this protein. Furthermore, during glucose starvation, Vid24p produced weak fluorescence upon studying the distribution of Vid24p by immunofluorescence microscopy. In contrast, Vid24p produced a stronger fluorescent signal following glucose replenishment for 30 to 60 min. Interesting, Vid24p was mostly distributed in punctate structures within cells. This suggested that Vid24p was associated with intracellular organelles, which were later determined to be the Vid vesicles. This indicates that Vid24p is a structural protein for the Vid vesicles. Furthermore, this also suggests that the *vid24-1* mutant belongs to the Class B category of mutants that accumulate FBPase in punctate structures. The above results highlight the requirement of Vid24p for the transport of FBPase from the Vid vesicles to the vacuole for degradation.

The next question pertains to the origin of the Vid vesicles. It has been proposed that the Vid vesicles may be derived from existing organelles and that they may be synthesized in cells even prior to glucose replenishment. Investigations surrounding the origins of the Vid vesicles have been hindered by the fact that Vid24p is only induced following 20-30 min of glucose replenishment. Therefore, events detailing the biogenesis of Vid vesicles during the first 20-30 min of glucose replenishment are difficult to examine with Vid24p. To circumvent this issue, an alternative strategy was designed that entailed the screening of mutants that failed to form Vid vesicles. This strategy would facilitate in assigning functions to mutants that were involved in specific steps of Vid vesicle biogenesis. In this manner, it was ascertained that the *UBC1* gene was required for Vid vesicle biogenesis [46]. As a matter of fact, the rate of FBPase degradation was observed to decrease in the null mutant of *UBC1*. Moreover, there was a diminished import of FBPase into the Vid vesicle fractions in the *Δubc1* mutant. As such, it could be inferred that in the *Δubc1* strain, there is a decrease in the level of Vid vesicles. For instance, Vid24p levels were enriched in the pellet fraction that was representative of Vid vesicles in wild-type cells. However, Vid24p levels were diminished in the pellet fraction in the *Δubc1* mutant, indicative of an impaired production of the Vid vesicles. At present, the mechanism by which *UBC1* is involved in the biogenesis of Vid vesicles has not been elucidated. Moreover, the formation of multi-ubiquitin chains has also been implicated in the degradation of FBPase in the Vid pathway. As such, yeast strains expressing the R48K/R63K ubiquitin mutant, which blocks multi-ubiquitin chain formation, resulted in inhibiting the degradation of FBPase in the Vid pathway. Interestingly, there was also a diminished amount of FBPase that was associated with the Vid vesicle fraction. Thus, these observations suggest that the *UBC1* gene and the formation of polyubiquitin chains are involved in the biogenesis of the Vid vesicles.

Another question is to understand how FBPase is imported into the Vid vesicles. To elucidate this, an *in vitro* system was developed to investigate the sequestration of FBPase into isolated Vid vesicles in the presence of the wild-type cytosol [47]. A wild-type strain in which the endogenous *FBP1* gene had been deleted for used for this *in vitro* assay. The Vid vesicles were isolated from this strain by differential centrifugation. Thereafter, the isolated Vid vesicles were incubated with a defined amount of purified FBPase in a reaction mixture that also contained wild-type cytosol, ATP and an ATP regenerating system. Proteinase K was added to the reaction mixture to degrade non-sequestered FBPase after 20 min of incubation. It was determined that 20-40% of the purified FBPase was protected from proteinase K digestion *in vitro*. Interestingly, addition of 2% Triton X-100 to permeabilize the membrane facilitated in the digestion of FBPase by proteinase K. As such, it can be inferred that FBPase is imported inside Vid vesicles, and that this import requires ATP and cytosol. In addition, our lab has also identified the cytosolic heat shock proteins Ssa1p and Ssa2p as being required for the import of FBPase into Vid vesicles [47].

Vid22p is a plasma membrane protein that was also determined to regulate FBPase sequestration into the Vid vesicles indirectly via the action of Cpr1p [43]. By using a transposon mutagenesis strategy, our lab identified the *VID22* gene. Following its synthesis in the cytosol, Vid22p is then targeted to the plasma membrane in a manner that is independent of the ER-Golgi transport pathway. It was determined that the null mutant of *VID22* inhibited the degradation of FBPase following glucose replenishment. Interestingly, FBPase was found to accumulate in the cytosol of the *Δvid22* mutant strain. This indicates that *VID22* may be required for the import of FBPase into the Vid vesicles. It was ascertained that FBPase sequestration into the Vid vesicles was inhibited upon combining the *Δvid22* mutant cytosol with the wild-type Vid vesicles using *in vitro* analysis. However, the wild-type FBPase import phenotype was rescued by incubating the wild-type cytosol with Vid vesicles from the *Δvid22* mutant. From these experiments, it can be inferred that the *Δvid22* mutant may contain functional Vid vesicles but have a defective cyotosolic environment. It has been determined that Vid22p, through its role in regulating the levels of Cpr1p, influences the degradation of FBPase. This is supported by the fact that the levels of Cpr1p in total lysates are diminished in the *Δvid22* mutant when compared to that observed in wild-type cells. However, this defect that is attributed to the absence of the *VID22* gene is rescued by the addition of purified Cpr1p *in vitro* or by overexpressing Cpr1p *in vivo*. As such, the Cpr1p protein, whose levels are regulated by Vid22p, directly promotes FBPase import into the Vid vesicles. At present, the mechanism by which Vid22p regulated Cpr1p levels has not been elucidated.

The peptidylprolyl isomerase cyclophilin A (Cpr1p) was identified as being required for the import of FBPase into Vid vesicles [48]. This cytosolic protein serves as a receptor for the immunosuppressant drug cyclosporin A. Our lab identified Cpr1p owing to its role as a mediator for the Vid protein Vid22p. By fractionating the wild-type cytosol by purification using ammonium sulfate precipitation, Superose 6 and G75 sizing chromatography, and DEAE ion exchange chromatography, our lab was able to isolate and identify Cpr1p. The role of Cpr1p in the degradation of FBPase was determined by using the *Δcpr1* mutant strain. It was ascertained that *in vitro* FBPase import and the subsequent degradation of FBPase was inhibited in the null mutant of *CPR1*. Furthermore, it was determined that the sequestration of FBPase into the wild-type Vid vesicles was impeded by the cytosol from the *Δcpr1* mutant. In contrast, import of FBPase into the Vid vesicles from *Δcpr1* mutants was not impaired when supplied with the wild-type cytosol. The role of Cpr1p in the involvement of FBPase import into the Vid vesicles was verified by adding increasing amounts of purified Cpr1p to an *in vitro* reaction mixture containing the Vid vesicles and cytosol from the null mutant of *CPR1*. A control experiment comprising of addition of BSA to the *in vitro* reaction mixture containing the Vid vesicles and cytosol from the null mutant of *CPR1* did not stimulate FBPase import. This suggests that Cpr1p has a direct involvement in the import of FBPase into the Vid vesicles.

7. The Vid pathway merges with the endocytic pathway to deliver cargo to the vacuole

In order to facilitate a better understanding of the biogenesis of Vid vesicles, Vid vesicles were isolated, purified and interacting proteins or those serving as structural components were identified using MALDI analysis. Interestingly, constituents of COPI vesicles such as Ret1p, Ret2p, Sec21p and Sec28p were identified on purified Vid vesicles [49]. As described previously, the COPI vesicles mediate transport of proteins from the Golgi to the ER [15,50]. It has been previously reported that COPI proteins have also been identified as components of endocytic compartments in both mammalian cells and in yeast [15,50]. Moreover, COPI proteins are involved in multivesicular body sorting in yeast and in endosomal trafficking in mammalian cells [15,50]. Our lab has demonstrated that COPI proteins associate with Vid vesicles [49]. This suggests that the COPI proteins may play a role in FBPase degradation. The *RET1, RET2, RET3, SEC26, SEC27, SEC21* and *SEC28* genes encode the different coatomer proteins in yeast. With the exception of *SEC28,* all the other genes are essential. As such, the role of the essential COPI genes in FBPase degradation was studied using temperature sensitive mutants. Following glucose replenishment of the null mutant of *SEC28* and the COPI temperature sensitive mutants, it was ascertained that FBPase degradation was impaired. Moreover, the *Δsec28* mutant and all of the temperature sensitive mutants of COPI genes inhibited the import of FBPase into the Vid vesicles. The *Δvam3* mutant served as a control in these experiments. The *VAM3* gene encodes a vacuolar t-SNARE that mediates fusion of intermediary vesicles with the vacuole. As such, the *Δvam3* mutant blocks FBPase degradation following its import into the Vid vesicles. These results suggest that the COPI genes are required for the import of FBPase into the Vid vesicles. The above results were verified by studying the distribution of FBPase in COPI mutants using sucrose density gradients. It was determined that FBPase distribution was enriched in the cytosolic fractions in these mutants and its levels were diminished in Vid vesicle fractions when compared to the *Δvam3* mutant. Intriguingly, the FBPase distribution in COPI mutants was similar to that observed in the *Δubc1* mutant. As these mutants inhibit the formation of Vid vesicles, this indicates that the COPI genes are also involved in Vid vesicle biogenesis. During glucose starvation, COPI proteins were observed to localize with the Vid vesicle marker Vid24p and the cargo FBPase. Interestingly, levels of COPI proteins in the Vid vesicle fractions displayed a transient increase and decrease following glucose replenishment. Furthermore, it was determined that COPI proteins associated with Vid24p forming a complex. This association was increased following glucose replenishment and was required for recruiting Vid24p to the Vid vesicles.

As the COPI genes have been previously reported to be involved in endocytosis in mammalian cells, it was important to determine whether endocytosis may be involved in our degradation pathway [49]. As a preliminary study, the kinetics of the uptake of the lipophilic dye FM4-64 was examined under our growth conditions. In wild-type cells, after its internalization, the FM4-64 dye stains the endocytic compartments before finally staining the vacuole membrane. Interestingly, the uptake of the FM dye differed upon studying its distribution in mutants that inhibited the degradation of FBPase in the Vid pathway. While mutants such as *Δvph1* displayed large FM circular distributions, other mutants such as *Δvam3* produced small FM-containing circles. Having identified Sec28p (COPI subunit of coatomer) as a structural protein of Vid vesicles, the distribution of this protein was studied as a means to monitor Vid vesicle trafficking. Sec28p was distributed in punctate structures following glucose replenishment of wild-type cells for 20-30 min. Following glucose replenishment, it was observed that Sec28p was localized to FM-containing endosomes in wild-type cells. In contrast, Sec28p failed to localize to FM-containing structures in the *Δvam3* mutant. As such, it can be inferred that the *VAM3* gene is required for the distribution of Sec28p to endosomes.

It has been previously determined that the *UBC1* gene is required for the biogenesis of Vid vesicles. In the null mutant of *UBC1*, FBPase is enriched in the cytosol and levels of Vid vesicles are diminished. The trafficking of Sec28p was also studied using a *Δubc1* mutant [49]. It was postulated that if COPI genes are involved in Vid vesicle biogenesis, then COPI proteins such as Sec28p may be discerningly distributed to structural precursors of Vid vesicles in the *Δubc1* mutant. Following glucose replenishment of the *Δubc1* mutant, it was observed that at the earlier time points, Sec28p was distributed at compartments that were stained by the FM dye. However, at later time points, Sec28p was distributed to the FM stained vacuole membrane. These results suggest that the *UBC1* gene is not required for the anterograde transport of Sec28p to the vacuole. As such, it can be inferred that the step following the delivery of Sec28p to vacuole membrane may require *UBC1*. It has been previously established that the biogenesis and budding of the COPI vesicles requires the assembly of COPI proteins at the budding site. Therefore, mutations of the COPI genes should result in altering the distribution of Sec28p to sites where the COPI vesicle buds from a precursor structure. Similarly, it was hypothesized that as Sec28p is a structural component of Vid vesicles, mutations of other COPI genes should affect the distribution of Sec28p to sites where the Vid vesicle is formed. To test this, the distribution of Sec28p was examined in a *ret2-1* mutant. In this mutant, the *ret2-1* gene encodes for a temperature sensitive protein which comprises the δ subunit of the COPI complex. Shortly after glucose replenishment, it was determined that Sec28p localized to FM containing endosomes in the *ret2-1* mutant. Interestingly, by 180 min following glucose replenishment, while FM had stained the vacuole membrane, Sec28p was observed as punctate dots near or on the vacuole membrane. This suggests that either Sec28p is a component of vesicles that are in the process of fusing with the vacuole or that Sec28p is budding from the vacuole as a component of retrograde vesicles. This was clarified by studying the distribution of Sec28p after pre-labeling the vacuole membrane with FM dye in the *ret2-1* strain. It was ascertained that Sec28p was distributed to buds that were forming on the vacuole membrane following glucose replenishment. Based on our results, it can be inferred that Sec28p containing vesicles are involved in both transport to and from the vacuole.

8. Early steps of endocytosis and actin polymerization are required for degradation of cargo to the Vid pathway

It has been previously ascertained that the Vid pathway merges with the endocytic pathway. An elucidation of this site of merger would afford a better understanding of the Vid pathway. According to one postulation, the Vid vesicles may originate from the plasma membrane or the vacuole. Alternatively, Vid vesicles may converge with endocytic vesicles that are forming on the plasma membrane. This may suggest that FBPase is also distributed near the plasma membrane. This was studied by examining, at the ultra-structural level, the distribution of FBPase in wild-type and *Δpep4* strains [51]. In these studies, following prolonged glucose starvation, the yeast strains were replenished with media containing fresh glucose for 20 min. Immuno-electron microscopy using affinity purified FBPase antibodies followed by secondary antibodies conjugated with 10 nm colloid gold particles facilitated in visualizing the FBPase distribution (Figure 1). It was determined that in both wild-type and *Δpep4* strains, a significant percentage of FBPase was distributed in irregularly shaped intracellular structures in the cytoplasm following glucose replenishment. Interestingly, FBPase was also found near the plasma membrane. This suggests that the early steps of the endocytic pathway are involved in the vacuole dependent degradation of FBPase. These irregularly shaped intracellular structures (containing FBPase) were purified by high speed centrifugation and passing the re-suspended pellet over a S-1000 column. In this manner, it was ascertained that these intracellular structures were enriched for the Vid vesicle marker Vid24p and the endosomal marker Pep12p. From this, it can be inferred that following glucose replenishment, Vid vesicles may associate with the endosomes to form large aggregates of FBPase containing structures.

Owing to the distribution of FBPase near the plasma membrane, this suggests that the early steps of endocytosis may be required for the Vid pathway. In yeast, it has been previously ascertained that the early steps of endocytosis is facilitated by actin polymerization [52-63]. Proteins involved in actin polymerization are recruited to the plasma membrane in a specific and orderedsequence (Figure 2). At the site of cargo internalization, coat module proteins and nucleation promotion factor (NPF) module proteins are recruited at the same time for shaping the membrane and for regulating actin assembly. Coat module proteins comprise of Sla1p, Lsb3p, Pan1p, and End3p. The NPF module proteins consist of Las17p, type I myosins Myo3p and Myo5p, and Vrp1p, Bzz1p and Bbc1p. With the exception of the type I myosins, it should be noted that the coat module proteins and the NPF module proteins are recruited independent of F-actin. Thereafter, the actin module proteins (consisting of 20 proteins) are recruited by F-actin to sites of actin assembly. The actin module proteins are involved in the organization and dynamics of the actin network. This module comprises of proteins such as Act1p, Arp2/3 protein complex, Abp1p, Cap1p, Cap2p, Sac6p and Aim3p among others. The Arp2/3 protein complex is involved in the nucleation of branched actin filaments. This protein complex is comprised of Arp2p, Arp3p, Arc15p, Arc18p, Arc19p, Arc35p and Arc40p. Additionally, the Las17p, Pan1p and Abp1p proteins are required for the activation of the Arp2/3 complex. Finally, the amphiphysin module proteins are recruited by F-actin to mediate scission of endocytic vesicles. This module comprises of Rvs161p and Rvs167p.

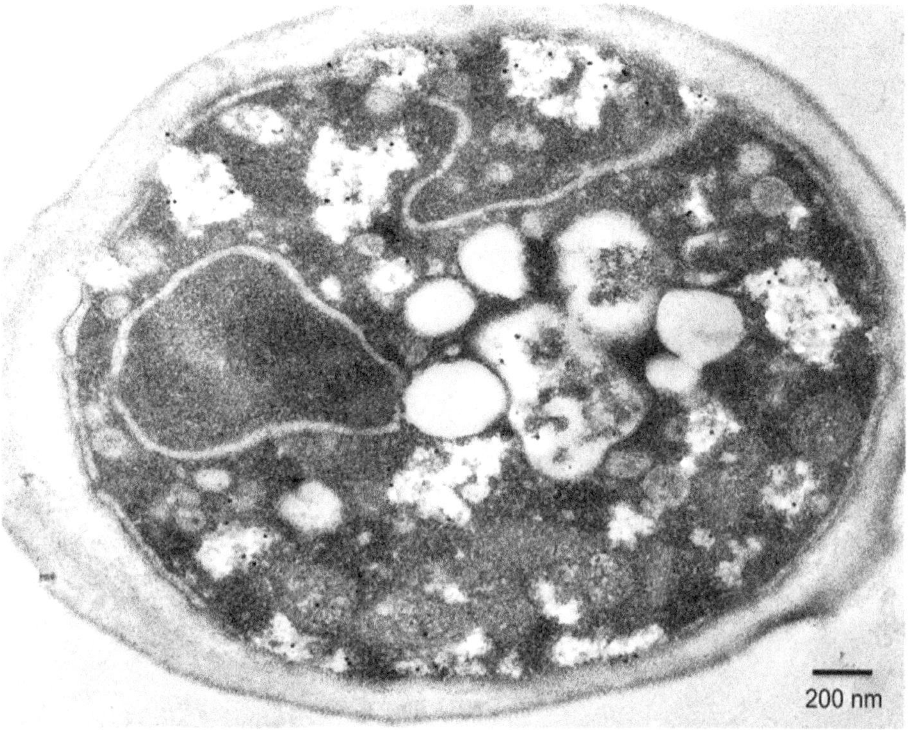

Figure 1. Ultra-structural distribution of FBPase in *Δpep4* cells following glucose replenishment for 20 min. FBPase was visualized using a purified primary antibody against FBPase and a secondary antibody conjugated with 10 nm colloid gold particles. This research was originally published in The Journal of Biological Chemistry (2010, vol. 285(2), pgs. 1516-1528). © the American Society for Biochemistry and Molecular Biology.

Under our growth conditions, upon examining the distribution of FBPase in a *Δend3* strain, it was determined that the distribution of FBPase in the plasma membrane, endosome and Vid vesicle fractions was diminished in comparison to the control *Δvph1* mutant [51]. This indicates that the early steps of endocytosis may be required for the association of FBPase with Vid vesicles. As it has been previously reported that actin polymerization facilitates the scission of endocytic vesicles from the plasma membrane, the degradation kinetics of FBPase were examined in mutants that blocked the different steps of actin polymerization. In this manner, it was ascertained that the null mutants of *END3* and *SLA1* served to inhibit the degradation of FBPase. Thus, it can be inferred that the actin polymerization genes are required for the association of FBPase with Vid vesicles. Next, fluorescent analysis was used to examine the distribution of proteins to actin patches (sites of actin polymerization). During glucose starvation, it was ascertained that there was a low percentage of co-localization of FBPase to actin patches in wild-type cells (Figure 3) [51,64]. Following glucose replenishment for up to 30 min, FBPase produced a high percentage of co-localization to actin patches. Interestingly, after 60 min of glucose replenishment, FBPase showed less co-localization to the actin patches. The distribution of MDH2 to actin patches also produced similar results. This indicates that the cargo proteins of the Vid pathway are targeted to the sites of actin polymerization on the plasma membrane.

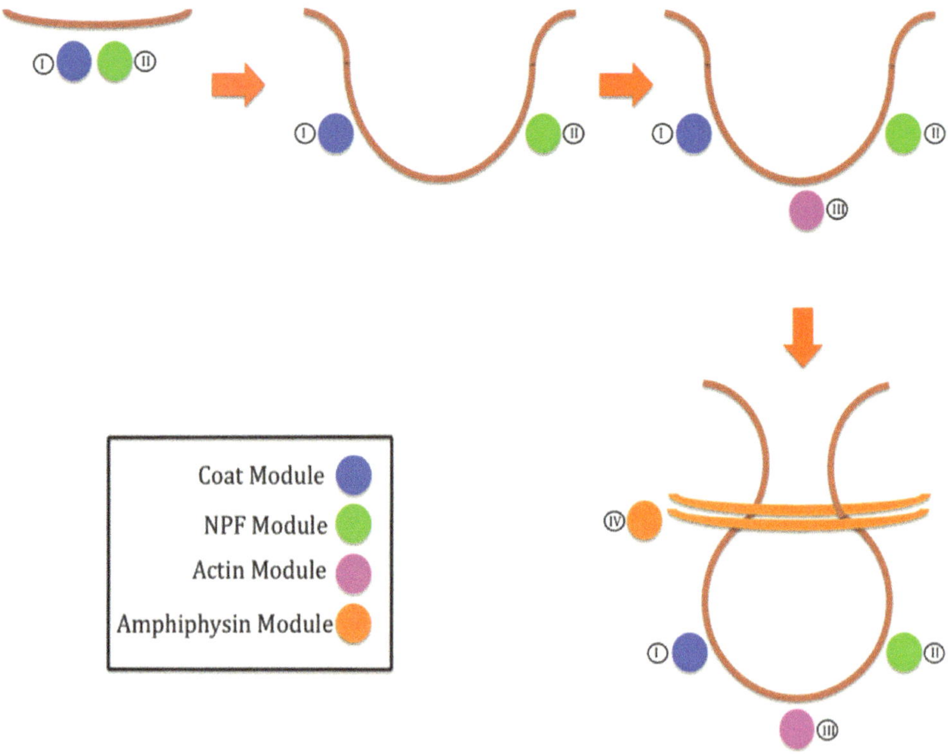

Figure 2. Actin polymerization assembly in yeast. (I, II) At the site of internalization, actin polymerization assembly recruits the coat module and nuclear promotion factor (NPF) module proteins for shaping the membrane. (III) The actin module proteins are then recruited for maintaining the integrity and the dynamics of actin assembly. (IV) The amphiphysin module proteins facilitate the scission of endocytic vesicles.

The distribution of the Vid24p to actin patches was next studied in wild-type cells as a means to determine whether the Vid vesicles are distributed to actin patches (Figure 4) [51,64]. During glucose starvation and following replenishment for up to 30 min, Vid24p was observed to be co-localized with actin patches. Intriguingly, by the 60 min time point, Vid24p demonstrated less co-localization with the actin patches. The distribution of Sec28p to actin patches also produced similar results. This suggests that during glucose starvation and following replenishment for up to 30 min, Vid vesicles associate with actin patches. In addition, in the *Δrvs167* strain, there is a prolonged association of Vid24p and Sec28p with actin patches. As such, it can be inferred that the actin patches mediate the scission of the Vid-endocytic vesicles from the plasma membrane.

To summarize the above results, we assert that Vid24p and Sec28p are distributed at the sites of actin polymerization (involved in the early steps of endocytosis) during glucose starvation. The gluconeogenic enzymes, FBPase and MDH2 are sequestered into free Vid vesicles and Vid vesicles that are aggregated at the site of actin polymerization. Following the scission of Vid-endocytic vesicles into the cytoplasm as small Vid-endosomes, these vesicles cluster and form large asymmetrically shaped structures. Therefore, the Vid-endosomes serve as intermediary carriers of cargo destined for degradation in the vacuole.

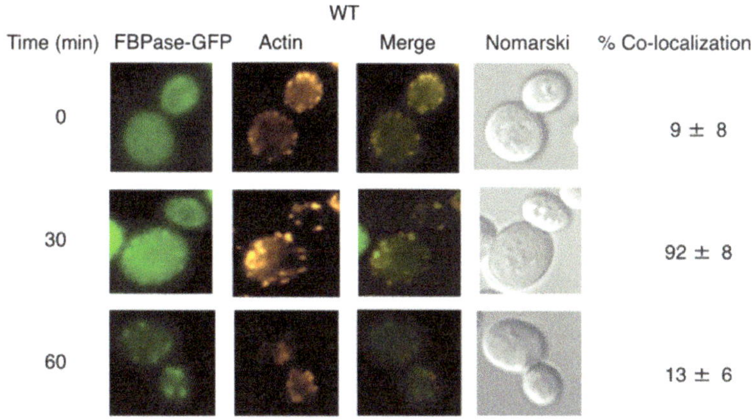

Figure 3. FBPase co-localizes with actin patches in wild-type cells. FBPase displays a low percentage of co-localization with actin patches in wild-type cells during glucose starvation. Following glucose replenishment for up to 30 min, FBPase displays a high percentage of co-localization with actin patches. Co-localization of FBPase with actin patches diminishes by the 60 min time point. This research was originally published in Autophagy (2012, vol. 8(1), pgs. 29-46). © Landes Bioscience.

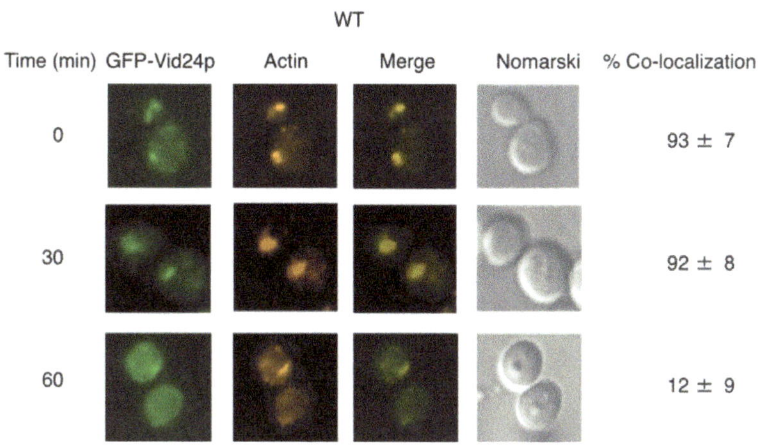

Figure 4. Vid24p co-localizes with actin patches in wild-type cells. Vid24p co-localizes to actin patches in wild-type cells during glucose starvation and for up to 30 min following glucose replenishment. Co-localization of Vid24p to actin patches diminishes by the 60 min time point. This research was originally published in Autophagy (2012, vol. 8(1), pgs. 29-46). © Landes Bioscience.

9. The association of Vid vesicles and actin patches requires *VID30*

As it has been previously determined that that the Vid pathway merges with the endocytic pathway, one could propose that association of Vid vesicles and actin patches may be a pivotal point of this integration. In that endeavor, the *VID30* gene was identified as a putative candidate involved in the Vid pathway using a transposon library screen [64]. This gene encodes a protein that has been previously reported to be involved in the proteasomal degradation of FBPase [65]. Vid30p forms a complex with Vid24p and serves as an E3 ligase in the ubiquitination of FBPase. The requirement of *VID30* in the Vid pathway was verified by examining FBPase degradation in both wild-type and *Δvid30* cells [64]. After glucose starvation for 3 days and following replenishment, FBPase was degraded in wild-type cells. In contrast, there was an inhibition of FBPase degradation in the *Δvid30* cells. This indicates that *VID30* is required for the vacuole dependent degradation of FBPase. In order to determine whether Vid30p was distribution to Vid vesicles, wild-type cells expressing Vid30p were glucose starved for 3 days followed by replenishment for up to 20 min. The cells were then subjected to differential centrifugation. Vid30p levels were enriched in the Vid vesicle enriched fraction. This infers that Vid30p is distributed to Vid vesicles.

Using pulldown assays, it was determined that Vid30p interacts with Vid24p and Sec28p under our growth conditions. Moreover, FBPase does not associate with this Vid30p-Vid24p complex. This further supports the notion that FBPase and Vid24p exist in topologically different environments. Thereafter, the effect of the absence of *SEC28* on the interaction of Vid30p and Vid24p was examined using pulldown assays. In this study, Vid30p was pulled down and the levels of Vid24p was examined the bound and unbound fractions. In the *Δsec28* mutant, the level of Vid24p in the bound fraction was diminished in comparison to that observed in wild-type cells. This indicates that Sec28p is required for the association of Vid30p with Vid24p. Furthermore, the absence of *VID24* also resulted in diminishing the interaction of Vid30p with Sec28p.

The co-localization of Vid30p with actin patches was studied using fluorescent miscroscopy. In wild-type cells, it was ascertained that Vid30p was co-localized with actin patches during glucose starvation and following glucose replenishment for up to 30 min (Figure 5) [64]. By the 60 min time point, the localization of Vid30p to actin patches began to diminish. In the absence of *VID24,* Vid30p co-localization with actin patches was prolonged following glucose replenishment (Figure 6) [64]. The absence of *SEC28* also prolonged the Vid30p co-localization to actin patches. This suggests that *SEC28* and *VID24* mediate the dissociation of Vid30p and actin patches. Interestingly, deletion of genes involved in the later steps of actin polymerization, such as *RVS161,* also resulted in prolonging the co-localization of Vid30p with actin patches.

Differential centrifugation was used to determine the step of the Vid pathway that requires the *VID30* gene. In this study, wild-type and *Δvid30* cells were glucose starved and replenished with glucose for 20 min. By differential centrifugation, it was determined that

FBPase was detected in both the Vid vesicle and cytosolic fractions in *Δvid30* cells. This is similar to what was observed for wild-type cells. Moreover, most of Vid24p was detected in the Vid vesicle fraction in the *Δvid30* mutant. This suggests that Vid vesicle formation occurs in the absence of the *VID30* gene. Even though Vid30p is distributed to multiple compartments, the deletion of this gene has no impact on the levels of FBPase and Vid24p in the Vid vesicle fraction. Moreover, it was determined that FBPase (Figure 7) and Vid24p (Figure 8) failed to co-localize with actin patches in *Δvid30* cells [64]. This suggests that Vid30p is required for the association of Vid vesicles and actin patches.

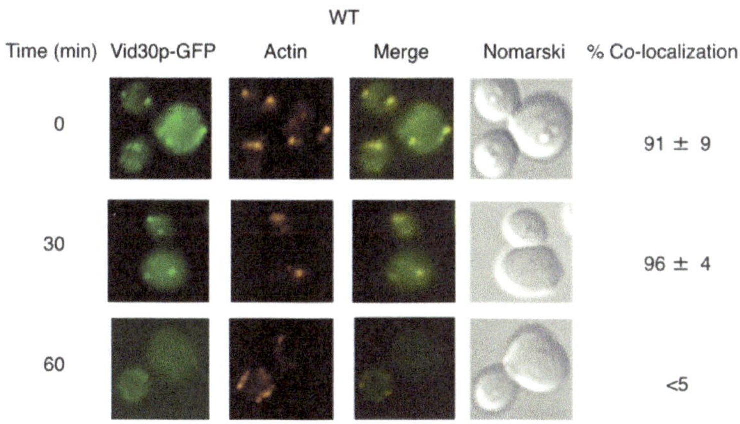

Figure 5. Vid30p co-localizes with actin patches in wild-type cells. Vid30p co-localizes to actin patches in wild-type cells during glucose starvation and for up to 30 min following glucose replenishment. Co-localization of Vid30p with actin patches diminishes by the 60 min time point. This research was originally published in Autophagy (2012, vol. 8(1), pgs. 29-46). © Landes Bioscience.

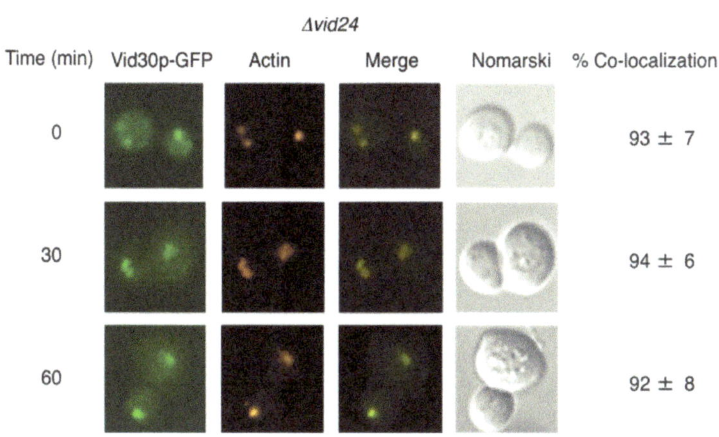

Figure 6. Vid30p co-localization with actin patches is prolonged in the null mutant of *VID24*. Vid30p is co-localized with actin patches during glucose starvation and for up to 60 min following glucose replenishment in the *Δvid24* strain. This research was originally published in Autophagy (2012, vol. 8(1), pgs. 29-46). © Landes Bioscience.

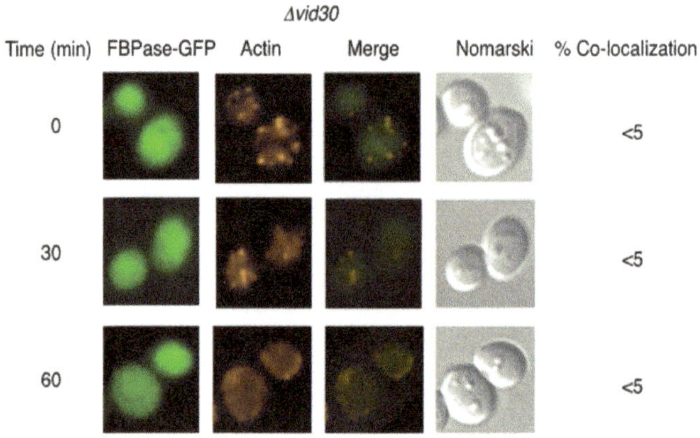

Figure 7. FBPase fails to co-localize with actin patches in *Δvid30* cells. During glucose starvation and following glucose replenishment for up to 60 min, FBPase fails to co-localize to actin patches in *Δvid30* cells. This research was originally published in Autophagy (2012, vol. 8(1), pgs. 29-46). © Landes Bioscience

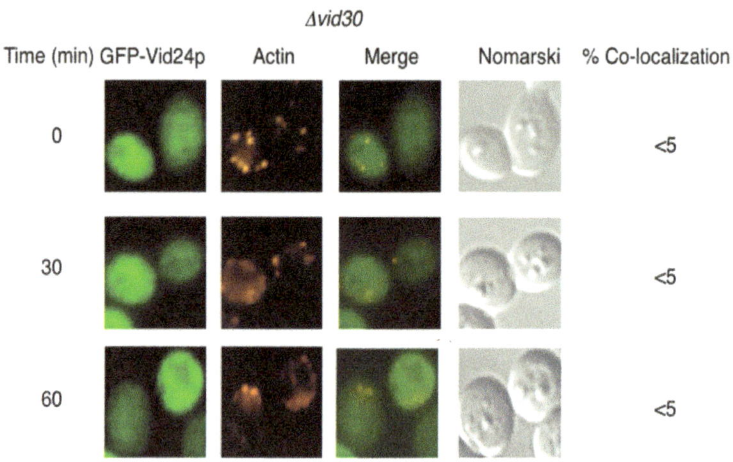

Figure 8. Vid24p fail to co-localize with patches in *Δvid30* cells. During glucose starvation and following glucose replenishment for up to 60 min, Vid24p fails to co-localize to actin patches in *Δvid30* cells. This research was originally published in Autophagy (2012, vol. 8(1), pgs. 29-46). © Landes Bioscience.

Vid30p contains two domains, a LisH (lissencephaly type 1-like homology) and CTLH (C-terminal to the LisH) domain. It has been previously reported that the *LIS1* gene is mutated in Miller-Dieker lissencephaly, a medical condition that contributes to retardation and premature mortality. In addition, the CTLH domain has been postulated to be involved in microtubule function. It was next determined if these domains play a role in the vacuole dependent degradation of FBPase. In either deletion of LisH or CTLH domain in the *VID30* gene, FBPase degradation was inhibited. Deletion of either domain also resulted in diminishing the association of Vid30p with Vid24p, and also with Sec28p. Vid30p, in which either LisH or CTLH domain had been deleted, was observed to be distributed to Vid vesicles and actin patches. Unlike in wild-type cells, the mutant Vid30p failed to be distributed to the vacuole membrane and was observed to aggregate in punctate structures. Similarly, upon deleting either domain, FBPase was also observed to localize to punctate structures. This indicates that the LisH and CTLH domains of Vid30p are involved in the later steps of the Vid pathway. Thus, in summation, *VID30* is required for the association of Vid vesicles and actin patches, and that the LisH and CTLH domains also required at a later step in the Vid pathway.

10. Proposed model for the Vid pathway

Based on the above findings, we postulate the following model for the vacuole import and degradation pathway (Figure 9). When *Saccharomyces cerevisiae* are grown under glucose starvation conditions, this induces synthesis of gluconeogenic enzymes such as FBPase, MDH2, Pck1p and Icl1p. Vid30p, Vid24p and Sec28p are present as a complex and are distributed on free Vid vesicles and on Vid vesicles aggregating around endocytic vesicles at the actin patch sites. Ubc1p has been implicated in the biogenesis of Vid vesicles. Moreover, Vid30p is required for the association of actin patches to Vid vesicles. In response to glucose, PKA facilitates in the phosphorylation of cargo proteins. Vid vesicles are distributed freely in the cytoplasm and also aggregate around the endocytic vesicles forming from the plasma membrane. The cargo proteins are sequestered into Vid vesicles and this step requires Vid22p and Cpr1p. Vid30p facilitates in the association of free Vid vesicles and actin patches. The amphiphysin module proteins (Rvs161p and Rvs167p) mediate the scission of Vid-endocytic vesicles and these are released as Vid-endosomes into the cytoplasm. Thereafter, the free Vid vesicles also accumulate around the Vid-endosomes to form large Vid-endosome clusters. The LisH and CTLH domains of Vid30p are required for the delivery of Vid-endosome clusters to the vacuole for degradation of cargo proteins. Overall, our current model highlights the association of Vid vesicles to actin patches as mediated by Vid30p as a crucial step in the degradation of gluconeogenic enzymes in the vacuole.

11. Future directions

Many questions remain to be answered concerning the degradation of cargo proteins by the Vid pathway. A pivotal question surrounding the mechanism that regulates the degradation of gluconeogenic enzymes by the proteasomal pathway versus the Vid pathway requires further elucidation. Previously, differential modification of cargo proteins following glucose replenishment has been attributed to dictate the site of degradation. For instance, proteins that are degraded in the proteasome are ubiquitinated prior to their degradation. In contrast, cargo proteins that are degraded in the vacuole are subject to phosphorylation by PKA before degradation [66]. Intriguingly, what signaling stimulus regulates this transition from degradation in the proteasome versus that in the vacuole? This warrants further elucidation.

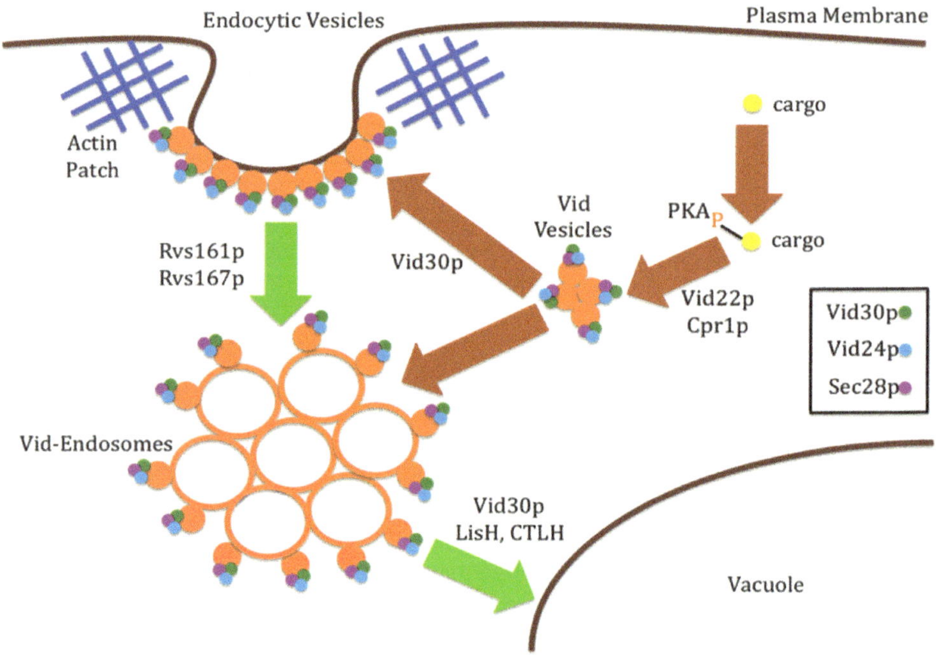

Figure 9. Current model of the Vid pathway. Growth of yeast cells under glucose starvation conditions induces the synthesis of gluconeogenic enzymes (cargo). Vid30p, Vid24p and Sec28p are present as a complex on free Vid vesicles and on those that are clustered around the endocytic vesicles at the site of actin patches. Following glucose replenishment, cargo proteins are phosphorylated by PKA and are subsequently imported into the Vid vesicles. Import of cargo into Vid vesicles requires Vid22p and Cpr1p. Vid30p mediates the association of free Vid vesicles and actin patches. Thereafter, Rvs161p and Rvs167p facilitate the scission of Vid-endocytic vesicles that are released into the cytoplasm as Vid-endosomes. The free Vid vesicles also aggregate with the Vid-endosomes to form larger clusters of Vid-endosomes. Finally, the Vid-endosome clusters deliver their cargo to the vacuole for degradation, and this step requires the LisH and CTLH domains of Vid30p.

A second concern pertains to the origin of Vid vesicles. According to one proposal, Vid vesicles may be derived from the plasma membrane or the sites of internalization. Alternatively, Vid vesicles may originate from the vacuole membrane as retrograde vesicles. Deletion of genes involved in plasma membrane internalization, such as *Δrvs161* or *Δrvs167*, contributed in prolonging the association of Vid vesicles to actin patches [51]. Perhaps Vid vesicles are components of endosomes. It is interesting to note that the Vid-endosome clusters that are formed following glucose replenishment share morphological similarities to multivesicular bodies. Moreover, the importance of the early steps of endocytosis and actin polymerization for the vacuole dependent degradation of cargo proteins requires further analysis. This could imply that cargo proteins are secreted out of the cells and then internalize at actin patch sites. As cargo proteins do not contain the ER-Golgi secretory signal sequence, this could facilitate in the understanding of the non-classical secretory pathway.

A comprehensive understanding of the Vid pathway could have significant implications in studying the etiology of diseases associated with abnormal gluconeogenesis in humans. And this could aid in developing therapeutics that regulate gluconeogenesis and treat the subsequent malady. For example, it has been previously reported that patients afflicted with Type II diabetes also suffer from an increase in levels of gluconeogenesis [67]. And an FBPase inhibitor called managlinat dialanetil has proven to be relatively successful in the treatment of Type II diabetes [67]. Another example is that FBPase may be attributed to cause clonorchiasis-associated hepatic fibrosis owing to the protein's secretion along with excretory products from *Clonorchis sinensis* adult worms [68]. Moreover, studies aimed at evaluating deterioration of the proximal renal tubules have identified FBPase as a crucial marker in this determination [69]. As such, seeking answers to the above questions will enable us in the development of treatments that will help improve the quality of life for the general public.

Author details

Abbas A. Alibhoy and Hui-Ling Chiang*
Department of Cellular and Molecular Physiology,
Penn State University College of Medicine, Hershey, USA

Acknowledgement

This work was supported by a NIH grant R01GM 59480 and a grant from the PA Tobacco Settlement Fund to Dr. Hui-Ling Chiang.

12. References

[1] Varshavsky A. Regulated protein degradation (2005) Trends Biochem. Sci. 30: 283-286
[2] Goldberg AL (2007) Functions of the proteasome: from protein degradation and immune surveillance to cancer therapy. Biochem. Soc. Trans. 35 (Part 1): 12-17.
[3] Mortimore GE, Hutson NJ, Surmacz CA (1983) Quantitative correlation between proteolysis and macro- and microautophagy in mouse hepatocytes during starvation and refeeding. Proc. Natl. Acad. Sci. USA 80 :2179-2183.
[4] Seglen PO, Gordon PB, Holen I (1990) Non-selective autophagy. Semin. Cell Biol. 1: 441-448.
[5] Mizushima N, Klionsky DJ (2007) Protein turnover via autophagy: implications for metabolism. *Annu. Rev. Nutr.* 7: 19-40.
[6] Wang CW, Klionsky DJ (2003) The molecular mechanism of autophagy. Mol. Med. 9: 65-76.

* Corresponding Author

[7] Klionsky DJ (2005) The molecular machinery of autophagy: unanswered questions. J. Cell Sci. 118: 7-18.
[8] Kaushik S, Cuervo AM (2008) Chaperone-mediated autophagy. *Methods Mol Biol* 445: 227-244.
[9] Massey A, Kiffin R, Cuervo AM (2004) Pathophysiology of chaperone-mediated autophagy. Int. J. Biochem. Cell Biol. 36: 2420-2434.
[10] Cuervo AM, Dice JF (2000) Age-related decline in chaperone-mediated autophagy. J. Biol. Chem 275: 31505-31513.
[11] Chiang, HL, Dice JF (1988) Peptide sequences that target proteins for enhanced degradation during serum withdrawal. J. Biol. Chem. 263: 6797-6805.
[12] Mizushima N, Levine B, Cuervo AM, Klionsky DJ (2008) Autophagy fights disease through cellular self-digestion. *Nature* 451: 1069-1075.
[13] Cuervo AM (2004) Autophagy: in sickness and in health. Trends Cell. Biol. 14: 70-77
[14] McMahon HT, Mills IG (2004) COP and clathrin-coated vesicle budding: different pathways, common approaches. Curr. Opin. Cell Biol. 16: 379-391.
[15] Gu F, Aniento F, Parton RG, Gruenberg J (1997) Functional dissection of COP-I subunits in the biogenesis of multivesicular endosomes. J. Cell Biol. 139: 1183-1195
[16] Duden R (2003) ER-to-Golgi transport: COP I and COP II function. Mol. Membr. Biol. 20: 197-207.
[17] Dunn WA Jr. (1990) Studies on the mechanisms of autophagy: formation of the autophagic vacuole. J. Cell Biol. 110: 1923-1933.
[18] Bryant NJ, Stevens TH (1998) Vacuole biogenesis in *Saccharomyces cerevisiae*: protein transport pathways to the yeast vacuole. Microbiol. Mol. Biol. Rev. 62: 230-247.
[19] Jones EW (1991) Three proteolytic systems in the yeast *Saccharomyces cerevisiae*. J. Biol. Chem. 266: 7963-7966.
[20] Robinson JS, Klionsky DJ, Banta LM, Emr SD (1988) Protein sorting in *Saccharomyces cerevisiae*: isolation of mutants defective in the delivery and processing of multiple vacuolar hydrolases. Mol. Cell. Biol. 8: 4936-4948.
[21] Nakatogawa H, Suzuki L, Kamada Y, Ohsumi Y (2009) Dynamics and diversity in autophagy mechanisms: lessons from yeast. Nat. Rev. Mol. Cell. Biol. 10: 458-467.
[22] Wedaman KP, Reinke A, Anderson S, Yates J 3rd, McCaffery JM, Powers T (2003) Tor kinases are distinct membrane-associated protein complexes in *Saccharomyces cerevisiae*. Mol. Biol. Cell 14(3): 1204-1220.
[23] Reinke A, Anderson S, McCaffery JM, Yates J 3rd, Aronova S, Chu S, Fairclough S, Iverson C, Wedaman KP, Powers T (2004) TOR complex 1 includes a novel component, Tco89p (YPL180w), and cooperates with Ssd1p to maintain cellular integrity in *Saccharomyces cerevisiae*. J. Biol. Chem. 279(15): 14752-14762.
[24] Yuan W, Stromhaug PE, Dunn WA (1999) Glucose-induced autophagy of peroxisomes in *Pichia pastoris* requires a unique E1-like protein. Mol. Biol. Cell. 10: 1353-1366.
[25] Chiang HL, Schekman R (1991) Regulated import and degradation of a cytosolic protein in the yeast vacuole. Nature 350: 313-318.

[26] Gancedo C (1971) Inactivation of fructose-1,6-bisphosphatase by glucose in yeast. J. Bacteriol. 107: 401-405.

[27] Holzer H (1989) Proteolytic catabolite inactivation in *Saccharomyces cerevisiae*. Revis. Biol. Celular. 21: 305-319.

[28] Minard KI, McAlister-Henn L (1992) Glucose-induced degradation of the MDH2 isozyme of malate dehydrogenase in yeast. J. Biol. Chem. 267: 17458-17464.

[29] Mazon MJ, Gancedo JM, Gancedo C (1982) Inactivation of yeast fructose-1,6-bisphosphatase. In vivo phosphorylation of the enzyme. J. Biol. Chem. 257(3): 1128-1130.

[30] Rittenhouse J, Harrsch PB, Kim JN, Marcus F (1986) Amino acid sequence of the phosphorylation site of yeast (*Saccharomyces cerevisiae*) fructose-1,6-bisphosphatase. J. Biol. Chem. 261(9): 3939-3943

[31] Jiang Y, Davis C, Broach JR (1998) Efficient transition to growth on fermentable carbon sources in *Saccharomyces cerevisiae* requires signaling through the Ras pathway. EMBO J. 17(23): 6942-6951.

[32] Schork SM, Thumm M, Wolf DH (1995) Catabolite inactivation of fructose-1,6-bisphosphatase of *Saccharomyces cerevisiae*. Degradation via the ubiquitin pathway. J. Biol. Chem. 270: 26446-26450.

[33] Schork SM, Bee G, Thumm M, Wolf DH (1994) Catabolite inactivation of fructose-1,6-bisphosphatase in yeast is mediated by the proteasome. FEBS Lett. 349: 270-274.

[34] Horak J, Wolf DH (2002) Two distinct proteolytic systems responsible for glucose induced degradation of fructose-1,6-bisphosphatase and the Gal2p transporter in the yeast *Saccharomyces cerevisiae* share the same protein components of the glucose signaling pathway. J. Biol. Chem. 77: 8248-8254.

[35] Regelmann J, Schule T, Josupeit FS, Horak J, Rose M, Entian KD, Thumm M, Wolf DH (2003) Catabolite degradation of fructose-1,6-bisphosphatase in the yeast *Saccharomyces cerevisiae*: a genome-wide screen identifies eight novel GID genes and indicates the existence of two degradation pathways. Mol. Biol. Cell 14: 1652-1663.

[36] Hammerle M, Bauer J, Rose M, Szallies A, Thumm M, Dusterhus S, Mecke D, Entian KD, Wolf DH (1998) Proteins of newly isolated mutants and the amino-terminal proline are essential for ubiquitin-proteasome catalyzed catabolite degradation of fructose-1,6-bisphosphatase of *Saccharomyces cerevisiae*. J. Biol. Chem. 273: 25000-25005.

[37] Chiang HL, Schekman R (1994) Site of catabolite inactivation. Nature 369: 284.

[38] Hung GC, Brown CR, Wolfe AB, Liu J, Chiang HL (2004) Degradation of the gluconeogenic enzymes fructose-1,6-bisphosphatase and malate dehydrogenase is mediated by distinct proteolytic pathways and signaling events. J. Biol. Chem. 279: 49138-49150.

[39] Chiang HL, Schekman R, Hamamoto S (1996) Selective uptake of cytosolic, peroxisomal and plasma membrane proteins by the yeast vacuole. J. Biol. Chem. 271: 9934-9941.

[40] Hoffman M, Chiang HL (1996) Isolation of degradation-deficient mutants defective in the targeting of fructose –1,6-bisphosphatase into the vacuole for degradation in *Saccharomyces cerevisiae*. Genetics 143: 1555-1566.

[41] Brown CR, Hung GC, Dunton D, Chiang HL (2010) The TOR complex 1 is distributed in endosomes and in retrograde vesicles that form from the vacuole membrane and plays an important role in the vacuole import and degradation pathway. J. Biol. Chem 285(30): 23359-23370.

[42] Egner R, Thumm M, Straub M, Simeon A, Schuller HJ, Wolf DH (1993) Tracing intracellular proteolytic pathways. Proteolysis of fatty acid synthase and other cytoplasmic proteins in the yeast *Saccharomyces cerevisiae*. J. Biol. Chem. 268(36): 27269-27276.

[43] Brown CR, McCann JA, Hung G, Elco C, Chiang HL (2002) Vid22p, a novel plasma membrane is required for the fructose-1,6-bisphosphatase degradation pathway. J. Cell Sci. 115: 655-666.

[44] Huang PH, Chiang HL (1997) Identification of novel vesicles in the cytosol to vacuole protein degradation pathway. J. Cell Biol. 136: 803-810.

[45] Chiang MC, Chiang HL (1998) Vid24p, a novel protein localized to the fructose-1,6-bisphosphatase containing vesicles, regulates targeting of fructose-1,6-bisphosphatase from the vesicles to the vacuole for degradation. J. Cell Biol. 140: 1347-1356.

[46] Shieh HL, Chen Y, Brown CR, Chiang HL (2001) Biochemical analysis of fructose-1,6-bisphosphatase import into vacuole import and degradation vesicles reveals a role for *UBC1* in vesicle biogenesis. J. Biol. Chem. 276: 10398-10406.

[47] Brown CR, McCann JA, Chiang HL (2000) The heat shock protein Ssa2p is required for import of fructose 1-6-bisphosphatase into Vid vesicles. J. Cell Biol. 150: 65-76.

[48] Brown CR, Cui DY, Hung GG, Chiang HL (2001) Cyclophilin A mediates Vid22p function in the import of fructose-1,6-bisphosphatase into Vid vesicles. J. Biol. Chem. 276: 48017-48026.

[49] Brown CR, Wolfe AB, Cui D, Chiang HL (2008) The vacuolar import and degradation pathway merges with the endocytic pathway to deliver fructose-1,6-bisphosphatase to the vacuole for degradation. J. Biol. Chem. 283: 26116-26127.

[50] Gabriely G, Kama R, Gerst JE (2007) Involvement of specific COPI subunits in protein sorting from the late endosome to the vacuole in yeast. Mol. Cell Biol. 27(2): 526-540.

[51] Brown CR, Dunton DD, Chiang HL (2010) The vacuole import and degradation pathway utilizes early steps of endocytosis and actin polymerization to deliver cargo proteins to the vacuole for degradation. J. Biol. Chem. 285(2): 1516-1528.

[52] Engqvist-Goldstein AE, Drubin DG (2003) Actin assembly and endocytosis: from yeast to mammals. Annu. Rev. Cell Dev. Biol. 19: 287-332.

[53] Toret CP, Drubin DG (2006) The budding yeast endocytic pathway. J. Cell Sci. 119: 4585-4587.

[54] Warren DT, Andrews PD, Gourlay CW, Ayscough KR (2002) Sla1p couples the yeast endocytic machinery to proteins regulating actin dynamics. J. Cell Sci. 115: 1703-1715.

[55] Howard JP, Hutton JL, Olson JM, Payne GS (2002) Sla1p serves as the targeting signal recognition factor for NPFX(1,2)D-mediated endocytosis. J. Cell Biol. 157: 315-326.

[56] Michelot A, Costanzo M, Sarkeshik A, Boone C, Yates JR III, Drubin DG (2010) Reconstitution and protein composition analysis of endocytic actin patches. Curr. Biol. 20: 1890-1899.

[57] Boettner DR, Chi RJ, Lemmon SK (2012) Lessons from yeast for clathrin-mediated endocytosis. Nat. Cell Biol. 14(1): 2-10.

[58] Burston HE, Maldonado-Baez L, Davey M, Montpetit B, Schluter C, Wendland B, Conibear E (2009) Regulators of yeast endocytosis identified by systematic quantitative analysis. J. Cell Biol. 185(6): 1097-1110.

[59] Liu J, Sun Y, Oster GF, Drubin DG (2010) Mechanochemical crosstalk during endocytic vesicle formation. Curr. Opin. Cell Biol. 22(1): 36-43.

[60] Kaksonen M, Sun Y, Drubin DG (2003) A pathway for association of receptors, adaptors, and actin during endocytic internalization. Cell 115(4): 475-487.

[61] Michelot A, Drubin DG (2011) Building distinct actin filament networks in a common cytoplasm. Curr. Biol. 21(14): R560-R569.

[62] Dewar H, Warren DT, Gardiner FC, Gourlay CG, Satish N, Richardson MR, Andrews PD, Ayscough KR (2002) Novel proteins linking the actin cytoskeleton to the endocytic machinery in *Saccharomyces cerevisiae*. Mol. Biol. Cell 13: 3646-3661.

[63] Kishimoto T, Sun Y, Buser C, Liu J, Michelot A, Drubin DG (2011) Determinants of endocytic membrane geometry, stability, and scission. Proc. Natl. Acad. Sci. U.S.A. 108(44): E979-988.

[64] Alibhoy AA, Giardina BJ, Dunton DD, Chiang HL (2012) Vid30 is required for the association of Vid vesicles and actin patches in the vacuole import and degradation pathway. Autophagy 8: 29-46.

[65] Santt O, Pfirrmann T, Braun B, Juretschke J, Kimmig P, Scheel H, Hofmann K, Thumm M, Wolf DH (2008) The yeast GID complex, a novel ubiquitin ligase (E3) involved in the regulation of carbohydrate metabolism. Mol. Biol. Cell 19(8): 3323-3333.

[66] Cui DY, Brown CR, Chiang HL (2004) The type 1 phosphatase Reg1p-Glc7p is required for the glucose-induced degradation of fructose-1,6-bisphosphatase in the vacuole. J. Biol. Chem. 279: 9713-9724.

[67] Wang Y, Tomlinson B (2007) Managlinat dialanetil, a fructose-1,6-bisphosphatase inhibitor for the treatment of type 2 diabetes. Curr. Opin. Investig. Drugs 8:849-858.

[68] Zheng M, Hu K, Liu W, Hu X, Hu F, Huang L, Wang P, Hu Y, Huang Y, Li W, Liang C, Yin X, He Q, Yu X (2011) Proteomic analysis of excretory secretory products from *Clonorchis sinensis* adult worms: molecular characterization and

serological reactivity of a excretory-secretory antigen-fructose-1,6-bisphosphatase. Parasit. Res. 109(3): 737-744.

[69] Kepka A, Dariusz Szajda S, Stypulkowska A, Waszkiewicz N, Jankowska A, Chojnowska S, Zwierz K (2008) Urinary fructose-1,6-bisphosphatase activity as a marker of the damage to the renal proximal tubules in children with idiopathic nephrotic syndrome. Clin. Chem. Lab. Med. 46:831-835.

Section 2

Using Microscopy to Study Endocytic Trafficking

Chapter 5

Imaging of Endocytosis in *Paramecium* by Confocal Microscopy

Paola Ramoino, Alberto Diaspro, Marco Fato and Cesare Usai

Additional information is available at the end of the chapter

1. Introduction

Endocytosis is the general term for internalization of fluid, solutes, macromolecules, plasma membrane components, and particles by the invagination of the plasma membrane and the formation of vesicles and vacuoles through membrane fusion. The means by which food material enters the body is to a great extent dependent on the size of the particles involved [1, 2]. One type of endocytosis is phagocytosis where large (>250 nm) particles are taken up by cells. In protozoa, phagocytosis is a feeding mechanism. Particles are brought into the cell in large endocytic vesicles called phagosomes (food vacuoles). The phagosomes fuse with lysosomes and digestion of the ingested particles occurs. In multicellular organisms, phagocytosis is a behavior seen only in certain specialized cells (for example, macrophages). The process is essentially the same whether it is phagocytosis of particles or other organisms or pinocytosis of molecules. Both endocytic processes are affected identically by inhibitors of aerobic metabolism and by low temperatures.

The various forms of endocytosis (or food uptake) really only differ in degree. They are closely linked in that the combined volume taken up by endocytosis is constant and critical. As phagocytosis increases, pinocytosis must decrease proportionally. In addition to the forms of bulk transport just considered, which involve invagination of the plasma membrane, other essential substances, dissolved nutrients of low molecular weight, enter the organism by facilitated diffusion or active transport through the plasma membrane.

Some protozoa secrete hydrolyzing enzymes into the external medium to degrade large nutritive molecules into smaller soluble units for transport through the plasma membrane. This facility for extracellular digestion is of value to facultative and obligate parasites and to other protozoa which live in a highly nutritive environment.

Ciliated protozoa collect or capture their food in a variety of ways, involving phagocytosis, fluid phase and receptor-mediated endocytosis

2. Phagocytosis

2.1. Phagocytosis in ciliates

Ciliates acquire their food as particles from the surrounding medium by a variety of means [1-5]. Filter feeders create water currents with special ciliary structures associated with the cytostome. The synchronized beating of these membranelles acts as a collecting sieve, where the food particles become trapped. Using this mode of feeding, ciliates can shift considerable volumes of water in relation to their size. Some of the most efficient filter feeders are the hymenostomes *Paramecium, Tetrahymena* and *Glaucoma*. All have a ventral buccal (oral) cavity containing well-developed ciliary membranelles and some form of paroral membrane on its right margin. Fluid and suspended food particles are collected in the buccal cavity and directed, by oral membranelle beating, downward to the cytopharynx and cytostome to form a new food vacuole [6, 7].

Herbivorous ciliates, instead, lack complex oral cilia and gather their food by a complex pharyngeal basket of rods and sheets of microtubules that forms an internal support to the cytostome. They ingest filamentous algae by grasping the filament, bending it like a hairpin, and drawing it into the cytopharynx, where it is broken up into fragments and enclosed in digestive vacuoles.

Gulper ciliates apprehend their prey with special structures called toxicysts, which are found in the oral region and release toxins that paralyze or kill the swimming prey organims. The paralyzed prey can then be ingested without difficulty. Indeed the oral area of gulpers can be extended greatly, as most carnivores take food at least as big as themselves. The initial contact may be due to chemotactic orientation or fortuitous contact. Some ciliates develop carnivorous tendencies only when their preferred food supply is exhausted. In the absence of bacteria *Blepharisma* eats its own kind and develops giant forms for self-protection [8].

Sophisticated organelles involved in ingestion are tentacles of suctoria. The predatory or parasitic suctorians are sessile ciliates. The ectocommensals on a wide variety of marine and freshwater hosts use the motile activities of their host or its feeding strategy to bring food to them. Carnivores have developed a special method of feeding in which the tips of the tentacles act as cytostomes. The contact between the prey and the tentacles of the predator triggers the stimulus for ingestion. The cell contents of the prey are transported up through the feeding tentacles into the suctorian, where digestive vacuoles are formed. The transporting mechanism is mediated by a complex array of microtubules within the tentacle.

A number of ciliates respond to exudates from animal tissues and have exploited this response by becoming active scavengers. Sheet-like membranous organelles associated with feeding are a feature of apostome ciliates which live in or on a variety of animal hosts, arthropods, echinoderms and sea anemones.

2.2. *Paramecium* as a model system for membrane trafficking

The ciliated protozoan *Paramecium* is easily cultured and manipulated. Therefore, it is especially useful to study *in vivo* vesicle formation, transport and fusion during the digestive process. It disposes of well-defined sites for formation of phagosomes (oral cavity, with cytostome and cytopharynx) [9, 10].

Food vacuoles undergo a series of sequential changes from their formation at the cytostome to their defecation at the cytoproct. Depending on their age, size, morphology, vacuolar pH, acid phosphatase activity and degree of bacteria digestion, they have been grouped into four stages [11-17].

The first stage, stage I, includes the nascent and newly formed food vacuoles which separate from the base of the cytopharynx and move toward the posterior end of the cell (Figure 1). During stage I, food vacuoles have no acid phosphatase activity and are bounded by acidosomes. Their content condenses, becoming progressively more acidic, and surplus vacuolar membrane and excess fluid are removed by pinocytosis of vesicles which migrate back to the cytopharynx. The condensed and acidic food vacuoles (stage II) are surrounded by enzyme-containing primary lysosomes. As lysosomes fuse and digestion proceeds, the vacuole enlarges again, becoming less acidic or even slightly alkaline (stage III). The breakdown products are pinched off as small vesicles (secondary lysosomes) to be transported where necessary. Following membrane and water elimination, the food vacuole decreases in size and tends to a neutral pH (stage IV); the active retrieval of lysosomal membrane may continue during this stage, but active acid phosphatase is not present. In the final stage, when the food vacuole becomes defecation-competent, it fuses with the plasma membrane at the cytoproct (a fixed spot on the ventral surface, posterior to the buccal cavity). The indigestible material is excreted, while the vacuolar membrane is retrieved and recycled as discoidal vesicles moving back to cytopharynx and providing the membrane to the nascent food vacuole.

2.3. Phagocytosis in *Paramecium* by confocal microscopy

In paramecia fed with indigestible particles, the duration of the digestive cycle is relatively short (20 to 60 minutes), and the digestive processes are synchronous enough and so temporally defined as to allow food vacuole selection in a specific digestion stage using a pulse-chase protocol. By immobilizing living cells pulsed with a food vacuole marker at succeeding times after a chase of unlabeled medium, it is possible visualize *in vivo* the intracellular movement of food vacuoles along an orderly path from their formation at the cytostome to their egestion at the cytoproct, as well as the flow of pinocytic vesicles from vacuolar membrane evagination to the fusion with other food vacuoles.

The sequence of appearance of the four vacuole stages in different regions of the cell follows the general path of cyclosis and indicates that the cytoplasm moves forward in the dorsal zone and backward in the ventral zone [6, 18]. Notwithstanding the fact that the digestive processes are sufficiently synchronous and separated in time, there is considerable overlap and variability in the length of the food vacuole stages from cell to cell, and possibly from food vacuole to food vacuole [16]. Food vacuoles less than 5 min old are vacuoles of stage I located in the posterior end of the cell and undergoing rapid acidification and condensation. Vacuoles between 5 to 10 min old are acidic and condensed food vacuoles of stage II; they are located near the oral region and around the macronucleus. The vacuoles of stage III range in age from 10 to about 20 min and are generally located in the anterior half of the cell, while stage IV food vacuoles are more than 21 min old and move toward the cytoproct. Vacuolar defecation has been shown to begin at about 20 min and is essentially completed by 60 min in axenically grown cells [14] as well as in bacterized cells.

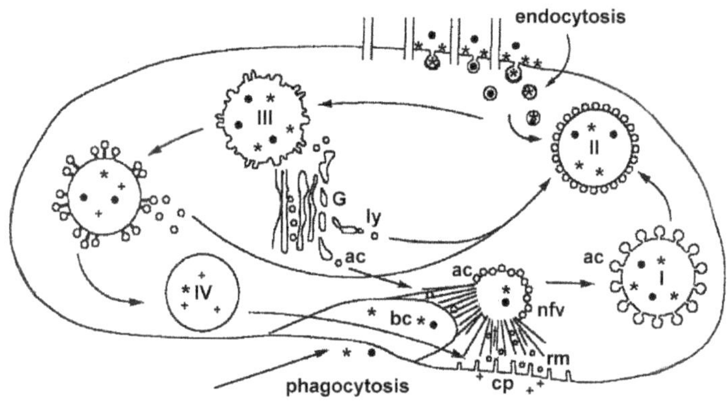

Figure 1. Schematic drawing of phagocytic and endocytic pathways of internalization in *Paramecium* based upon WGA-FITC and dextran-TXR staining. Without phagocytosis inhibition the ingested material is directed by the oral membranelle beating into the buccal cavity (bc), the cytopharynx and, at last, into the nascent food vacuole (nfv). The newly formed food vacuole (I) is surrounded by acidosomes (ac) discharging their content into it. The vacuole reduces its size by eliminating water and membrane trough small pinocytic vesicles. The condensed vacuole (II) receives the enzymes contained into lysosomes (ly). The cargo is digested (III) and pinocytic vesicles containing digestion products are pinched off. At last, the indigestible material (IV) is excreted at the cytoproct (cp), while the vacuolar membrane is retrieved (rm). The endocytosis of WGA-FITC and dextran-TXR also occurs at the parasomal sacs located next the ciliar basal bodies. Exogenous fluid and plasma membrane components are internalized by vesicles which fuse with food vacuoles. G = Golgi apparatus; * = WGA-FTC; ° = dextran-TXR; + = degraded material; ↔ = flow direction (modified from Allen et al. [17]).

In order to characterize the cytoplasmic distribution and movement of food vacuoles and pynocytic vesicles, living cells were continuously fed with BSA-FITC and latex particles (LP) in culture medium for a time period ranging from 30 sec to 30 min, washed, immobilized with $NiCl_2$ [19]. $NiCl_2$ inhibits both locomotive activity [20] and formation of food vacuoles with solid particle content [21], without affecting cytoplasmic streaming [21]. To demonstrate the reutilization of pinocytic vesicles and vesicles formed by the membrane retrieved from spent vacuoles at the cytoproct, cells first fed with BSA-FITC and LP for 30 min, then washed in sterile filtered culture medium for 20 to 30 min, were labeled with carmine particles or BSA-Texas red for 1 min, washed and immobilized at various times during chase [19].

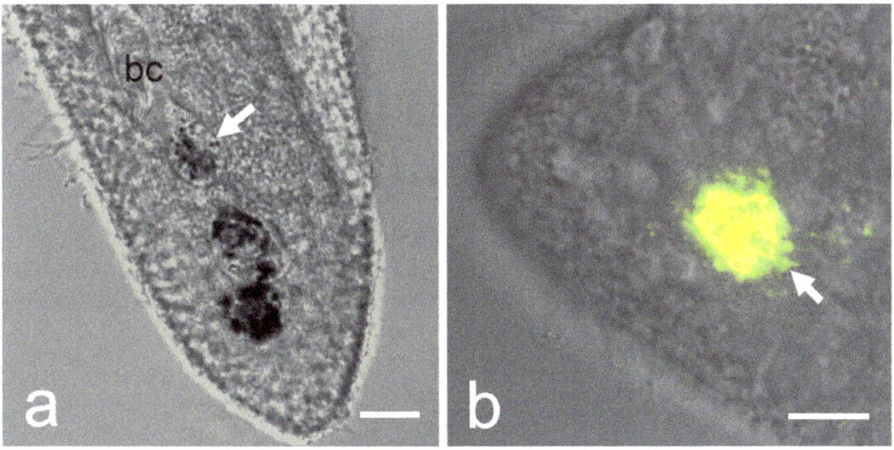

Figure 2. (a) One nascent food vacuole (arrow) at the bottom of buccal cavity (bc) and two newly formed food vacuole in a cell fed (carmine particles). (b) In cells fed with BSA-TXR, newly formed food vacuoles which fused with BSA-FITC labeled recycled vesicles (arrow), contain both fluorochromes. Bars, 10 μm.

In cells fed with carmine particles for 30 sec, only 1-2 food vacuoles are labeled (Figure 2a). To enlarge the membrane surrounding the nascent food vacuoles, cells utilize recycled vesicles that are pinched off from food vacuoles during the digestive process and the vesicles deriving from the membrane of the spent vacuoles at the cytoproct. So, in cells fed with BSA-Texas red newly formed food vacuoles, which fused with BSA-FITC labeled recycled vesicles, contain both fluorochromes (Figure 2b). Food vacuoles separating from the cytopharynx and moving toward the posterior end of the cell, are surrounded by small acidic vesicles (acidosomes) and by lysosomes (containing hydrolytic enzymes) which fuse with vacuolar membrane. To analyze the distribution of acidosomes and lysosomes during food vacuole migration, living cells pulsed with LP for 1-2 min, were labeled with acridine orange (AO) at various times after the chase in unlabeled medium, and immobilized. AO is a fluorescent tertiary amine that accumulates in the acidic compartment of living cells and is commonly used to stain lysosomes [23-25]. The fluorescence around the newly formed food vacuoles (<5 min old) is due to acidosomes [25], whereas the fluorescence around and within >5 min-old food vacuoles is due to lysosomes. The AO-stained granules produce a punctuate pattern throughout the cell and around certain food vacuoles. Fluorescent granules, including both primary and secondary lysosomes, form a relatively thin rim around 6 min-old food vacuoles containing LP (Figure 3a). The rim becomes quite prominent in other food vacuoles (10 min old), as the lysosomes fuse with the vacuolar membrane and discharge their content into the vacuole (Figure 3b).

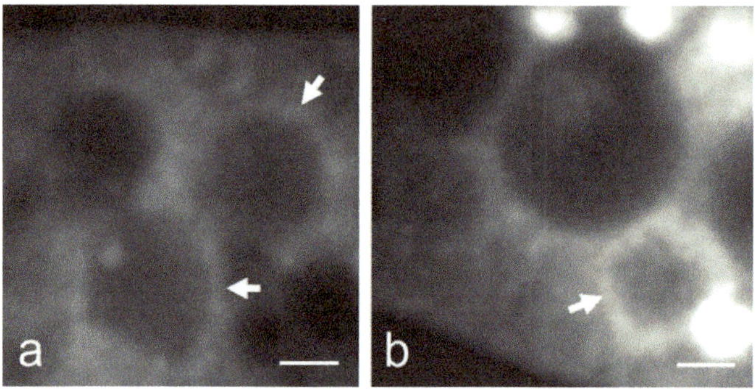

Figure 3. (a) Stage II LP-labeled food vacuoles (non-fluorescent) surrounded by AO-labeled lysosomes (arrows). (b) Stage III LP-labeled food vacuole (more fluorescent in the surface area, arrow): AO-labeled lysosomes are fusing and discharging their content into the vacuole. Bars, 10 μm .(modified from Ramoino et al. [19])

As the vacuolar content is digested by lysosomal enzymes, the breakdown substances pass into the cytoplasm through vacuolar membrane or by way of small pinocytic vesicles. The vesicles evaginated from the membrane, go away from the food vacuole and move in the cytoplasm toward the cytopharynx where they enlarge the membrane of the nascent food vacuole. These vesicles can also fuse with stage II food vacuoles (Figure 4), when the vacuoles of stage II increase their size, changing from an acidic to an alkaline status. For better visualization of the movement of the pinocytic vesicles, cells are fed with BSA-FITC and LP for 30 sec. and washed in a sterile culture medium. So, only 1 or 2 labeled vacuoles are formed and few fluorescent vesicles move in the cytoplasm. When the cells are immobilized between 15 to 20 min after chase in sterile medium, the food vacuoles are in the digestion stage and small pinocytic fluorescent vesicles pinch off. The multimodal image analysis utilizing the pseudo-color technique [26] shows changes on the direction of movement of the vesicles going away from the vacuole (Figure 5a).

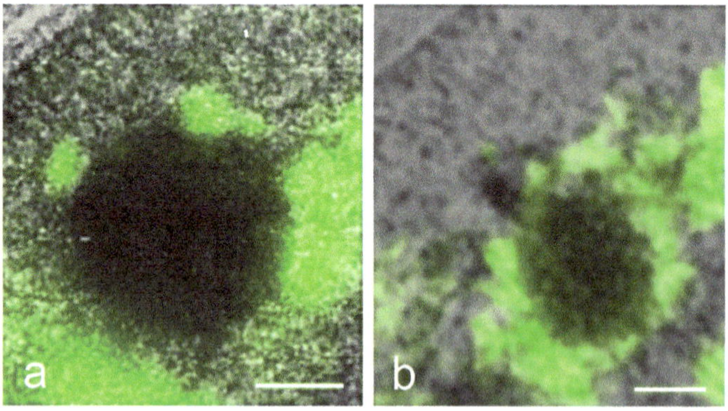

Figure 4. (a) Carmine-labeled food vacuole (non-fluorescent) surrounded by FITC- labeled fluorescent vesicles. Carmine-labeled food vacuole (more fluorescent in the superficial zone) surrounded by FITC-labeled fluorescent vesicles. Bars, 5 μm (modified from Ramoino et al. [19])

The utilization of recycled vesicles from stage II and early stage III food vacuoles is evidenced by labeling food vacuoles either with carmine particles or with a second fluorescent probe. When carmine particles are utilized, food vacuoles in cells immobilized after a 5-min. chase (stage II) are initially non-fluorescent and surrounded with a lot of fluorescent small vesicles (Figure 4a), then, as the vesicles fuse with the food vacuole (stage III), their content becomes fluorescent. At the beginning fluorescence is only located in the surface area (Figure 4b), then, it increases as the food vacuole age progresses and more vesicles fuse discharging their labeled content inside the vacuole. Vesicles, which fuse with food vacuoles, move apparently in unidirectional manner (Figure 5b).

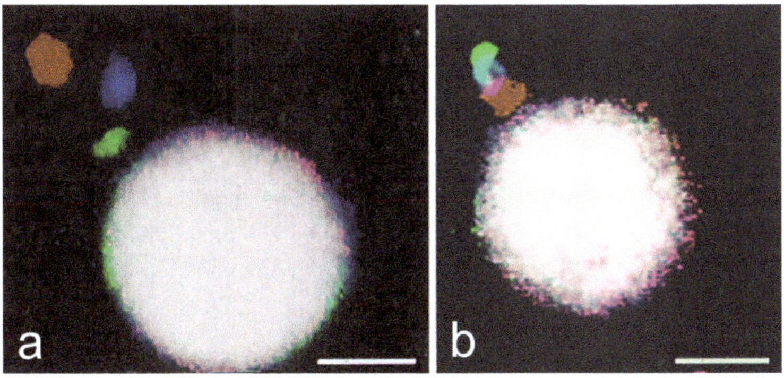

Figure 5. Composite false-color images showing the global vesicle movement of one FITC-labeled vesicle moving away from a food vacuole (a) as well as of one small FITC-labeled vesicle moving toward the vacuole (b). A different color, green, blue and red, was associated with three successive temporal images t_1, t_2, t_3, respectively. Bar, 10 μm. (modified from Ramoino et al. [19])

In *Paramecium,* a saltatory movement with changes in direction and velocity, stops and starts, was described for cell organelles such as mitochondria and trichocysts in subcortical regions of the cell [27] and for all motions within the cytoplasmic streaming [28], whereas a smooth and continuous unidirectional movement along the microtubules joined to the cytopharynx was reported for acidosomes and discoidal vesicles [29].

After a digestion period, the food vacuole becomes defecation-competent and fuses with plasma membrane; the indigestible material is excreted. Figure 6 is a composite picture of phase-contrast CLSM images demonstrating the temporal and spatial movement and egestion of food vacuoles at the cytoproct.

In *P. primaurelia* food vacuole formation depends on membrane material supply [30]. By using different solid particle concentrations in unbacterized culture medium it is shown that a given amount of membrane material is available for food vacuole formation. This membrane amount is utilized more rapidly if the concentration of particles is higher, where the food vacuole size is larger, than for a lower concentration, where the food vacuole size is smaller. After the utilization of the membrane made available for the cell, a decrease in the food vacuole number occurs. Furthermore, the rate of food ingestion decreases in starved cells pressed continuously to form food vacuoles because of particles suspended in the culture medium. The kinetics of food vacuole formation does not differ when cells are fed on latex particles in bacterized and unbacterized medium for short periods of time. For longer labeling periods with high particle concentrations, the food vacuole number decreases after the maximum value more rapidly in cells stained with particles in non-nutrient medium than in cells fed with particles diluted in bacterized medium. The spent vacuole membrane is insufficient to keep the food vacuole number at a high level. Failing new syntheses, the vacuolar membrane amount goes on depressing.

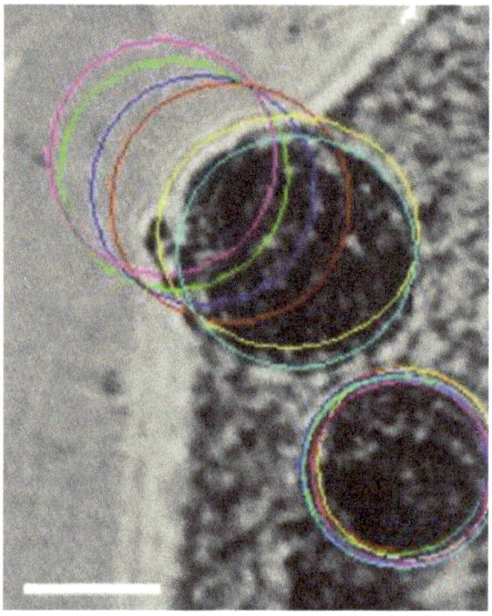

Figure 6. Composite image showing the movement of a food vacuole egesting its content at the cytoproct. A different color cyan, yellow, red, blue, green and magenta was associated with six successive temporal images t_1, t_2, t_3, t_4, t_5 and t_6, respectively. Bar, 10 μm.

3. Endocytosis

3.1. Endocytosis in ciliates

Fluid phase and receptor-mediated endocytosis has been extensively studied in mammalian cells [31-34] and most of what is understood about the endocytic process in protozoa comes from the studies of pinocytosis in amoeba [35-37]. In an amoeba pinocytotic vesicles are formed at the bases of long narrow invaginations, pinocytotic channels. Small vesicles are pinched off at the base of a channel deep in the cytoplasm and are passed into the interior. But fluid phase endocytosis does not necessarily involve the development of channels. The parasitic *Opalina ranarum,* which must take up nutrients through its plasma membrane, pinches off small vesicles in the grooves between the folds in its pellicle [38]. Once inside, vesicles coalesce to form larger vacuoles bounded by unit membranes, vacuoles which are not markedly different from endocytic vacuoles produced by phagocytosis. Fluid phase endocytosis may occur simultaneously all over the surface as in amoebae, or it may be restricted to clearly defined regions, such as the walls of the flagellar pocket of some trypanosomes.

The limited data about endocytosis in ciliates is also due to the difficulty of visualizing this process at an optical level. Indeed, in ciliates only defined areas on cell surface are potential sites for endocytic uptake since most of the surface is covered internally by an extensive system of alveoli and an underlying fibrous epiplasm [39]. This system is interrupted only at the cytopharynx, the cytoproct, contractile vacuole pores and along the junctions of the abutting units of alveolar membrane sacs. Only the punctuate indentations of the plasma membrane, called parasomal sacs, and pellicular pores are potential endocytic entry ports of all fluid phase and putative receptor-mediated endocytosis [17, 40-42].

Detailed morphological and tracer studies on endocytosis carried out by electron microscopy showed that in *Paramecium multimicronucleatum* fluid phase markers such as horseradish peroxidase (HRP) and in *Tetrahymena pyriformis* receptor-mediated markers such as cationized ferritin are internalized via coated pits and are found in coated vesicles [40, 41]. Both coated pits and vesicles are also labeled in fixed cells when a monoclonal antibody against the plasma membrane of *P. multimicronucleatum* (C6 antigen) is applied to cryosections, suggesting that both membrane-bound and fluid phase markers are internalized at the coated pits [40]. Most endocytic sites are clathrin coated pits, however there is increasing evidence for mammalian cells for clathrin-independent pathways, mediated by caveolae or non-coated vesicles [43].

In *Paramecium* the fluorescence amount internalized by endocytosis is less than that internalized by phagocytosis even if an increased endocytic rate is obtained when food vacuole formation is blocked. Indeed, evidence was provided by electron microscopy studies that the number of endocytic vesicles increased when food vacuole formation was blocked by trifluoperazine, a calmodulin antagonist [41]. In addition, by means of a quantitative analysis, it was evidenced more specifically that the HRP influx rate increased twofold when phagocytosis was blocked by propranolol, a β-adrenergic antagonist [44]. Wyroba [44] suggested that in *Paramecium* the increased fluid phase uptake indicates that the two pathways, though independent, may be limited by a membrane pool and/or energy requirements. Indeed, forskolin and phorbol ester, powerful stimulants of *Paramecium* phagocytosis [45], reduce the HRP uptake rate.

3.2. Endocytosis in *Paramecium* by confocal microscopy

Endocytosis in *P. primaurelia* was studied using WGA (*Triticum vulgaris agglutinin*) and $GABA_B$ receptor antibodies, which bind to surface constituents of fixed [46, 47] and living cells, as markers for membrane transport and dextran as a marker for fluid phase endocytosis.

Endocytosis markers are internalized via food vacuoles formed at the cytopharynx when they are added to the cell incubation medium without phagocytosis inhibition [48]. Cells pulsed with WGA-FITC for 3 min show some food vacuoles at the posterior pole of the body. After a 10-min chase in unlabeled medium the number of fluorescent food vacuoles increases (data not shown). The increase in labeled food vacuoles in a fluorochrome-free medium is due to the fact that the ingested lectins are degraded and pass into the cytoplasm by small vesicles which then fuse with other food vacuoles [19]. Increasing the chase in unlabeled medium increases the fluorescence inside the cytoplasm, which is found later in the vesicles of the phagosome-lysosome system and at the plasma membrane level. Conversely, when phagocytosis is blocked by trifluoperazine, the fluorescence is initially found, in 3 min pulsed cells, on the plasma membrane and cilia and inside the cell into small cytoplasmic vesicles (Figure 7a). After a 5-10 min chase in unlabeled medium, fluorescent vesicles fuse with some food vacuoles (Figure 7b), and after 20-30 min the labeled food vacuoles increase in number and small vesicles throughout the cytoplasm fluoresce (Figure 7c). Therefore, the digestion inside the vacuoles of lectins internalized via endocytosis begins later with respect to lectins internalized via food vacuole formation (phagocytosis). Moreover, a very weak fluorescence is detectable on plasma membrane after longer time periods compared with lectin internalization via food vacuole formation.

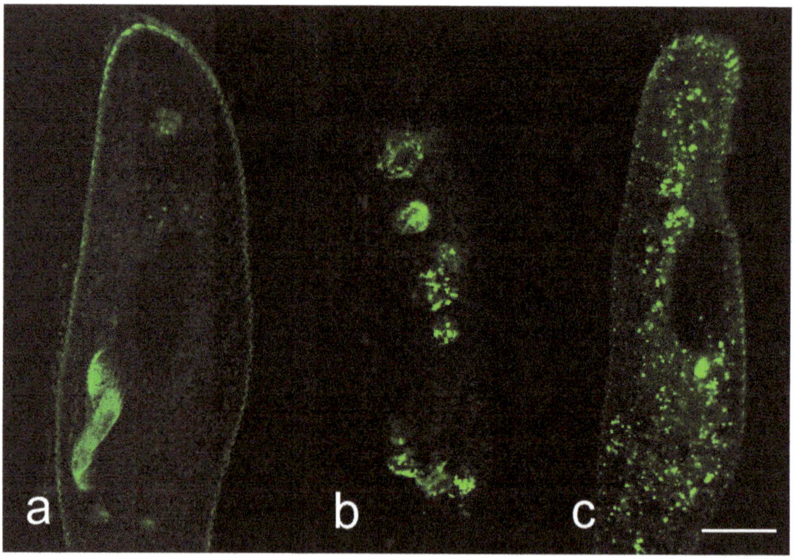

Figure 7. WGA internalization and intracellular flow. Cells labeled with WGA-FITC for 3 min. Plasma membrane and small vesicles inside the cell fluoresce (a). After a 10 min chase in unlabeled medium (b) fluorescence is visible in a few food vacuoles whereas after 30 min (c) small vesicles throughout the cytoplasm fluoresce. Bar, 20 μm.

The fusion of endocytic vesicles with food vacuoles is evidenced by a double-labeling experiment in which the vesicles are dyed with WGA-FITC and the food vacuoles with BSA-TXR (Figure 8).

Similar results were obtained in cells blocked in the phagocytic activity, incubated at 25°C in a culture medium containing an antibody anti-GABAB R1 receptor for 15-30 minutes and then fixed and processed for immunolabeling.

Dextran-TXR, a fluid phase endocytosis marker, does not label the plasma membrane and enters the cell via small vesicles initially localized at the cortical level (Figure 9a). The vesicles later migrate in the cytoplasm and fuse with other endocytic vesicles and with food vacuoles (Figure 9b). The number of labeled food vacuoles increases as the dextran-labeled vesicles fuse with food vacuoles (Figure 9c) and then decreases when the vacuolar content is digested and food vacuoles containing the indigestible material are ejected at the cytoproct.

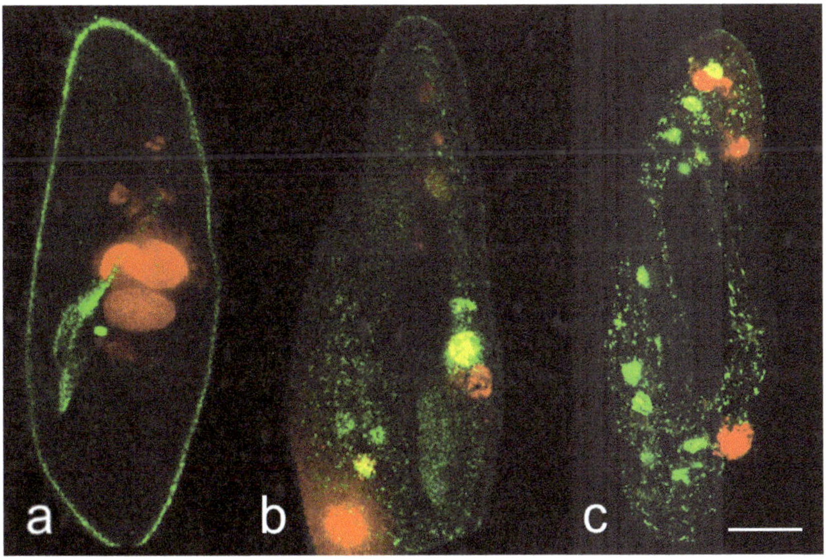

Figure 8. Fusion of endocytic vesicles with food vacuoles. Cells fed with BSA-TXR for 20 min, washed, and labeled with WGA-FITC for 5 min. At first, green fluorescence is localized on plasma membrane (a); after 20 (b) and 30 min (c) of chase in unlabeled medium green fluorescence is also present inside the food vacuoles. Bar, 20 μm.

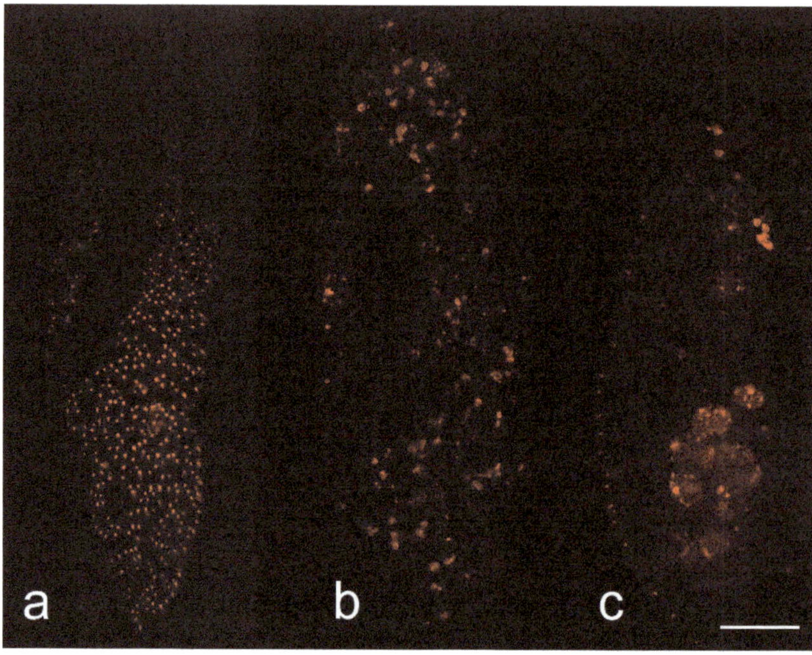

Figure 9. Dextran internalization and intracellular flow. In cells labeled with dextran-TXR for 3 min fluorescence is visible in small vesicles located in the cortex under the plasma membrane (a). After a 10 min chase in unlabeled medium (b), vesicles decrease in number and increase in size. After 30 min several food vacuoles are labeled (c). Bar, 20 μm.

The relationships between the two different routes of internalization, membrane transport and fluid phase endocytosis, are clearly shown when cells blocked in their phagocytic activity are simultaneously fed with WGA-FITC and dextran-TXR (Figure 10). The data obtained by confocal microscopy suggest that WGA and dextran are present in different endocytic vesicles soon after initiation of uptake (< 10 min). The two probes probably partly join prior to their fusion with the phago-lysosomal compartment. From these data we can assume that dextran-TXR and WGA-FITC enter the cells through two different vesicle populations, which then can fuse together or with food vacuoles.

In order to understand if the two markers are internalized through two separate pathways, cells were incubated either in a hypertonic medium or in acetic acid. Indeed, subjecting mammalian cells to either incubation in media containing sucrose [49] or cytosol acidification with acetic acid [50, 51] has been shown to inhibit clathrin-mediated endocytosis by interfering with clathrin-adaptor interactions [52], or by altering the structure of clathrin itself [53-55]. In *P. primaurelia* 0.20 M sucrose incubation completely blocks the internalization of WGA, which stops at the plasma membrane (Figure 11a). It also reduces dextran uptake, which is localized in small vesicles in the cortical part of the cell and in a few vesicles throughout the cytoplasm. Through cytosol acidification by 10 mM acetic acid, pH 5.0, WGA fluorescence is localized at the plasma membrane level whereas small red vesicles containing dextran are localized both in the cortex, under the plasma membrane, and throughout the cytoplasm (Figure 11b). A similar fluorescent pattern was seen by using chlorpromazine (data not shown), a cationic amphiphilic drug which inhibits clathrin-dependent receptor mediated endocytosis by reducing the number of coated pit-associated receptors at the cell surface [56, 57].

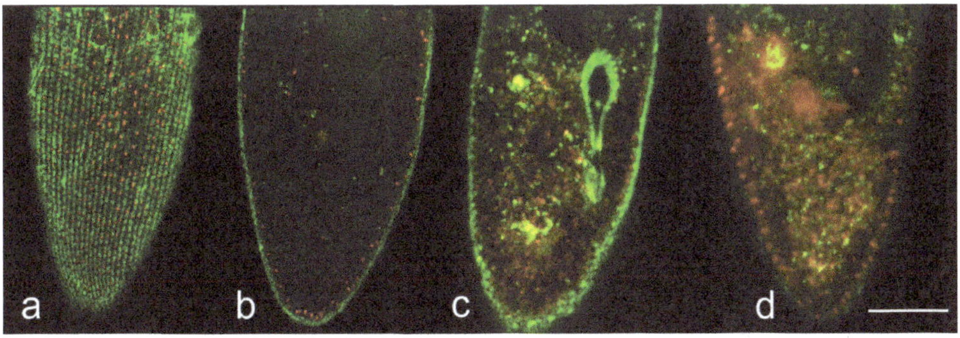

Figure 10. Double labeling with WGA and dextran. Cells are fed with WGA-FITC and dextran-TXR for 5 min (a, b), and fixed after 10 (c) and 20 min (d) of chase in unlabeled medium. (a) and (b) are images of the same cell acquired at different focus planes. WGA and dextran are present in different endocytic vesicles soon after initiation of uptake (< 10 min), then the two probes partly join prior to their fusion with food vacuoles. Bar, 20 μm.

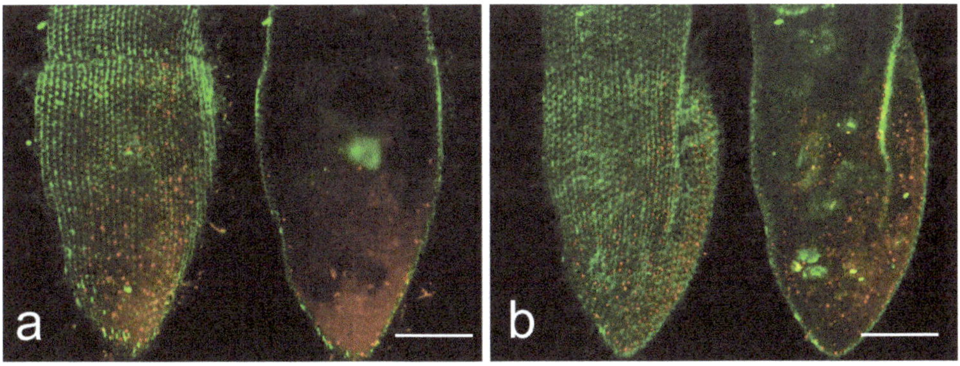

Figure 11. Fig. 11. Inhibition of clathrin-mediated endocytosis. (a) Effect of hypertonic medium on fluid phase and membrane mediated transport. Cells blocked in phagocytic activity are incubated in 0.20 M sucrose, WGA-FITC and dextran-TXR for 10 min. Sucrose inhibits WGA internalization and reduces dextran internalization. (b) Effect of cytosol acidification on fluid phase and membrane mediated transport. Cells blocked in phagocytic activity are incubated in 10 mM acetic acid, pH 5.0, WGA-FITC and dextran-TXR for 5 min. Green fluorescence is localized on the plasma membrane and red fluorescence in vesicles in both the cortical region and throughout the cytoplasm. (a, b) Two images of the same cell acquired at different focus planes from the dorsal side (left) to the internal cytoplasm (right). Bars, 20 μm.

Conversely, dextran internalization is blocked by filipin and nystatin (Figure 12), sterol-binding agents that disrupt caveolar structure and function [58].

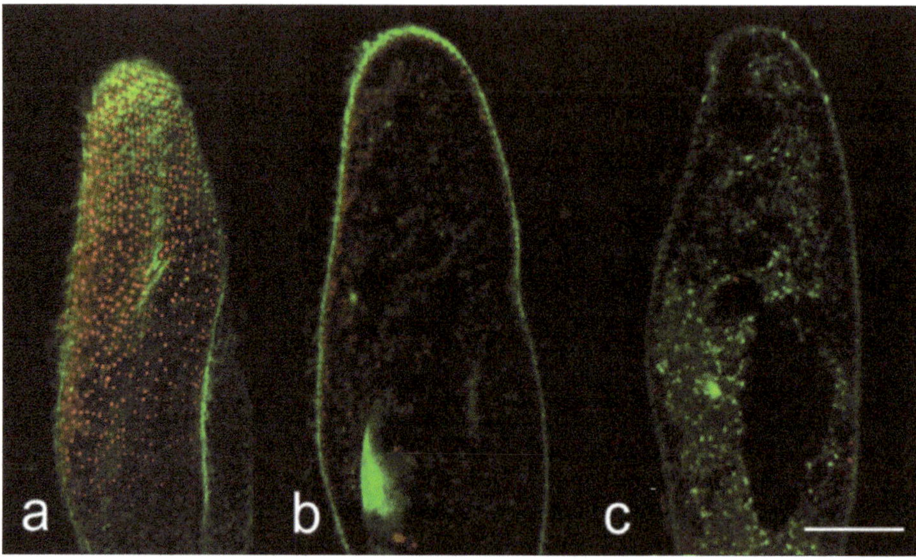

Figure 12. Inhibition of clathrin-independent endocytosis. Effect of nystatin on fluid phase and membrane mediated transport. Cells blocked in phagocytic activity are incubated in 2μg/ml nystatin, WGA-FITC and dextran-TXR for 5 (a, b) and 20 min (c) min. (a) and (b) are images of the same cell acquired at different focus planes. Green fluorescence locate on the plasma membrane and in vesicles through the cytoplasm; no red vesicles are seen inside the cell. Bar, 20 μm.

4. Receptor internalization and recycling

4.1. Receptor endocytosis

The classical paradigm of receptor function assumes that receptors localize on the cell surface and are activated by the binding of agonist ligands. After activation, most receptors are endocytosed from cell surface and travel to low pH endosomes, allowing the ligand to detach before the receptor is recycled back to the cell surface or sent through late endosomes to lysosomes for degradation [59]. Increasing evidence shows that some G protein-coupled receptors are not totally inactive in the absence of ligands but exhibit a constitutive activity, too, with elevated basal levels of intracellular signaling [60, 61]. It was found that receptor internalization from the neuronal surface occurring both constitutively and in response to agonist exposure is mediated by clathrin-dependent endocytosis [62-64]. Clathrin-coated vesicles are the initial vehicles for sequestration of surface receptors, which are ultimately degraded or recycled. Endocytosis of such membrane proteins involves a series of steps beginning with the clustering of receptors at specific sites of the plasma membrane, regions that later turn into clathrin-coated pits. Receptors do this by recruiting cytosolic AP2 adaptor complexes through their cytoplasmic tails.

AP2 is a key component of the endocytic machinery that links cargo membrane proteins to the clathrin lattice, selects molecules for sorting into clathrin-coated vesicles and recruits clathrin to the plasma membrane [65-69]. It is composed of subunits: α, $\beta 2$, $\mu 2$, and σ. The $\mu 2$ subunit (AP50) binds the endocytic sequence motif of cargo proteins, whereas the $\beta 2$ subunit binds to clathrin and the α region interacts via distinct domains with amphiphysin, AP180 and eps15. In addition to the AP2 adaptor complex, amphiphysin interacts with dynamin and the disruption of dynamin-amphiphysin interaction by recombinant amphiphysin src homology 3 (SH3) domain *in vivo* leads to a potent block in clathrin-mediated endocytosis [70, 71]. Dynamin, a large GTP-binding protein, pinches off vesicles at constricted clathrin-coated pits by forming a ring-like structure collaring the neck of the vesicle that is thought to drive vesicle separation. Eps15 binds the C-terminal domain of the AP2 adaptor α-subunit and mediates the interaction of AP2 with proteins such as epsin, CALM/AP180 and synaptojanin, implicated in regulation of receptor-mediated endocytosis. Eps15 function in clathrin-dependent endocytosis seems to be restricted to the early events leading to clathrin-coated pit formation: indeed eps15 is not present in clathrin-coated vesicles [72].

It has been shown that endocytosis of receptors may also occur through other membrane structures, including noncoated membrane invaginations [73, 74] and caveolae [75]. The β_2-adrenergic receptor, which is endocytosed by clathrin-coated pits in several cell types [76, 77], is endocytosed by membrane invaginations resembling to caveolae in other cells [74, 75]. Cholecystokinin receptors have been observed in both clathrin-coated pits and caveolae in the same cells [73]. Caveolae are cholesterol- and sphingolipid-rich smooth invaginations of the plasma membrane that partition into raft fractions and the expression of which is associated with caveolin 1.

A clathrin- and dynamim-dependent mechanism in the β_2-adrenergic receptor internalization has been already shown in *Paramecium* [78, 79]. An homologue of dynamin, a protein present in mammalian cells with three isoforms generating more than 25 possible spliced variants expressed in a tissue-specific manner, was identified in *Paramecium*, too [80]. A gene fragment of this dynamin reveals 74% similarity to human dynamin 2 mRNA and the deduced amino acid sequence shows 61.1% homology in a 175 amino acid overlap to the N-terminal region of human, mouse and rat dynamin [81]. Endocytosis in *Tetrahymena* also involves a protein in the dynamin family [82].

4.2. Receptor trafficking after internalization

Endocytosis of receptors can contribute to functional resensitization of signal transduction by promoting dephosphorylation and recycling of receptors to the plasma membrane [83] as well as to down-regulation of receptors, a process that leads to functional desensitization of signal transduction by reducing the number of receptors present in the plasma membrane and promoting degradation of receptors in lysosomes [73, 84, 85].

These processes of receptor regulation are thought to involve membrane trafficking of receptors via distinct recycling or degradative pathways and can mediate opposite effects on the regulation of functional signal transduction [83, 86]. Golgi-derived vesicles provide newly synthesized receptors to the cell surface, whereas clathrin coated vesicles are the initial vehicles for sequestration of surface receptors, which are ultimately degraded or recycled back to the plasma membrane, either directly or through the recycling endosomes [87-89]. These processes are mediated by a continuous traffic of vesicular and tubular intermediates which needs to be coordinated to ensure proper progression of cargo through the different compartments. Several rab family members have been localized to distinct compartments of the endocytic pathway and play different roles in endocytosis and recycling [90-93]. Rab5 and rab4 are both localized to early endosomes but exert opposite effects on the uptake of membrane-bound proteins. Rab5 plays a role in the formation of clathrin-coated vesicles at the plasma membrane [94], their subsequent fusion with early endosomes, in the homotypic fusion between early endosomes [95, 96] and in the interaction of early endosomes with microtubules [97]. Rab4 has been implicated in the regulation of membrane recycling from the early endosomes to the recycling endosomes or directly to the plasma membrane [98].

In accordance with this functional diversity, rab5 lies at the center of a complex machinery comprising several effector proteins [99]. Of these proteins, EEA1 was identified as a core component of the homotypic endosome docking and fusion machinery and was shown to play a role in the docking/tethering of the endosome membranes [99]. EEA1 is predominantly localized to the early endosomes and is regarded as a specific marker of this compartment. Because of this localization and given its function in endosome membrane docking [99] it has been proposed that EEA1 may confer directionality to rab5-dependent vesicular transport to the early endosomes. Another effector protein for rab5 is rabaptin-5. Rabaptin-5 binds directly to the GTP-bound form of rab5 and is recruited to early endosomes by rab5 in a GTP-dependent manner [100], stabilizes rab5 in the GTP-bound active form by down-regulating GTP hydrolysis [101] and, finally, it is required for the homotypic fusion between early endosomes as well as for the heterotypic fusion of clathrin-coated vesicles with early endosomes in vitro [100, 102]. Rabadpin-5 also interacts, via a distinct structural unrelated N-terminal RBD, with GTP-bound rab4 but does not appear to interact with rab11, a GTPase that is highly enriched on the recycling endosome and whose activity is required for receptor recycling through this compartment [89]. Thus the same effector interacts with the two rab proteins which act sequentially in transport through the early endosomes. Furthermore, the lysosomal sorting of receptors is dependent upon rab 7 activity [103].

Small GTPase rab is a widely conserved molecular switch among eukaryotes and regulates membrane trafficking, also in ciliates. In the *T. thermophila* genome 56 different rab protein genes were identified [104]. These do not include 17 putative rabs previously reported [82]. This is a remarkable number, considering that somewhat over 63 rabs have been identified in humans [105]. Some of them are very conserved and some others are ciliate specific [104, 106]. Endocytic compartments were found to be associated with a large number of rabs, including both conserved endocytic rabs but also a roughly equal number of divergent rabs. One of the conserved rabs did not fall into any of the proposed core clades. The animal rabs in this clade are associated with transport of lysosome-related organelles, while the

Tetrahymena protein localized to phagosomes. The remaining 14 conserved rabs in *Tetrahymena* fall within five of the proposed core pathways: ER-to Golgi, endocytosis/recycling, endocytosis, retrograde Golgi and late endocytosis [104, 106].

86 rab genes in the *Tetrahymena* genome and 229 rab genes in *P. tetraurelia* were found by Saito-Nakano et al. [107]. By comparing the amino acid sequence of rabs in humans and the budding yeast *Saccharomyces cerevisiae,* 42 conventional and 44 species-specific rabs were categorized in *Tetrahymena* and 157 conventional and 72 species-specific rabs in *Paramecium*. Among them, nine *Paramecium* rab genes showed high homology to seven *Tetrahymena* rabs, suggesting the conservation of ciliate-specific rab [107].

4.3. Investigating the $GABA_B$ receptor trafficking pathway in *Paramecium* using confocal microscopy

In our studies we are interested in understanding the endocytic properties of $GABA_B$ receptors in *Paramecium* [108, 109]. Although most G protein-coupled receptors undergo endocytosis, the conditions and mechanisms of this process vary from receptor to receptor. Many of them are endocytosed via clathrin-coated pits, but some are not [110, 111]. Some have an agonist-induced endocytosis, some are continuously endocytosed even in the absence of stimulation, while others exhibit both a constitutive and a stimulated endocytosis [110, 112, 113]. Currently, very little is known about the targeting and trafficking mechanisms of $GABA_B$ receptor in cells. In the past years, attention has mainly been focused on endocytosis of ionotropic GABA ($GABA_A$) receptors. It has been shown that $GABA_A$ receptors are internalized by a clathrin-coated pit-mediated process in hippocampal neurons and in A293 cells [114] and in a clathrin independent manner in HEK-293 cells [115]. Using a dominant-negative dynamin construct K44A Herring et al. [116] showed that constitutive endocytosis of $GABA_A$ receptors in HEK-293 cells is dynamin-mediated, while Cinar and Barnes [115] found that it is dynamin-independent. It was also shown that both recombinant and neuronal $GABA_A$ receptors can constitutively recycle between the cell surface and an intracellular endosomal compartment [117]. In *Paramecium* a dynamin- and clathrin-dependent pathway has been already observed [78, 79].

Constitutive internalization and intracellular trafficking of receptors in *P. primaurelia* was visualized by multiple immunofluorescence analysis using $GABA_B$ receptors as marker. $GABA_B$ receptors display a dotted vesicular pattern dispersed on the cell surface and throughout the cytoplasm (Figure 13a), and are internalized via clathrin-dependent and -independent endocytosis. Indeed, $GABA_B$ receptors colocalize with the adaptin complex AP2, which is implicated in the selective recruitment of integral membrane proteins to clathrin-coated vesicles, and with caveolin 1, which is associated with uncoated membrane invaginations [109].

Cells were double labeled with a guinea pig anti-$GABA_B$ receptor R1 subunit antibody and with a monoclonal anti-clathrin or anti-caveolin 1 antibody and visualized with Alexa Fluor 594-conjugated anti-guinea pig and Alexa Fluor 488-conjugated anti-mouse secondary antibodies, respectively. Staining with anti-clathrin or antI-caveolin antibody led to a punctuate pattern throughout the cytoplasm representing endocytic vesicles. The expression of GABA$_B$ receptors and clathrin- or caveolin-coated vesicles exhibited a clustered distribution on the cell membrane and inside the cytoplasm (Figure13). Importantly, GABA$_B$ receptor and clathrin- or caveolin-coated vesicle clusters were partly colocalized (yellow fluorescence). Furthermore, GABA$_B$ receptors colocalize with β_2 adaptin in a number of sites on the plasma membrane [109].

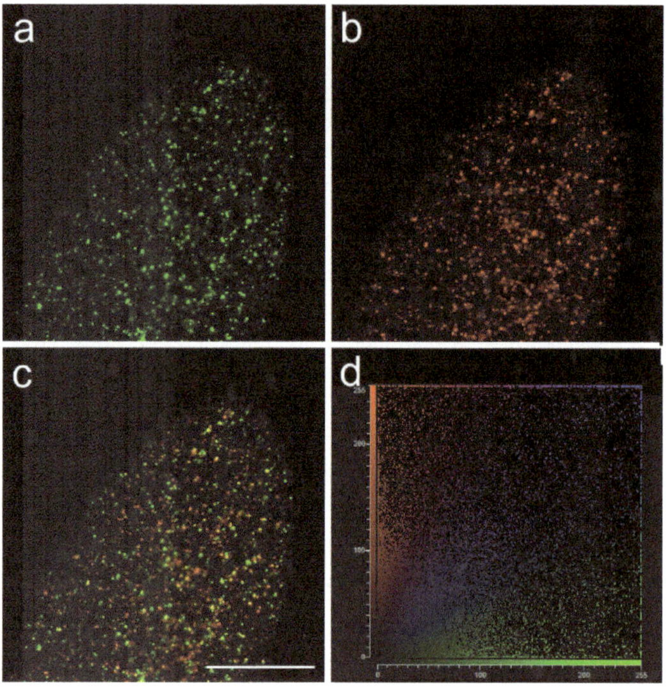

Figure 13. Colocalization of GABA$_B$ receptors and clathrin. In cells labeled with a polyclonal antibody against GABA$_B$ receptor (b) and a monoclonal antibody against clathrin HC (a), a clustered distribution of fluorescence is detected on the plasma membrane and inside the cytoplasm. GABA$_B$ receptors and clathrin vesicles are partly colocalized (c, yellow fluorescence). Bar, 20 μm. (d) 2D cytofluorogram: colocalized pixels are clustered along the diagonal line (visualized in blue).

In addition, we have shown that GABA$_B$ receptors are removed from the plasma membrane by clathrin-dependent and -independent endocytosis by blocking receptors internalization by hypertonic sucrose. However, it has recently been found that sucrose inhibits GABA$_A$ receptor endocytosis that is not mediated by clathrin-coated pits [115]. Therefore, we have also used cytosol acidification with acetic acid for clathrin-mediated endocytosis inhibition [50]. Furthermore, GABA$_B$ receptor internalization in *Paramecium* is blocked by filipin and nystatin, cholesterol binding drugs. The sensitivity of endocytosis to nonacute cholesterol depletion with agents such as filipin and nystatin distinguishes caveolae and raft pathways from clathrin-dependent and constitutive pinocytosis pathways [118].

Treatment of cells with 150 mM sucrose or cytosol acidification significantly inhibited the internalization of receptors, as shown by the considerable reduction in receptors inside the cytoplasm (Figure 14) as compared to the control (Figure14b). This observation strongly suggests that GABAB receptor internalization in *Paramecium* is mediated by clathrin-dependent endocytosis. In these experiments phagocytosis was blocked by trifluoperazine.

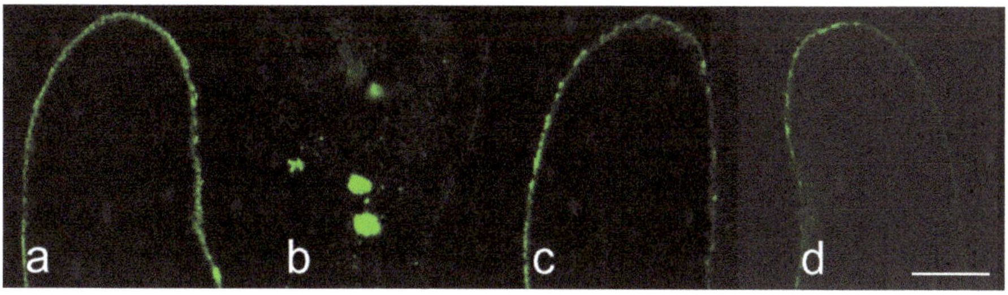

Figure 14. GABAB-receptor internalization is mediated by clathrin-coated vesicles. In cells whose phagocytic activity is blocked by trifluoperazine, 20-minute treatment with 150 mM sucrose (c) or citosol acidification (d) inhibits receptor internalization, which can be seen by receptor accumulation on the cell membrane and receptor reduction inside the cytoplasm. Controls are cells incubated with the anti-GABAB receptor antibody for 1 (a) and 20 min (b) in the absence of inhibitors; the antibody is localized in endosomes and phagosomes. Incubation temperature, 25°C. Bar, 20 μm.

Moreover, when endocytosis was blocked by filipin or by nystatin the receptor internalization decreased (Figure 15). In these experiments cells were incubated in the anti-GABAB receptor antibody for 30 minutes at 4°C (temperature inhibiting phagosome and endosome formation, [16], so that receptors were accumulated on the cell membrane (Figure 15a). After removal of the excess of antibody, cells were incubated at 25°C. 84% receptors were internalized in untreated cells after 20 minutes incubation at 25°C, as shown both by the reduction of cell membrane fluorescence intensity and by the fluorescence localization into endosomes and phagosomes (Figure 15b). Only 37% and 46% fluorescence was internalized in filipin ($p<0.01$) and nystatin-treated cells ($p<0.01$), respectively (Figure 16).

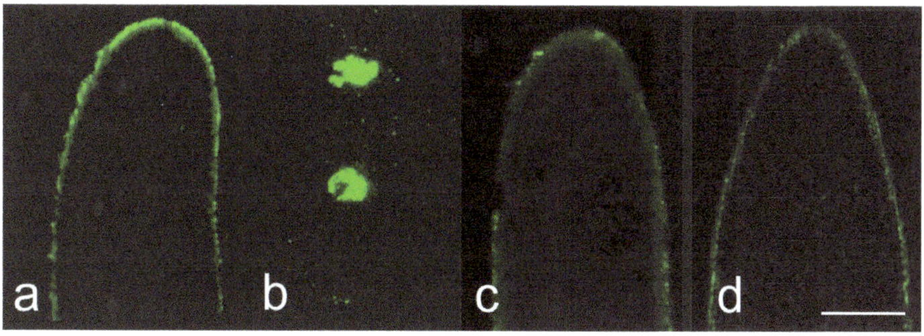

Figure 15. GABAB receptors are internalized by non-coated endocytosis. Cells preincubated at 4°C for 30 minutes and labeled with anti-GABAB receptor antibody for 30 minutes (a) are fixed after a 20-minute chase at 25°C in the absence (b) or in the presence of non-coated-pit endocytosis inhibitors filipin (c) and nystatin (d). Bar, 20 μm.

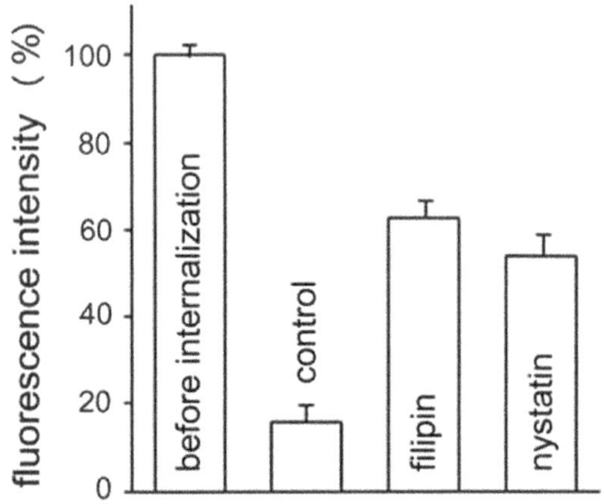

Figure 16. Measurement of the internalization shown in Figure 15. Constitutive receptor internalization is partially inhibited by filipin and nystatin (38% and 47%, respectively); $P < 0.01$, Student's *t*-test). Data were normalized to cells before internalization at 25°C (shown in Figure15a).

Colocalization values reported in Table 1 and experiments carried out in living cells suggest that GABAB receptors are internalized through the two pathways in a similar quantity. Quantification of cell membrane fluorescence was performed by ImageJ 1.46b software (Wayne Rasband, Nat. Inst. of Health, USA).

Red	Green	Red/green (%)	Green/red (%)
GABAB receptor	Clathrin	18 ± 3	19 ± 3
GABAB receptor	Caveolin 1	21 ± 6	25 ± 4

Every colocalization value is the average from four optical sections of ten cells. Data were calculated as the mean ± s.e. and are given in percent.

Table 1. Colocalization of GABAB receptor labeling with proteins involved in endocytosis

After internalization by clathrin-coated vesicles and by caveolae GABAB receptors are transported by rab5-linked vesicles to early endosomes, characterized by the EEA1 marker. Receptors are then partly recycled back to cell membrane and partly degraded. The recycling of GABAB receptors is evidenced by the overlapping of their immunolocalization with both rab4 and rab11 immunostaining. Rab4 controls the rapid recycling of cargo proteins directly back to the cell surface from rab4/rab5 positive endosomal structures, and the slow recycling of cargo via rab11 positive recycling endosomes. The traffic of GABAB receptors to Golgi apparatus is evidenced by colocalization of GABAB with TGN38 immunoreactivity. The communication between contiguous rab-domains and thereby the sequential transport of receptors from one intracellular compartment to another is regulated by rab effector rabaptin-5. Furthermore a fraction of GABAB R1 seemed to be directed to lysosomes, as shown by GABAB R1 and LAMP1 (lysosomal marker) immunocolocalization, and to phagosomes for degradation. An immunolocalization of rab7 on phagosomal membrane was also reported in *Paramecium* [119-122].

Colocalization of GABA$_B$ receptors with proteins involved in the endocytosis and recycling was demonstrated by both the colocalized pixels in a 2D cytofluorogram and the similarity of green and red profiles along the *z*-axes of fluorescence intensity of double-stained vesicles [109]. 2D cytofluorogram (Figure 14d) was generated using ICA plugin of ImageJ. Colocalization along the *z*-axes of double-stained vesicles is also demonstrated by the similarity of green and red profiles of their fluorescence intensity of double-labeled vesicles from a stack of 30 images (total thickness 2 μm). For three different focal planes a sample of a vesicle that shows colocalization (yellow fluorescence) was selected (Figure 17).

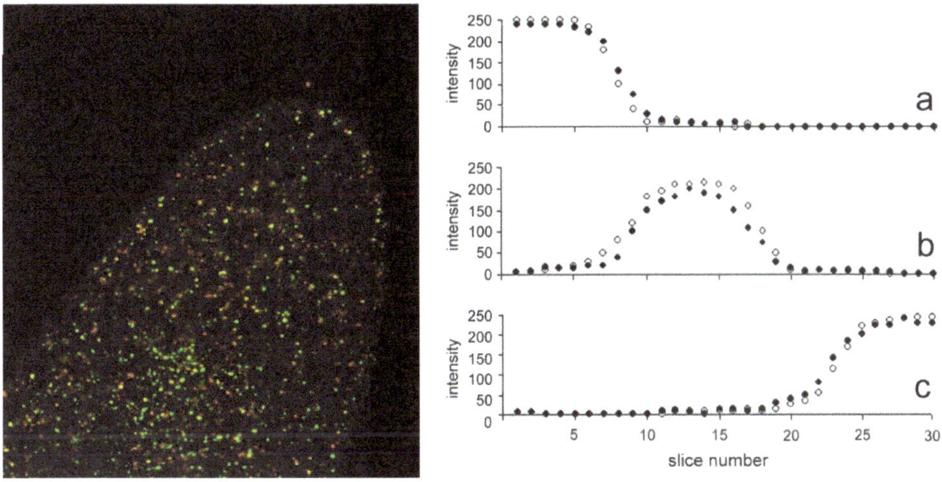

Figure 17. *z*-Stack profile of fluorescence intensity of double-labeled vesicles. The left side of the figure shows a optical plane from a stack of 30 images (total thickness 2 μm). The right side of the figure shows fluorescence-intensity distribution along the *z*-axis of three yellow-labeled vesicles selected from different optical planes (a, top; b, middle; c, bottom) (green, open circle; red, filled circle).

The quantitative estimation of colocalized proteins in immunocytochemical studies has been performed calculating the colocalization coefficients [123] from the red- and green-channel scatterplot. Colocalization coefficients express the fraction of colocalizing molecular species in each component of a dual-color image and are based on the Pearson's correlation coefficient, a standard procedure for matching one image with another in pattern recognition [124]. If two molecular species are colocalized, the overlay of their spatial distributions has a correlation value higher than what would be expected by chance alone. Costes et al. [125] developed an automated procedure to evaluate correlation between the green and red channels with a significance level >95%. The same procedure automatically determines an intensity threshold for each color channel based on a linear least-square fit of the green and red intensities in the image's 2D correlation cytofluorogram. Costes' approach has been accomplished by macro routines integrated as plug-ins (WCIF Co-localization Plugins, Wright Cell Imaging Facility, Toronto Western Research Institute, Canada) in the ImageJ 1.46b image-analysis software (Wayne Rasband, NIH, USA).

5. Conclusion

In ciliates, essential components of membrane trafficking during endocytosis have been identified, based on work mainly with *Paramecium* and *Tetrahymena*. We used *P. primaurelia* and laser scanning microscopy to show *in vivo* vesicle formation, movement and fusion. The retrieval of membrane of stage III and stage IV food vacuoles, and recycling back both to nascent food vacuoles as small vesicles and to the acidified food vacuoles (stage II) as secondary lysosomes have been clearly and dynamically documented. Furthermore, the multimodal analysis using the pseudo-color technique enabled us to observe the changes in the direction of movement of pinocytic vesicles after evagination from food vacuoles.

Using endocytosis markers and confocal microscopy we have also shown that WGA and dextran enter the cell via two distinct vesicle populations and that in *Paramecium*, as in mammalian cells, fluid phase endocytosis is unaffected by treatments that arrest coated pit-mediated endocytosis, indicating that fluid phase endocytosis is primarily clathrin-independent. So, plasma membrane components are internalized by endosomes, which are first localized in the cortical region of the cell, transported in the most internal cytoplasmic portion and fused with other endosomal compartments, until their content is transferred to the phagosomes.

Furthermore, $GABA_B$ receptors are removed from the plasma membrane by clathrin-dependent and -independent endocytosis. Indeed, internalization of receptors is blocked by hypertonic sucrose and cytosol acidification, classic inhibitors of clathrin-mediated endocytosis, as well as by nystatin and filipin, sterol-binding agents that disrupt caveolar structure and function.

Using standard immunomarkers for early endosomes, recycling vesicles and lysosomes, and comparing our data with those obtained in mammalian cells relating to the internalization and recycling of some other receptors, we inferred that also in *Paramecium* $GABA_B$ receptors are partly recycled to cell plasma membrane and partly degraded into lysosomes. So, using immunohistochemical methods we demonstrated that in the single-celled organism *Paramecium*, as in mammalian cells, rab-like proteins are involved in the vesicle transport from one compartment to another.

Author details

Paola Ramoino*
Department of Earth, Environment and Life Sciences (DISTAV), University of Genoa, Genova, Italy

Alberto Diaspro
The Italian Institute of Technology (IIT), Nanophysics Unit, Genova, Italy

*Corresponding Author

Marco Fato
Department of Communication, Computer and System Sciences (DIST), University of Genoa, Genova, Italy

Cesare Usai
Institute of Biophysics, CNR Genoa, Genova, Italy

6. References

[1] Hall RP (1965) Protozoan Nutrition. Blaisdell Publishing Company, New York.
[2] Nisbet B (1984) Nutrition and Feeding Strategies in Protozoa. Croom Helm, London.
[3] Radek R, Hausmann K (1996) Phagotrophy of ciliates. In: Hausmann K, Bradbury PC, editors. Ciliates: Cells as Organisms. Gustav Fischer Verlag, Stuttgart, pp. 197-219.
[4] Verni F, Gualtieri P (1997) Feeding behaviour in ciliated protists. Micron 28: 487–504.
[5] Hausmann K (2002) Food acquisition, food ingestion and food digestion by protists. Jpn. J. Protozool. 35: 85–95.
[6] Mast SO (1947) The food-vacuole in *Paramecium*. Biol. Bull. 92: 31-72.
[7] Fenchel T (1980) Suspension feeding in ciliated protozoa: structure and function of feeding organelles. Arch. Protistenkd. 123: 239-260.
[8] Giese AC (1973) Cannibalism and gigantism in *Blepharisma*. Trans. Amer. Micros. Soc. 57: 245-255.
[9] Allen RD (1974) Food vacuole membrane growth with microtubule-associated membrane transport in *Paramecium*. J. Cell Biol. 63: 904-922.
[10] Allen RD, Wolf RW (1974) The cytoproct of *Paramecium caudatum*: structure and function, microtubules, and fate of food vacuole membranes. J. Cell Sci. 14: 611–631.
[11] Müller M., Törö I (1962) Studies on feeding and digestion in protozoa. III. Acid phosphatase activity in food vacuoles of *Paramecium multimicronucleatum*. J. Protozool. 9: 98-102.
[12] Müller M, Röhlich P, Toth J, Törö I (1963) Fine structure and enzymic activity of protozoan food vacuoles. Ciba Found. Symp. Lysosomes, Churchill, London, pp. 201-216.
[13] Allen RD, Staehelin LA (1981) Digestive system membranes: freeze-fracture evidence for differentiation and flow in *Paramecium*. J. Cell Biol. 89: 9-20.
[14] Fok AK, Lee Y, Allen RD (1982) The correlation of digestive vacuole pH and size with the digestive cycle in *Paramecium caudatum*. J. Protozool. 29: 409-414.
[15] Allen RD, Fok AK (1984) Retrieval of lysosomal membrane and acid phosphatase from phagolysosomes of *Paramecium caudatum*. J. Cell Biol. 99: 1955-1959.
[16] Fok AK, Muraoka JH, Allen RD (1984) Acid phosphatase in the digestive vacuoles and lysosomes of *Paramecium caudatum*: a timed study. J. Protozool. 31: 216-220.
[17] Allen RD, Fok AK (2000) Membrane trafficking and processing in *Paramecium*. Int. Rev. Cytol. 198: 277-318.
[18] Kuznicki L, Sikora J (1971) Cytoplasmic streaming within *Paramecium aurelia*. I. Movements of crystals after immobilization by antiserum. Acta Protozool. 8: 439-445.

[19] Ramoino P, Beltrame F, Diaspro A, Fato M (1996) Time-variant analysis of organelle and vesicle movement during phagocytosis in *Paramecium primaurelia* by means of fluorescence confocal laser scanning microscopy. Microsc. Res. Tech. 35: 377-384.
[20] Kuznicki L (1963) Reversible immobilization of *Paramecium caudatum* evoked by nickel ions. Acta Protozool. 1: 301-312.
[21] Brutkowska M (1967) Immobilization effect of $NiCl_2$ and food vacuole formation in *Paramecium caudatum*. Boll. Acad. Pol. Sci. Cl II, 15: 119-122.
[22] Sikora J, Wasik A (1978) Cytoplasmic streaming with Ni^{2+} immobilized *Paramecium aurelia*. Acta Protozool. 17: 389-397.
[23] Robbins E, Marcus P, Gonotas NK (1964) Dynamics of acridine orange-cell interaction. II. Dye-induced ultrastructural changes in multivesicular bodies (acridine orange particles). J. Cell Biol. 21: 49-62.
[24] Zelenin AV (1966) Fluorescence microscopy of lysosomes and related structures in living cells. Nature 212: 425-426.
[25] Allen RD, Fok AK.(1983) Nonlysosomal vesicles (acidosomes) are involved in phagosome acidification in *Paramecium*. J. Cell Biol. 97: 566-570.
[26] Beltrame F, Diaspro A, Fato M, Martin I, Ramoino P, Sobel I (1995) Use of stereo vision and 24-bit false colour imagering to enhance visualisation of multimodal confocal images. Proc. SPIE 2412: 222-229.
[27] Aufderheide KJ (1977) Saltatory motility of uninserted trichocysts and mitochondria in *Paramecium tetraurelia*. Science 198: 299-300.
[28] Sikora J (1981) Cytoplasmic streaming in *Paramecium*. Protoplasma 109: 57-77.
[29] Schroeder CC, Fok AK, Allen AD (1990) Vesicle transport along microtubular ribbons and isolation of cytoplasmic dynein from *Paramecium*. J. Cell Biol. 111: 2553-2562.
[30] Ramoino P (1996/97) Membrane supply and food vacuole formation in *Paramecium primaurelia*. Arch. Protistenkd. 147: 323-329.
[31] Mellman I, Howe C, Helenius A (1987) The control of membrane traffic on the endocytic pathway. Curr .Top. Membr. Transp. 29: 255-288.
[32] Courtoy PJ (1989) Dissection of endosomes. In: Steer C, Hanover J editors Intracellular Trafficking of Proteins, Cambridge University Press, New York, pp 103-156.
[33] Gruenberg J, Howell KE (1989) Membrane traffic in endocytosis: insights from cell-free assay. Annu. Rev. Cell Biol. 5: 453-481.
[34] van Deurs B, Petersen OW, Olsnes S, Sandvig K (1989) The ways of endocytosis. Int. Rev. Cytol. 171: 131-176.
[35] Chapman-Andresen C (1962) Studies on pinocytosis in *Amoebae*. Comp. Rend. Trav. Lab. Carlsberg 33: 73-264.
[36] Stockem W, Christofidou-Solomidou M (1993) Food uptake and digestion in amoebae. In: H. Plattner, editor. Membrane Traffic in Protozoa. JAI Press Inc., Greenwich (Conn.), USA, pp. 371-407.
[37] Stockem W (1977) Endocytosis. In: Jamieson GA, Robinson DM, editors. Mammalian Cell Membranes, vol. 5. Butterworths, London, pp 151-195.
[38] Munch R. (1970): Food uptake by endocytosis in *Opalina ranarum*. Cytobiologie 2: 108-122.

[39] Allen RD (1988) Cytology. In: Görtz H-D, editor. Paramecium, Springer-Verlag, Berlin Heidelberg, pp 4-40.
[40] Nilsson JR, van Deurs P (1983) Coated pits and pinocytosis in *Tetrahymena*. J. Cell Sci. 63: 209-222.
[41] Allen RD, Schroeder CC, Fok AK (1992) Endosomal system of *Paramecium*: coated pits to early endosomes. J. Cell Sci. 101: 449-461.
[42] Allen RD, Fok AK (1993) Endosomal membrane traffic of ciliates. In: Plattner H, editor. Advances in Cell and Molecular Biology of Membranes, Membrane Traffic in Protozoa, JAI Press Inc, Greenwich, CT, pp 283-309.
[43] Lamaze C, Schmid SL (1995) The emergence of clathrin-independent pinocytic pathways. Curr. Opin. Cell Biol. 7: 573-580.
[44] Wyroba E (1991) Quantitation of fluid phase uptake in *Paramecium*. 1. Kinetics in the cells blocked in phagocytic activity. Cell Biol. Int. Rep. 15: 1207-1216.
[45] Wyroba E (1987) Stimulation of *Paramecium* phagocytosis by phorbol ester and forskolin. Cell Biol. Int. Rep. 11: 657-664.
[46] Allen RD, Ueno MS, Fok AK (1988) A survey of lectin binding in *Paramecium*. J. Protozool. 35: 400-407.
[47] Ramoino P (1997) Lectin-binding glycoconjugates in *Paramecium primaurelia*: changes with cellular age and starvation. Histochem. Cell Biol. 107: 321-329.
[48] Ramoino P, Fronte P, Diaspro A, Fato M, Beltrame F, Robello M (2001) Fluid phase and receptor mediated endocytosis in *Paramecium*. Europ. Biophys. J. Biophys. 30: 305-312.
[49] Daukas G, Zigmond SH (1985) Inhibition of receptor-mediated but not fluid-phase endocytosis in polymorphonuclear leukocytes. J. Cell Biol. 101: 1673-1679.
[50] Davoust J, Gruenberg J, Howell K E (1987) Two threshold values of low pH block endocytosis at different stages. EMBO J. 6: 3601-3609.
[51] Sandvig K, Olsnes S, Petersen OW, van Deurs B (1987) Acidification of the cytosol inhibits endocytosis from coated pits. J. Cell Biol. 105: 679-689.
[52] Hansen SH, Sandvig K, van Deurs B (1993) Molecules internalized by clathrin-independent endocytosis are delivered to endosomes containing transferrin receptors. J. Cell Biol. 123: 89-97.
[53] Hansen SH, Sandvig K, van Deurs B (1993) Clathrin and HA2 adaptors: effects of potassium depletion, hypertonic medium, and cytosol acidification. J. Cell Biol. 121: 61-72.
[54] Heuser J (1989) Effects of cytoplasmic acidification on clathrin lattice morphology. J. Cell Biol. 108: 401-411.
[55] Heuser JE, Anderson RG (1989) Hypertonic media inhibit receptor-mediated endocytosis by blocking clathrin-coated pit formation. J. Cell Biol. 108: 389-400.
[56] Wang L-H, Rothberg KG, Anderson RGW (1993) Mis-assembly of clathrin lattices on endosomes reveals regulatory switch for coated pit formation. J. Cell Biol. 123: 1107-1117.
[57] Sofer A, Futherman AH (1995) Cationic amphiphilic drugs inhibit the internalization of colera toxin to the Golgi apparatus and the subsequent elevation of cyclic AMP. J. Biol. Chem. 270: 12117-12122.

[58] Schnitzer JE, Oh P, Pinney E, Allard J (1994) Filipin-sensitive caveolae-mediated transport in endotelium: reduced transcytosis, scavenger endocytosis, and capillary permeability of selected macromolecules. J. Cell Biol. 127: 1217-1232.
[59] Ferguson SS (2001) Evolving concepts in G protein-coupled receptor endocytosis: the role in receptor desensitization and signaling. Pharmacol. Rev. 53: 1–24.
[60] Seifert R, Wenzel-Seifert K (2002) Constitutive activity of G-protein-coupled receptors: cause of disease and common property of wild-type receptors. Naunyn. Schmiedebergs Arch. Pharmacol. 366: 381-416.
[61] Royle SJ, Murrell-Lagnado RD (2003) Constitutive cycling: a general mechanism to regulate cell surface proteins. BioEssays 25: 39-46.
[62] Tehrani MHJ, Barnes EM, Jr (1997) Sequestration of gamma-aminobutyric acid A receptors on clathrin-coated vesicles during chronic benzodiazepine administration in vivo. J. Pharmacol. Exp. Ther. 283: 384-390.
[63] Grampp T, Notz V, Broll I, Fischer N, Benke D (2008) Constitutive, agonist-accelerated, recycling and lysosomal degradation of $GABA_B$ receptors in cortical neurons. Mol. Cell Neurosci. 39: 628-637.
[64] Traub LM (2010) Tickets to ride: selecting cargo for clathrin-regulated internalization. Nat. Rev. Mol. Cell Biol. 10: 583-596.
[65] Kirchhausen T (1999) Adaptors for clathrin-mediated traffic. Annu. Rev. Cell. Dev. Biol. 15: 705-732.
[66] Takei K, Haucke V (2001) Clathrin-mediated endocytosis: membrane factors pull the trigger. Trends Cell Biol. 11: 385-391.
[67] Traub LM (2003) Sorting it out: AP-2 and alternate clathrin adaptors in endocytic cargo selection. J. Cell Biol. 163: 203-208.
[68] Owen DJ, Collins BM, Evans PR (2004) Adaptors for clathrin coats: structure and function. Annu. Rev. Cell. Dev. Biol. 20: 153-191.
[69] Rodemer C, Haucke V (2008) Clathrin/AP-2-dependent endocytosis: a novel playground for the pharmacological toolbox? Handb. Exp. Pharmacol. 186: 105-122.
[70] Shupliakov O, Low P, Grabs D, Gad H, Chen H, David C, Takei K, De Camilli P, Brodin L (1997) Synaptic vesicle endocytosis impaired by distruption of dynamin-SH3 domain interactions. Science 276: 259-263.
[71] Wigge P, Kohler K, Vallis Y, Doyle CA, Owen D, Hunt SP, McMahon HT (1997) Amphiphysin heterodimers: potential role in clathrin-mediated endocytosis. Mol. Biol. Cell. 8: 2003-2015.
[72] Cupers P, Jadhav AP, Kirchhausen T (1998) Assembly of clathrin coats disrupts the association between Eps15 and AP-2 adaptors. J. Biol. Chem. 273: 1847-1850.
[73] Roettger BF, Rentsch RU, Pinon D, Holicky E, Hadac E, Larkin JM, Miller IJ (1995) Dual pathways of internalization of the cholecystokinin receptor. J. Cell Biol. 128: 1029-1041.
[74] Raposo, G., Dunia, I., Delavier-Klutchko, C., Kaveri, S., Strosberg, A. D. and Benetti, E.L. (1989) Internalization of beta–adrenergic receptor in A431 cells involves non-coated vesicles. Eur. J. Cell Biol. 50: 340-352.
[75] Dupree P, Parton RG, Raposo G, Kurzchalia TV, Simons K (1993) Caveolae and sorting in the trans-Golgi network of epithelial cells. EMBO J. 12: 1597-1605.

[76] von Zastrow M, Kobilka BK (1994) Antagonist-dependent and -independent steps in the mechanism of adrenergic receptor internalization. Biol. Chem. 269: 18448-18452.
[77] Zhang J, Ferguson SS, Barak LS, Menard L, Caron MG (1996) Dynamin and beta-arrestin reveal distinct mechanisms for G protein-coupled receptor internalization. J. Biol. Chem. 271: 18302-18305.
[78] Wiejak J, Surmacz L, Wyroba E (2004) Dynamin- and clathrin-dependent endocytic pathway in unicellular eukaryote *Paramecium*. Biochem. Cell Biol. 82: 547-558.
[79] Wiejak J, Surmacz L, Wyroba E (2004) Dynamin-association with agonist-mediated sequestration of beta-adrenergic receptor in single-cell eukaryote *Paramecium*. J. Exp. Biol. 207: 1625-1632.
[80] Wiejak J, Wyroba E (2002) Dynamin: characteristics, mechanisms of action and function. Cell Biol. Int. Rep. 13: 667-678.
[81] Surmacz L, Wiejak J, Wyroba E (2001) Alterations in the protein pattern of subcellular fractions isolated from *Paramecium* cells suppressed in phagocytosis. Folia Histochem. Cytobiol. 39: 301-305.
[82] Elde NC, Morgan G, Winey M, Sperling L, Turkewitz AP (2005) Elucidation of clathrin-mediated endocytosis in *Tetrahymena* reveals an evolutionarily convergent recruitment of dynamin. PLoS Genet. 1: 52.
[83] Ferguson SS, Zhang J, Barak LS, Caron MG (1998) Molecular mechanisms of G protein-coupled receptor desensitization and resensitization. Life Sci. 62: 1561-1565.
[84] Hoxie JA, Ahuja M, Belmonte E, Pizarro S, Parton R, Brass LF (1993) Internalization and recycling of activated thrombin receptors. J. Biol. Chem. 268: 13756-13763.
[85] Kallal L, Gagnon AW, Penn RB, Benovic JL (1998) Visualization of agonist-induced sequestration and down-regulation of a green fluorescent protein-tagged beta2-adrenergic receptor. J. Biol. Chem. 273: 322-328.
[86] Bohm SK, Khitin LM, Grady EF, Aponte G, Payan DG, Bunnett NW (1996) Mechanisms of desensitization and resensitization of proteinase-activated receptor-2. J. Biol. Chem. 271: 22003-22016.
[87] Yamashiro DJ, Maxfield FR (1984) Acidification of endocytic compartments and the intracellular pathways of ligands and receptors. J. Cell Biochem. 26: 231-246.
[88] Hopkins CR, Gibson A, Shipman M, Strickland DK, Trowbridge IS (1994) In migrating fibroblasts, recycling receptors are concentrated in narrow tubules in the pericentriolar area, and then routed to the plasma membrane of the leading lamella. J. Cell Biol. 125: 1265-1274.
[89] Ullrich O, Reinsch S, Urbe S, Zerial M, Parton RG. 1996. Rab11 regulates recycling through the pericentriolar recycling endosome. J. Cell Biol. 135: 913-924.
[90] Chavrier P, Parton RG, Hauri HP, Simons K, Zerial M (1990) Localization of low molecular weight GTP binding proteins to exocytic and endocytic compartments. Cell 62: 317-329.
[91] Van Der Sluijs P, Hull M, Zahraoui A, Tavitian A, Goud B, Mellman I (1991) The small GTP-binding protein rab4 is associated with early endosomes. Proc. Natl. Acad. Sci. USA 88: 6313-6317.

[92] Olkkonen VM, Dupree P, Killisch I, Lutcke A, Zerial M, Simons K (1993) Molecular cloning and subcellular localization of three GTP-binding proteins of the rab subfamily. J. Cell Sci. 106: 1249-1261.
[93] Stenmark H (2009) Rab GTPases as coordinators of vesicle traffic. Nat. Rev. Mol. Cell Biol. 10: 513-525.
[94] McLauchlan H, Newell J, Morrice N, Osborne A, West M, Smythe E (1998) A novel role for Rab5-GDI in ligand sequestration into clathrin-coated pits. Curr. Biol. 8: 34-45.
[95] Gorvel JP, Chavrier P, Zerial M, Gruenberg J (1991) Rab5 controls early endosome fusion in vitro. Cell 64: 915-925.
[96] Bucci C, Parton RG, Mather IH, Stunnenberg H, Simons K, Hoflack B, Zerial M (1992) The small GTPase rab5 functions as a regulatory factor in the early endocytic pathway. Cell 70: 715-728.
[97] Nielsen E, Severin F, Backer JM, Hyman AA, Zerial M (1999) Rab5 regulates motility of early endosomes on microtubules. Nat. Cell Biol. 1: 376-382.
[98] Daro E, van der Sluijs P, Galli T, Mellman I (1996) Rab4 and cellubrevin define different early endosome populations on the pathway of transferrin receptor recycling. Proc. Natl. Acad. Sci. USA 93: 9559-9564.
[99] Christoforidis S, McBride HM, Burgoyne RD, Zerial M (1999) The Rab5 effector EEA1 is a core component of endosome docking. Nature 397: 621-625.
[100] Stenmark H, Vitale G, Ullrich O, Zerial M (1995) Rabaptin-5 is a direct effector of the small GTPase Rab5 in endocytic membrane fusion. Cell 83: 423-432.
[101] Rybin V, Ullrich O, Rubino M, Alexandrov K, Simon I, Seabra MC, Goody R, Zerial M (1996) GTPase activity of Rab5 acts as a timer for endocytic membrane fusion. Nature 383: 266-269.
[102] Horiuchi H, Lippe R, McBride HM, Rubino M, Woodman P, Stenmark H, Rybin V, Wilm M, Ashman K, Mann M, Zerial M (1997) A novel Rab5 GDP/GTP exchange factor complexed to Rabaptin-5 links nucleotide exchange to effector recruitment and function. Cell 90: 1149-1159.
[103] Marchese A, Paing MM, Temple BR, Trejo J (2008) G protein-coupled receptor sorting to endosomes and lysosomes. Annu. Rev. Pharmacol. Toxicol. 48: 601–629.
[104] Bright LJ, Kambesis N, Nelson SB, Jeong B, Turkewitz AP (2010) Comprehensive analysis reveals dynamic and evolutionary plasticity of Rab GTPases and membrane traffic in *Tetrahymena thermophila*. PLoS Genet. 6: 1001155.
[105] Lal K, Field MC, Carlton JM, Warwicker J, Hirt RP (2005) Identification of a very large Rab GTPase family in the parasitic protozoan Trichomonas vaginalis. Mol. Biochem. Parasitol. 143: 226-235.
[106] Turkewitz AP, Bright LJ (2011) A Rab-based view of membrane traffic in the ciliate *Tetrahymena thermophila*. Small GTPases 2: 222-226.
[107] Saito-Nakano Y, Nakahara T, Nakano K, Nozaki T, Numata O (2010) Marked amplification and diversification of products of ras genes from rat brain, Rab GTPases, in the ciliates *Tetrahymena thermophila* and *Paramecium tetraurelia*. J. Eukaryot. Microbiol. 57: 389-99;

[108] Ramoino P, Usai C, Beltrame F, Fato M, Gallus L, Tagliaferro G, Magrassi R, Diaspro A (2005) GABA$_B$ receptor intracellular trafficking after internalization in *Paramecium*. Microsc. Res. Tech. 68: 290-295.

[109] Ramoino P, Gallus L, Beltrame F, Diaspro A, Fato M, Rubini P, Stigliani S, Bonanno G, Usai C (2006) Endocytosis of GABA$_B$ receptors modulates membrane excitability in the single-celled organism *Paramecium*. J. Cell Sci. 119: 2056-2064.

[110] Nichols BJ, Lippincott-Schwartz J (2001) Endocytosis without clathrin coats. Trends. Cell Biol. 11: 406-412.

[111] Johannes L, Lamaze C (2002) Clathrin-dependent or not: is it still the question? Traffic 3: 443-451.

[112] Dale LB, Bhattacharya M, Seachrist JL, Anborgh PH, Ferguson SSG (2001) Agonist-stimulated and tonic internalization of metabotropic glutamate receptor 1a in human embryonic kidney 293 cells: agonist-stimulated endocytosis is β-arrestin1 isoform-specific. Mol. Pharmacol. 60: 1243-1253.

[113] Tsao PI, von Zastrow M (2001) Diversity and specifity in the regulated endocytic membrane trafficking of G-protein-coupled receptors. Pharmacol. Ther. 89: 39-147.

[114] Kittler JT, Delmas P, Jovanovic JN, Brown DA, Smart TG, Moss SJ (2000) Constitutive endocytosis of GABA$_A$ receptors in association with adaptin AP2 complex modulates inhibitory synaptic currents in hippocampal neurons. J. Neurosci. 20: 7972-7977.

[115] Cinar H, Barnes EM Jr (2001) Clathrin-independent endocytosis of GABA(A) receptors in HEK 293 cells. Biochemistry 40: 14030-14036.

[116] Herring D, Huang R, Singh M, Robinson LC, Dillon GH, Leidenheimer NJ (2003). Constitutive GABA$_A$ receptor endocytosis is dynamin-mediated and dependent on a dileucine AP2 adaptin-binding motif within the {beta}2 subunit of the receptor. J Biol. Chem. 278: 4046-24052.

[117] Connolly CN, Kittler JT, Thomas P, Uren .M, Brandon NJ, Smart TG, Moss S. J. (1999). Cell surface stability of gamma-aminobutyric acid type A receptors. J. Biol. Chem. 274: 36565-36572.

[118] Nabi IR, Le PH (2003) Caveolae/raft-dependent endocytosis. J. Cell Biol. 161: 673-677.

[119] Surmacz L, Wiejak J, Wyroba E (2003) Evolutionary conservancy of the endocytic machinery in the unicellular eukaryote *Paramecium*. Biol. Cell 95: 69-74.

[120] Surmacz L, Wiejak J, Wyroba E (2006) Cloning of two genes encoding Rab7 in *Paramecium*. Acta Biochim. Biol. 53: 149-156.

[121] Wyroba E, Surmacz L, Osinska M, Wiejak J (2007) Phagosome maturation in unicellular eukaryote *Paramecium*: the presence of RILP, Rab7 and LAMP-2 homologues. Eur. J. Histochem. 51: 163–172.

[122] Osinska M, Wiejak J, Wypych E, Bilski H, Bartosiewicz R, Wyroba E (2011) Distinct expression, localization and function of two Rab7 proteins encoded by paralogous genes in a free-living model eukaryote. Acta Biochim. Pol. 58: 597-607.

[123] Manders EM, Verbeek FJ, Aten JA (1993) Measurement of co-localization of objects in dualcolour confocal images. J. Microscopy 169: 375-382.

[124] Gonzalez RC, Wintz P (1987) Digital Image Processing, 2nd edn., Addison Wesley Publication Company, Mass.USA.

[125] Costes SV, Daelemans D, Cho EH, Dobbin Z, Pavlakis G, Lockett S (2004) Automatic and quantitative measurement of protein-protein colocalization in live cells. Biophys. J. 86: 3993-4003.

Chapter 6

Advanced Optical Imaging of Endocytosis

Jesse S. Aaron and Jerilyn A. Timlin

Additional information is available at the end of the chapter

1. Introduction

Endocytosis is the highly controlled and complex process by which a portion of the plasma membrane, including its lipids, proteins, and local extracellular fluid becomes internalized in a cell. Endocytosis serves to mediate a multitude of interactions between a cell and its environment, including nutrient uptake, mitosis, motility, as well as adaptive and innate immune response, among many others. There are multiple routes of endocytotic uptake into cells, with the most studied being clathrin mediated endocytosis (CME). Although CME differs significantly on a molecular level from the clathrin-independent endocytosis mechanisms (e.g. macropinocytosis, phagocytosis), all of the endocytic mechanisms involve a sequence of changes in morphology, molecular composition, and protein interactions at the plasma membrane, as well as throughout the bulk of the cell. Further, each of these changes is tightly regulated in space and time. To fully characterize endocytic pathways and their intertwined relationship to other signalling pathways, there is a need to visualize the dynamics of multiple species at the plasma membrane and within the cell with high three-dimensional spatial resolution.

Traditional biochemical and genetic approaches have provided, and will continue to provide, a wealth of information about the cellular pathways and key molecules involved in endocytosis. However, such bulk assays are only able to provide ensemble measurements. Thus, they cannot shed light on the important and stochastic sub-cellular spatiotemporal information that is inherent to endocytosis. High resolution electron microscopy studies can address this limitation with exquisite spatial detail approaching atomic resolution, but cannot easily capture the dynamics of endocytosis. For this reason, live cell fluorescence microscopy has been exploited to provide vital information at the subcellular and single molecule level about the localization of components involved in individual endocytosis events.

Very early optical microscopy investigations used traditional organic fluorophores and widefield fluorescence imaging to follow membrane associations during endocytosis.

Widefield fluoresce microscopy is readily available, simple to conduct, and provides lateral spatial resolution of 200-400 nm when using visible excitation light and axial spatial resolution on the order of 1 μm with optimal objective and microscope configurations. This resolution is sufficient to understand the overall arrangement of proteins on the cellular membrane and determine uptake into cells as demonstrated by Leserman et al. (Leserman, et al., 1980); but, the presence of interfering signal from throughout the depth of the cell limits the ability to visualize single events using this approach. The addition of the confocal pinhole into the fluorescence microscope serves to reject much of the out-of-focus light, providing a significant improvement in axial spatial resolution, and signal to noise (SNR); confocal fluorescence microscopy has demonstrated wide success in following endocytotic processes within living cells in three dimensions (Betz, et al., 1996; Muller, 2006). The reader is referred to Stephens and Allan for a review of the basics of widefield and confocal fluorescence microscopy technologies for live cell imaging (Stephens & Allan, 2003).

In this chapter, we will summarize optical imaging methodologies beyond the simple transmission optical, widefield fluorescence, and confocal fluorescence microscopes. The advanced techniques presented here have significant advantages in spatial, spectral, and/or temporal resolution when compared to traditional microscopy methods and are well-suited for real-time tracking of individual endocytotic events in living cells. We will cover: (1) total internal reflection fluorescence microscopy, which has become a dominant technology for endocytosis dynamics due to its specificity for the plasma membrane, (2) super-resolution microscopy, whose exquisite spatial resolution has led to emerging applications in the field of endocytosis, and (3) spectral imaging, which exploits the spectral properties of fluorophores and spectral deconvolution to extend fluorescence microscopy much further into the multiplexed regime. In each of these areas, we will introduce the basic concepts of the measurement technique, present important developments in analysis algorithms, and highlight recent studies with regard to endocytosis. It is important to note that although the focus of this chapter is advanced optical imaging methodologies for following endocytosis in living cells, the techniques presented here demonstrate the potential utility in visualizing exocytotic processes and the various vesicle trafficking events that are critical to cell function.

2. Total internal reflection fluorescence microscopy

2.1. Basic principles

Total internal reflectance fluorescence (TIRF) microscopy offers a unique approach for selective imaging of biological components and events very near (typically <200nm) to the plasma membrane in cells (Axelrod, 2001). This technique avoids much of the background signal emanating from fluorophores within the cytoplasm, thereby increasing detection sensitivity over traditional widefield or confocal microscopy. TIRF was first demonstrated in living cells by Axelrod and colleagues for visualization of acetyl choline receptors, and as a sensitive measure of membrane topology (Axelrod, 1981). It has since been widely adopted in biological laboratories for a large range of applications, particularly after the introduction of commercially available objective-based TIRF systems (Mattheyses & Axelrod, 2006). Demonstrations have included characterizations of cell receptor distributions (N. L. Thompson, et al., 1997), and other membrane bound biomolecules (Sund & Axelrod, 2000), dynamic imaging of exocytic/secretory vesicle trafficking and fusion (Schmoranzer, et al., 2000), as well as single molecule (Tokunaga, et al., 1997) and single nanoparticle (Aaron, et al., 2011) 2D tracking within the membrane. In addition, TIRF has allowed for enhanced biophysical characterizations of endocytotic events, as discussed in the following section.

The principle behind TIRF relies on the creation of evanescent excitation. Snell's law accurately predicts the angle of light refraction through media of various refractive indices. However, it can be shown that in cases where light propagates from a higher refractive index material (such as glass) to a lower refractive index material (such as air or water), there exists a critical angle, θ_c, above which refraction cannot occur. Mathematically, this is represented by:

$$\theta_c = \sin^{-1}\left(\frac{n_1}{n_2}\right) \tag{1}$$

where n_1 and n_2 correspond to the lower and higher refractive indices, respectively. At angles that exceed this value, total internal reflection occurs, and light does not propagate through the lower refractive index material, but rather is reflected away from the interface in the opposite direction. Interestingly, a more detailed analysis using Maxwell's equations reveals that a portion of the impinging light's energy extends slightly into the lower refractive index material. This is referred to as an evanescent wave, which propagates parallel to the interface, and decays quickly in the perpendicular direction. The decrease in intensity from the surface can be described by an exponential function, with characteristic decay constant, d, given by:

$$d = \frac{\lambda}{4\pi}\left(n_2^2 \sin^2\theta - n_1^2\right)^{-\frac{1}{2}} \tag{2}$$

where λ refers to the excitation wavelength, θ is the angle at which the light impinges normal to the interface ($\theta > \theta_c$), and n_1 and n_2 are as described above. Equation (2) indicates that the penetration depth of the evanescent field will typically extend a distance less than the wavelength of light used, and will decrease with increasing illumination angle. For instance, excitation at 532nm, passing from a glass coverslip (n_1 = 1.52) to an aqueous environment (n_2 = 1.33), at an angle of 68° to the surface, will exhibit an evanescent decay length of only 165nm. This represents more than a 3-fold smaller distance than the axial resolution of a typical confocal microscope. As such, TIRF has become a widely used modality to study events very near the cell membrane, including the myriad of endocytosis mechanisms. The following sections focus on three phenomena where TIRF imaging has dramatically impacted current knowledge of internalization-related phenomena: (1) clathrin mediated endocytosis, (2) cellular uptake of viruses, and (3) internalization of engineered nanoparticles.

2.2. Understanding clathrin mediated endocytosis via TIRF microscopy

CME is perhaps the best characterized pathway for internalizing receptor-specific biomolecules and is conserved among nearly all eukaryotic cells (Rappoport, et al., 2004). Many of the earliest studies examining CME relied on electron microscopy due to the ultra-high resolving power of that modality (Kirchhausen, et al., 1986). However, the advent of fluorescence imaging has permitted the important advantage of capturing dynamic, molecular-specific behavior in living cells. When combined with fluorescent protein constructs or other labels, new insights into this complex process have been gleaned using time-resolved live cell microscopy.

Earlier wide-field fluorescence imaging studies reported previously unseen behavior of clathrin-coated pits (CCPs) on or near the plasma membrane (Gaidarov, et al., 1999). Interestingly, distinct patterns were observed within a cell-wide population of CCPs. A majority of CCPs displayed limited/random, or no lateral motion within the membrane, and were generally termed "static". With the introduction of TIRF microscopy, a subset of CCPs was observed to be motile – exhibiting rapid active transport motion (Keyel, et al., 2004). This latter observation has led to a shift in the overall model of the CME pathway. While clathrin was originally thought to only participate in vesicle formation, Rappoport et al. showed that some clathrin coated vesicles (CCVs) persisted beyond the initial plasma membrane-bound state, and were transferred to microtubules parallel to the membrane (Rappoport, Taha, & Simon, 2003). This rapid motion was correlated with CCV internalization. However, still others were shown to disappear or re-appear from the TIRF field of view without active transport-like motion. This suggested that some disassembly or reassembly of the clathrin triskelia is concomitant with their internalization away from the evanescent field, without transport by motor proteins/microtubule network (Merrifield, et al., 2002; Merrifield, et al., 2005).

Matters were complicated further when later data suggested that a single CCP could give rise to multiple vesicles (Rappoport, 2008). Single particle analysis of TIRF data showed that some CCV were initiated *de novo* – that is, the assembly of a single CCP resulted in complete disappearance of clathrin-associated signal into an internalized vesicle. However, other CCPs were seen to separate into sub-structures, only a portion of which were seen to internalize, while other CCPs were seen to merge into larger structures (often termed clathrin coated plaques). As an example, Figure 1 shows TIRF images and analysis representing a single event of the latter type, taken from (Rappoport, 2008). Close inspection of the indicated point spread function from a single sub-diffraction sized CCP shows a broadening and eventual separation into two distinct features. This was followed by the disappearance of the newly isolated CCP, presumably as it is internalized as a vesicle. This model of CCV formation has been referred to as iterative budding. The relative contributions of *de novo* formation and iterative budding mechanisms to overall CME behavior have shown to be highly cell-line dependent. Swiss 3T3 fibroblasts exhibited 59%:41% iterative:*de novo* behavior in one study (Merrifield, et al., 2005), while BSC1 cells showed exclusively *de novo* CCV formation in another (Ehrlich, et al., 2004).

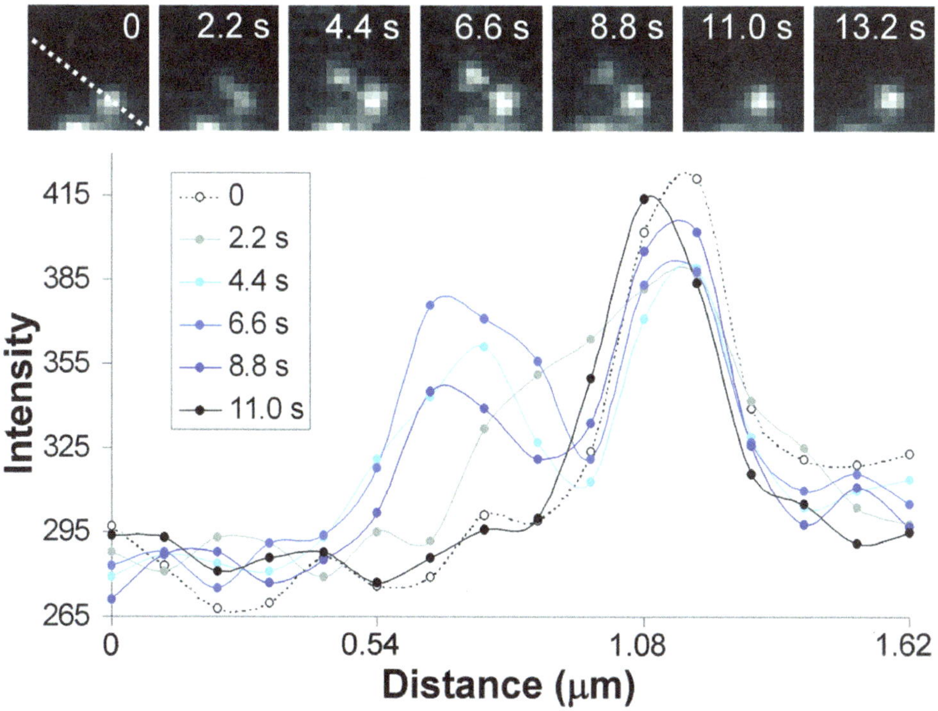

Figure 1. Iterative budding of a CCV from a larger CCP. At top, an image sequence illustrates TIRF images of CME. At bottom, signal intensity profiles indicate the progressive widening of a single CCP, and eventual partial separation, resulting in the iterative budding of a single CCV (Rappoport, 2008). Reproduced with permission, from Rappoport, (2008), *Biochem. J.*, **412**, 415-523. © The Biochemical Society.

The complexity and mechanistic diversity of CME has been postulated to arise in part from variety of adapter and accessory proteins that are expressed in a given cell. However, there does appear to be a "core" group of molecular players present in nearly all forms of CME. Arguably the most ubiquitous accessory proteins are dynamin and actin. Dynamin fulfills a plethora of roles within many cell signaling pathways (particularly dynamin-2). One of its most prominent functions is in aiding the initial formation of CCVs, and their scission from the plasma membrane. Despite its importance, the precise sequence of events surrounding the role of dynamin is still uncertain. Merrifield et al. initially showed, via dual-color TIRF microscopy, an increase in dynamin-associated fluorescence immediately prior to clathrin internalization, and a synchronized decrease during CCV internalization (Merrifield et.al, 2002). However, the underlying reason for the increase in fluorescence is still not clear. Alternate models predict either a recruitment of dynamin from the cytosol directly to the "neck" of the newly formed CCV, or recruitment from the cytosol to the whole vesicle, and then translocation to the point of invagination.

Along with dynamin, actin also plays a role in CCV formation as an accessory protein in many cells (Kaksonen, et al., 2006; Merrifield, et al., 2002). Merrifield et al. showed that EGFP-actin signal displayed a transient increase near CCPs during internalization

(Merrifield, et al., 2002). Interestingly, using TIRF microscopy to compare the kinetics of dynamin and actin recruitment over a number of cells clearly indicated that dynamin recruitment consistently precedes local actin polymerization (in addition to clathrin internalization), thus giving valuable insight to the sequence of events in the CME process.

TIRF microscopy has also been instrumental in disentangling the complex interactions of adapter proteins involved in CME (Rappoport, et al., 2006). The first identified, and most studied, of these are the adaptins (Boehm & Bonifacino, 2001). These proteins can form a tetrameric complex referred to as AP-2, and act as an intermediate between cell surface receptors and the endocytic machinery by concentrating cargo bound for internalization into a CCP. However, the role for AP-2 during the post-internalization phase has been controversial. Rappoport et al. initially showed that AP-2 is lost from the CCV during internalization (Rappoport, Taha, Lemeer, et al., 2003), although Keyel et al. later proposed that AP-2 accompanies CCVs into the cytosol, suggesting its possible regulatory role the downstream sorting machinery (Keyel, et al., 2004). However, more detailed TIRF image analysis confirmed the former hypothesis, and showed that while AP-2 was co-localized to static CCPs in the membrane, it was absent from those CCPs observed to disappear into the cytosol (Rappoport, et al., 2005).

2.3. Tracking single viruses and endocytosis in living cells with TIRF imaging

Simliar to its utility in understanding the kinetics of endogenous protein-protein interactions during CME, TIRF microscopy has also been highly useful to probe the mechanisms of pathogen invasion via similar routes. Of particular interest is the mechanism by which viruses enter their host cells (Brandenburg & Zhuang, 2007). Although some viruses (such as HIV) replicate via direct genome injection through the plasma membrane, most have evolved a multitude of methods to gain entry to cells via endocytosis; specific mechanisms include CME, macropinocytosis, and caveolin-dependent internalization (Sieczkarski & Whittaker, 2002a). As a prominent example, this section will highlight studies of Influenza A viral entry by CME, as elucidated by TIRF microscopy.

Zhuang and colleagues were among the first to visualize the interactions between single influenza A viruses and host cells using both widefield (Lakadamyali, et al., 2003) and TIRF microscopy (Floyd, et al., 2008; Rust, et al., 2004). Influenza A is an enveloped, single-stranded RNA virus thought to enter cells via CME, although more recent data indicate it may also utilize a clathrin-independent pathway (Sieczkarski & Whittaker, 2002b). One of the first TIRF imaging studies revealed that influenza A particles were internalized via *de novo* CCP formation and internalization to CCV, as shown in Figure 2 (Rust, et al., 2004).

Images indicate that Influenza viruses, which were labeled with a lipophilic fluorescent tracer (DiD, shown in red), bind to the surface of live BSC1 cells. A few minutes after binding, GFP-tagged clathrin is seen to accumulate around a subset of viral particles, as shown in (B). Following this, the velocity of the viral particle dramatically increases, indicative of attachment to microtubules. The increase in viral velocity is then correlated with a complete disappearance of both DiD and GFP fluorescence signal, suggesting directed internalization within a single, *de novo* CCV. Interestingly, however, some viral particles were seen to exhibit the aforementioned velocity increase without apparent recruitment of clathrin. While that may simply suggest the presence of non-fluorescent clathrin, it may also point to the presence of alternate endocytosis mechanisms. This later hypothesis is supported by other data that demonstrate little change in influenza infectivity in the presence of inhibitors of CME (Sieczkarski & Whittaker, 2002b). Nevertheless, CME does seem to be a route that is well-exploited by viral pathogens, albeit not always in the classical sense. For instance, Johannsdottir, et al. used single particle tracking TIRF microscopy to show that while dynamin-2 was required for Vesicular stomatis virus (VSV) internalization, AP-2 was not (Johannsdottir, et al., 2009).

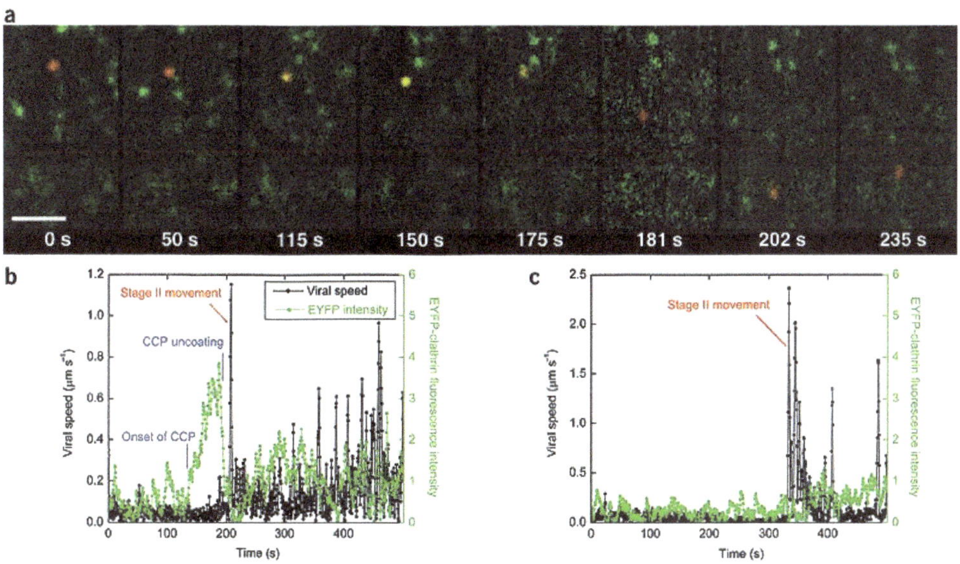

Figure 2. Dual color TIRF microscopy reveals that Influenza virus is endocytosed via de novo CCV formation. In (a), a single influenza virus (shown in red) binds to the cell surface, and is eventually colocalized to EGFP-tagged clathrin. After colocalization, the clathrin/virus complex displays enhanced motility, and eventual disappearance from the evanescent field. In (b), intensity profiles (in green) show the recruitment of clathrin near the single influenza virus. Black traces indicate viral velocity, indictative of active transport into the cytoplasm. In (c), an example of viral internalization is shown that does not indicate dependance on clathrin (Rust, et al., 2004). *Adapted by permission from Macmillan Publishers Ltd:* Nature Structural and Molecular Biology *M.J. Rust, et al.,* ***11****(5), 567-573, © 2004.*

TIRF microscopy has been used to not only visualize initial viral entry, but also its behavior later in the endocytosis pathway. When enveloped viruses are labeled with high density of a fluorescent dye such as DiD or lipophilic Rhodamine, fluorescence self-quenching occurs such that viral particles are relatively non-fluorescent. Upon fusion of the virus to the endosomal membrane, the density of fluorophore decreases, resulting in a dramatic increase in detectable signal (Hoekstra, et al., 1984; van der Schaar, et al., 2007). Furthermore, viral particles can independently or simultaneously be loaded with a tracer molecule to measure release of genomic material into the cytosol, subsequent to fusion (Brandenburg, et al., 2007). For instance, Floyd, et al. used a dual-labeling approach to gain new insights into influenza fusion using a supported lipid bilayer model under TIRF interrogation as shown in Figure 3 (Floyd, et al., 2008).

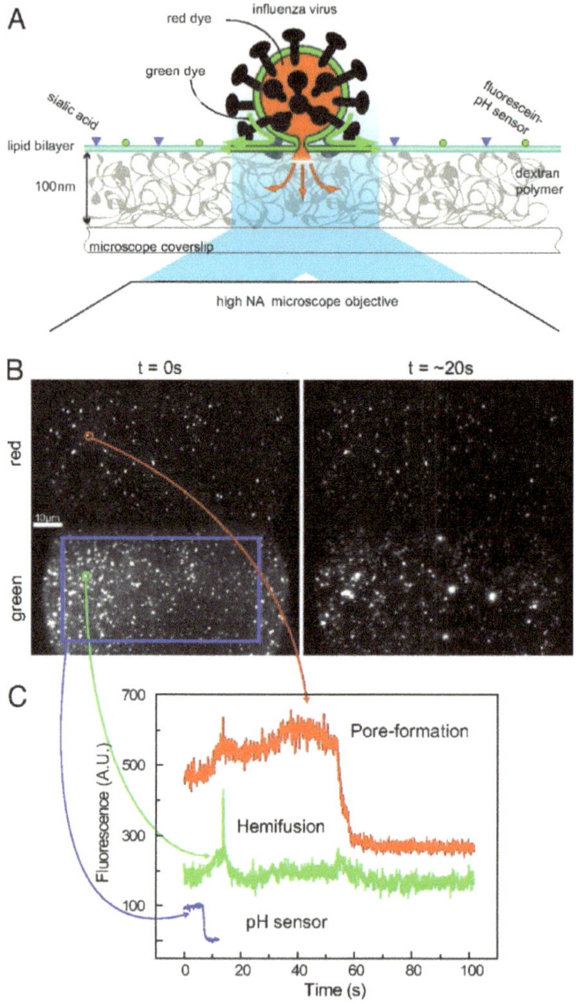

Figure 3. In vitro TIRF characterization of influenza fusion and content mixing. In (A), the experimental setup is illustrated, showing influenza virus binding to a dextran-supported lipid bilayer to model the endosomal membrane. Viral envelopes are labeled with lipophilic rhodamine (green), and the interior is loaded with sulforhodamine B (SRB) (red). Images in (B) show representative dual color TIRF images. In (C), fluorescence intensity plots show a sharp spike in the lipophilic rhodamine signal (green), indicative of viral hemifusion, while the SRB (red) shows the dispersal of viral cargo into the sub-bilayer space after pore formation (Floyd, et al., 2008). *Reproduced with permission, from D.L. Floyd, et al.,* Proc. Nat. Acad. Sci., **105***(40), 15382-15387, ©2008 by the National Academy of Sciences.*

In this study, influenza viruses were labeled first with a lipophilic Rhodamine derivative, which exhibits a fusion dependent signal increase under these conditins. Additionally, the interior of the virus was labeled with sulforhodamine B (SRB). While only 30% of virus contained both dyes in sufficient quantities, the sequence of viral docking, fusion, and content release was able to be monitored in near real time, and corresponding rate constants, number of intermediate states, and lag times between fusion and pore formation were all able to be calculated. Figure 3 shows TIRF experimental setup (A) and data (B-C) obtained from influenza virus interactions with a liquid supported bilayer. In (C), dual color TIRF microscopy shows pore formation (red) vs. hemifusion (green) as a function of time. Hemifusion can be detected by a sharp, transient increase in fluorescence intensity as the lipophilic dye is released from its self-quenching state, and diffuses throughout the bilayer. The viral content release is assessed by the decay in red fluorescence as the SRB enters the sub-bilayer space and diffuses away.

As can be seen, TIRF microscopy has produced notable insight into pathogenic infection mechanisms, particularly with regard to their endocytosis by host cells. As opposed to bulk studies, imaging approaches allow for a "single cell", and even "single virus" quantification of behavior. This capability has far-reaching consequences in understanding fundamental molecular mechanisms. For instance, the data above was used to model the kinetics of viral hemifusion with endosomal membranes to clearly reveal that three intermediate stages exist in this process, thereby opening avenues for potential, specific therapeutic targets for Influenza infection.

2.4. TIRF microscopy for studying endocytosis of engineered nanomaterials

In addition to monitoring the internalization of pathogens, TIRF microscopy has also been instrumental in characterizing the uptake of engineered nanoparticles aimed at therapeutic or diagnostic applications (West & Halas, 2003), as well as in an effort to assess possible toxicological consequences of these materials (Marquis, et al., 2009). Engineered nanoparticles comprised of porous silica (Slowing, et al., 2008), liposomes (Hashida, et al., 2005), and other polymer materials (Panyam & Labhasetwar, 2003) have been widely successful as therapeutic carriers for both drug and gene delivery. Much of these approaches depend on the endocytic uptake and release of the material in question into the cytoplasm. Thus, quantitative characterizations of endocytosis are imperative in order to assess diagnostic/therapeutic effect.

Among the first and most widely used nanomaterials for diagnostic use include quantum dots (QDs). QDs are typically <20nm in diameter, and are comprised of various heavy metal/semiconductor materials such as CdSe, CdTe, or PbS, among others. This configuration results in an electronic bandgap that typically falls in the optical/NIR wavelength range. As such, QDs make highly attractive optical imaging probes with narrow emission bandwidth, broad absorption spectra, and relative resistance to photobleaching (Medintz, et al., 2005). Nie and colleagues were the first to demonstrate their utility as molecular imaging probes (Chan & Nie, 1998), and they have since gained wide-spread acceptance in this regard, including their use as *in vivo* diagnostic agents (Gao, et al., 2005).

Nevertheless, questions remain about the possible cytotoxic effects of semiconductor/heavy metal nanomaterials. In this regard, TIRF microscopy has been utilized to quantify the uptake properties of various CdSe QDs in immune cells. Aaron et al. have shown that while QD diameter is not a good predictor of cellular uptake (both in terms of uptake rate and extent), there does appear to be a correlation with QD shape (Aaron, et al., 2011). Figure 4 shows representative TIRF images of QD (emitting at 605nm, referred to as QD605) uptake in RBL mast cells over time. Similar to studies examining CME, this approach relied on the gradual disappearance of QD-associated signal as particles are transported to the cytoplasm, and away from the evanescent field.

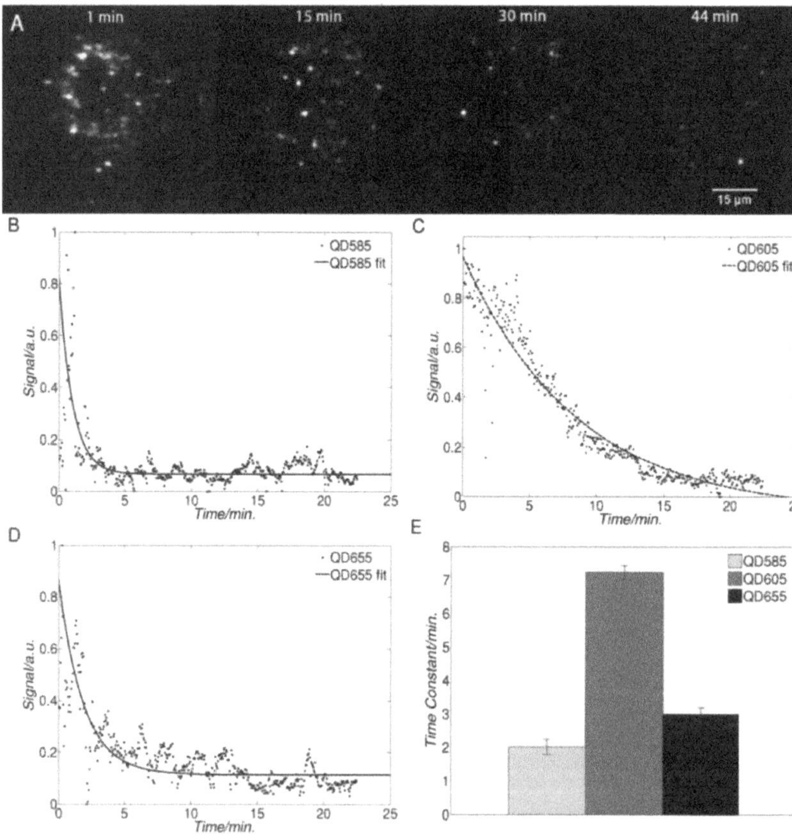

Figure 4. TIRF microscopy to quantify internalization rates of various shape/size quantum dots (QDs) in an RBL mast cell line. In (A), fluorescence signal from QDs are seen to gradually disappear over 30-60 minutes, as they move beyond the TIRF evanescent field into the cytoplasm. The total QD signal can be plotted vs. time to calculate a characteristic endocytosis time (B-D) for various sized/shaped QDs. These values are represented in (E) for two spheroidal QDs (QD585 and QD655, in light and dark grey, respectively), and one rod-shaped QD (QD605, medium grey). Data suggest that particle shape regulates internalization, with uptake times of rod-shaped particles nearly two-fold longer than spheroidal QDs (Aaron, et al., 2011).

Live cell time-course TIRF imaging of QD uptake (shown in A) allows for a measure of endocytosis rates for these materials. By simply plotting fluorescence signal as a function of time, a typical decay behavior becomes evident. Uptake measurements taken for QDs emitting at 585nm, 605nm, and 655nm (B, C, and D, respectively) showed marked differences in their rates of internalization, summarized in (E). Interestingly, the uptake rates did not correlated with size (average diameter increases with increasing emission peak), but upon closer inspection, was found to be related to QD shape. While QD585 and QD655 displayed relatively small aspect ratios (1.2 and 1.6, respectively), QD605 were found to have more rod-shaped character, with aspect ratio of 2.0. This suggests that spheriodal particles are internalized at a higher rate than elliptical particles, giving insight into the shape and size effects on nanoparticle-cell interactions.

2.5. Emerging TIRF microscopy methods

TIRF microscopy is currently in the midst of another renaissance, as more advanced methodologies are being developed to better extract meaningful, quantitative information about events at the plasma membrane. Two such approaches are directly applicable to imaging of endocytosis, and include polarization sensitive and multi-angle TIRF imaging.

Polarization-sensitive TIRF (pTIRF) microscopy had been proposed for some time (N. L. Thompson, et al., 1984), yet only in the last several years have these concepts been applied in biological samples (Anantharam, et al., 2010; Sund, et al., 1999). This method is based on the observation that, at the surface of a cell, endocytosis events create a localized birefringent environment, as illustrated in Figure 5. As can be seen, during endocytosis (or exocytosis), the deformation of the plasma membrane creates portions of the membrane that are parallel and perpendicular (as illustrated by arrows) to the s- and p-polarizations of the evanescent field, respectively. Therefore, a polarized detection scheme will be sensitive to separate regions within the nascent vesicle, provided that fluorescent dyes (such as DiD) are all oriented similarly with respect to the lipid bilayer. Resulting images may include a "doughnut" appearance at sites of membrane invagination, due to the alternative parallel and perpendicular orientations of the membrane with respect to the evanescent field polarization. While still in its infancy, this method has sensitively detected exocytosis of neuronal vesicles (Anantharam, et al., 2010), as well as fusion of SNARE-bearing vesicles on a supported lipid bilayer (Kiessling, et al., 2010). Further studies combining pTIRF with atomic force microscopy have shed light on fundamental mechanisms of protein-mediated membrane disordering (Oreopoulos & Yip, 2009).

While technically more complex, pTIRF may also be an ideal method for imaging endocytosis pathways that do not have well identified proteomic markers. For instance, a growing body of evidence has shown that Influenza A virus may make use of multiple endocytic pathways for infection of host cells, with at least one route being both clathrin and caveolin-independent. Using pTIRF to image the structural and kinetic properties of this cryptic pathway may lead to ultimately elucidating its origin.

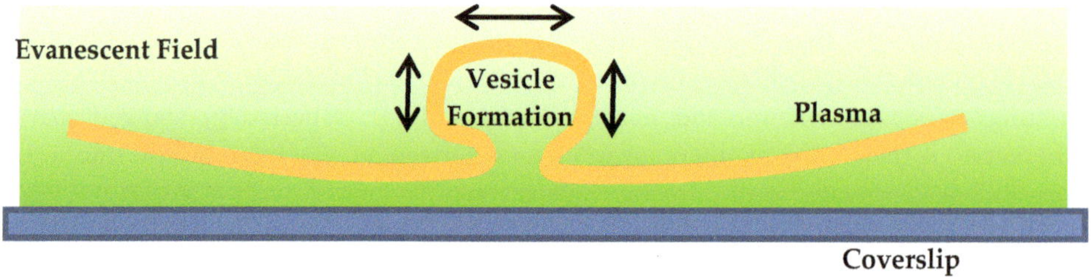

Figure 5. Principle of pTIRF. During membrane reorganization during endocytosis, fluorophores become oriented perpendicular and parallel to the incoming evanescent field polarization. Using cross-polarized detection, endosomal vesicles appear as alternating dark and light rings.

Another promising emerging strategy is the implementation of multi-angle TIRF microscopy. Hypothesized more than 20 years ago (Reichert, et al., 1987), it has only more recently been applied to quantitatively determining axial distributions of biomolecules at the nanoscale. Recall that Equation 2 in section 2.1 illustrates how the evanescent field depth is a sensitive function of illumination angle, with decreasing field penetration with increasing beam angle. This offers the intriguing possibility of optical sectioning at various axial positions near the sample/coverslip interface, far below the optical diffraction limit by systematically varying the TIRF angle, θ. This strategy has been successfully implemented in a compact design (Stock, et al., 2003), and utilized for a number of applications, including mapping cell membrane topology relative to the cytoplasm (Olveczky, et al., 1997), viewing exocytosis of secretory granules in Chromaffin cells (Oheim, et al., 1998), as well as detecting sub-diffraction axial movements of surface-immobilized DNA molecules, all with accuracies of less than 50nm along the optical axis (He, et al., 2005). Doubtless, the application of this methodology may shed new lights into endocytic mechanisms as well.

3. Super-resolution microscopy

3.1. History and background

Until the last decade, interrogation of cells and cellular processes with a microscope was limited by diffraction. Practically speaking, this meant that cellular features could only be distinguished if they were laterally separated by at least half the wavelength of the illumination light, as elegantly described by Ernest Abbe in in the late 19th century (Abbe, 1873). Under visible wavelength excitation, this means that cellular features and structures must be at least 200-350 nm apart in order to be resolved in X and Y. Unfortunately, this resolution limit is more than an order of magnitude larger than the spatial scale on which most biochemical processes occur. To address this, scientists have developed specialty optical microscopy techniques over the years to achieve information on a spatial scale below the limits of optical diffraction in living organisms. The most well-known of these are Förster resonance energy transfer (FRET), fluorescence correlation spectroscopy (FCS), and TIRF microscopy which as discussed above can provide axial resolution of *ca.* 100 nm, but is limited by diffraction in the lateral dimension.

More recent advances have produced several new methodologies, collectively termed "super-resolution" microscopy or "nanoscopy" that effectively break the traditional diffraction barrier in all three spatial dimensions. These resolutions are more aligned with the spatial scales on which biomolecular processes occur, and have potential to re-define the state-of-the art in biological imaging. Though a complete review of all the super-resolution microscopies is outside of the scope of this chapter, this section will discuss two major approaches with demonstrated applications in endocytosis: (1) localization microscopy and (2) stimulated emission depletion microscopy (STED). The reader is referred to recent review articles and the references within for additional information on super-resolution microscopy (B. Huang, et al., 2009; Schermelleh, et al., 2010).

3.2. Localization microscopy

3.2.1. Fundamentals of localization microscopy

If a single molecule within a diffraction-limited volume can be imaged independently from any other nearby emitters, localization techniques (R. E. Thompson, et al., 2002) can be employed to determine that molecule's location with precision of approximately

$$\Delta x \approx \frac{PSF}{\sqrt{N}} \tag{3}$$

where Δx is the localization precision, PSF is the size of the point spread function, and N is the number of detected photons from a single chromophore. With laser excitation and modern detectors, this accuracy can routinely be accomplished with <50nm precision. Interestingly, if this localization procedure could be repeated for many molecules, then an image could be constructed from the sum of all the localizations, with lateral resolution nearly 10-fold less than the diffraction limit. The practical challenge of this approach is rendering the vast majority of fluorophores in a sample in a "dark" state, only allowing a small subset to be visible at any given time. Indeed, conventional immunofluorescence labelling may result in thousands of visible fluorophores within a diffraction limited volume. However, several methods based on wide-field imaging of subpopulations of molecules activated in a stochastic fashion have emerged, including STORM (Rust, et al., 2006), PALM (Betzig, et al., 2006), and FPALM (Hess, et al., 2006). These approaches, collectively termed localization microscopies, each differ in the photophysics and photochemistry through which the single molecule activation and deactivation is achieved, but in general, make use of some form of "photoswitching" to turn individual fluorescent dyes to/from an on/off state. Then, the localization procedure remains essentially the same as described previously. In each approach, a delicate optimization is necessary between dye choice, imaging buffer solutions, labelling density, excitation wavelength and intensity, emission wavelength, and acquisition speed to produce images of the highest quality. A comprehensive review of fluorophore characteristics for use in localization microscopy has been recently published (Dempsey, et al., 2011).

To visualize biological processes like endocytosis which occur in three-dimensions, it is important to develop techniques that improve spatial resolution in x, y, and z and are compatible with imaging of living cells. More recent methods based on Astigmatism (Huang, Wang, et al., 2008) and dual-focal plane imaging (Juette, et al., 2008) have achieved axial localization precisions of 50 nm and 75 nm respectively, over depths of several hundreds of nanometers. Additionally, sub- 25 nm axial localization precision has been demonstrated using interferometric methods (Shtengel, et al., 2009). Z-scanning and single-particle tracking can be combined with these methods to extend the depth to several micrometers permitting imaging throughout the cell. (Huang, Jones, et al., 2008; Juette, et al., 2008). Though localization microscopies were originally limited to imaging fixed cells due to the conditions necessary to provide the stochastic photoswitching and the need to have no movement during the lengthy acquisition times, current methods are compatible with live-cell imaging (Manley, et al., 2008; Shroff, et al., 2008) while still maintaining axial spatial resolutions in the 50-60 nm range. This is an active area of research and further advancements are anticipated to eventually permit visualization of endocytotic dynamics at sub-50 nm resolution in all three dimensions.

3.2.2. Current applications of localization microscopy in endocytosis

The past five years have seen a flurry of localization-based superresolution microscopy studies related to endocytic processes, in a number of contexts. For instance, Betzig et al. successfully detected lysosomal membrane-associated proteins using a PALM approach with better than 10nm lateral resolution in fixed cell sections (Betzig, et al., 2006). Furthermore, Zhuang and colleagues were able to construct exquisite 3D images of both microtubule networks and clathrin coated vesicles (CCVs) in intact samples using the astigmatism-based STORM approach described above (Huang, Wang, et al., 2008). In addition, highly multiplexed studies using another localization variant, ground state depletion followed by individual molecule return (GSDIM), have shed light on the interactions between clathrin, tubulin, actin, and peroxisomes (Testa, et al., 2010). The interactions of HIV with the host cell cofactor tethrin (a protein implicated in preventing virus internalization) were revealed with excellent detail using a combination of photoactivatable proteins and photoswitchable organic dyes (Lehmann, et al., 2011). Another excellent example of super-resolution imaging applied toward endocytosis mechanisms includes a study by Subach, et al. In this case, novel photoactivatable proteins were exploited to acquire dual-color PALM images to visualize the clustering of transferrin receptors into clathrin coated pits (CCPs) at 25nm spatial resolution, as illustrated in Figure 6 (Subach, et al., 2009).

Data in Figure 6 show the substantial increase in image detail afforded by super-resolution imaging (b, e, h) over TIRF microscopy (a, d, g), with enlarged areas (indicated by white boxes), shown in (c, f, i). Clathrin (green) is generally co-localized to clusters of transferrin (red), although a large background of isolated/non-colocalized receptors is also apparent.

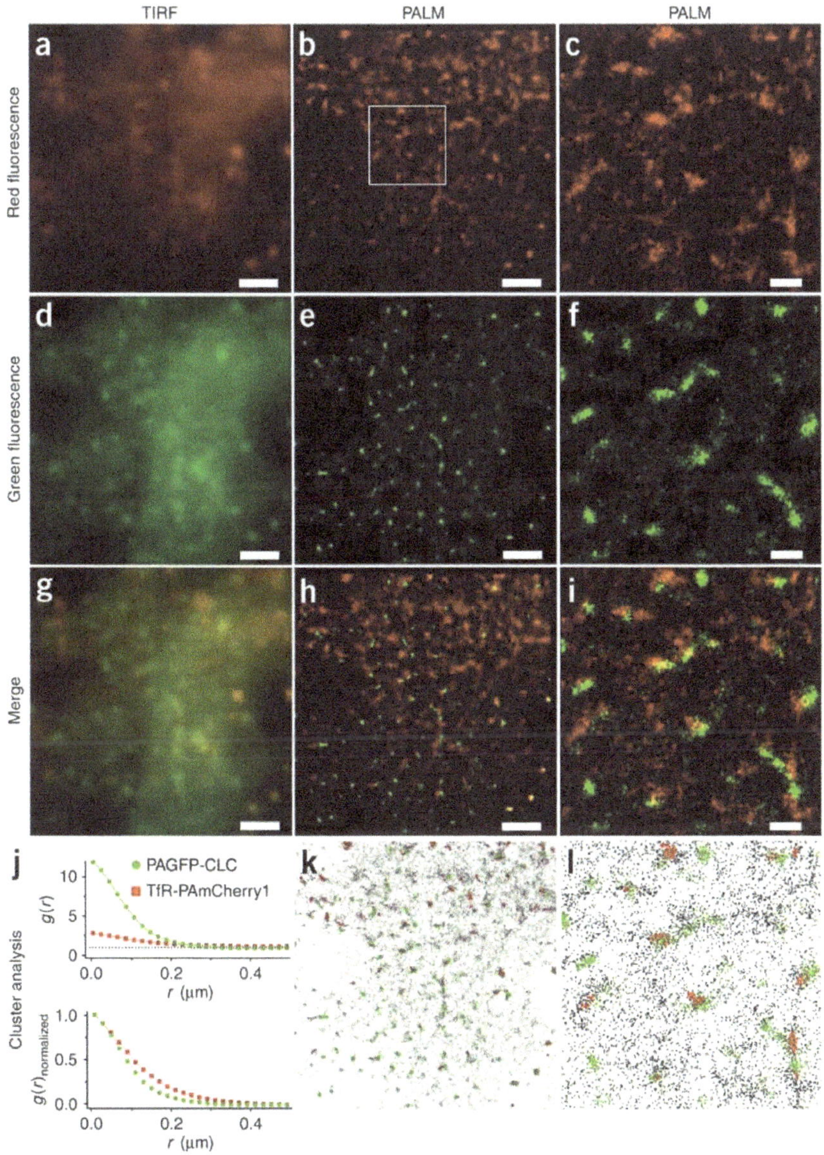

Figure 6. Localization-based super-resolution microscopy of transferrin receptor co-localization to CCPs. (a, d, g) indicate conventional TIRF images of receptor, clathrin, and overlay image, respectively. (b,e,h) illustrate the large increase in image detail after super-resolution localization is preformed, with zoomed in regions displayed in (c, f, and i). Co-cluster analysis was performed on areas where transferrin receptor/clathrin density was greater than 5-fold the mean (k, l). Correlation functions indicate a characteristic cluster size of approximately 200nm, below the Abbe limit (j) (Subach, et al., 2009). *Reprinted by permission from Macmillan Publishers Ltd:* Nature Methods *F.V. Subach, et al.,* **6**(2), *153-159, ©2009.*

As this data suggests, super-resolution microscopy also allows for image analysis with greatly increased precision over diffraction-limited imaging. In (k-j) receptor/clathrin clusters are analysed such that only areas where receptor density is five-fold greater than the image-wide average are considered (k-i). Spatial pair-correlation analysis (j) gives a measure of cluster diameter in the 200nm range, representing a detailed optical image analysis below the Abbe limit. This is important, as the above treatment demonstrates that while super-resolution imaging can provide images with exquisite detail and multiplexed capability, perhaps its greatest utility is its ability to enable improved quantification of biomolecular behaviour *in situ*. As more demonstrations are reported, new biological insights will doubtless be gained with the ability to monitor changes in biomolecular localization and dynamics at the nanoscale.

3.3. Stimulated emission depletion microscopy

3.3.1. Fundamentals of STED microscopy

In contrast to the localization-based super-resolution methods described in sections 3.1-3.2, stimulated emission-depletion (STED) microscopy relies on a different mechanism, and falls into the category of illumination-based techniques. Instead of localizing many random fields of single fluorophors to form a complete image, Illumination-based methods rely on a carefully engineered point spread function (PSF) that effectively limits fluorescence emission to a small, sub-diffraction volume. This modified PSF is subsequently scanned across a field of view in order to construct an image via confocal detection.

By far the most common way to accomplish a restriction in the PSF is to make use of two, superimposed beams of light, as shown in Figure 7. The first "excitation" beam (green) is a conventionally focused laser spot, whose diameter is subject to the diffraction limit. The second "depletion" beam (orange) is also diffraction limited, but a phase function is imparted such that it forms an optical vortex or "doughnut" when focused on the sample. When superimposed, the depletion beam prevents conventional fluorescence emission except for a small area near the center of the vortex. Using this method, optical resolutions approaching 7nm have been achieved (Rittweger, et al., 2009).

The concept of stimulated emission as a means to break the diffraction barrier extends from Stephan Hell's seminal paper exploring the theoretical basis (Hell & Wichmann, 1994), with experimental demonstration following (Klar, et al., 2000). The original implementation of this concept invovled complex, expensive instrumentation, including pairs of highly synchronized, femtosecond pulsed laser sources, in addition to other non-trivial timing electronics. However, subsequent simplifications were made such that STED could be accomplished with a single light source (Wildanger, et al., 2008), as well without any pulsed light sources (Willig, et al., 2007).

STED microscopy remains a very active area of development, with applications demonstrated in a wide variety of fields (Nägerl, et al., 2008; Rittweger, et al., 2009; Willig, Kellner, et al., 2006; Willig, Rizzoli, et al., 2006). Endocytosis stands as a enticeing area in which to apply STED microscopy, due to the intricate interplay between proteomic mediators and the sensitive spatiotemporally varying nature of cargo internalization.

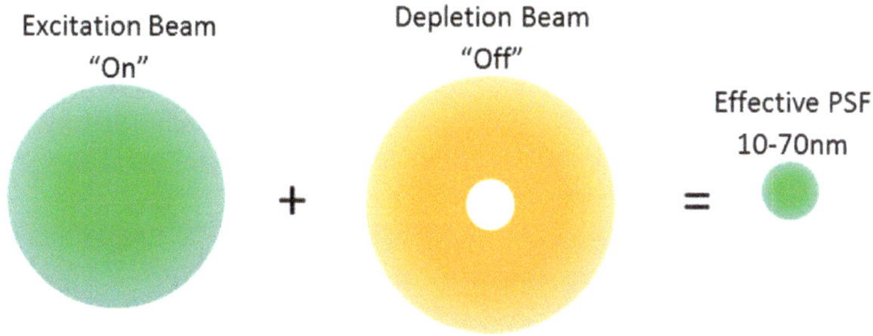

Figure 7. Principle of Stimulated Emission Depletion (STED) Microscopy. A conventional exictation spot (green) is overlayed with a optical vortex depletion beam (orange) to confine fluorescence emission to a sub-diffraction volume, and then scanned across a sample to create an image.

3.3.2. Current applications of STED microscopy in endocytosis

Illumination-based super-resolution techniques such as STED generally require more complex instrumentation as compared to localization approaches such as PALM/STORM. However, STED offers the advantage of more facile dynamic imaging. Although STED-based methods have somewhat lagged in their application toward the understanding of endocytosis as compared to localization techniques, several studies have begun to bring the considerable power of STED microscopy to bear on a number of pathways that are relevant in this regard. For instance, Schneider, et al. utilized STED microscopy to gain insight into the function of flotillin proteins in the context of Alzheimer's disease. Flotillins have been implicated in non-clathrin/caveolin mediated endocytosis as a mediator of amyloid regulation. Via knockdown models, they were able to show, with convincing image detail, that amyloid precursor protein (APP) internalization was reduced in the absence of flotillin-2. Furthermore, the increased resolution also permitted measurement of membrane-bound APP clusters with 70nm precision, and revealed that flotillin knockdown significantly reduced APP cluster size (Schneider, et al., 2008).

Additionally, Barrantes and colleagues successfully probed the nanoscale arrangement of acetylcholine receptors using STED microscopy (Kellner, et al., 2007). Perturbations in plasma membrane cholesterol via methyl-β-cyclodextrin resulted in significant, yet sub-diffraction changes in receptor behaviour, with clear implications for their regulation via endocytosis (Barrantes, 2007).

In combination with the development of video rate STED microscopy (Westphal, et al., 2008), Hell and colleagues were able to dynamically image synaptic vesicle trafficking in neurons at 40-60nm resolution (Willig, Rizzoli, et al., 2006). While synaptic transmission is often treated as an exocytic phenomenon, these results indicated that synaptotagmin remains clustered after exocytic vesicle fusion with the neuronal plasma membrane. This observation has clear implications for neurotransmitter re-endocytosis, as the precise mechanism by which endosomal recycling controls neurotransmitter release is still under investigation. These results indicate that membrane re-sorting of neurotransmitters may not be necessary for their recycling back to the cytoplasm. Via multi-temperature immunolabeling, combined with appropriate blocking and permeabilization, these data indicated re-endocytosis of synaptotagmin occurred within seconds of their initial exocytosis. Below, Figure 8 shows the increase in image quality afforded by STED when imaging re-endocytosed synaptic vesicles, taken from (Willig, Rizzoli, et al., 2006).

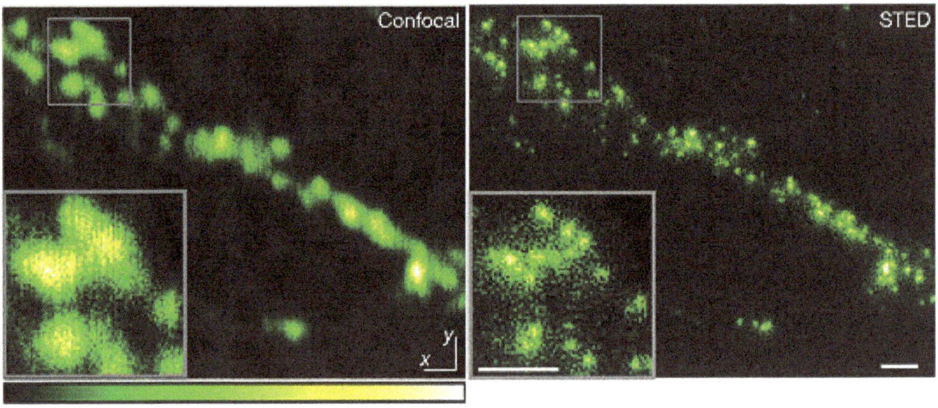

Figure 8. Confocal (left) and STED (right) microscopy images of synaptotagmin clustering on the surface of neuronal cells. STED microscopy affords a nearly order of magnitude increase in image resolution, allowing better quantifaction of neurotransmitter clustering, thereby giving better insight into the role of re-endocytosis as a mediator of synaptic transmission (Willig, Rizzoli, et al., 2006). *Reprinted by permission from Macmillan Publishers Ltd:* Nature *K.I. Willig et al.,* **440**, *935-939, ©2006*

4. Spectral imaging

4.1. Basic principles

Traditionally, optical microscopy (including the confocal and TIRF modalities applied to applications in endocytosis), and even superresolution imaging have been accomplished using a set of one or more filters to select a specific range of emission wavelengths to pass on to the detector. Filter-based microscopy is readily commercially available and can be extremely fast, producing high quality images at frame rates of up to hundreds or thousands of Hz with modern detectors. Filter-based microscopy requires that multiple chromophores of interest have well-separated emissions in order to avoid a phenomenon known as spectral channel crosstalk or spectral bleed through (SBT). Thoughtful choices of fluorophore labels can permit two- or three-color imaging in well characterized systems with filter-based microscopes; however in most live cell applications, filter-based microscopy is further limited by the presence of cellular autofluorescence. Cellular autofluorescence typically displays a broad emission that can span most of the visible wavelengths, and its spectral characteristics and intensity can vary widely across cell types and even within cells. In many applications where sensitivity is not a limiting factor, thresholding is used to minimize the SBT effect of cellular autofluorescence. Unfortunately, thresholding approaches may not be suitable for imaging of endocytotic processes if the signal at the single event level is often very near the intensity level of the cellular autofluorescence, since this will greatly confound quantitative analyses.

Spectral imaging is an alternative to filter-based microscopy whereby an entire emission spectrum is collected at each image pixel (2D) or voxel (3D) (Garini, et al., 2006; Zimmerman, et al., 2003). Spectral imaging has been implemented in a variety of optical modalities for biological applications including visible reflectance (Zuzak, et al., 2002), fluorescence (Michalet, et al., 2003) and vibrational spectroscopies such IR absorption (Levin & Bhargava, 2005), Raman scattering (Christensen & Morris, 1998), and surface-enhanced Raman (SERS) (Sharonov, et al., 1994)), as well as in non-optical methods like mass spectrometry (Fletcher, et al., 2008). In practice, higher degrees of multiplexing, higher accuracy, and lower detection limits are achievable with spectral imaging due to the ability to implement multivariate analysis methods to identify and/or classify spectral signatures even in the presence of high degrees of spectral overlap from other labels and cellular autofluorescence (Mansfield, et al., 2005). The trade-off is usually a sacrifice in speed, however microscope designs have been recently introduced that are competitive with current filter-based microscope acquisition rates (Sinclair, et al., 2006). In addition, further advances in the speed of acquisition are possible and anticipated given the latest detector technology (Coates, 2011; Fowler, et al., 2010).

Of the spectral imaging modalities, fluorescence and Raman-based spectral imaging are of particular interest to the field of endocytosis due to their demonstrated success in increasing the degree of multiplexing and providing label-free molecular specificity, respectively. Lerner provides a comprehensive tutorial covering the general principles of imaging spectrometers applicable to both fluorescence and Raman modalities and the reader is referred there for additional information (Lerner, 2006).

Fluorescence spectral imaging, also termed hyperspectral fluorescence microscopy, can be implemented in a wide variety of formats that differ predominantly in the way the spectral information is obtained. Hyperspectral fluorescence microscopes typically use one of three approaches to generate spectrally-resolved information: (1) a prism or grating to disperse the fluorescence emission onto a linear detector array or a charge-coupled device (CCD) detector in point-scanning (Sinclair, et al., 2006) or line-scanning (Sinclair, et al., 2004) formats, (2) interferometric methods that measure the intensity as a function of optical path length difference and glean spectral information through Fourier analysis (Malik, et al., 1996), (3) sequential, narrow bandpass filter scanning of discrete wavelength regions using acousto-optical or liquid crystal tunable filter (Gat, 2000).

4.2. Applications of fluorescence spectral imaging in endocytosis

Due to their relatively large excitation cross section, size-determined emission properties, and improved photostability as compared with traditional organic fluorophores, semi-conductor quantum dots (QDs) are becoming increasingly popular for biomedical research, with applications including targeted therapeutics and disease diagnostics. However the behavior and ultimate fate of these and other engineered nanoparticles in living systems has yet to be fully characterized. To this end, Aaron and co-authors took advantage of the multiplexing capabilities of hyperspectral confocal fluorescence microscopy and multivariate curve resolution (MCR), a constrained alternating least squares method, to identify and localize three colors of quantum dots and a lysosome-specific dye simultaneously (Aaron, et al., 2011), as shown in Figure 9. This work revealed unanticipated compartmentalization of the QDs on the plasma membrane (B and D) of a non-phagocytotic immune cell line (RBL cells), as well as an accurate measure of the relative fraction of QDs located within the lysosomes following endocytosis (C). These data were acquired with high precision, despite the significant spectral overlap between the various QDs and the lysosome-specific tracer dye, as shown in (A).

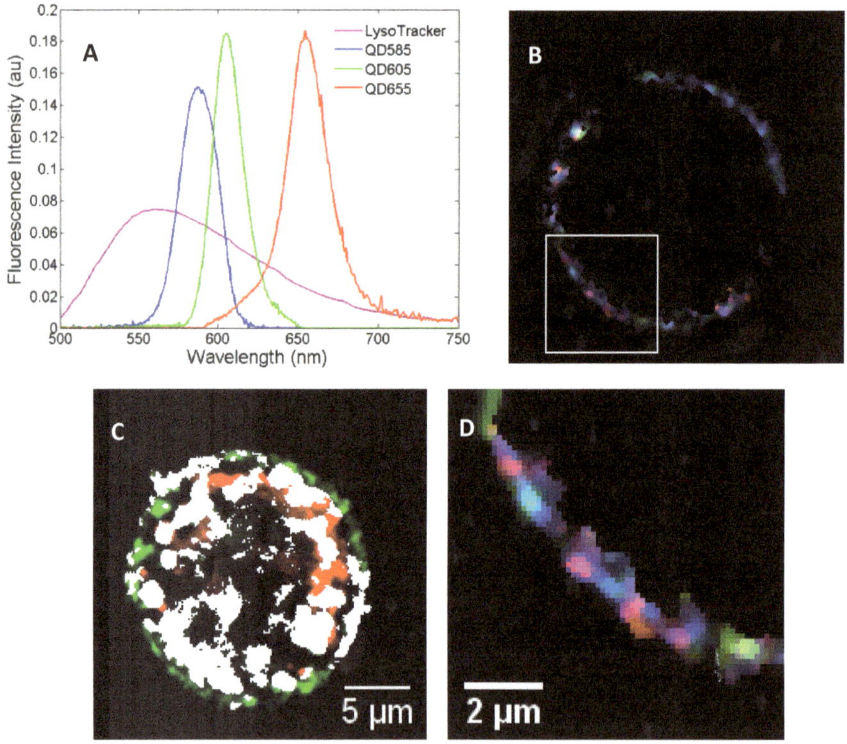

Figure 9. Hyperspectral imaging of quantum dot (QD) endocytosis. In (A), pure component spectra for three sized/shaped of QDs and a lysosome-specific fluorescent tracer are calculated without the need for any a priori information, and despite signficant spectral overlap. (B) indicates QDs not present in lysosomes, with the white box denoting the enlarged region shown in (D). Images indicate a compartmentalization of similar sized/ shaped QDs into distinct regions near the membrane, rather than a random distribution. In (C), QD signal is shown in green, while lysosome-specific dye is indicated in red. Areas of QD/lysosome co-localization are shown in white.

Huth and colleagues have also demonstrated the power of fluorescence hyperspectral imaging for visualizing uptake and intracellular trafficking of liposomes (Huth, et al., 2004). This work has particular relevance to understanding and manipulating the mechanisms of drug delivery via liposomal vehicles. They utilized Fourier-transform based spectral imaging technology to generate hyperspectral images of five fluorescent dyes in COS-7 cells. With the help of multivariate analysis algorithms, they were able to determine vesicle distribution throughout the cell relative to membrane lipids, lysosomes, and nuclear compartments. Their work clearly shows the multiplexing and accuracy advantages of spectral imaging for visualizing multiple subcellular compartments, while following the distribution of endocytosed cargo.

These highlighted applications illustrate the suitability of hyperspectral fluorescence microscopy for fundamental research into endocytotic mechanisms and make it easy to imagine future work employing hyperspectral fluorescence microscopy to follow the distributions of many of the cellular factors listed in Table 1 of Mercer (Mercer, et al., 2010), as well as potential cargo with diffraction-limited spatial and moderate temporal (10-100's of frame/sec) resolution. Studies of this type would provide information unavailable with other techniques.

4.3. Applications of Raman spectral imaging in endocytosis

Unlike fluorescence spectral imaging, Raman spectral imaging does not typically utilize exogenous labels to generate image contrast (Lewis & Edwards, 2001). Instead, the technique relies on the interaction of excitation light with the native molecular vibrations that are characteristic of distinct molecular components within the sample. These molecular "signatures" provide a label-free detection method for many important biomolecules, including proteins, nucleic acids, lipids, phospholipids, and carbohydrates. Though Raman spectral signatures are much weaker than fluorescence emission spectra, it is possible to perform Raman spectral imaging at the single cell level with modern detection technologies. Hyperspectral Raman microscopy can be implemented in a variety of formats similar to those described for hyperspectral fluorescence microscopy (Christensen & Morris, 1998; Govil, et al., 1993; Morris, et al., 1996), however the most commonly utilized for visualizing endocytosis in living cells has been the confocal point-scanning method, due to its availability, high sensitivity, optical sectioning capability, and speed.

Chernenko and colleagues applied hyperspectral confocal Raman microscopy to noninvasively query the distribution of cellular organelles relative to two biodegradable polymeric nanoparticle delivery systems (Chernenko, et al., 2009). It is very important to characterize the biocompatibility, cellular uptake and intracellular trafficking of these and other nanoparticle vehicles for drug delivery. Typically this is accomplished through the use of 1-2 fluorescent labels at a time and as such is likely to be inefficient and can suffer problems with label stability and interference with the nanocarrier. Importantly, in addition to the multiplexed advantage demonstrated with fluorescence-based spectral imaging, Raman spectra are exquisitely sensitive to changes in the local biochemical environment. This gives the added ability to detect and monitor changes that are associated with nanoparticle degradation (such as endosomal acidification). The authors employed a multivariate analysis algorithm known as Vertex Component Analysis to decompose spectra into their individual components (also called endmembers) (Nascimento & Dias, 2005). The resulting data were able to represent the spatial distribution of proteins, nanoparticles, lipid/phospholipids rich organelle membranes, and endosomal vesicles all without the need for exogenous labels.

Toward similar goals of characterizing the endocytotic uptake and trafficking of gold nanoparticles for applications in biomedical diagnostics and targeted gene/drug delivery, Park et al. used surface-enhanced Raman scattering (SERS) and dark field microscopy to visualize gold nanoparticles conjugated to transferrin protein (Park, et al., 2011). This work demonstrates the additional sensitivity offered by SERS over traditional Raman spectroscopy, and the potential of this method to not only follow nanoparticle distribution in three dimensions in a single living cell, but also make use of SERS spectral changes to indicate alterations in protein conjugation due to biochemical reactions.

5. Conclusion

A fuller understanding of endocytosis processes and the signalling cascades that regulate them is critically important for developing diagnostics, therapeutics, and vaccines. In this chapter we have presented three advanced optical imaging methodologies that have demonstrated advantages in spatial, temporal, and/or spectral resolution over traditional microscopy for interrogation of the processes involved in cellular uptake and trafficking. The ability to visualize dynamics of multiple species within living cells with high 3D spatial and temporal resolution provides unique information about molecular level interactions and their heterogeneity, both within and between cells, that is unavailable with other techniques. The examples we highlighted from recent literature illustrate how these tools are being engaged to address unanswered questions about the roles of key biomolecules including actin, dynamin, and others in the field of endocytosis as well as the sequence of biomolecular events during cellular response.

Yet, the potential of advanced imaging for studying endocytosis-related processes has not been fully realized. Recent developments in super-resolution microscopy, spectral imaging, and specialized TIRF modalities have extended imaging into a realm where multiple biomolecules involved in individual endocytic events can be visualized with never before seen clarity, detail, and precision. Future efforts will doubtlessly focus on continued improvements to these enabling technologies individually, as well as on coupling the aforementioned approaches. Progress towards both ends will provide more complete visualizations that are necessary to complement bulk biochemical and genetic approaches, and thus better characterize endocytosis pathways in the living cell.

Author details

Jesse S. Aaron and Jerilyn A. Timlin
Sandia National Laboratories, USA

Acknowledgement

Sandia National Laboratories is a multi-program laboratory managed and operated by Sandia Corporation, a wholly owned subsidiary of Lockheed Martin Corporation, for the U.S. Department of Energy's National Nuclear Security Administration under contract DE-AC04-94AL85000. This work was supported by the National Institutes of Health Director's New Innovator Award Program, 1-DP2-OD006673-01 (JSA and JAT). We would also like to acknowledge Ms. Michelle Raymer for helpful assistance in preparing this manuscript.

6. References

Aaron, J.S., Greene, A.C., Kotula, P.G., Bachand, G.D., & Timlin, J.A. (2011). Advanced Optical Imaging Reveals the Dependence of Particle Geometry on Interactions Between CdSe Quantum Dots and Immune Cells. *Small,*Vol.7, No.3, pp. 334-341. 1613-6829

Abbe, E. (1873). Beiträge zur Theorie des Mikroskops und der mikroskopischen Wahrnehmung. *Archiv für Mikroskopische Anatomie,*Vol.9, No.1, pp. 413-418. 0176-7364

Anantharam, A., Onoa, B., Edwards, R.H., Holz, R.W., & Axelrod, D. (2010). Localized topological changes of the plasma membrane upon exocytosis visualized by polarized TIRFM. *The Journal of Cell Biology,*Vol.188, No.3, February 8, 2010, pp. 415-428.

Axelrod, D. (1981). Cell-substrate contacts illuminated by total internal reflection fluorescence. *The Journal of Cell Biology,*Vol.89, No.1, April 1, 1981, pp. 141-145.

Axelrod, D. (2001). Total Internal Reflection Fluorescence Microscopy in Cell Biology. *Traffic,*Vol.2, No.11, pp. 764-774. 1600-0854

Barrantes, F.J. (2007). Cholesterol effects on nicotinic acetylcholine receptor. *Journal of Neurochemistry,*Vol.103, No.pp. 72-80. 1471-4159

Betz, W.J., Mao, F., & Smith, C.B. (1996). Imaging exocytosis and endocytosis. *Current Opinion in Neurobiology,*Vol.6, No.3, pp. 365-371. 0959-4388

Betzig, E., Patterson, G.H., Sougrat, R., Lindwasser, O.W., Olenych, S., Bonifacino, J.S., et al. (2006). Imaging intracellular fluorescent proteins at nanometer resolution. *Science,*Vol.313, No.5793, Sep, pp. 1642-1645.

Boehm, M., & Bonifacino, J.S. (2001). Adaptins. *Molecular Biology of the Cell,*Vol.12, No.10, October 1, 2001, pp. 2907-2920.

Brandenburg, B., Lee, L.Y., Lakadamyali, M., Rust, M.J., Zhuang, X., & Hogle, J.M. (2007). Imaging Poliovirus Entry in Live Cells. *PLoS Biology,*Vol.5, No.7, pp. e183.

Brandenburg, B., & Zhuang, X. (2007). Virus trafficking - learning from single-virus tracking. [10.1038/nrmicro1615]. *Nature Reviews Microbiology,*Vol.5, No.3, pp. 197-208. 1740-1526

Chan, W.C.W., & Nie, S. (1998). Quantum Dot Bioconjugates for Ultrasensitive Nonisotopic Detection. *Science,*Vol.281, No.5385, September 25, 1998, pp. 2016-2018.

Chernenko, T., Matthäus, C., Milane, L., Quintero, L., Amiji, M., & Diem, M. (2009). Label-Free Raman Spectral Imaging of Intracellular Delivery and Degradation of Polymeric Nanoparticle Systems. *ACS Nano,*Vol.3, No.11, 2009/11/24, pp. 3552-3559. 1936-0851

Christensen, K.A., & Morris, M.D. (1998). Hyperspectral Raman microscopic imaging using Powell lens line illumination. *Applied Spectroscopy,*Vol.52, No.9, 1998, pp. 1145-1147.

Coates, C. (2011). New sCMOS vs. Current Microscopy Cameras: Andor Technolgy.

Dempsey, G.T., Vaughan, J.C., Chen, K.H., Bates, M., & Zhuang, X. (2011). Evaluation of fluorophores for optimal performance in localization-based super-resolution imaging. [10.1038/nmeth.1768]. *Nature Methods,*Vol.8, No.12, pp. 1027-1036. 1548-7091

Ehrlich, M., Boll, W., van Oijen, A., Hariharan, R., Chandran, K., Nibert, M.L., et al. (2004). Endocytosis by Random Initiation and Stabilization of Clathrin-Coated Pits. *Cell,*Vol.118, No.5, pp. 591-605. 0092-8674

Fletcher, J.S., Rabbani, S., Henderson, A., Blenkinsopp, P., Thompson, S.P., Lockyer, N.P., et al. (2008). A New Dynamic in Mass Spectral Imaging of Single Biological Cells. *Analytical Chemistry,*Vol.80, No.23, pp. 9058-9064.

Floyd, D.L., Ragains, J.R., Skehel, J.J., Harrison, S.C., & van Oijen, A.M. (2008). Single-particle kinetics of influenza virus membrane fusion. *Proceedings of the National Academy of Sciences,*Vol.105, No.40, October 7, 2008, pp. 15382-15387.

Fowler, B., Liu, C., Mims, S., Balicki, J., Li, W., Do, H., et al. (2010). A 5.5Mpixel 100 frames/sec wide dynamic range low noise CMOS image sensor for scientific applications. *Proceedings of SPIE,* 9780819479297, San Jose, CA, Jan 2010.

Gaidarov, I., Santini, F., Warren, R.A., & Keen, J.H. (1999). Spatial control of coated-pit dynamics in living cells. [10.1038/8971]. *Nature Cell Biology,*Vol.1, No.1, pp. 1-7. 1465-7392

Gao, X., Yang, L., Petros, J.A., Marshall, F.F., Simons, J.W., & Nie, S. (2005). In vivo molecular and cellular imaging with quantum dots. *Current Opinion in Biotechnology,*Vol.16, No.1, pp. 63-72. 0958-1669

Garini, Y., Young, I.T., & McNamara, G. (2006). Spectral imaging: Principles and applications. *Cytometry Part A,*Vol.69A, No.8, pp. 735-747. 1552-4930

Gat, N. (2000). Imaging Spectroscopy Using Tunable Filters: A Review. *Proceedings of Wavelet Applications VII.*

Govil, A., Pallister, D.M., & Morris, M.D. (1993). Three-dimensional confocal Raman microscopy. *Applied Spectroscopy,*Vol.47, No.1, 1993, pp. 75-79.

Hashida, M., Kawakami, S., & Yamashita, F. (2005). Lipid Carrier Systems for Targeted Drug and Gene Delivery. *Chemical and Pharmaceutical Bulletin,*Vol.53, No.8, pp. 871-880.

He, Y., Li, H.-W., & Yeung, E.S. (2005). Motion of Single DNA Molecules at a Liquid−Solid Interface As Revealed by Variable-Angle Evanescent-Field Microscopy. *The Journal of Physical Chemistry B,*Vol.109, No.18, 2005/05/01, pp. 8820-8832. 1520-6106

Hell, S.W., & Wichmann, J. (1994). Breaking the diffraction resolution limit by stimulated emission: stimulated-emission-depletion fluorescence microscopy. *Optics Letters,*Vol.19, No.11, pp. 780-782.

Hess, S.T., Girirajan, T.P.K., & Mason, M.D. (2006). Ultra-High Resolution Imaging by Fluorescence Photoactivation Localization Microscopy. *Biophysical Journal*,Vol.91, No.11, December 1, 2006, pp. 4258-4272.

Hoekstra, D., De Boer, T., Klappe, K., & Wilschut, J. (1984). Fluorescence method for measuring the kinetics of fusion between biological membranes. *Biochemistry*,Vol.23, No.24, 1984/11/01, pp. 5675-5681. 0006-2960

Huang, B., Bates, M., & Zhuang, X. (2009). Super-resolution fluorescence microscopy. *Annual Review of Biochemistry*,Vol.78, No.pp. 993-1016. 1545-4509 (Electronic)

Huang, B., Jones, S.A., Brandenburg, B., & Zhuang, X. (2008). Whole-cell 3D STORM reveals interactions between cellular structures with nanometer-scale resolution. [10.1038/nmeth.1274]. *Nature Methods*,Vol.5, No.12, pp. 1047-1052. 1548-7091

Huang, B., Wang, W., Bates, M., & Zhuang, X. (2008). Three-Dimensional Super-Resolution Imaging by Stochastic Optical Reconstruction Microscopy. *Science*,Vol.319, No.5864, February 8, 2008, pp. 810-813.

Huth, U., Wieschollek, A., Garini, Y., Schubert, R., & Peschka-Süss, R. (2004). Fourier transformed spectral bio-imaging for studying the intracellular fate of liposomes. *Cytometry Part A*,Vol.57A, No.1, pp. 10-21. 1552-4930

Johannsdottir, H.K., Mancini, R., Kartenbeck, J., Amato, L., & Helenius, A. (2009). Host Cell Factors and Functions Involved in Vesicular Stomatitis Virus Entry. *Journal of Virology*,Vol.83, No.1, January 1, 2009, pp. 440-453.

Juette, M.F., Gould, T.J., Lessard, M.D., Mlodzianoski, M.J., Nagpure, B.S., Bennett, B.T., et al. (2008). Three-dimensional sub-100 nm resolution fluorescence microscopy of thick samples. [10.1038/nmeth.1211]. *Nature Methods*,Vol.5, No.6, pp. 527-529. 1548-7091

Kaksonen, M., Toret, C.P., & Drubin, D.G. (2006). Harnessing actin dynamics for clathrin-mediated endocytosis. [10.1038/nrm1940]. *Nature Reviews Molecular Cell Biology*,Vol.7, No.6, pp. 404-414. 1471-0072

Kellner, R.R., Baier, C.J., Willig, K.I., Hell, S.W., & Barrantes, F.J. (2007). Nanoscale organization of nicotinic acetylcholine receptors revealed by stimulated emission depletion microscopy. *Neuroscience*,Vol.144, No.1, pp. 135-143. 0306-4522

Keyel, P.A., Watkins, S.C., & Traub, L.M. (2004). Endocytic Adaptor Molecules Reveal an Endosomal Population of Clathrin by Total Internal Reflection Fluorescence Microscopy. *Journal of Biological Chemistry*,Vol.279, No.13, March 26, 2004, pp. 13190-13204.

Kiessling, V., Domanska, M.K., & Tamm, L.K. (2010). Single SNARE-Mediated Vesicle Fusion Observed In Vitro by Polarized TIRFM. *Biophysical Journal*,Vol.99, No.12, pp. 4047-4055. 0006-3495

Kirchhausen, T., Harrison, S.C., & Heuser, J. (1986). Configuration of clathrin trimers: Evidence from electron microscopy. *Journal of Ultrastructure and Molecular Structure Research*,Vol.94, No.3, pp. 199-208. 0889-1605

Klar, T.A., Jakobs, S., Dyba, M., Egner, A., & Hell, S.W. (2000). Fluorescence microscopy with diffraction resolution barrier broken by stimulated emission. *Proceedings of the National Academy of Sciences*,Vol.97, No.15, July 18, 2000, pp. 8206-8210.

Lakadamyali, M., Rust, M.J., Babcock, H.P., & Zhuang, X. (2003). Visualizing infection of individual influenza viruses. *Proceedings of the National Academy of Sciences,*Vol.100, No.16, August 5, 2003, pp. 9280-9285.

Lehmann, M., Rocha, S., Mangeat, B., Blanchet, F., Uji-i, H., Hofkens, J., et al. (2011). Quantitative Multicolor Super-Resolution Microscopy Reveals Tetherin HIV-1 Interaction. *PLoS Pathog,*Vol.7, No.12, pp. e1002456.

Lerner, J.M. (2006). Imaging spectrometer fundamentals for researchers in the biosciences—A tutorial. *Cytometry Part A,*Vol.69A, No.8, pp. 712-734. 1552-4930

Leserman, L.D., Weinstein, J.N., Blumenthal, R., & Terry, W.D. (1980). Receptor-mediated endocytosis of antibody-opsonized liposomes by tumor cells. *Proceedings of the National Academy of Sciences,*Vol.77, No.7, July 1, 1980, pp. 4089-4093.

Levin, I.W., & Bhargava, R. (2005). Fourier Transform Infrared Vibrational Spectroscopic Imaging: Integrating Microscopy and Molecular Recognition. *Annual Review of Physical Chemistry,*Vol.56, No.1, pp. 429-474.

Lewis, I.R., & Edwards, H.G.M. (2001). Handbook of Raman Spectroscopy *Practical Spectroscpy Series* (pp. 1074). New York: Marcel Dekker, Inc.

Malik, Z., Cabib, D., Buckwald, R.A., Talmi, A., Garini, Y., & Lipson, S.G. (1996). Fourier transform multipixel spectroscopy for quantitative cytology. *Journal of Microscopy,*Vol.182, No.2, pp. 133-140. 1365-2818

Manley, S., Gillette, J.M., Patterson, G.H., Shroff, H., Hess, H.F., Betzig, E., et al. (2008). High-density mapping of single-molecule trajectories with photoactivated localization microscopy. [10.1038/nmeth.1176]. *Nature Methods,*Vol.5, No.2, pp. 155-157. 1548-7091

Mansfield, J.R., Gossage, K.W., Hoyt, C.C., & Levenson, R.M. (2005). Autofluorescence removal, multiplexing, and automated analysis methods for in-vivo fluorescence imaging. *Journal of Biomedical Optics,*Vol.10, No.4, pp.

Marquis, B.J., Love, S.A., Braun, K.L., & Haynes, C.L. (2009). Analytical methods to assess nanoparticle toxicity. *Analyst,*Vol.134, No.3, pp. 425-439. 0003-2654

Mattheyses, A.L., & Axelrod, D. (2006). Direct measurement of the evanescent field profile produced by objective-based total internal reflection fluorescence. *Journal of Biomedical Optics,*Vol.11, No.1, pp. 014006-014007.

Medintz, I.L., Uyeda, H.T., Goldman, E.R., & Mattoussi, H. (2005). Quantum dot bioconjugates for imaging, labelling and sensing. [10.1038/nmat1390]. *Nature Materials,*Vol.4, No.6, pp. 435-446. 1476-1122

Mercer, J., Schelhaas, M., & Helenius, A. (2010). Virus Entry by Endocytosis. *Annual Reviews of Biochemistry,*Vol.79, No.pp. 803-833.

Merrifield, C.J., Feldman, M.E., Wan, L., & Almers, W. (2002). Imaging actin and dynamin recruitment during invagination of single clathrin-coated pits. [10.1038/ncb837]. *Nature Cell Biology,*Vol.4, No.9, pp. 691-698. 1465-7392

Merrifield, C.J., Perrais, D., & Zenisek, D. (2005). Coupling between Clathrin-Coated-Pit Invagination, Cortactin Recruitment, and Membrane Scission Observed in Live Cells. *Cell,*Vol.121, No.4, pp. 593-606. 0092-8674

Michalet, X., Kapanidis, A.N., Laurence, T., Pinaud, F., Doose, S., Pfughoefft, M., et al. (2003). The power and prospects of fluorescence microscopies and spectroscopies. *Annual Review of Biophysics and Biomeolecular Structure,*Vol.32, No.pp. 161-182.

Morris, H.R., Hoyt, C.C., Miller, P., & Treado, P.J. (1996). Liquid Crystal Tunable Filter Raman Chemical Imaging. *Applied Spectroscopy,*Vol.50, No.6, pp. 805-811.

Muller, M. (2006). *Introduction to Confocal Microscopy* (second ed. Vol. TT69).SPIE Press, Bellingham, Washington.

Nägerl, U.V., Willig, K.I., Hein, B., Hell, S.W., & Bonhoeffer, T. (2008). Live-cell imaging of dendritic spines by STED microscopy. *Proceedings of the National Academy of Sciences,*Vol.105, No.48, December 2, 2008, pp. 18982-18987.

Nascimento, J.M.P., & Dias, J.M.B. (2005). Vertex Component Analysis: A Fast Algorithm to Unmix Hypersepctral Data. *IEEE Transactions on Geoscience and Remote Sensing,*Vol.43, No.4, pp. 898-910.

Oheim, M., Loerke, D., Stühmer, W., & Chow, R.H. (1998). The last few milliseconds in the life of a secretory granule. *European Biophysics Journal,*Vol.27, No.2, pp. 83-98. 0175-7571

Olveczky, B.P., Periasamy, N., & Verkman, A.S. (1997). Mapping fluorophore distributions in three dimensions by quantitative multiple angle-total internal reflection fluorescence microscopy. *Biophysical Journal,*Vol.73, No.5, pp. 2836-2847. 0006-3495

Oreopoulos, J., & Yip, C.M. (2009). Probing Membrane Order and Topography in Supported Lipid Bilayers by Combined Polarized Total Internal Reflection Fluorescence-Atomic Force Microscopy. *Biophysical Journal,*Vol.96, No.5, pp. 1970-1984. 0006-3495

Panyam, J., & Labhasetwar, V. (2003). Biodegradable nanoparticles for drug and gene delivery to cells and tissue. *Advanced Drug Delivery Reviews,*Vol.55, No.3, pp. 329-347. 0169-409X

Park, J.-H., Park, J., Dembereldorj, U., Cho, K., Lee, K., Yang, S., et al. (2011). Raman detection of localized transferrin-coated gold nanoparticles inside a single cell. *Analytical and Bioanalytical Chemistry,*Vol.401, No.5, pp. 1635-1643. 1618-2642

Rappoport, J.Z. (2008). Focusing on clathrin-mediated endocytosis. *Biochemical Journal,*Vol.412, No.pp. 415-523.

Rappoport, J.Z., Benmerah, A., & Simon, S.M. (2005). Analysis of the AP-2 Adaptor Complex and Cargo During Clathrin-Mediated Endocytosis. *Traffic,*Vol.6, No.7, pp. 539-547. 1600-0854

Rappoport, J.Z., Kemal, S., Benmerah, A., & Simon, S.M. (2006). Dynamics of clathrin and adaptor proteins during endocytosis. *American Journal of Physiology - Cell Physiology,*Vol.291, No.5, November 1, 2006, pp. C1072-C1081.

Rappoport, J.Z., Simon, S., & Benmerah, A. (2004). Review Understanding Living Clathrin-Coated Pits. [Article]. *Traffic,*Vol.5, No.5, pp. 327-337. 13989219

Rappoport, J.Z., Taha, B.W., Lemeer, S., Benmerah, A., & Simon, S.M. (2003). The AP-2 Complex Is Excluded from the Dynamic Population of Plasma Membrane-associated Clathrin. *Journal of Biological Chemistry,*Vol.278, No.48, November 28, 2003, pp. 47357-47360.

Rappoport, J.Z., Taha, B.W., & Simon, S.M. (2003). Movement of Plasma-Membrane-Associated Clathrin Spots Along the Microtubule Cytoskeleton. *Traffic,*Vol.4, No.7, pp. 460-467. 1600-0854

Reichert, W.M., Suci, P.A., Ives, J.T., & Andrade, J.D. (1987). Evanescent Detection of Adsorbed Protein Concentration-Distance Profiles: Fit of Simple Models to Variable-Angle Total Internal Reflection Fluorescence Data. *Applied Spectroscopy,*Vol.41, No.3, pp. 503-508.

Rittweger, E., Han, K.Y., Irvine, S.E., Eggeling, C., & Hell, S.W. (2009). STED microscopy reveals crystal colour centres with nanometric resolution. [10.1038/nphoton.2009.2]. *Nature Photonics,*Vol.3, No.3, pp. 144-147. 1749-4885

Rust, M.J., Bates, M., & Zhuang, X. (2006). Sub-diffraction-limit imaging by stochastic optical reconstruction microscopy (STORM). [10.1038/nmeth929]. *Nature Methods,*Vol.3, No.10, pp. 793-796. 1548-7091

Rust, M.J., Lakadamyali, M., Zhang, F., & Zhuang, X. (2004). Assembly of endocytic machinery around individual influenza viruses during viral entry. [10.1038/nsmb769]. *Nature Structural Molecular Biology,*Vol.11, No.6, pp. 567-573. 1545-9993

Schermelleh, L., Heintzmann, R., & Leonhardt, H. (2010). A guide to super-resolution fluorescence microscopy. *The Journal of Cell Biology,*Vol.190, No.2, July 26, 2010, pp. 165-175.

Schmoranzer, J., Goulian, M., Axelrod, D., & Simon, S.M. (2000). Imaging Constitutive Exocytosis with Total Internal Reflection Fluorescence Microscopy. *The Journal of Cell Biology,*Vol.149, No.1, April 3, 2000, pp. 23-32.

Schneider, A., Rajendran, L., Honsho, M., Gralle, M., Donnert, G., Wouters, F., et al. (2008). Flotillin-Dependent Clustering of the Amyloid Precursor Protein Regulates Its Endocytosis and Amyloidogenic Processing in Neurons. *The Journal of Neuroscience,*Vol.28, No.11, March 12, 2008, pp. 2874-2882.

Sharonov, S., Nabiev, I., Chourpa, I., Feofanov, A., Valisa, P., & Manfait, M. (1994). Confocal three-dimensional scanning laser Raman–SERS–fluorescence microprobe. Spectral imaging and high-resolution applications. *Journal of Raman Spectroscopy,*Vol.25, No.7-8, pp. 699-707. 1097-4555

Shroff, H., Galbraith, C.G., Galbraith, J.A., & Betzig, E. (2008). Live-cell photoactivated localization microscopy of nanoscale adhesion dynamics. [10.1038/nmeth.1202]. *Nature Methods,*Vol.5, No.5, pp. 417-423. 1548-7091

Shtengel, G., Galbraith, J.A., Galbraith, C.G., Lippincott-Schwartz, J., Gillette, J.M., Manley, S., et al. (2009). Interferometric fluorescent super-resolution microscopy resolves 3D cellular ultrastructure. *Proceedings of the National Academy of Sciences,*Vol.106, No.9, March 3, 2009, pp. 3125-3130.

Sieczkarski, S.B., & Whittaker, G.R. (2002a). Dissecting virus entry via endocytosis. *Journal of General Virology,*Vol.83, No.7, July 1, 2002, pp. 1535-1545.

Sieczkarski, S.B., & Whittaker, G.R. (2002b). Influenza Virus Can Enter and Infect Cells in the Absence of Clathrin-Mediated Endocytosis. *Journal of Virology,*Vol.76, No.20, October 15, 2002, pp. 10455-10464.

Sinclair, M.B., Haaland, D.M., Timlin, J.A., & Jones, H.D.T. (2006). Hyperspectral confocal microscope. *Applied Optics,*Vol.45, No.24, pp. 3283-3291.

Sinclair, M.B., Timlin, J.A., Haaland, D.M., & Werner-Washburne, M. (2004). Design, construction, characterization, and application of a hyperspectral microarray scanner. *Applied Optics,*Vol.43, No.10, pp. 2079-2089.

Slowing, I.I., Vivero-Escoto, J.L., Wu, C.-W., & Lin, V.S.Y. (2008). Mesoporous silica nanoparticles as controlled release drug delivery and gene transfection carriers. *Advanced Drug Delivery Reviews,*Vol.60, No.11, pp. 1278-1288. 0169-409X

Stephens, D.J., & Allan, V.J. (2003). Light Microscopy Techniques for Live Cell Imaging. *Science,*Vol.300, No.5616, April 4, 2003, pp. 82-86.

Stock, K., Sailer, R., Strauss, W.S.L., Lyttek, M., Steiner, R., & Schneckenburger, H. (2003). Variable-angle total internal reflection fluorescence microscopy (VA-TIRFM): realization and application of a compact illumination device. *Journal of Microscopy,*Vol.211, No.1, pp. 19-29. 1365-2818

Subach, F.V., Patterson, G.H., Manley, S., Gillette, J.M., Lippincott-Schwartz, J., & Verkhusha, V.V. (2009). Photoactivatable mCherry for high-resolution two-color fluorescence microscopy. [10.1038/nmeth.1298]. *Nature Methods,*Vol.6, No.2, pp. 153-159. 1548-7091

Sund, S.E., & Axelrod, D. (2000). Actin Dynamics at the Living Cell Submembrane Imaged by Total Internal Reflection Fluorescence Photobleaching. *Biophysical Journal,*Vol.79, No.3, pp. 1655-1669. 0006-3495

Sund, S.E., Swanson, J.A., & Axelrod, D. (1999). Cell Membrane Orientation Visualized by Polarized Total Internal Reflection Fluorescence. *Biophysical Journal,*Vol.77, No.4, pp. 2266-2283. 0006-3495

Testa, I., Wurm, C.A., Medda, R., Rothermel, E., von Middendorf, C., Fölling, J., et al. (2010). Multicolor Fluorescence Nanoscopy in Fixed and Living Cells by Exciting Conventional Fluorophores with a Single Wavelength. *Biophysical Journal,*Vol.99, No.8, pp. 2686-2694. 0006-3495

Thompson, N.L., Drake, A.W., Chen, L., & Broek, W.V. (1997). Equilibrium, Kinetics, Diffusion and Self-Association of Proteins at Membrane Surfaces: Measurement by Total Internal Reflection Fluorescence Microscopy. *Photochemistry and Photobiology,*Vol.65, No.1, pp. 39-46. 1751-1097

Thompson, N.L., McConnell, H.M., & Burhardt, T.P. (1984). Order in supported phospholipid monolayers detected by the dichroism of fluorescence excited with polarized evanescent illumination. *Biophysical Journal,*Vol.46, No.6, pp. 739-747. 0006-3495

Thompson, R.E., Larson, D.R., & Webb, W.W. (2002). Precise Nanometer Localization Analysis for Individual Fluorescent Probes. *Biophysical Journal,*Vol.82, No.5, pp. 2775-2783. 0006-3495

Tokunaga, M., Kitamura, K., Saito, K., Iwane, A.H., & Yanagida, T. (1997). Single Molecule Imaging of Fluorophores and Enzymatic Reactions Achieved by Objective-Type Total Internal Reflection Fluorescence Microscopy. *Biochemical and Biophysical Research Communications,*Vol.235, No.1, pp. 47-53. 0006-291X van der Schaar, H.M., Rust, M.J., Waarts, B.-L., van der Ende-Metselaar, H., Kuhn, R.J., Wilschut, J., et al. (2007). Characterization of the Early Events in Dengue Virus Cell Entry by Biochemical Assays and Single-Virus Tracking. *Journal of Virology,*Vol.81, No.21, November 1, 2007, pp. 12019-12028.

West, J.L., & Halas, N.J. (2003). Engineered Nanomaterials for Biophotonics Applications: Improving Sensing, Imaging, and Therapeutics. *Annual Review of Biomedical Engineering,*Vol.5, No.1, pp. 285-292.

Westphal, V., Rizzoli, S.O., Lauterbach, M.A., Kamin, D., Jahn, R., & Hell, S.W. (2008). Video-Rate Far-Field Optical Nanoscopy Dissects Synaptic Vesicle Movement. *Science,*Vol.320, No.5873, April 11, 2008, pp. 246-249.

Wildanger, D., Rittweger, E., Kastrup, L., & Hell, S.W. (2008). STED microscopy with a supercontinuum laser source. *Opt. Express,*Vol.16, No.13, pp. 9614-9621.

Willig, K.I., Harke, B., Medda, R., & Hell, S.W. (2007). STED microscopy with continuous wave beams. [10.1038/nmeth1108]. *Nature Methods,*Vol.4, No.11, pp. 915-918. 1548-7091

Willig, K.I., Kellner, R.R., Medda, R., Hein, B., Jakobs, S., & Hell, S.W. (2006). Nanoscale resolution in GFP-based microscopy. [10.1038/nmeth922]. *Nature Methods,*Vol.3, No.9, pp. 721-723. 1548-7091

Willig, K.I., Rizzoli, S.O., Westphal, V., Jahn, R., & Hell, S.W. (2006). STED microscopy reveals that synaptotagmin remains clustered after synaptic vesicle exocytosis. [10.1038/nature04592]. *Nature,*Vol.440, No.7086, pp. 935-939. 0028-0836

Zimmerman, T., Rietdorf, J., & Pepperkok, R. (2003). Spectral imaging and its applications in live cell microscopy. [minireview]. *Federation of European Biochemical Societies Letters,*Vol.546, No.pp. 87-92.

Zuzak, K.J., Schaeberle, M.D., Lewis, E.N., & Levin, I.W. (2002). Visible reflectance hyperspectral imaging: Characterization of a noninvasive, in vivo system for determining tissue perfusion. *Analytical Chemistry,*Vol.74, No.9, pp. 2021-2028.

Section 3

Viruses

Chapter 7

Clathrin-Associated Endocytosis as a Route of Entry into Cells for Parvoviruses

F. Brent Johnson and Enkhmart Dudleenamjil

Additional information is available at the end of the chapter

1. Introduction

For successful infection, viruses must solve several problems and overcome barriers that would impede or prevent infection. Some of these barriers include finding a suitable host, circumventing the immune system, interacting with cells within their host range, attaching to the cell surface, penetrating to the interior of the cell and finding ways for transcription, synthesis of their gene products, genome replication, assembly of the particle components, and escape from the infected cell. As a whole, viruses are paradoxically, fascinatingly simple yet complex. Some viruses, such as those that are enveloped, enter the target cell by attacking the cell membrane with a fusion protein, or a fusion domain of an attachment protein, that invades the cell membrane and promotes fusion of the viral envelope with the cell membrane. This fusion event may occur either at the plasma membrane level or at a membrane site in the endocytic vesicle. The viral genome then is free of its envelope and can enter the cell interior. Nonenveloped viruses generally do not enter by fusion at the plasma membrane level but are commonly taken into the cell by endocytosis, then, by some mechanism, are able to invade the cell cytoplasm by escaping from confinement in the endosome. The pathways for virus entry along with the entry subtleties are nearly as diverse as the virus families themselves. Many of the parvoviruses are known to employ the clathrin-associated acid mediated endocytotic pathway for penetration into their host cell and this chapter deals with parvovirus entry.

The virus family *Parvoviridae* is so named because of the small size of the virus particles. As a group, they are approximately 22-26 nm in diameter, have T=1 capsid structure, and contain a single stranded DNA genome. Six genera have been named. They infect a variety of mammalian, avian, non-mammalian vertebrate, and invertebrate hosts (Table 1). Some of these viruses will be described in the discussion of endocytic entry pathways that follow. Parvovirus host range and entry was reviewed by Cotmore and Tattersall (2007) emphasizing viruses in the genus *Parvovirus*. For purposes of illustration of parvovirus entry, bovine parvovirus 1 (BPV-1) will be emphasized in the current review. In subsequent descriptions this virus is referred to simply as BPV. BPV is a member of the genus *Bocavirus*. Other Bocaviruses are canine minute virus (CnMV) and human bocavirus (HBoV). BPV-1 is the prototype virus of this genus. It is a small (22 nm), non-enveloped virus with icosahedral structure. BPV is a pathogen of cattle found worldwide that causes severe gastroenteritis in calves, mild respiratory infection, and may cause reproductive failure (Manteufel & Truyen, 2008; Sandals et al., 1995). It is a contamination risk for commercially prepared bovine serum and bovine-derived products. The genome consists of single stranded DNA that is composed of about 5515 nucleotides (Chen et al., 1986; Qiu et al., 2007; Sun, et al., 2009). The genome contains non-identical palindromic sequences at the two ends. The palindromic sequences have signals that are important for genome replication and packaging (Berns, 1990; Shull et al., 1988). The genome has three open reading frames (ORFs): the left ORF encodes the nonstructural proteins NS1 and NS2; the central ORF encodes the nonstructural protein NP-1; and the right ORF encodes two or three structural proteins VP1, VP2, and VP3 (Johnson & Hoggan, 1973; Qiu et al., 2007). According to parvovirus structural protein nomenclature, the viruses with two proteins (VP1 and VP2) designate the largest protein VP1. The two proteins are co-terminal at the carboxyterminal ends but a unique sequence (called VP1u or the VP1 unique sequence) is present on the larger protein. Similar nomenclature applies to AAV except there are three structural proteins and the unique ends are the VP1/VP2u sequences. Parvoviruses do not encode polymerases. These enzymes are provided by the host cell. BPV genome replication relies on host cell DNA polymerase and replication factors found in S-phase cells (Berns, 1990). Moreover, transcription is carried out by cell RNA polymerase II which requires a double-stranded transcription template. Thus, genome replication is a necessary precursor to transcription. The genomic organization of HBoV, the second human-pathogenic parvovirus known (discovered after Parvovirus B19), closely resembles the other known bocaviruses BPV and CnMV (Allander et al., 2005; Kaplan et al., 2006; Ma et al., 2006; Sun et al., 2009). The mid-ORF product of HBoV is homologous to NP-1 of BPV and CnMV, and these proteins have 47% amino acid identity. The proteins of the two major ORFs (NS and VP) of HBoV have 42-43% homology with NS1, VP1, and VP2 proteins of BPV and CnMV (Allander et al., 2005; Bi et al., 2007).

Most parvoviruses, as understood to date, utilize several different subtle variations of a general strategy to deliver their genome to the cell nucleus, the site of virus replication. These are late penetrating viruses, in the sense that they are pH-dependent and persist longer through the endocytic transport system than earlier penetrating viruses (Greber, 2011). The rugged, mature, extracellular virion undergoes multistep conformational changes that are time- and cell compartment-dependent. In general, several discrete steps have been recognized that entering parvoviruses follow. These are interaction with a cell surface receptor, trafficking through the endosomal pathway to the late endosome or lysosome, or through macropinocytosis for porcine parvovirus (Boisvert et al., 2010), escape from the endosomal pathway using the newly deployed phospholipase 2 (PLA_2) domain of the capsid protein, and cytoskeletal-controlled transport of the modified particle to the nucleus.

Techniques employed in the laboratory for the elucidation of viral entry pathways involve both direct observational detection of virus in cells and indirect detection of virus movement by blocking specific functions in the process with chemical inhibitors. Electron microscopy is used as a method for showing virus associated with cell structures giving "snap shot" views of virus over time intervals in the entry process. Parvoviruses, for example, can be seen associated with clathrin coated pits and in endosomes (Dudleenamjil et al., 2010; Parker & Parrish, 2000; Vendeville et al., 2009). Chemical inhibitors are assumed to block the entry pathway at certain specific points and incomplete virus replication in the presence of inhibitor is presumed to indicate viral penetration using the pathway blocked by the drug. Commonly used drugs are chlorpromazine (prevents assembly and disassembly of clathrin lattices at the cell surface and on endosomes by inhibiting clathrin polymerization which blocks assembly of coated pits and is also a phospholipase A_2 inhibitor, Blanchard et al., 2006); bafilamycin A1 (inhibits vacuolar proton ATPases inhibiting acidification of endosomes by blocking transport of protons into the vesicle, Clague et al., 1994; Drose & Altendorf, 1997; Wassmer et al., 2005); ammonium chloride (penetrates into the endosome increasing endosomal pH, Jin et al., 2002; Liebl et al., 2006); chloroquine (blocks assembly of clathrin coated pits and raises endosomal pH, Mani et al., 2006; Ros et al., 2002). Brefeldin A has no effect on the early endosome but inhibits vesicle transport and early-to-late endosome transition (Clague et al., 1994; Nebenfuhr et al., 2002).

Some drugs block virus entry through caveolae and can be used to help distinguish virus entry through caveolae or clathrin-associated endosomes. The antifungal drug nystatin complexes with and sequesters cholesterol inhibiting lipid rafts and caveolae (Chazal & Gerlier, 2003; Damm et al., 2005; Sieczkarski & Whittaker, 2002). Phorbol-12-myristate-13-acetate, a mitogen and a tumor promoter, decreases membrane caveolin-1 (Smart et al., 1994). Methyl-β-cyclodextrin disrupts detergent-resistant lipid rafts (Beer et al., 2005). Genistein blocks phosphorylation of tyrosine kinase which is involved in the formation of caveosomes (Pelkmans, 2005a). Other drugs block transport of virions (or modified virions) through the cell cytoplasm. Some viruses depend on microtubules for transport and some associate with actin filaments. Nocodazole blocks microtubule polymerization (Brandenburg & Zhuang, 2007; Pelkmans & Helenius, 2003), vanadate is a dynein inhibitor (Beckerle & Porter, 1982), erythro-9-3-(2hydroxynonyl)adenine also is a dynein inhibitor (Krietensson et al., 1986). Cytochalasin D disrupts actin microfilaments and blocks actin polymerization (Vendeville et al., 2009), and latrunculin A also inhibits actin polymerization (Damm et al., 2005; Forest et al., 2005).

There are some important concerns about the interpretation of the results obtained in studies that employ inhibitors. Concentrations of the drugs must be below the cell toxicity levels to preserve virus-dependent cell activities. In most cases complete shut-off of virus replication does not occur. So, the possibilities of pathway leakiness or the presence of more than one entrance pathway must be considered. None of the drugs possesses absolute specificity, therefore possible side effects may occur that affect interpretation. In studies like these, the use of several chemical inhibitors having different modes of action help to somewhat mitigate the problem of cloudy interpretation if the results are mutually supportive.

Genus	Species	Pathogenesis[1]
Parvovirus	*Canine parvovirus*	Gastroenteritis, cardiovascular failure
	Chicken parvovirus	Enteric disease, cerebellar hypoplasia
	Feline panleukopenia virus	Gastroenteritis, leukopenia, cerebellar hypoplasia
	Goose parvovirus	Gastroenteritis, anorexia, nasal discharge, death
	Hamster parvovirus	Osteolytic infection, facial and tooth malformation
	H-1 parvovirus	A hamster osteolytic virus
	Human PARV 4	Viremia, ?other disease
	Kilham rat virus	Mostly asymptomatic, small litters, runting
	Lapine parvovirus	Enteritis
	LUIII virus	Systemic infection in hamsters
	Mink enteritis virus	Gastroenteritis, leukopenia
	Mouse minute virus	Asymptomatic, MMVi is immunosuppressive
	Murine parvovius 1	No known disease
	Porcine parvovirus	Reproductive failure, embryonic death
	Raccoon parvovirus	Enteritis
	Rat minute virus 1a, 1b, 1c	Asymptomatic
	Rat parvovirus	(Also known as Kilham rat virus)
Erythrovirus	*Human parvovirus B19*	Erythema infectiosum (Fifth disease), exanthema, hydrops fetalis
Bocavirus	*Bovine parvovirus 1*	Gastroenteritis (calf scours), dyspnea, stillborn calves
	Bovine parvovirus 2	Same as BPV-1
	Bovine parvovirus 3	Same as BPV-1
	Canine minute virus	Diarrhea, anorexia, dyspnea
	Human bocavirus	Upper respiratory, coryza
Amdovirus	*Aleutian mink disease virus*	Immune complex disease, glomerulonephritis, death
Dependovirus	*Adeno-associated virus* (types 1-11)	None known, latency in chromosome 19
Densovirus	*Brevidensovirus*	Infects mosquito larvae
	Densovirus	GmDNV kills moth larvae
	Iteravirus	Infects silkworm larvae
	Pefudensovirus	Host is the cockroach

Table 1. Members of family *Parvoviridae*. Other presumptive parvovirus isolates are not listed. [1]In some instances virus infection in adult animals is largely asymptomatic but in neonates some pattern of symptomatic infection has been recognized.

Immunofluoresceinated virus particles have been used to show association of virus with cell receptors and clathrin-rich early endosomes (Freistadt & Eberle, 2006). Purified virus particles are directly labeled with the fluorescein fluor by treatment with N-hydroxysuccinimide-fluorescein. Attachment and uptake of virus by cells is tracked by flow cytometry, fluorescence microscopy, or confocal microscopy. Most viruses, including the very small parvoviruses, cannot be seen as individual particles by light microscopy, but clumps of viruses can be seen. Also, accumulated virus particles present in cell organelles such as late endosomes-lysosomes can be seen. Among the Rab proteins in cells, Rab5 is primarily associated with early endosomes, Rab7 with late endosomes, and Rab 11 with recycling endosomes. Co-localization of virus with specific Rab proteins provides evidence of virus trafficking through certain endosomal compartments.

A recent report provided cryoelectron microscopic images of the HBoV particle (Gurda et al., 2010). Other reports show the small, T=1 capsid of parvoviruses displays amino acid side chains to the inner capsid surface which bond to the bases of the single stranded genome (Agbandje-McKenna & Chapman, 2006; Chapman & Agbandje-McKenna, 2006). Such interactions occur as the genome is packaged into the preformed shell. Several features including the small size, compact capsid, interior molecular interactions, and lack of an envelope combine to make this virus remarkably environmentally stable. These viruses undergo several essential protein conformation shifts during the trafficking process but are able to remain essentially intact after attachment, engulfment, endosomal transit, travel through the cytoplasmic environment, association with the nuclear membrane, and travel across the nuclear membrane. Although there are some capsid alterations during this process (Cotmore & Tattersall, 2007), the viral genome remains capsid-associated as it enters the nuclear compartment.

2. Attachment of virus to the cell

The term *receptor* to indicate the cell surface structure to which the virus attaches and the term *antireceptor* to indicate the virus-associated component that binds to the cell receptor are used to describe the virus-cell interaction that leads to sticking of virus to cell. Several studies have been reported that show the orientations of the capsid proteins and the capsid surface topography (Agbandje-McKenna & Chapman, 2006; Agbandje et al., 1998; Chapman & Agbandje-McKenna, 2006; Chapman & Rossmann, 1993; Tsao et al., 1991) and show the major structural protein forming ridges and valleys on the faces and edges with a pore at each vertex. In the parvoviruses studied for surface structure, there is a surface "spike" positioned on the icosahedral face or axis of three-fold rotational symmetry. The antireceptor is located with this icosahedral face spike (Cotmore & Tattersall, 2007).

Parvoviruses are known to utilize a variety of cell surface molecules as their receptors including glycoproteins, glycolipids, and glycans (Cotmore & Tattersall, 2007; Harbison et al., 2008). Members of the genus *Parvovirus* in the FPV serotype (Feline and canine parvoviruses) can use transferrin receptors (TfR) for attachment and entry (Cotmore &

Tattersall, 2007; Hueffer et al., 2004; Parker et al., 2001; Suikkanen, 2003). These FPV viruses also bind to neuraminidase-sensitive N-glycolyl neuraminic acid side chains present on some cells, but these probably only act as attachment receptors and not entry receptors because infectious entry is insensitive to neuraminidase and infectious entry occurs on TfR. The block to FPV infection in canine cells is largely due to the lack of a functional cell surface receptor. Feline panleukopenia virus (FPV) and canine parvovirus (CPV) both bind feline TfR and use it to infect cat cells, but CPV preferentially binds canine TfR and infects dog cells (Palermo et al., 2006). In contrast to the FPV entry program, MVM binds to sialoglycoprotein receptors and both binding and entry are neuraminidase sensitive suggesting these receptors provide two functions, attachment and entry. Porcine parvovirus (PPV) is a major etiologic agent of reproductive failure in swine. PPV binds to sialic acid receptors on surface glycoproteins (Boisvert et al., 2010). These authors found pre-treatment of cells with neuraminidase prevented infection. Resialation on sialidase-treated cells with either α-2,3-*O*-sialyltransferase or α-2,3-*N*-sialyltransferase partially restored infectivity suggesting both O-linked and N-linked forms of carbohydrate moieties may act as receptors, but leaves open the possibility of other sialic acids functioning in the attachment process.

The mammalian adeno-associated viruses (AAVs) have been extensively studied for use as gene vectors for therapy of diseases resulting from genetic defects. Cellular receptors and coreceptors are an area of crucial interest in viral vector-mediated gene therapy because receptor preference and receptor tissue distribution may dictate vector choice for a given organ. By understanding the mechanisms of viral entry into target cells it may be possible to manipulate the gene vector in order to target a cell type of interest. Heparan sulfate proteoglycan mediates attachment of AAV-2 to susceptible cell lines (Summerford & Samulski, 1998), and other cell entry receptors or co-receptors for AAV-2 include human fibroblast growth factor receptor-1 (FGFR-1) (Qing et al., 1999), $\alpha V\beta 5$ integrin (Sanlioglu et al., 2000; Summerford et al., 1999), and hepatocyte growth factor receptor (c-Met) (Kashiwakura et al., 2005). It has been suggested that heparin sulfate may play a role in AAV-2 infection as a low affinity attachment molecule (Qiu et al., 2000). Reports indicate that a group of basic amino acids that contribute to heparin binding are clustered in three positions on the three-fold spike of the AAV-2 capsid (Kern et al., 2003). Additionally, AAV-5 reportedly binds to cell surface 2,3-linked sialic acids (Walters et al., 2001) and the AAV-5 receptor for hemagglutination and transduction is α2,3-N-linked sialic acid and that for AAV-4 is α2,3-O-linked sialic acid (Kaludov et al., 2001). AAV-5 also binds to the platelet derived growth factor receptor (Di Pasquale et al., 2003). AAV type 3 can use fibroblast growth factor receptor 1 for binding (Blackburn et al., 2006).

A characteristic feature of BPV is its ability to hemagglutinate erythrocytes, hence its original name, hemadsorbing enteric virus (HADEN) (Abinanti & Warfield, 1961). This aspect of virus-cell interaction is possible because this virus attaches to receptors on the red cell membrane. Glycophorin A is an abundant transmembrane glycoprotein found in the erythrocyte membrane (Tomita et al., 1978). The glycophorin A monomer is composed of a 131 amino acid sequence found in three domains that form the hydrophilic cytoplasmic domain, the hydrophobic transmembrane domain, and the aminoterminal glycosylated external domain. Glycophorin A naturally exists as a homodimer with extensive O-linked oligosaccharide glycosylation on the external domain. Exposure of the red cell surface to proteolytic enzymes or to neuraminidase destroyed the erythrocyte receptors for BPV indicating the receptor consists of a sialylglycoprotein (Thacker & Johnson, 1998). Virus probes on western blots and virus attachment to purified glycophorin A on dot blots confirmed virus attachment to this glycoprotein. Moreover, purified glycophorin A completely competed out virus attachment to the natural receptor. Further, Blackburn et al. (2005) showed BPV binding to α-2,3-linked sialic acid located on the O-linked oligosaccharides of the glycophorin A molecule. Treatment of glycophorin A with α2,3,-6,-8 neuraminidase eliminated binding of virus to this receptor. Beta-elimination of O-linked sialic acids on glycophorin A also eliminated binding while removal of N-linked carbohydrates using the N-glycosidase PNGase failed to eliminate virus binding. After enzymatic removal of the receptors, virus binding could be restored by reconstitution of the O-linked α2,3 neuraminic acids. On nucleated bovine host cells BPV attachment occurs on both α-2,3-O-linked and α-2,3-N-linked sialic acids (Johnson et al., 2004).

In studies on PPV infection, prior treatment of cells with neuraminidase prevented infection by eliminating PPV receptors (Boisvert et al., 2010). Resialation with α-2,3-*O*-sialyltransferase or α-2,3-*N*-sialyltransferase, or with a combination of the two enzymes partially restored infectivity. Therefore, the sialylglycoprotein receptors on the cell surface for PPV appeared to consist of both O- and N-linked sialic acids. Possibly other sialic acid receptor moieties exist in addition to these because total reconstitution did not occur but it was unclear whether completion of the reconstitution reaction could occur under the experimental conditions.

Host range is a property of the virus-host interaction that provides a suitable environment for complete virus replication resulting in the production of virus progeny. Only certain organs, tissues, and cells are within the host range of viruses. Commonly, host range is thought of relating to the availability of cell surface receptors. However, determination of host range and its resultant tissue tropism can occur at many levels including attachment to compatible receptors, the entry pathway, uncoating, the transcriptional environment, genome duplication, translational processes, assembly of virus particles, and mechanisms of escape from the host cell. At one level, host range for autonomous parvoviruses is determined by the viral requirement for S-phase cells. Frequently, in examples of parvovirus disease, the tissues involved are those which have an abundance of mitotic cells. In the case of FPV and CPV, related parvoviruses, attachment then entry by endocytosis can occur in many different kinds of non-permissive cells, indicating that their host range can be determined by events after cell entry. Amino acid residues 359 to 375 found in a flexible capsid protein loop were found to exhibit differential conformation when exposed to various concentrations of protons and Ca^{++} (Simpson et al., 2000). It was found that this region was functionally associated with both hemagglutinating activity and host range determinants providing continuance of successful replication.

3. Engulfment

Several possible pathways of receptor-mediated endocytosis are recognized: clathrin- and caveolae-mediated endocytosis, macropinocytosis, and novel nonclathrin/noncaveolae pathways (Brindley & Maury, 2008; Damm et al., 2005; Dimitrov, 2004; Kee et al., 2004; Marsh & Helenius, 2006; Meier & Greber, 2004; Meier et al., 2002; Mercer & Helenius, 2009; Pelkmans & Helenius, 2003; Pelkmans et al. 2004; Sieczkarski & Whittaker, 2002; Stuart & Brown, 2006). To date, for viral infections that have been studied, viruses mostly take advantage of clathrin-mediated endocytosis for internalization (Marsh & Helenius, 2006; Pelkmans & Helenius, 2003; Sieczkarski & Whittaker, 2002). During clathrin-mediated endocytosis, transport vesicles are surrounded by a clathrin coat, which is a three-dimensional array of triskelia. A triskelion is composed of three clathrin heavy chains (CHCs, approximately 190-kDa) and three light chains (CLCs, about 25-29 kDa), and has three-fold rotational symmetry (Edeling et al., 2006; Merrifield et al., 2005). Ligands that are to be transported to the cytosol, including viruses, are concentrated on the cell surface, and the concentrated ligands, as a patch, trigger recruitment of clathrin-adaptor proteins to the cytoplasmic side of the plasma membrane. Clathrin-adaptor complexes (APs) include the main distinct complexes, AP1 and AP2. A third protein, AP180, is used mainly for synaptic vesicles. The AP-2 complex consistes of α-adaptin, β2-adaptin, μ2-chain, and σ2-chain. APs bind to membranes by recognizing phosphoinositides and link clathrin to the membrane. Therefore, clathrin coated vesicles (CCVs) are three-layered: 1) the inner, membrane layer with its embedded receptor/ligand complex, 2) the middle layer that is composed of APs and other regulatory proteins for clathrin assembly, and 3) the outer clathrin shell (Edling et al., 2006). CCVs comprise one of the most common and well defined coated transport vesicles. The internal pH in the CCV is around 6.5. Once fission of the pit has occurred through the action of accessory proteins including dynamin and other proteins forming CCVs, the clathrin coat must be rapidly shed to allow fusion of the vesicle with its target membrane. Uncoating of clathrin is resolved by auxilin and the molecular chaperone Hsc70 (heat shock protein 70). Auxilin interacts with assembled clathrin and binds to Hsc70 via its carboxyl-terminal J domain triggering Hsc70's ATPase activity. Hsc 70 then interrupts clathrin-clathrin interactions, causing shedding of the clathrin coat. Disassembled clathrin, accessory proteins and the endocytic recycling compartment (internal pH about 6.5) are recycled and promote the clathrin-coated vesicle cycle (Brandenburg & Zhuang, 2007; Dawsen et al., 2006; DeTulleo & Kirhchausen, 1998; Doxsey et al., 1987; Edling et al., 2006; Heuser & Anderson, 1989; Huang et al., 2004; Lemmon, 2001; Pu & Zhang, 2008; Sun et al., 2002). The cargo-containing vesicle then matures to the early endosome with an acidic environment (pH 6.5 to 6.0). In addition to the pH, markers for the early endosome include Rab5-GDP and Early Endosome Associated Protein-1. Virus-engaged receptors are uncoupled from their ligands at this mildly acidic environment of the early endosomes and ligand molecules are recycled back to the plasma membrane (Van der Goot & Gruenberg, 2006). The early endosomes are major sorting stations where the endosome is recycled back to the cell membrane, or where endocytosed cargo, including some viruses, can be released to the cytoplasm or can progress farther into the endosomal pathway to more acidic late endosomes (pH 6.0 to 5.0) and lysosomes. The acidification of endosomes is required for release of virus into the cytoplasm (Damm & Pelkmans, 2006; Gagescu et al., 2000; Lakadamyali et al., 2006; Marsh & Helenius, 2006; Pelkmans & Helenius, 2003; Russell et al., 2006; Sieczkarski & Whittaker, 2002; Smith & Helenius, 2004; Van der Goot & Gruenberg, 2006). The late endosome becomes a degradative body with lower pH. It acquires the marker for mannose-6-phosphate receptor (MPR+) and the Rab7-GDP marker. They also acquire the unusual lipid lysobisphosphatidic acid (LBPA).

After engulfment the genome or nucleocapsids of viruses are released into the cytosol by fusion of the viral envelope with the endosomal membrane for enveloped viruses or, for non-enveloped viruses, capsid disassembly occurs in the endosome for some viruses with subsequent genomic escape to the cytosol (Brandenburg et al., 2007; Stidwill & Greber, 2000). The acidic pH of endosomes plays an essential role to trigger these events. Further, some non-enveloped viruses begin the uncoating process in the late endosome, but complete uncoating is delayed and it is a nuclear event. For example, in some instances such as adenovirus, AAV and canine parvovirus infections, the genome together with modified capsid components translocate to the nuclear membrane (Meier & Greber, 2004; Sonntag et al., 2006; Vihinen-Ranta et al., 2002) where final uncoating occurs for adenovirus but the small parvovirus capsid crosses the nuclear membrane before final uncoating within the nucleus. Digestion of material enclosed in the late endosome may not be complete and fusion with lysosomes may occur forming a hybrid organelle with an internal pH of about 5.0 and are MPR-negative. Viruses that exploit clathrin-dependent acid-mediated entry are sensitive to the inhibitors of endosomal acidification. Thus, inhibition of virus replication by endosomal pH inhibitors is taken as evidence for virus tracking through an acid-mediated endocytosis pathway.

The transport of endosomes is mediated by microtubules and proceeds toward the microtubule organization center which is found in the perinuclear area of the cell. Thus, transport to the late endosomes/lysosomes is beneficial for both virion conformation adjustment and for transport to the nucleus. Once the virus has escaped from the endosomal compartment the virus particle itself may interact with microtubule motors or with actin filaments to complete the journey to the nuclear membrane. In the case of parvoviruses, proteosomal digestion can help or hinder virus infection. AAV particles are degraded by the proteosome causing an aborted infection (Douar et al., 2001) while proteosome processing is required for MVM infection (Ros et al., 2002).

Electron microscopic images of BPV-infected cells show vacuoles consistent with CCVs containing virus particles (Dudleenamjil et al., 2010) at 15 minutes post-infection. In a separate report on canine parvovirus (Parker & Parrish, 2000), virus particles were found in endosomes at 5 minutes and 15 minutes post-infection. Densonucleosis virus was shown by EM in CCVs of *Lepidoptera* cells at 5 minutes post infection (Vendeville et al., 2009). These images in three different host cells were very similar and characteristic of CCVs bound with membrane-linked clathrin. An electron micrograph (Fig. 1) illustrates the possible invagination and pinching off process and shows a CCV that contains a virus-like particle.

In contrast to clathrin-mediated pH-dependent endocytosis, caveolae-mediated entry is an event triggered by binding of, for example, virus particles to receptor molecules on the cell surface that induces the clustering of lipid rafts with a high content of cholesterol and sphingolipids (Chazal & Gerlier, 2003; Hommelgaard et al., 2005; Pelkmans, 2005a). The area of the plasma membrane with the clusters invaginates to the cytosol, and the vesicle is surrounded by caveolins, the most characterized proteins of caveolae-mediated entry. These caveolin coated vesicles formed at the cholesterol-rich microdomains at the plasma membrane are the caveolae. Caveolins stabilize caveolae, and they are remarkably static in caveosomes (Marsh & Helenius, 2006; Pelkmans, 2005b). Accessory proteins involved in caveolae-mediated entry are dynamin and actin, and they are recruited by tyrosine kinase activites (Dimitrov, 2004; Marsh & Helenius, 2006; Pelkmans, 2005a; Pelkmans & Helenius, 2003; Pelkmans et al., 2005; Smith & Helenius, 2004). The caveolae containing the virus/receptor complex, close, pinch off from the cell membrane, and fuse together forming caveosomes. Caveosomes are part of the endocytic organelles with a neutral pH and the absence of markers for early, recycling, and late endosomes (Pelkmans, 2005a; Pelkmans & Helenius, 2003). Nevertheless, caveosomes connect with the smooth endoplasmic reticulum (ER), early and late endosomes, and the cell membrane. Release of virus taken up by the caveolar-raft system can occur from caveosomes (Echo 1), the ER (SV40), and endosomes (polyomaviruses and BK virus) (Eash et al., 2004; Marsh & Helenius, 2006; Pelkmans & Helenius, 2003). The interaction with endosomes may be crucial for some viruses that are taken up by caveolae-mediated endocytosis but require a low-pH environment for escape to the cytoplasm. Chemical inhibitors, targeted for a certain part of interconnected organelles of caveolae entry, are extensively exploited for examination of caveolae pathways used for virus entry. Results of studies utilizing inhibitors targeting caveolae found no evidence for BPV entry through these vesicles (Dudleenamjil et al., 2010) nor for PPV (Boisvert et al., 2010).

Active cell entry routes of some members of the Parvoviridae family have been described. Parvoviruses are known to utilize a variety of cell surface molecules as their receptor including glycoproteins (Blackburn et al., 2005; Thacker & Johnson, 1998), glycolipids, and glycans (Cotmore & Tattersall, 2007). It has been reported that parvoviruses with known cell entry routes enter into CCVs and establish successful infections (Basak & Compans, 1989; Cotmore & Tattersall, 2007; Harbison et al., 2008; Op De Beeck & Caillet-Fauquet, 1997; Parker & Parrish, 2000; Ros et al., 2002; Vendeville et al., 2009). Most studied for illumination of this process are MVM, CPV, PPV, DNV and AAV. Adeno-associated virus (AAV)'s entry into the host cell is mediated by clathrin coated pits and then routes to the late endosomes (Bartlett et al., 2000). The virus particles then escape to the cytoplasm where they are partially degraded by the proteasome and delivered to the nucleus for replication (Douar et al., 2001). The canine parvovirus, having used the transferrin receptor (TfR) for attachment (Parker et al., 2001), enters through the CCVs, and localizes in endosomes (Hueffer et al., 2004; Parker & Parrish, 2000; Vihinen-Ranta et al., 2002). Both CPV and FPV bind to TfR but species-specific binding controls host range. CPV binds to the filopodia of canine cells while FPV infects cats binding to the TfR on feline cells. FPV does not bind the canine TfR, does not infect dogs, or infect cultured canine cells (Harbison et al., 2009). Conversely, CPV can infect feline cells by binding to TfR on the cell body. Minute virus of mice (MVM)'s cell entry and following events in the cytoplasm were investigated analyzing the effects of drugs that interfere with the endosomal acidification and ubiquitin-proteasome activities. Results suggested that MVM's entry is pH-dependent, and the interaction with the ubiquitin-proteasome system is required for MVM replication (Ros et al., 2002). The relatively rapid endocytic uptake of parvoviruses appears to be followed by slower traffic along the endocytic compartments toward the nucleus. The endosomal pathway undertaken by parvoviruses appears to be complex and depends on the virus, its concentration, and likely the cell type (Dorsch et al., 2002; Mani et al., 2006; Sonntag et al., 2006; Suikkanen et al., 2003; Yuan & Parrish, 2001). Conformational alterations in capsid structure probably occur in the endosomal compartment facilitating uncoating and transport to the nucleus.

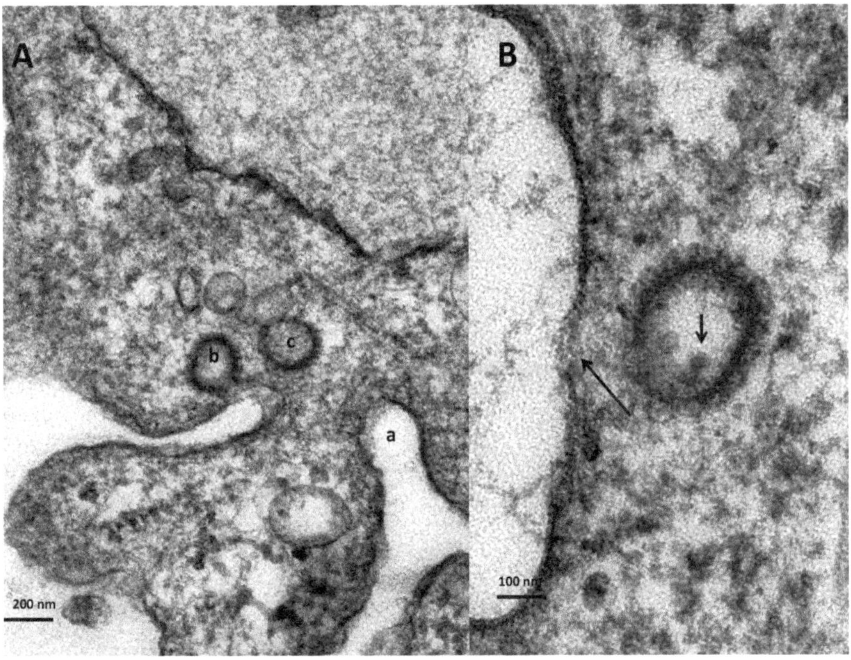

Figure 1. Electron micrograph of a clathrin coated vesicle-associated parvovirus-like particle. A) Negative-stained uninfected EBTr cell thin section showing invagination of membrane (a); (b) a nearly closed clathrin coated vesicle (CCV); (c) an enclosed CCV. B) A bovine parvovirus-infected cell (at 15 minutes post-infection) showing the cytoplasmic membrane site of CCV formation (long arrow), and a CCV containing unidentified material and a parvovirus-like particle (short arrow). A puddle of negative stain lies next to the particle slightly obscuring the capsid edge.

The vacuolar proton ATPase (V-ATPase) is a multisubunit enzyme complex, and it is responsible for the acidification of membrane-bounded organelles like endosomes. V-ATPase transports H^+ over membranes against an electrochemical potential under ATP hydrolysis, and H^+ ions acidify endosomal environments. Baf A1 blocks the V-ATPase activity causing neutralization of the acidic environment of endosomes. Baf A1 is routinely used as a suppressor of V-ATPases, and it is used to test whether specific viral entry is reliant on endosomal acidification (Clague et al., 1994; Drose & Altendorf, 1997; Jin et al., 2005; Wassmer et al., 2005). It has been widely used as a probe to study acid-mediated viral trafficking using a variety of virus models. Entry pathway studies of mouse polyomavirus (Liebl et al., 2006), feline calicivirus (Stuart & Brown, 2006), bovine viral diarrhea virus (Lecot et al., 2005), AAV (Bartlett et al., 2000), MVM (Mani et al., 2006; Ros et al., 2002), mouse hepatitis virus type 2 (Pu & Zhang, 2008), poliovirus (Brandenburg et al., 2007), baculovirus (Long et al., 2006), hepatitis C virus (Blanchard et al., 2006), influenza virus (Guinea & Carrasco, 1995; Sieczkarski et al., 2003), HIV-1 (Fredericksen et al., 2002), and human rhinovirus 14 (Bayer et al., 1999) have analyzed the effects of Baf A1 on virus trafficking and examined the role of endosomal acidification in a route that leads to viral replication. For studies of parvovirus entry Baf A1 may be an inhibitor preferred over chlorpromazine because chlorpromazine is not only an inhibitor of clathrin lattice processing, but is also an inhibitor of phospholipase A_2. Because parvoviruses use their own version of PLA_2 for entry, confusion may arise regarding the point of inhibition.

4. Virus in the early endosome

In the process of acid mediated endocytosis acidification begins in the early endosome reaching a pH of 6.5 to 6.0. Proton transport continues during the transition to the late endosome which develops a pH of about 5.0 within this vesicle. The intravesicular environment of the lysosome is also characterized by low pH. Early endosomes are considered an initial sorting station where cargos for degradation are distinguished from those for recycling and this sorting process begins in clathrin coated vesicles, depends on microtubule motility, and appears to involve endocytosis adaptors (Lakadamyali et al., 2006). Acidification of endosomes is known to be essential for viruses which internalize within CCVs. Studies on parvovirus entry have reported that BPV, MVM, AAV, and CPV internalization require endosomal acidification, and the endosomal acidic environment may induce capsid conformational changes vital for viral release from endosomes to the cytoplasm (Basak & Compans, 1989; Douar et al., 2001; Dudleenamjil et al., 2010; Mani et al., 2006; Ros et al., 2002). Because the proton concentration increases during transition from early to late endosomes and high acidity is maintained in the lysosome, it is possible that acid-dependent uncoating of viruses, specifically pH-dependent capsid protein conformational shifts, may occur at various points within the early endosome, in the late endosome, or after exposure to the harsh environment in the lysosome complex.

The mechanisms of capsid endosomal processing in a low pH environment are poorly understood but are under investigation. The N termini of AAV VP1 and VP2 (VP1/VP2u), like the N terminus of MVM, CPV, and PPV VP1 (VP1u), contain motifs homologous to PLA_2 and nuclear localization signals. AAV mutants that are consistent with a nuclear localization signal-deficient phenotype, traffick through the early endosome as the recombinant AAV2 control does, but forms a more diffuse accumulation pattern in the perinuclear area consistent with an NLS defect (Johnson et al., 2010). In a recent study, AAV-8 was exposed to pHs ranging from 7.5 to 4.0 and crystal structures of empty particles and green fluorescent protein gene-packaged particles were analyzed (Nam et al., 2011). The capsid surface topologies of particles exposed to various pHs were similar except changes located close to the two-fold depression and significant amino acid side chain conformational changes were seen on the interior surface of the capsid under the three-fold axis. The three-fold change is consistent with low pH-associated release of genomic DNA from an ordered state on the interior surface of the capsid and likely indicate capsid transitions that ultimately result in genome uncoating. The surface alteration results in disruption of VP-VP interface interactions along with a decrease in buried surface area between VP monomers. This destabilization may lead to activation of the PLA_2 activity for endosomal escape and the NLS for nuclear targeting. As noted, it was observed by Nam et al. (2011) that at pH 4.0 specific interactions between the capsid and the packaged DNA genome are weakened but it is insufficient for genome uncoating. Rather, the intracapsid genome may be compacted. This process could be a step toward genome release or shift of the VP1u to externalization. These authors further point out that current thinking postulates that the five-fold pore of AAV is the externalization portal for the AAV VP1 and VP2 N-termini, the other parvovirus' VP1u, and the packaged DNA. But there is only a small difference in diameter at the top of the channel between pH 7.5 and pH 4.0. Therefore, in addition to acid pH other cellular factors such as proteolytic enzymes likely operate to facilitate capsid dynamic events to externalize these VP1u and VP1/VP2 N-terminal domains and ultimately genome release after trafficking from the early endosome.

A contrast exists between the deployment times of VP1u of some autonomous parvoviruses with that of Parvovirus B19. As noted above VP1u deployment occurs during trafficking through the acidified endosomal compartments. It was found that B19 attachment to human erythrocytes caused early accessibility and activity of PLA_2 without entry into a nucleated cell (Bönsch et al., 2008). Thus, VP1u is displayed as the virus remains attached to the surface of erythrocytes. The phospholipase does not cause lysis of the cells but does cause increased osmotic fragility. In an earlier study on VPu of B19 virus, it was reported that the VPu motif is internally oriented but becomes exposed in heat-treated particles and in particles exposed to low pH (Ros et al., 2006).

CPV capsids labeled with fluorescent markers were seen in Rab-5 positive endosomes within minutes of uptake (Harbison et al., 2009). Capsids were also seen in Rab7- and Rab11-positive endosomal compartments by 10-15 minutes after infection. Gradually the virus accumulated near the microtubule organizing center. The CA form of Rab5 induces large, ring-like vesicles in cells and a high proportion of the CPV capsids entered these vesicles and remained there for a period of time of one hour or more. Many of the particles remained attached to the vesicle wall probably in association with the receptors.

In Lepidoptera cells densonucleosis virus (DNV) particles are rapidly internalized and are found in CCVs. They then traffic slowly within early endosomes, then to late endosomes (Vendeville et al., 2009). An alternative route from the endosome is to the multivesicular body (MVB) which contains numerous intraluminal vesicles (ILVs). Fusion of the MVB with the plasma membrane releases small exosomes to the extracellular environment. Exosomes are microvesicles 40-100 nm in size that exhibit a cup-like morphology and appear to be released ILVs (Meckes & Raab-Traub, 2011). The size of an exosome would accomodate one or more parvovirus particles but this pathway would be a detour diverting virus from the entry process leading to nuclear penetration which is essential for virus replication. It is not clear at this time if parvovirus particles are extruded from the cell in exosomes. It would be feasible for cells outside of the viral host range to expel endocytosed particles in this manner, but such a resistance pathway for parvoviruses is yet to be demonstrated.

5. Transition to the late endosome

It is thought that four general biological alterations accompany the early-to-late endosomal transition: acidification of the endosomal lumen, formation of luminal vesicles, the switch of Rab GTPases, and microtubule-mediated transport between the organelles (Greber & Cosset, 2011). Some viruses escape to the cytoplasm under the effects of the acidic environment (pH 6.5 to 6.0) of the early endosome. Sensitivity to endosomal acidification raises a question regarding duration of the virus within the endosome. That is, whether the virus particles are directly released from the early endosomal compartment to the cytosol or are routed farther into the late endocytic compartment or even to the late endosome-lysosome complex. If virus does not escape from the early endosomal station, it would follow the transition pathway to the late endosome. In this transition, the early endosome becomes a transport intermediate recruiting ADP ribosylation factor-1 (Arf1)-dependent coatomer proteins (COPI, clathrin, and AP-1) and converts to the late endosome. Arf 1 or small GTPase's activities are catalyzed by Sec7-type GTP-exchange factors (GEFs). GEFs are primarily targets of BFA. Therefore, tubulation of maturation of early endosomes is delayed, and virus transition to the late endosome is blocked (Brandenburg et al., 2007; Nebenfuhr et al., 2002; Stuart & Brown, 2006). Studies reported that BFA also affects transport between Golgi and endoplasmic reticulum with the same mechanism through Arf1. BFA was chosen as an inhibitor of pH-dependent endocytosis to investigate early events of murine polyomavirus, SV40, AAV, DNV, MVM, PV, and FCV interaction with the host cell (Blanchard et al., 2006; Damm et al., 2005; Douar et al., 2001; Gilbert & Benjamin, 2004; Guinea & Carrasco, 1995; Mani et al., 2006; Ros et al., 2002; Stuart & Brown, 2006; Vendeville et al., 2009). That BFA blocks infection by these viruses is suggestive of late endosomal involvement in the entry process.

Rab proteins are small GTPases that regulate vesicular transport in endocytosis and exocytosis. They are considered master regulators of transport. Early endosomes are converted to late endosomes as a shift occurs in their linked Rab GTPases from the early endosome-associated Rab5 to the late endosome-associated Rab7 (Cabrera & Ungermann, 2010; Rink et al., 2005; Rodman & Wandinger-Ness, 2000). Rab conversion is the mechanism by which cargo moves from early to late endosomes. The Rab7 domain grows on the early endosome and converts the Rab5-positive endosome to a Rab7-positive endosome (Poteryaev et al., 2010). Participating in this process are the cofactors SAND-1 and Mon1

(Nordmann et al., 2010; Poteryaev et al., 2007). The Rab7 complex is activated by the Class C VPS/HOPS complex (vacuolar protein sorting/homotypic fusion and vacuole protein sorting) complex which is a GEF (guanine nucleotide exchange factor) for Rab7. It interacts with Rab5 and is required for the Rab5 to Rab7 conversion (Rink et al., 2005). Activated Rab5 is important for sequestering ligands into clathrin-coated pits and subsequent fusion of these vesicles with early endosomes. The actin cytoskeleton plays a prominent role in both the early stages of endocytosis and the late Rab5 function (Rodman & Wandinger-Ness, 2000) and actin facilitates fusion among late endosomes and between late endosomes and phagosomes (Kjeken et al., 2004). Materials destined for degradation are delivered to early endosomes then segregated for transport to late endosomes, then to lysosomes (Rodman &Wandinger-Ness, 2000). The transition of the Rab5 early endosome to the Rab 7 late endosome, mediated by the Class C VPS/HOPS complex, is facilitated by Syntaxin-7 which is localized to the late endosome and is required for late endosome and lysosome fusion (Kim et al., 2001). Moreover, the endosomal membrane protein Ema interacts with Class C VPS/HOPS to promote endosomal maturation (Kim et al., 2010).

ADP-ribosylation factor (Arf) in association with Sec7 (Arf GDP/Sec7) is a GEF used in membrane traffic at the Golgi. Arf GDP/Sec7 is phosphorylated to the GTP level which may be used for activation of Class C VPS/HOPS-GDP to Class C VPS/HOPS-GTP that is used in the Rab5 to Rab7 conversion. BFA is a drug with specificity for the Arf-GDP/Sec7 complex and by binding at the interface between Arf-GDP and the Sec7 domains acts as an uncompetitive inhibitor of Arf activation and freezes (stabilizes) the complex that cannot proceed to nucleotide dissociation (Cherfils & Melancon, 2005; Zeghouf et al., 2005). BFA disrupts maturation of the early endosome to the late endosome (Douar et al., 2001; Vieira et al., 2002), but it also interferes with the secretory pathway (Greber & Way, 2006). However, inhibition of parvovirus infection with BFA is not due to interference with the secretory pathway as no parvoviruses are known to exit the infected cell in that way. So, viral susceptibility to BFA in the parvovirus replication cycle is most likely due to blockage of the early-to-late endosome transition. Thus, using BFA inhibition of Arf GTPase employs a strategy for disrupting early to late endosome traffic (Vieira et al., 2002).

A possible mechanism for BFA inhibition of late endosome formation is shown in Fig. 2. Illustrated in the figure is the molecular interaction between the ArfGDP and Sec7 subunits of the ArfGDP/Sec7 GEF complex. This interaction stabilizes the molecular complex preventing its activity as a GEF. Altogether, biochemical and structural data using isolated Sec7 domains provide a consistent explanation for the action of BFA, that its only target is the Arf-GDP/Sec7 interface (Cherfils & Melancon, 2005). This, in turn, inhibits the pathway responsible for the Rab5 to Rab7 conversion which is necessary for endosome maturation. The bovine parvovirus entry pathway goes through clathrin-associated endocytosis (Dudleenamjil et al., 2010) and it may go through extended compartments in this pathway. The results of inhibitor studies using BFA in BPV entry were consistent with virus persistence within the endosome until transition to late endosome is complete (Dudleenamjil & Johnson, unpublished data).

After internalization, generally both virus particles and vesicles that carry virus particles are able to interact with cell cytoskeletal structures and utilize their activities to reach specific sites in the cytoplasm or nucleus for replication. Virus-associated trafficking routes may involve actin filaments together with myosin motors or microtubules with their dynein and kinesin motors. Globular (G-) actin is polymerized to filamentous (F-) actin during synthesis of actin fibers. Synthesis begins with a loose association of three to four G-actin monomers to an unstable oligomer, a process called nucleation. Filaments are then elongated by addition of G-actin monomers. F-actin is involved in both cell movement and in the movement of cell organelles. Endocytic vesicles move at the tips of actin tails and appear to be pushed through the cytosol (Merrifield et al., 1999) and late endosomes can nucleate F-actin whereas early endosomes cannot (Kjeken et al., 2004). LAT A is a natural toxin secreted by red sea sponges, for example *Latrunculia magnifica* (Coue' et al., 1987). Lat A binds to G-actin and prevents it from adding to a filament end during synthesis of F-actin (Yarmola et al., 2000). Growing evidence has suggested a tight interaction between the actin network and acid-mediated endocytosis at the level of the late endosome (Kjeken et al., 2004; Rodman & Wandinger-Ness, 2000). Thus, inhibition of virus movement through the endocytic compartment by LAT A would be evidence that viral transit through the late endosome is essential in the process of getting the virus to the cell nucleus. Cells treated with increasing noncytotoxic concentrations of the inhibitor LAT A reduced bovine parvovirus infectivity (Dudleenamjil & Johnson, unpublished observations). The reduction of virus infection by this drug is evidence that acid-mediated endocytosis is a functional route of BPV internalization into the host cell and utilizes actin filaments in the trafficking of ligands contained in the late endosome.

6. Lysosomal interaction

In the viral entry process followed by some non-parvoviruses, exposure of stable capsids to low pH may allow proteolysis of capsid protein to occur creating a metastable configuration displaying sequences for membrane penetration (Cotmore & Tattersall, 2007). CPV enters the cell in association with its receptor, TfR, and the virus along with the TfRs is transported to late endosomes-lysosomes before escape into the cytoplasm (Suikkanen, 2003). In the pathway followed by MVM, a low pH environment fosters proteololysis of the VP2 N-termini which results in enhanced stability at low pH (Cotmore & Tatersall, 2007). Thus, these modified particles may be required to return to a pH-neutral environment before they undergo the structural transition that exposes the PLA_2 activity required for membrane penetration. However, it would be expected that after late endosome-lysosome fusion this hazardous compartment would result in extensive viral polypeptide hydrolysis and DNA damage. Results indicate that CPV does not face this same requirement. CPV exposed to low pH *in vitro* develops PLA_2 activity which persists when returned to neutral pH (Suikkanen et al., 2003). It seems clear that CPV requires PLA_2 activity as PLA_2 inhibitors block viral replication (Suikkanen, 2003).

Cells infected with viable parvovirus, empty capsids, or entry-defective mutants accumulate virus in large, crescent-shaped, peri-nuclear vesicular clusters that are probably microtubule organizing centers. The vesicles appear to be late endosomes-lysosomes as the processing of early endosomes to these late structures utilizes microtubule transport. Nocodazole, a microtubule depolymerizing drug, inhibits CPV infection and leaves vesicles containing CPV near the cell surface (Vihinen-Ranta et al., 1998) and they fail to accumulate in the perinuclear crescents. Likewise, in the MVM model, microtubule polymerization moves virus to the perinuclear late endosome-lysosome complex and depolymerization of microtubules scatters virus toward the cell periphery. Upon repolymerization of the microtubules the virus returns to crescent complexes (Cotmore &Tattersall, 2007). Similarly, BPV infection is sensitive to nocodazole treatment.

Some viruses that infect the gastrointestinal tract, notably rotavirus, require proteolytic activation in order to promote virus entry and infection. The proteolytic enzymes in the gastrointestinal tract provide this service to the virus. Influenza A viruses require proteolytic cleavage of the hemagglutinin (H) molecule to separate HA1 and HA2 exposing the fusion peptide located on the N-terminal end of HA2 which is required for bridging the endosomal membrane. Also, the acidic pH of the endosome promotes conformational shift in the HA structure resulting in functional availability of the hydrophobic fusion domain for penetration into the endosomal membrane. Many parvoviruses target the enteric tract and/or the respiratory tract of their natural hosts. Although parvoviruses, many of which may be exposed to the proteolytic enzymes in the gastrointestinal tract or to respiratory proteolysis such as that mediated by tryptase clara, may not require proteolysis for cell entry, but one considers a possible role for proteolysis as well as low pH conformational shifts in capsid protein structure for enhancement of infection relating to the display and activity of the PLA_2 and NLS motifs. On the other hand, exposure to low pH may circumvent the infectious pathway. Acid pHs promote virus aggregation and crystal formation (see Fig. 3). The pH of the respiratory tract is acidic resulting from the CO_2 reaction with H_2O to form carbonic acid. Parvoviruses that infect the respiratory tract such as Parvovirus B19, HuBoV, BPV, AAV and others may aggregate in the respiratory tract. Viruses infecting the gastrointestinal tract such as PPV, MVM, and others may aggregate in the gastric environment. In infected cells, intranuclear newly assembled virions clearly form large crystalline arrays, then, upon escape from the infected cell these newly produced aggregates may not be easily dispersed even at physiological pH. Moreover, regarding entry, it is possible multiple virions may be engulfed in single CCVs and transport together within the endosomal system and upon acidification form aggregates. Because aggregates become too large to penetrate the nucleus as is the case for individual particles, the fate of the virions within the aggregate may not include nuclear entry but they may be digested in the lysosome complex. As demonstrated for PPV, viral aggregates enter by macropinocytosis. The outcome of this pathway may be virus destruction rather than nuclear entry. In this regard, most viral progeny in parvoviral infections may not end up as infectious units but may be cleared and destroyed in the lysosome. More work will be required to decipher the probability of multiple routes of cell entry, and multiple intra-cellular outcomes.

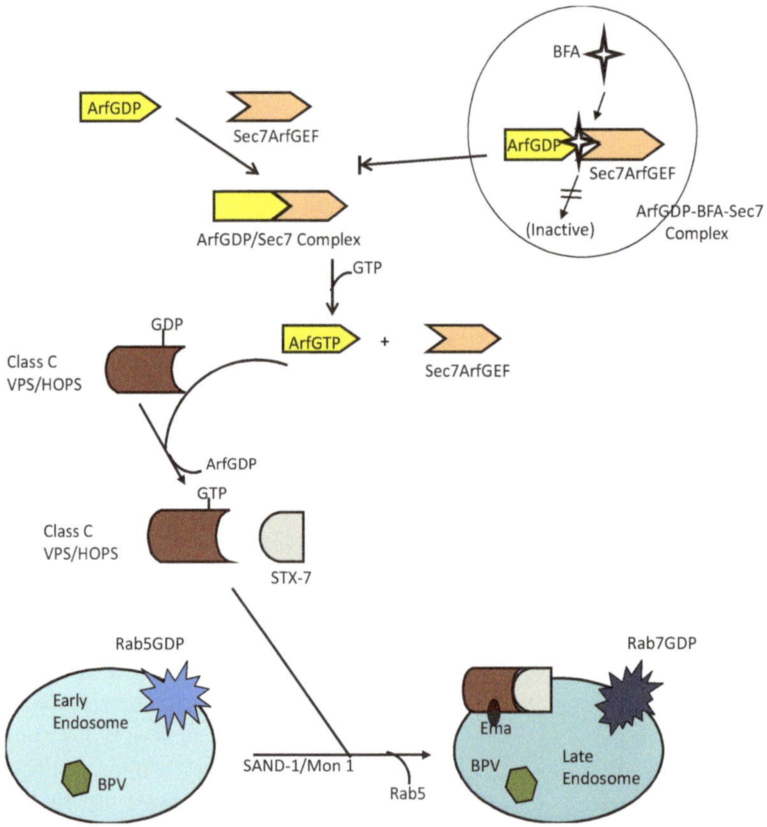

Figure 2. A possible mechanism of inhibition of the early to late endosome transition by brefeldin A (BFA). Rab5GDP is a marker on the early endosome. It is phosphorylated and gradually replaced by Rab7GDP which is a marker of the late endosome. The Mon1-Ccz 1 complex with SAND-1 control this process (Cabrera & Ungermann, 2010; Poteryaev et al., 2010). During this event, Class C VPS/HOPS is activated by phosphorylation, complexes with STX-7, and binds to the late endosome with Ema. The guanine exchange factor ArfGTP may activate the Class C VPS/HOPS complex but it can perform this function only through the activation of the ArfGDP/Sec7 complex. It is at the ArfGDP/Sec7 complex where BFA works. The drug binds to the ArfGDP-Sec7 interface and stabilizes the complex making it inactive as an ArfGDP/Sec complex preventing Arf activation (Cherfils & Melancon, 2005; Zeghouf et al., 2005), thus inhibiting the early-to-late endosome conversion. **Abbreviations:** Arf (ADP-ribosylation factor, a GTPase). ArfGDP/Sec 7 (a guanine nucleotide exchange factor, the target of BFA). BFA (brefeldin A). GEF (guanine nucleotide exchange factor, a GTPase). Class C VPS/HOPS (vacuolar protein sorting/homotypic fusion and vacuolar protein sorting, a GEF for Rab7). STX-7 (Syntaxin-7, a vsnare). Ema (an endosomal membrane protein).

7. Virus escape from endosomes

Paradoxically, parvoviruses must have a capsid shell rugged enough and stable enough to protect their single stranded DNA genomes from damage and degradation during transit from host to host and from portal of entry to ultimate target tissue, yet sensitive enough to navigate the cell-entry process leading to final uncoating within the proper viral replication compartment. Intracellular navigation, in the case of parvoviruses, requires subtle reorientations of capsid structural proteins exposing functional domains on the VP1u motif (Cotmore & Tattersall, 2007). Conceptually, a T=1 virus capsid would be constructed of 60 copies of a structural polypeptide. MVM empty capsids are constructed, on average, of 50 molecules of VP2 and 10 copies of VP1. In parvoviruses, the major (most abundant) structural polypeptide is the smallest structural protein (VP2 or VP3). Altogether, as noted above, the capsids contain either two or three related structural polypeptides that are coterminal at their carboxyterminals and have small unique sequences on their amino terminals. Unique to the larger versions of the polypeptides are elements on the VP1u region that are required for trafficking through the host's entry pathways (Cotmore & Tattersall, 2007). VP1 is dispensable for capsid assembly and genome packaging, but is absolutely required for infectivity. The VP1u sequence, which is within the N-terminal unique region, contains both the PLA_2 phospholipase domain and nuclear localization signals. The proteins are translated from a large mRNA and initiate either by leaky scanning or after differential splicing so that the upstream initiation codon is removed and leaves a downstream codon available for starting translation of the smaller capsid protein. One of these elements, PLA_2 which is a lipolytic enzyme is employed to breach the endosomal membrane allowing virus escape from the endocytic organelles, releasing it into the cytosol. An MVMp PLA_2 mutant deficient in this enzyme is unable to escape from its vacuolar confine and these particles accumulate in the endosomes. Once the virus is released, the virus targets the nucleus presumably by cellular factors that facilitate transport to the nuclear membrane, penetration across the nuclear membrane, capsid disassembly, exposure of the genome, and movement to an appropriate intranuclear replication compartment. Preliminary capsid conformational change, which may allow eventual endosomal escape, may occur in the early or late endosome but remains to be clarified for the various parvoviruses. It's likely that parvoviruses as a group utilize their phospholipase activity to escape the endosomal pathway. The parvovirus capsid is structurally dynamic undergoing multiple conformational changes during its replication cycle including the externalization of the VP1 N terminus during entry. In such a condition, VP1 remains tethered to the viral shell and appears to be active in particle escape from the endosome. Interestingly, an active particle may also operate in *trans* allowing for escape of VP1u-deficient particles if contained in the same endosome (Farr et al., 2005). It has also been shown that parvovirus B19, the human virus that causes erythema infectiosum (fifth disease) has a VP1-unique region that contains PLA_2 activity (Dorsch et al., 2002) and presumably operates in a manner similar to that described for MVM.

When studies use high multiplicities of virus and track the particles in infected cells, many of the particles enter dead-end pathways and never enter the nucleus. Moreover, in cases where ratios of numbers of physical particles to infectious particles have been studied, the ratio is quite high, 300:1 for MVMp and 1000:1 for CPV (Cotmore & Tattersall, 2007). Since only a small number of particles actually arrive at their nuclear destination, it makes the interpretation of vacuolar transport studies that use high viral multiplicities somewhat difficult. Many of these particles may be flagged for destruction and not infection. In studies on BPV entry, virus was labeled with fluorescein, cells exposed to high multiplicities of virus, then observed microscopically over time to assess virus transport through the cell. Labeled virus appeared in vacuolar structures but whether virus was visible in the nucleus was not conclusive (Dudleenamjil & Johnson, unpublished results).

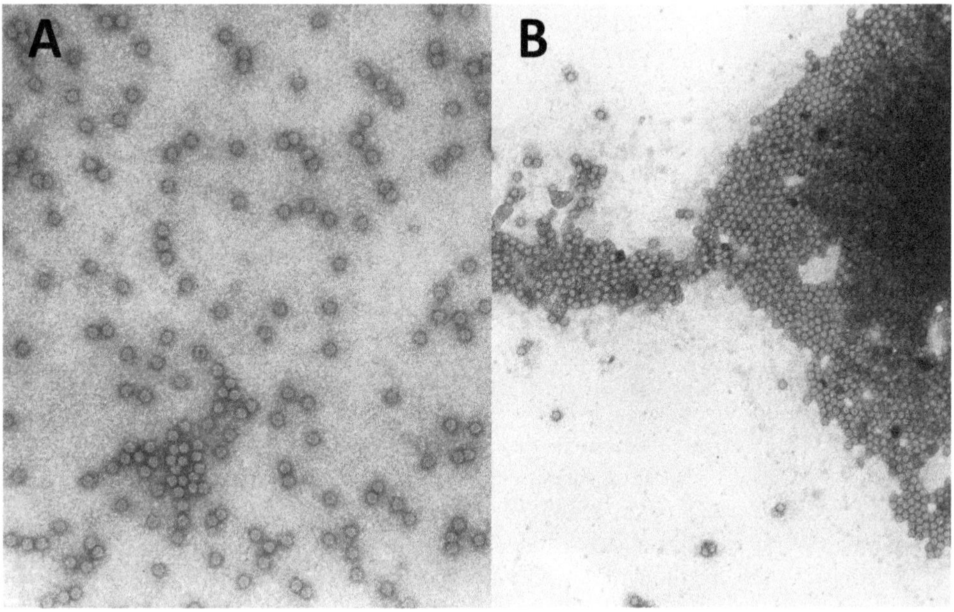

Figure 3. Aggregation of parvovirus particles at acid pH. A) Adeno-associated Virus Type-3 at pH 7.4 negatively stained with uranyl acetate. B) AAV-3 particles shifted to pH 6.1 also negatively stained with uranyl acetate, showing extensive particle aggregation and formation of crystalline array. Original micrograph magnification: 141,000x.

8. Transport to the nucleus and nuclear invasion

In the parvovirus life cycle, structural proteins must be transported at least twice from cytoplasm to the nucleus: following structural protein synthesis so that nuclear capsid assembly can be accommodated and also after virus entry during infection for invasion of the host cell nucleus. Both MVM and CPV have nuclear localization signals (NLS) on their capsid proteins (Cotmore & Tattersall, 2007; Lombardo et al., 2002). Because capsid-associated VP1 undergoes structural rearrangement during entry exposing functional signals, it is possible that the VP1 NLS motif participates in penetration of the invading capsid across the nuclear membrane. Thus, a life-cycle advantage that parvoviruses have over other nuclear-dependent viruses is their small size, small enough at 26 nm to be able to import into the nucleus as an intact (although modified) virion controlled by a bifunctional set of NLS motifs. In support of capsid transport across the nuclear membrane, CPV virions microinjected into cells were found to translocate into the nucleus intact and initiate gene expression (Suikkanen et al., 2003; Vihinen-Ranta et al., 2000). Entry of the microinjected virions circumvented the natural endocytic pathway and placed them in position to penetrate the nucleus, but it is unclear if the NLS motifs were active in the process. As an alternative mechanism of nuclear entry, MVM nuclear penetration in mouse fibroblast cells was tracked by fluorescence microscopy and electron microscopy (Cohen et al., 2006). It was found that this virus caused marked changes in nuclear shape, alterations of nuclear lamin immunostaining and breaks in the nuclear membrane. These changes may allow direct physical access to the nuclear interior for the virus.

Microtubule-associated activity in parvoviral trafficking is involved with dynein-dependent endosomal trafficking of CPV capsids before escape. Further, dynein dependency is seen in movement of CPV capsids after escape through the cytoplasm to the nucleus (Suikkanen, 2003). It was reported that AAV-2 trafficking to the nucleus utilizes the PI13 kinase activation cascade directing virus along microtubules and microfilaments (Sanlioglu et al., 2000). Surprisingly, one study found AAV's trafficking to the nucleus appears to be independent from the microtubule network (Hirosue et al., 2007). These authors found that overexpression of dynamitin which results in a functional inhibition of the minus-end-directed micotubule motor protein dynein did not inhibit transduction of rAAV2. Treatment of HeLa cells with nocadazole or vinblastine disrupted microtubules but did not significantly affect virus transduction. In contrast, high concentrations of taxol resulted in microtubule stabilization and high vinblastine concentrations caused formation of tubulin paracrystals both reducing rAAV2 transduction. These authors concluded that these results demonstrate that AAV2 can infect HeLa cells independent of dynein function or an intact microtubule network. In another report (Johnson & Samulski, 2009), a population of rAAV2 virions entered the nucleus and accumulated in the nucleolus after infection but empty capsids were excluded from nuclear entry. Interestingly, virions trafficked to the nucleolus were found to retain infectivity in secondary infections. Thus, in the case of AAV, mobilization from the nucleolar site to nucleoplasmic locations likely permits uncoating and gene expression. Also in this study proteosome inhibitors were found to potentiate nucleolar accumulation.

Proteosome activity is essential for PPV infection (Boisvert et al., 2010). In the presence of lactacystin and MG-132, two commonly used proteosome inhibitors, the virus remains in a more diffused state in the perinuclear area of the cell and low-level replication occurs. Supportive data showed that PPV capsid proteins were ubiquitinated early in infection. These observations suggest proteosomal interaction during the last stages of transport prior to nuclear entry. Interaction with the proteosome has also been demonstrated for MVM (Ros et al., 2002; Ros & Kempf, 2004). The exact role for ubiquitination and proteosomal interaction for these viruses remains to be elucidated.

Because the autonomously replicating parvoviruses require the S-phase nuclear environment for replication, after they penetrate the nucleus they remain in a genomic nonsynthetic state until S-phase occurs. The Dependoviruses do not have this constraint as the helper virus provides the synthetic environment, yet these viruses have a constraint of their own, the necessary co-infection with the helper virus. Once in the nucleus, AAV appears to undergo subnuclear mobilization accumulating in the nucleolus (Johnson et al., 2010), uncoating and genome replication. Although the purpose of this review is to only take the entering virus to the intranuclear environment, it is interesting to note that in CPV infection the nuclear replication compartment expands and is accompanied by chromatin marginalization to the vicinity of the nuclear membrane, virus capsids move by passive diffusion, intranuclear structure and dynamics are extensively affected enlarging the interchromosomal domain which contains viral proteins, genomes, and capsids (Ihalainen et al., 2009). After parvovirus assembly and maturation within the nucleus of infected cells, they egress by processes that include apoptosis (Poole et al., 2004; Poole et al., 2006) and cell necrosis (Abdel-Latif et al., 2006).

9. Conclusion

Not all parvoviruses have been examined in detail for their entry and trafficking pathways, but the model viruses that have been studied reveal some commonalities shared by the viruses in this family. Among these are:

1. Parvoviruses penetrate their host cell through receptor-mediated endocytosis, interacting with a host cell receptor, and entering in clathrin-coated vesicles.
2. The viruses are processed in the early to late endosome and possibly lysosome prior to nuclear entry.
3. In preparation for vacuolar escape, the low pH environment induces capsid protein conformational shifts which display the viral VP1 PLA_2 domain and nuclear location signals.
4. Viral escape from the endosome is mediated by PLA_2.
5. Microtubules and actin generally play a role in virus particle transport to the nuclear membrane where capsid penetration to the intranuclear environment occurs.
6. Intranuclear mobility occurs by random diffusion for some and for others (AAV) by tracking to the nucleolus.

Alternative pathways may exist for some of these viruses, as seen by macropinocytosis for PPV. There is yet much to be learned about the virus-cell interactions utilized by parvoviruses, mechanisms of bridging the nuclear membrane, intranuclear localization and the uncoating process.

Author details

F. Brent Johnson and Enkhmart Dudleenamjil
Brigham Young University, Provo, Utah, USA

10. References

Abdel-Latif, L., Murray, B. K., Renberg, R. L., O'Neill, K. L., Porter, H., Jensen, J. B., & Johnson, F. B. (2006). Cell death in parvovirus-infected embryonic bovine tracheal cells is mediated by necrosis rather than apoptosis. *J. Gen. Virol.*, Vol.87, No.9, (Sep 2006), pp. 2539-2548, ISSN 0022-1317

Abinanti, F. R., & Warfield, M. S. (1961). Recovery of a hemadsorbing virus (HADEN) from the gastrointestinal tract of calves. *Virology*, Vol.14, No.2, (Jun 1961), pp. 288-289, ISSN 0042-6822

Agbandje-McKenna, M., & Chapman, M. S. (2006). Correlating structure with function in the viral capsid, In: *The Parvoviruses*, Kerr, J. R., Cotmore, S. F., Bloom, M. E., Linden, R. M., & Parrish, C. R, pp. 125-140, Hodder Arnold, ISBN 9780340811986, London

Agbandje-McKenna, M., Llamas-Saiz, A. L., Wang, F., Tattersall, P., & Rossmann, M. G. (1998). Functional implications of the structure of the murine parvovirus, minute virus of mice. *Structure*, Vol.6, No.11, (Nov 1998), pp. 1369-1381, ISSN 0969-2126

Allander, T., Tammi, M.T., Eriksson, M., Bjerkner, A., Tiveljung-Lindell, A., & Andersson, B. (2005). Cloning of a human parvovirus by molecular screening of respiratory tract samples. *PNAS*, Vol.102, No.36, (Sep 2005), pp. 12891-12896, ISSN 1091-6490

Bartlett, J. S., Wilcher, R., & Samulski, R. J. (2000). Infectious entry pathway of adeno-associated virus and adeno-associated virus vectors. *J. Virol.*, Vol.74, No.6, (Mar 2000), pp. 2777-2785, ISSN 0022-538X

Basak, S., & Compans, R. W. (1989). Polarized entry of canine parvovirus in an epithelial cell line. *J. Virol.*, Vol.63, No.7, (Jul 1989), pp. 3165-3167, ISSN 0022-538X

Bayer, N., Prchla, E., Schwab, M., Blaas, D., & Fuchs, R. (1999). Human rhinovirus HRV14 uncoats from early endosomes in the presence of bafilomycin. *FEBS Letters*, Vol.463, No.1, (Dec 1999), pp. 175-178, ISSN 0014-5793

Beckerle, M. C., & Porter, K. R. (1982). Inhibitors of dynein activity block intracellular transport in erythrophores. *Nature*, Vol. 295, No. 5851, (Feb 1982), pp. 701-703, ISSN 0028-0836

Beer, C., Andersen, S. D., Rojek, A., & Pedersen, L., (2005). Caveola-dependent endocytic entry of amphotropic murine leukemia virus. *J. Virol.*, Vol. 79, No. 16, (Aug 2005), pp. 10776-10787, ISSN 0022-538X

Berns, K. I. (1990). Parvovirus replication. *Microbiol. Mol. Biol. Rev.*, Vol.54, No.4, (Dec 1990), pp. 316-329, 1092-2172, ISSN 1092-2172

Bi, Z., Formenty, P. B. H., & Roth, C. E. (2007). Human bocavirus, a real respiratory tract pathogen. *Afr. J. Microbiol. Res.*, Vol.1, No.5, (Oct 2007), pp. 051-056, ISSN 1996-0808

Blackburn, S. D., Cline, S. E., Hemming, J. P., & Johnson, F. B. (2005). Attachment of bovine parvovirus to O-linked alpha 2,3 neuraminic acid on glycophorin A. *Arch. Virol.*, Vol.150, No.7, (Jul 2005), pp. 1477-1484, ISSN 0304-8608

Blackburn, S. D., Steadman, R. A., & Johnson, F. B. (2006). Attachment of adeno-associated virus type 3H to fibroblast growth factor receptor 1. *Arch. Virol.*, Vol.151, No.3, (Mar 2006), pp. 617-623, ISSN 0304-8608

Blanchard, E., Belouzard, S., Goueslain, L., Wakita, T., Dubuisson, J., Wychowski, C., & Rouille, Y. (2006). Hepatitis C virus entry depends on clathrin-mediated endocytosis. *J. Virol.*, Vol.80, No.14, (Jul 2006), pp. 6964-6972, ISSN 0022-538X

Bönsch, C., Kemph, C., & Ros, C. (2008). Interaction of Parvovirus B19 human erythrocytes alters virus structure and cell membrane integrity. *J. Virol.*, Vol.82, No.23, (Dec 2008), pp. 11784-11791, ISSN 0022-538X

Boisvert, M., Fernandes, S., & Tijssen, P. (2010). Multiple pathways involved in porcine parvovirus cellular entry and trafficking toward the nucleus. *J. Virol.*, Vol.84, No.15, (Aug 2010), pp. 7782-7792, ISSN 0022-538X

Brandenburg, B., Lee, L. Y., Lakadamyali, M., Rust, M. J., Zhuang, X., & Hogle, M. J. (2007). Imaging poliovirus entry in live cells. *PLoS Biol.*, Vol.5, No.7, (Jul 2007), pp. 1543-1555, ISSN 1544-9173

Brandenburg, B., & Zhuang, X. (2007). Virus trafficking-learning from single-virus tracking. *Nat. Rev. Microbiol.*, Vol.5, No.3, (Mar 2007), pp. 197-208, ISSN 1740-1526

Brindley, M. A., & Maury, W. (2008). Equine infectious anemia virus entry occurs through clathrin-mediated endocytosis. *J. Virol.*, Vol.82, No.4, (Feb 2008), pp. 1628-1637, ISSN 0022-538X

Cabrera, M., & Ungermann, C. (2010). Guiding endosomal maturation. *Cell*, Vol.141, No.3, (Apr 2010), pp. 404-406, ISSN 0092-8674

Chapman, M. S., & Agbandje-McKenna, M. (2006). Atomic structure of viral particles, In: *The Parvoviruses*, Kerr, J. R., Cotmore, S. F., Bloom, M. E., Linden, R. M., & Parrish, C. R, pp. 107-124, Hodder Arnold, ISBN 9780340811986, London

Chapman, M. S., & Rossmann, M. G. (1993). Structure, sequence, and function correlations among parvoviruses. *Virology*, Vol.194, No.2, (Jun 1993), pp. 491-508, ISSN 0042-6822

Chazal, N., & Gerlier, D. (2003). Virus entry, assembly, budding, and membrane rafts. *Microbiol. Mol. Biol. Rev.*, Vol.67, No.2, (Jun 2003), pp. 226-237, ISSN 1092-2172

Chen, K. C., Shull, B. C., Moses, E. A., Lederman, M., Stout, E. R., & Bates, R. C. (1986). Complete nucleotide sequence and genome organization of bovine parvovirus. *J. Virol.*, Vol.60, No.3, (Dec 1986), pp. 1085-1097, ISSN 0022-538X

Cherfils, J., & Melancon, P. (2005). On the action of brefeldin A on Sec7-stimulated membrane-recruitment and GDP/GTP exchange of Arf proteins. *Biochem. Soc. Trans.*, Vol.33, No.Pt4, (Jun 2005), pp. 635-638, ISSN 0300-5127

Clague, M. J., Urbe, S., Aniento, F., & Gruenberg, J. (1994). Vacuolar ATPase activity is required for endosomal carrier vesicle formation. *J. Biol. Chem.*, Vol.269, No.11, (Jan 1994), pp. 21-24, ISSN 0021-9258

Cohen, S., Behzad, A. R., Carroll, J. B., & Pante, N. (2006). Parvoviral nuclear import: bypassing the host nuclear-transport machinery. *J. Gen. Virol.*, Vol.87, No.11, (Nov 2006), pp. 3209-3213, ISSN 0022-1317

Cotmore, S. F., & Tattersall, P. (2007). Parvoviral host range and cell entry mechanisms. *Adv. Virus. Res.*, Vol.70, (2007), pp. 183-232, ISSN 0065-3527

Coue', M., Brenner, S. L., Spector, I., & Korn, E. D. (1987). Inhibition of actin polymerization by latrunculin A. *FEBS Lett.* Vol.213, No.2, (Mar 1987), pp. 316-318

Damm, E., & Pelkmans, L. (2006). Systems biology of virus entry in mammalian cells. *Cell. Microbiol.*, Vol.8, No.8, (Aug 2006), pp. 1219-1227, ISSN 1462-5814

Damm, E. M., Pelkmans, L., Kartenbeck, J., Mezzacasa, A., Kurzchalia, T., & Herenius, A. (2005). Clathrin- and caveolin-1-independent endocytosis: entry of simian virus 40 into cells devoid of caveolae. *J. Cell. Biol.*, Vol.168, No.3, (Jan 2005), pp. 477-488, ISSN 0021-9525

Dawson, J.C., Legg, J. A., & Machesky, L. M. (2006). Bar domain proteins: a role in tabulation, scission, and actin assembly in clathrin-mediated endocytosis. *Trends Cell Biol.*, Vol.16, No.10, (Oct 2006), pp. 493-498, ISSN 0962-8924

DeTulleo, L., & Kirhchausen, T. (1998). The clathrin endocytic pathway in viral infection. *The EMBO J.*, Vol.17, No.16, (Aug 1998), pp. 4585-4593, ISSN 0261-4189

Dimitrov, D. S. (2004). Virus entry: molecular mechanisms and biomedical applications. *Nat. Rev. Microbiol.*, Vol.2, No.81, (Feb 2004), pp. 111-122, ISSN 1740-1526

DiPasquale, G., & Chiorini, J. A. (2003). PKA/PrKX activity is a modulator of AAV/adenovirus interaction. *EMBO J.*, Vol.22, No.7, (Apr 2003), pp. 1716-1724, ISSN 0261-4189

Dorsch, S., Liebisch, G., Kaufmann, B., von Landenberg, P., Hoffmann, J. H., Drobnik, W., & Modrow, S. (2002). The VP1 unique region of parvovirus B19 and its constituent phospholipase A2-like activity. *J. Virol.*, Vol.76, No.4, (Dec 2002), pp. 2014-2018, ISSN 0022-538X

Douar, A. M., Poulard, K., Stockholm, D., & Danos, O. (2001). Intracellular trafficking of adeno-associated virus vectors: routing to the late endosomal compartment and proteasome degradation. *J.Virol.*, Vol.75, No.4, (Feb 2001), pp. 1824-1833, ISSN 0022-538X

Doxsey, S. J., Brodsky, F. M., Blank, G. S., & Helenius, A. (1987). Inhibition of endocytosis by anti-clathrin antibodies. *Cell*, Vol.50, No.3, (Nov 1987), pp. 453-463, ISSN 0092-8674

Drose, S., & Altendorf, K. (1997). Bafilomycins and concanamycins as inhibitors of V-ATPases and P-ATPases. *J. Exp. Biol.*, Vol.200, No.1, (Jan 1997), pp. 1-8, ISSN 0022-0949

Dudleenamjil, E., Lin, C-Y., Dredge, D., Murray, B. K., Robison, R. A., Johnson, F. B. (2010). Bovine parvovirus uses clathrin-mediated endocytosis for entry. *J. Gen. Virol.*, Vol.91, No.12, (Dec 2010), pp. 3032-3042, ISSN 0022-1317

Eash, S., Querbes, W., & Atwood, W. J. (2004). Infection of vero cells by BK virus is dependent on caveolae. *J. Virol.*, Vol.78, No.21, (Nov 2004), pp. 11583-11590, ISSN 0022-538X

Edeling, M. A., Smith, C., & Owen, D. (2006). Life of a clathrin coat: insights from clathrin and AP structures. *Nat. Rev. Mol. Cell Biol.*, Vol.7, No.1, (Jan 2006), pp. 32-44, ISSN 1471-0072

Farr, G. A., Zhang, L-G., & Tattersall, P. (2005) Parvoviral virions deploy a capsid tethered lipolytic enzyme to breach the endosomal membrane during cell entry. *PNAS*, Vol.102, No.47, (Nov 2005), pp. 17148-17153, ISSN 0027-8424

Forest, T., Barnard, S., & Baines, J. D. (2005). Active intranuclear movement of herpesvirus capsids. *Nat. Cell Biol.*, Vol. 7, No. 4, (Apr 2005), pp. 429-431, ISSN 1465-7392

Fredericksen, B. L., Wei, B. L., Yao, J., Luo, T., & Garcia, V. J. (2002). Inhibition of endosomal/lysosomal degradation increases the infectivity of human immunodeficiency virus. *J. Virol.*, Vol.76, No.22, (Nov 2002), pp. 11440-11446, ISSN 0022-538X

Freistadt, M. S., & Eberle, K. E. (2006). Fluorescent poliovirus for flow cytometric cell surface binding studies. *J. Virol. Methods*, Vol.134, No.1-2, (Jun 2006), pp. 1-7, ISSN 0166-0934

Gagescu, R., Demaurex, N., Parton, R. G., Hunziker, W., Huber, L. A., & Gruenberg, J. (2000). The recycling endosome of madin-darby canine kidney cells is a mildly acidic compartment rich in raft components. *Mol. Biol. Cell*, Vol.11, No.8, (Aug 2000), pp. 2775-2791, ISSN 1059-1524

Gilbert, J., & Benjamin, T. (2004). Uptake pathway of polyomavirus via ganglioside GD1a. *J.Virol.*, Vol.78, No.22, (Nov 2004), pp. 12259-12267, ISSN 0022-538X

Greber, U. F., & Cosset, F. L. (2011). Too smart to fail-how viruses exploit the complexity of host cells during entry. *Curr. Opin.Virol.*, Vol.1, No.1, (Jul 2011), pp.3-5, ISSN 1879-6257

Greber, U. F., & Way, M. (2006). A superhighway to virus infection. *Cell*, Vol.124, No.4, (Nov 2006), pp. 741-754, ISSN 0092-8674

Guinea, R., & Carrasco, L. (1995). Requirement for vacuolar proton-ATPase activity during entry of influenza virus into cells. *J. Virol.*, Vol.69, No.4, (Apr 1995), pp. 2306-2312, ISSN 0022-538X

Gurda, B. L., Parent, K. N., Bladek, H., Sinkovits, R. S., DiMattia, M. A., Rence, C., Castro, A., McKenna, R., Olson, N., Brown, K., Baker, T. S., & Agbandje-McKenna, M. (2010). Human bocavirus capsid structure: Insights into the structural repertoire of the *parvoviridae*. *J. Virol.*, Vol.84, No.12, (Jun 2010), pp. 5880-5889, ISSN 0022-538X

Harbison, C. E., Chiorini, J. A., & Parrish, C. R. (2008). The parvovirus capsid odyssey: from the cell surface to the nucleus. *Trends Microbiol.*, Vol.16, No.5, (May 2008), pp. 208-214, ISSN 0966-842X

Harbison, C. E., Lyi, S. M., Weichert, W. S., & Parrish, C. R. (2009). Early steps in cell infection by Parvoviruses: Host-specific differences in cell receptor binding but similar endosomal trafficking. *J. Virol.*, Vol.83, No.20, (Oct 2009), pp. 10504-10514. ISSN 0022-538X

Heuser, J. E., & Anderson, R. G. W. (1989). Hypertonic media inhibit receptor-mediated endocytosis by blocking clathrin-coated pit formation. *J. Cell Biol.*, Vol.108, No.2, (Feb 1989), pp. 389-400, ISSN 0021-9525

Hirosue, S., Senn, K., Clement, N., Nonnenmacher, M., Gigout, L., Linden, R. M., & Weber, T. (2007). Effect of inhibition of dynein function and microtubule-altering drugs on AAV2 transduction. *Virology*, Vol.367, No.1, (Oct 2007), pp. 10-18, ISSN 0042-6822,

Hommelgaard, A. M., Roepstorff, K., Vilhardt, F., Torgersen, M. L., Sandvig, K., & Deurs, B. V. (2005). Caveolae: stable membrane domains with a potential for internalization. *Traffic*, Vol.6, No.9, (Sep 2005), pp. 720-724, ISSN 1600-0854

Huang, F., Khvorova, A., Marshall, W., & Sorkin, A. (2004). Analysis of clathrin-mediated endocytosis of epidermal growth factor receptor by RNA interference. *J. Biol. Chem.*, Vol.279, No.16, (Apr 2004), pp. 16657-16661, ISSN 0021-9258

Hueffer, K., Palermo, L. M., & Parrish, C. R. (2004). Parvovirus infection of cells by using variants of the feline transferrin receptor altering clathrin-mediated endocytosis, membrane domain localization, and capsid-binding domains. *J. Virol.*, Vol.78, No.11, (Jun 2004), pp. 5601-5611, ISSN 0022-538X

Ihalainen, T. O., Niskanen, E. A., Jylhävä, J., Paloheimo, O., Dross, N., Smolander, H., Langowski, J., Timonen, J., & Vihinen-Ranta, M. (2009). Parvovirus induced alterations in nuclear architecture and dynamics. *pLoS One*, Vol.4, No.6, (Jun 2009), pp. 1-16, ISSN 1932-6203

Jin, M., Park, J., Lee, S., Park, B., Shin, J., Song, K. J., Ahn, T. I., Hwang, S. Y., Ahn, B. Y., & Ahn, K. (2002). Hantaan virus enters cells by clathrin-dependent receptor-mediated endocytosis. *Virology*, Vol. 294, No. 1, (Mar 2002), pp. 60-69, ISSN 0042-6822

Jin, S., Zhang, B., Weisz, O. A., & Monteralo, R. C. (2005). Receptor-mediated entry by equine infectious anemia virus utilizes a pH-dependent endocytic pathway. *J. Virol.*, Vol.79, No.23, (Dec 2005), pp. 14489-14497, ISSN 0022-538X

Johnson, F. B., Fenn, L. B., Owens, T. J., Faucheux, L. J., & Blackburn, S. D. (2004). Attachment of bovine parvovirus to sialic acids on bovine cell membranes. *J. Gen. Virol.*, Vol.85, No.8, (Aug 2004), pp. 2199-2207, ISSN 0022-1317

Johnson, F. B., & Hoggan, M. D. (1973). Structural proteins of HADEN virus. *Virology*, Vol.51, No.1, (Jan 1973), pp. 129-137, ISSN 0042-6822

Johnson, J. S., Li, C., DiPrimio, N., Weinberg, M. S., McCown, T. J., Samulski, R. J. (2010). Mutagenesis of adeno-associated virus type2 capsid protein VP1 uncovers new roles for basic amino acids in trafficking and cell-specific transduction. *J. Virol.*, Vol.84, No.17, (Sep 2010), pp. 8888-902, ISSN 0022-538X

Johnson, J. S., & Samulski, R. J. (2009). Enhancement of Adeno-associated virus infection by mobilizing capsids into and out of the nucleolus. *J. Virol.*, Vol.83, No.6, (Mar 2009), pp. 2632-2644, ISSN 0022-538X

Kaludov, N., Brown, K. E., Walters, R. W., Zabner, J., & Chiorini, J. A. (2001). Adeno-associated virus serotype 4 (AAV4) and AAV5 both require sialic acid binding for hemagglutination and efficient transduction but differ in sialic acid linkage specificity. *J. Virol.*, Vol.75, No.15, (Aug 2001), pp. 6884-6893, ISSN 0022-538X

Kaplan, N. M., Dove, W., AbuZeid, A. F., Shamoon, H. E., Abd-Eldayem, S. A., & Hart, A. C. (2006). Human bocavirus infection among children, Jordan. *Emerg. Infect. Dis.*, Vol.12, No.9, (Sep 2006), pp. 1418-1420, ISSN 1080-6040

Kashiwakura, Y., Tamayose, K., Iwabuchi, K., Hirai, Y., Shimada, T., Matsumoto, K., Nakamura, T., Watanabe, M., Oshimi, K., & Daida, H. (2005). Hepatocyte growth factor receptor is a coreceptor for adeno-associated virus type 2 infection. *J. Virol.*, Vol.79, No.1, (Jan 2005), pp. 609-614, ISSN 0022-538X

Kee, S., Cho, E., Song, J., Park, K. S., Baek, L., & Song, K. (2004). Effects of endocytosis inhibitory drugs on rubella virus entry into veroE6 cells. *Microbiol. Immunol.*, Vol.48, No.11, (Dec 2004), pp. 823-829, ISSN 0385-5600

Kern, A., Schmidt, K., Leder, C., Müller, O. J., Wobus, C. E., Bettinger, K., Von der Lieth, C. W., King, J. A., & Kleinschmidt, J. A. (2003). Identification of a heparin-binding motif on adeno-associated virus type 2 capsids. *J. Virol.*, Vol.77, No.20, (Oct 2003), pp. 11072-11081, ISSN 0022-538X

Kim, B. Y., Kramer, H., Yamamoto, A., Kominami, E., Kohsaka, S., & Akazawa, C. (2001). Molecular characterization of mammalian homologues of class C Vps proteins that interact with Syntaxin-7. *J. Biol. Chem.*, Vol.276, No.31, (Aug 2001), pp. 29393-29402, ISSN 0021-9258

Kim, S., Waikar, Y. P., Daniels, R. W., & DiAntonio, A. (2010). The novel endosomal membrane protein Ema interacts with the class C Vps-HOPS complex to promote endosomal maturation. *J. Cell Biol.*, Vol.188, No.5, (Mar 2010), pp. 717-734, ISSN 0021-9525

Kjeken, R., Edeberg, M., Habermann, A., Kuehnel, M., Peyron, P., Floetenmeyer, M., Walther, P., Jahrus, A., Defacque, H., Kuznetsov, S. A., & Griffiths, G. (2004). Fusion between phagosomes, early and late endosomes: a role for actin in fusion between late, but not early endocytic organelles. *Mol. Biol. Cell*, Vol.15, No.1, (Jan 2004), pp. 345-358, ISSN 1059-1524

Krietensson, K., Lycre, E., Roytta, M., Svennerholm, B., & Vahlne, A. (1986). Neuritic transport of herpes simplex virus in rat sensory neurons in vitro. Effects of substances interacting with microtubular function and axonal flow [nocodazole, taxol, and erythro-9-3(2-hydroxynonyl)adenine]. *J. Gen. Virol.*, Vol. 67, No. 9, (Sep 1986), pp. 2023-2028, ISSN 0022-1317

Lakadamyali, M., Rust, M. J., & Zhuang, X. (2006). Ligands for clathrin-mediated endocytosis are differentially sorted into distinct populations of early endosomes. *Cell*, Vol.124, No.5, (Mar 2006), pp. 997-1009, ISSN 0092-8674

Lecot, S., Belouzard, S., Dubuisson, J., & Rouille, Y. (2005). Bovine viral diarrhea virus entry is dependent on clathrin-mediated endocytosis. *J. Virol*, Vol.79, No.16, (Aug 2005), pp. 10826-10829, ISSN 0022-538X

Lemmon, S. K. (2001). Clathrin uncoating: Auxilin comes to life. *Curr. Biol.*, Vol.11, No.2, (Jan 2001), pp. R49-R52, ISSN 0960-9822

Liebl, D., Difato, F., Hornikova, L., Mannova, P., Strokrova, J., & Forstova, J. (2006). Mouse polyomavirus enters early endosomes, requires their acidic pH for productive infection, and meets transferrin cargo in Rab11-positive endosomes. *J. Virol.*, Vol.80, No.9, (May 2006), pp. 4610-4622, ISSN 0022-538X

Lombardo, E., Ramirez, J. C., Garcia, J., & Almendral, J. M. (2002). Complementary roles of multiple nuclear targeting signals in the capsid proteins of the parvovirus minute virus of mice during assembly and onset infection. *J. Virol.*, Vol.76, No.14, (Jul 2002), pp. 7049-7059, ISSN 0022-538X

Long, G., Pan, X., Kormelink, R., & Vlak, J. M. (2006). Functional entry of baculovirus into insect and mammalian cells is dependent on clathrin-mediated endocytosis. *J. Virol.*, Vol.80, No.17, (Sep 2006), pp. 8830-8833, ISSN 0022-538X

Ma, X., Endo, R., Ishiguro, N., Ebihara, T., Ishiko, A. T., & Kikuta, H. (2006). Detection of human Bocavirus in Japanese children with lower respiratory tract infections. *J. Clin. Microbiol.*, Vol.44, No.3, (Mar 2006), pp. 1132-1134, ISSN 0095-1137

Mani, B., Baltzer, C., Valle, N., Almendral, J. A., Kempf, C., & Ros, C. (2006). Low pH-dependent endosomal processing of the incoming parvovirus minute virus of mice virion leads to externalization of the VP1 N-terminal sequence (N-VP1), N-VP2 cleavage, and uncoating of the full-length genome. *J. Virol.*, Vol.80, No.2, (Jan 2006), pp. 1015-1024, ISSN 0022-538X

Manteufel, J., & Truyen, U. (2008). Animal bocaviruses: A Brief review. *Intervirology*, Vol.51, No.5, (Nov 2008), pp. 328-334, ISSN 0300-5526

Marsh, M., & Helenius, A. (2006). Virus entry: Open Sesame. *Cell*, Vol.124, No.4, (Feb 2006), pp. 729-740, ISSN 0092-8674

Meckes, D. G, Jr., & Raab-Traub, N. (2011). Minireview. Microvesicles and viral infection. *J. Virol.*, Vol.85, No.24 , (Dec 2011), pp. 12844-12854, ISSN 0022-538X

Meier, O., Boucke, K., Hammer, S. V., Keller, S., Stidwill, P. R., Hemmi, S., & Greber, U. F. (2002). Adenovirus triggers macropynocytosis and endosomal leakage together with its clathrin-mediated uptake. *J. Cell Biol.*, Vol.158, No.6, (Sep 2002), pp. 1119-1131, ISSN 0021-9525

Meier, O., & Greber, U. F. (2004). Adenovirus endocytosis. *J. Gene Med*, Vol.6, No.S1, (Feb 2004), pp. S152-S163, ISSN 1099-498X

Mercer, J., & Helenius, A. (2009). Virus entry by macropynocytosis. *Nat. Cell Biol.*, Vol.11, No.5, (May 2009), pp. 510-520, ISSN 1465-7392

Merrifield, C. J., Moss, S. E., Ballestrem, C., Imhof, B. A., Giese, G., Wunderlich, I., & Almers, W. (1999). Endocytic vesicles move at the tips of actin tails in cultured mast cells. *Nat. Cell Biol.*, Vol.1, No.1, (May 1999), pp. 72-74, ISSN 1465-7392

Merrifield, C. J., Perrais, D., & Zenisek, D. (2005). Coupling between clathrin-coated pit invagination, contractin recruitment, and membrane scission observed in live cells. *Cell*, Vol.121, No.5, (May 2005), pp. 593-606, ISSN 0092-8674

Nam, H. J., Gurda, B. L., Mckenna, R., Potter, M., Byrne, B., Salganik, M., Muzyczka, N., & Agbandje-McKenna, M. (2011). Structural studies of Adeno-associated virus serotype 8 capsid transitions associated with endosomal trafficking. *J. Virol.*, Vol.85, No.22, (Nov 2011), pp. 11791-11799, ISSN 0022-538X

Nebenfuhr, A., Ritsenthaler, C., & Robinson, D. G. (2002). Brefeldin A: deciphering an enigmatic inhibitor of secretion. *Plant Physiol.*, Vol.130, No.3, (Nov 2002), pp. 1102-1108, ISSN 0032-0889

Nordmann, M., Cabrera, M., Perz, A., Brocker, C., Ostrowicz, C., Engelbrecht-Vandre, S., & Undermann, C. (2010). The Mon1-Ccz1 complex is the GEF of the late endosomal

Rab7 homolog Ypt. *Curr. Biol.*, Vol.20, No.18, (Aug 2010), pp. 1654-1659, ISSN 0960-9822

Op De Beeck, A., & Caillet-Fauquet, P. (1997). Viruses and the cell cycle. *Prog. Cell Cycle Res.*, Vol.3, No.1, (1997), pp. 1-19, ISSN 1087-2957

Palermo, L. M., Hafenstein, S. L., & Parrish, C. R. (2006). Purified feline and canine transferrin receptors reveal complex interactions with the capsid of Canine and Feline Parvoviruses that correspond to their host ranges. *J. Virol.*, Vol.80, No.17, (Sep 2006), pp. 8482-8492, ISSN 0022-538X

Parker, J. S. L., Murphy, W. J., Wang, D., O'Brien, S. J., & Parrish, C. R. (2001). Canine and feline parvoviruses can use human or feline transferrin receptors to bind, enter, and infect cells. *J. Virol.*, Vol.75, No.8, (Apr 2001), pp. 3896-3902, ISSN 0022-538X

Parker, J. S., & Parrish, C. R. (2000). Cellular uptake and infection by canine parvovirus involves rapid dynamin-regulated clathrin-mediated endocytosis, followed by slower intracellular trafficking. *J. Virol.*, Vol.74, No.4, (Feb 2000), pp. 1919-1930, ISSN 0022-538X

Pelkmans, L. (2005a). Secrets of caveolae- and lipid raft-mediated endocytosis revealed by mammalian viruses. *Biochim. Biophys. Acta*, Vol.1746, No.3, (Dec 2005), pp. 295-304, ISSN 0006-3002

Pelkmans, L. (2005b). Viruses as probes for systems analysis of cellular signaling, cytoskeleton reorganization, and endocytosis. *Curr. Opin. Microbiol.*, Vol.8, No.3, (Jun 2005), pp. 331-337, ISSN 1369-5274

Pelkmans, L., & Helenius, A. (2003). Insider information: what viruses tell us about endocytosis. *Curr. Opin. Cell Biol.*, Vol.15, No.4, (Aug 2003), pp. 414-422, ISSN 0955-0674

Pelkmans, L., Burli, T., Zerial, M., & Helenius, A. (2004). Caveolin-stabilized membrane domains as multifunctional transport and sorting devices in endocytic membrane traffic. *Cell*, Vol.118, No.6, (Sep 2004), pp. 767-780, ISSN 0092-8674

Pelkmans, L., Fava, E., Grabner, H., Hannus, M., Habermann, B., Krausz, E., & Zerial, M. (2005). Genome-wide analysis of human kinases in clathrin- and caveolae/raft-mediated endocytosis. *Nature*, Vol.436, No.7, (Jul 2005), pp.78-86, ISSN 0028-0836

Poole, B. D., Karetnyi, Y. V., & Naides, S. J. (2004). Parvovirus B19-induced apoptosis of hepatocytes. *J. Virol.*, Vol.78, No.14, (Jul 2004), pp. 7775-7783, ISSN 0022-538X

Poole, B. D., Zhou, J., Grote, A., Shiffenbauer, A., & Naides, S. J. (2006). Apoptosis of liver-derived cells induced by parvovirus B19 nonstructural protein. *J. Virol.*, Vol.80, No.8, (Apr 2006), pp. 4114-4121, ISSN 0022-538X

Poteryaev, D., Datta, D., Ackema, K., Zerial, M., & Sprang, A. (2010). Identification of the switch in early-to-late endosome transition. *Cell*, Vol.141, No.3, (Apr 2010), pp. 497-508, ISSN 0092-8674

Poteryaev, D., Fares, H., Bowerman, B., & Sprang, A. (2007). Caenorhabditis elegans SAND-1 is essential for RAB-7 functionis endosomal traffic. *EMBO J.*, Vol.26, No.2, (Jan 2007), pp. 301-312, ISSN 0261-4189

Pu, Y., & Zhang, X. (2008). Mouse Hepatitis Virus Type 2 enters cells through a clathrin-mediated endocytosis pathway independent of Eps15Δ. *J. Virol.*, Vol.82, No.16, (Aug 2008), pp. 8112-8123, ISSN 0022-538X

Qing, K., Mah, C., Hansen, J., Zhou, S., Dwarki, V., & Srivastava, A. (1999). Human fibroblast growth factor receptor 1 is a coreceptor for infection by adeno-associated virus 2. *Nat. Med.*, Vol.5, No.1, (Jan 1999), pp. 71-78

Qiu, J., Cheng, F., Johnson, B. F., & Pintel, D. (2007). The transcription profile of the Bocavirus bovine parvovirus is unlike those of previously characterized parvoviruses. *J. Virol.*, Vol.81, No.21, (Nov 2007), pp. 12080-12085, ISSN 0022-538X

Qiu, J., Handa, A., Kirby, M., & Brown, K. E. (2000). The interaction of heparin sulfate and adeno-associated virus 2. *Virology*, Vol.269, No.1, (Mar 2000), pp. 137-147, ISSN 0042-6822

Ramanathan, H., & Jonsson, C. B. (2008). New and old world hantaviruses differentially utilize host cytoskeletal components during their life cycles. *Virology*, Vol.374, No.1, (Apr 2008), pp. 138-150, ISSN 0042-6822

Rink, J., Ghigo, E., Kalaidzidis, Y., & Zerial, M. (2005). Rab conversation as a mechanism of progression from early to late endosomes. *Cell*, Vol.122, No.5, (Sep 2005), pp. 735-749, ISSN 0092-8674

Rodman, J. S., & Wandinger-Ness, A. (2000). Rab GTPases coordinate endocytosis. *J. Cell Sci.*, Vol.113, No.Pt2, (Jan 2000), pp. 183-192, ISSN 0021-9533

Ros, C., Burckhardt, C. J., & Kempf, C. (2002). Cytoplasmic trafficking of minute virus of mice: Low-pH requirement, routing to late endosomes and proteasome interaction. *J. Virol*, Vol.76, No.24, (Dec 2002), pp. 12634-12645, ISSN 0022-538X

Ros, C., Gerber, M., & Kempf, C. (2006). Conformational changes in the VP1-unique region of native human parvovirus B19 lead to exposure of internal sequences that play a role in virus internalization and infectivity. *J. Virol.*, Vol.80, No.24, (Dec 2006), pp. 12017-12024, ISSN 0022-538X

Ros, C., & Kempf, C. (2004). The ubiquitin-proteasome machinery is essential for nuclear translocation of incoming minute virus of mice. *Virology*, Vol.324, No.2, (Jul 2004), pp. 350-360, ISSN 0042-6822

Russell, M. R. G., Nickerson, D. P., & Odorozzi, G. (2006). Molecular mechanisms of late endosome morphology, identity, and sorting. *Curr. Opin. Cell. Biol.*, Vol.18, No.4, (Aug 2006), pp. 422-428, ISSN 0955-0674

Sandals, W. C. D., Rovey, R. C., & Meek, A. H. (1995). Prevalence of bovine parvovirus in Ontario dairy cattle. *Can. J. Vet. Res.*, Vol.59, No.2, (Apr 1995), pp. 81-86, ISSN 0830-9000

Sanlioglu, S., Benson, P. K., Yang, J., Atkinson, E. M., Reynolds, T., & Engelhardt, J. F. (2000). Endocytosis and nuclear trafficking of Adeno-associated virus type 2 are controlled by Rac1 and phosphatidylinositol-3 kinase activation. *J. Virol.*, Vol.74, No.19, (Oct 2000), pp. 9184-9196, ISSN 0022-538X

Shull, B. C., Chen, K. C., Lederman, M., Stout, E. R., & Bates, R. C. (1988). Genomic clones of bovine parvovirus: construction and effect of deletions and terminal sequence inversions on infectivity. *J. Virol.*, Vol.62, No.2, (Feb 1988), pp. 417-426, ISSN 0022-538X

Sieczkarski, S. B., Brown, H. A., & Whittaker, G. R. (2003). Role of protein kinase C ßII in influenza virus entry via late endosomes. *J. Virol.*, Vol.77, No.1, (Jan 2003), pp. 460-469, ISSN 0022-538X

Sieczkarski, S. B., & Whittaker, G. R. (2002). Dissecting virus entry via endocytosis. *J. Gen. Virol.*, Vol.83, No.7, (Jul 2002), pp. 1535-1545, ISSN 0022-1317

Simpson, A. A., Chandrasekar, V., Hėbert, B., Sullivan, G. M., Rossmann, M. G., & Parrish, C. R. (2000). Host range and variability of calcium binding by surface loops in the capsids of Canine and Feline Parvoviruses. *J. Mol. Biol.*, Vol.300, No.3, (Jul 2000), pp. 597-610, ISSN 0022-2836

Smart, E. J., Foster, D. C., Ying, Y. S., Kamen, B. A., & Anderson, R. G. W. (1994). Protein kinase C activators inhibit receptor-mediated potocytosis by preventing internalization of caveolae. *J. Cell Biol.*, Vol. 124, No. 3, (Feb 1994), pp. 307-313, ISSN 0021-9525

Smith, A. E., & Helenius, A. (2004). How viruses enter animal cells. *Sci.*, Vol.304, No.9, (Apr 2004), pp. 237-241, ISSN 0036-8075

Sonntag, F., Bleker, S., Leuchs, B., Fischer, R., & Kleinschmidt, J. A. (2006). Adeno-associated virus type 2 capsids with externalized VP1/VP2 trafficking domains are generated prior to passage through the cytoplasm and are maintained until uncoating occurs in the nucleus. *J. Virol.*, Vol.80, No.22, (Nov 2006), pp. 11040-11054, ISSN 0022-538X

Stidwill, R. P., & Greber, U. F. (2000). Intracellular virus trafficking reveals physiological characteristics of the cytoskeleton. *News Physiol. Sci*, Vol.15, No.2, (Apr 2000), pp. 67-71, ISSN 0886-1714

Stuart, A. D., & Brown, D. K. T. (2006). Entry of feline calicivirus is dependent on clathrin-mediated endocytosis and acidification in endosomes. *J. Virol.*, Vol.80, No.15, (Aug 2006), pp. 7500-7509, ISSN 0022-538X

Suikkanen, S. (2003). Cell biology of canine parvovirus entry, Jan 11, 2012, Available from:< https://jyx.jyu.fi/dspace/bitstream/handle/123456789/13139/9513917851.pdf?sequence=1>

Suikkanen, S., Antila, M., Jaatinen, A., Vihinen-Ranta, M., & Vuento, M. (2003). Release of canine parvovirus from endocytic vesicles. *Virology*, Vol.316, No.2, (Nov 2003), pp. 267-280, ISSN 0042-6822

Summerford, C., Bartlett, J. S., & Samulski, R. J. (1999). AlphaVbeta5 intergin: a co-receptor for adeno-associated virus type 2 infection. *Nat. Med.*, Vol.5, No.1, (Jan 1999), pp. 78-82, ISSN 1078-8956

Summerford, C., & Samulski, R. J. (1998). Membrane-associated heparin sulfate proteoglycan is a receptor for adeno-associated virus type 2 virions. *J. Virol.*, Vol.72, No.2, (Feb 1998), pp. 1438-1445, ISSN 0022-538X

Sun, T. X., Hoek, A. V., Huang, Y., Bouley, R., McLaughlin, M., & Brown, D. (2002). Aquaporin-2 localization in clathrin-coated pits: inhibition of endocytosis by dominant-negative dynamin. *Am. J. Physiol. Renal Physiol.*, Vol.282, No.6, (Jun 2002), pp. F998-F1011, ISSN 1931-857X

Sun, Y., Chen, A. Y., Cheng, F., Guan, W., Johnson, F. B., & Qiu, J. (2009). Molecular characterization of infectious clones of the minute virus canines reveals unique features of bocaviruses. *J. Virol.,* Vol.83, No.8, (Apr 2009), pp. 3956-3967, ISSN 0022-538X

Thacker, T. C., & Johnson, F. B. (1998). Binding of bovine parvovirus to erythrocyte membrane sialylglycoproteins. *J. Gen. Virol.,* Vol.79, No.9, (Sep 1998), pp. 2163-2169, ISSN 0022-1317

Tomita, M., Furthmayr, H., & Marchesi, V. T. (1978). Primary structure of human erythrocyte glycophorin A. Isolation and characterization of peptides and complete amino acid sequence. *Biochemistry,* Vol.17, No.22, (Oct 1978), pp. 4756-4770

Tsao, J., Chapman, M. S., Agbandje, M., Keller, W., Smith, K., Wu, H., Luo, M., Smith, T. J., Rossmann, M. G., Compans, R. W., & Parrish, C. R. (1991). The three-dimensional structure of canine parvovirus and its functional implications. *Sci,* Vol.251, No.5000, (Mar 1991), pp. 1456-1464, ISSN 0036-8075

Van der Goot, F. G., & Gruenberg, J. (2006). Intra-endosomal membrane traffic. *Trends Cell Biol.,* Vol.16, No.10, (Oct 2006), pp. 514-521, ISSN 0962-8924

Vendeville, A., Ravallec, M., Jousset, F. X., Devise, M., Mutuel, D., Lopez-Ferber, M., Fournier, P., Dupressoir, T., & Ogliastro, M. (2009). Densovirus infectious pathway requires clathrin-mediated endocytosis followed by trafficking to the nucleus. *J. Virol.,* Vol.83, No.9, (May 2009), pp. 4678-4689, ISSN 0022-538X

Viera, O. V., Botelho, R. J., & Grinstein, S. (2002). Phagosome maturation: aging gracefully. *Biochem. J.,* Vol.366, No.Pt3, (Sep 2002), pp. 689-704, ISSN 02646021

Vihinen-Ranta, M., Kalela, A., Makinen, P., Kakkola, L., Marjomaki, V., & Vuento, M. (1998). Intracellular route of canine parvovirus entry. *J. Virol.,* Vol.72, No.1, (Jan 1998), pp. 802-806, ISSN 0022-538X

Vihinen-Ranta, M., Wang, D., Weichert, W. S., & Parrish, C. R. (2002). The VP1 N-terminal sequence of canine parvovirus affects nuclear transport of capsids and efficient cell infection. *J. Virol.,* Vol.76, No.4, (Feb 2002), pp. 1884-1891, ISSN 0022-538X

Vihinen-Ranta, M., Yuan, W., & Parrish, C. R. (2000). Cytoplasmic trafficking of the Canine Parvovirus capsid and its role in infection and nuclear transport. *J. Virol.,* Vol.74, No.10, (May 2000), pp. 4853-4859, ISSN 0022-538X

Walters, R. W., Yi, S. M. P., Keshavjee, S., Brown, K. E., Welsh, M. J., Chiorini, J. A., & Zabner, J. (2001). Binding of adeno-associated virus type 5 to 2,3-linked sialic acid is required for gene transfer. *J. Biol. Chem.,* Vol.276, No.23, (Mar 2001), pp. 20610-20616, ISSN 0021-9258

Wassmer, T., Froissard, M., Plattner, H., Kissmehl, R., & Cohen, J. (2005). The vacuolar proton-ATPase plays a major role in several membrane-bounded organelles in Paramecium. *J. Cell. Sci.,* Vol.118, No.13, (Jul 2005), pp. 2813-2825, ISSN 0021-9533

Yarmola, E. G., Somasundaram, T., Boring, T. A., Spector, I., & Bubb, M. R. (2000). Actin-latrunculin A structure and function. *J. Biol. Chem.,* Vol.275, No.36, (Sept 2000), pp. 28120-28127

Yuan, W., & Parrish, C. R. (2001). Canine parvovirus capsid assembly and differences in mammalian and insect cells. *Virology,* Vol.279, No.2, (Jan 2001), pp. 546-557, ISSN 0042-6822

Zeghouf, M., Guibert, B., Zeeh, J-C., & Cherfils, J. (2005). Arf, Sec7, and brefeldin A: a model towards the therapeutic inhibition of guanine nucleotide-exchange factors. *Biochem. Soc. Trans.*, Vol.33, No.Pt6, (Dec 2005), pp. 1265-1268, ISSN 0300-5127

Chapter 8

Caveolae-Dependent Endocytosis in Viral Infection

Norica Branza-Nichita, Alina Macovei and Catalin Lazar

Additional information is available at the end of the chapter

1. Introduction

1.1. Caveolae structure and morphogenesis

Morphologically described as small "caves" in the plasma membrane, caveolae are highly specialized membrane domains with important roles in cell endocytosis, lipid metabolism, and signaling. Since their discovery sixty years ago [1,2], the functional relevance of caveolae has challenged many scientists, raising numerous debates and controversies. The electron microscopy images of caveolae show a rather cell-type-dependent appearance. In endothelial cells, caveolae opening is more constricted [3], while in epithelial cells they appear open to the extracellular medium and smaller in size [4]. In muscle cells, multiple caveolae units cluster together forming T-like tubules invaginating from sarcolemma [5]. Regardless their shape, caveolae appear as immobile structures, in tight connection with the cortical actin cytoskeleton underlying the plasma membrane. Video microscopy and fluorescence recovery after photobleaching analysis have shown that caveolae detach from the membrane only upon ligand binding and specific signaling [6].

The discovery of caveolins (Cav), the structural proteins enveloping caveolae in a spike-like coat [7] marked a significant breakthrough in understanding the nature and importance of these organelles.

Three members of the Cav family have been described in mammalian cells, to date: Cav-1, -2, -3, which share a significant homology and are conserved throughout evolution [8]. Cav-1 and -2 are relatively ubiquitous, with highest distribution in fibroblasts, adipocytes, endothelial cells, and pneumocytes, being co-expressed in most cells types [9]. Cav-3 is expressed independently of Cav-1 and -2 and is limited to skeletal muscle fibers and cardiac myocytes [10-12]. Over-expression of Cav-1 in caveolae-deficient cells is necessary and apparently sufficient to drive caveolae biogenesis [13]. Moreover, Cav-1 expression is required for the membrane localization and stability of Cav-2. Although unable to drive caveolae biogenesis on its own, Cav-2 may however influence it, at least in several polarized, epithelial cells [14,15]. The capacity to modulate caveolae assembly, shape and size has been shown to depend on Cav-2 phosphorylation status [16]. Similar to Cav-1, Cav-3 protein is sufficient to drive formation of caveolae in muscle cells [17].

Most of the molecular data available on caveolins refer to Cav-1; therefore the following discussion will focus on this protein, as a representative of the caveolin family, which is shown schematically in Fig.1.

Cav-1 is an integral membrane protein of 21kDa with an unusual topology. Both the N- and C-termini are cytoplasmically oriented and connected by a central hydrophobic domain, comprising approximately residues 102-134, inserted into, but not spanning the membrane bilayer, in a hairpin (or U bent, or horseshoe) configuration [18,19]. Interestingly, a peptide corresponding to the last 20 residues of the hydrophilic N-domain, (amino acids 82–101) enriched in aromatic residues, can also bind to membranes independently [20,21]. This so-called caveolin scaffolding domain (CSD) is a highly conserved region responsible for many functions associated with Cav-1.

In silico analysis of the conformation of this hydrophobic domain, showed that mutation of a single residue, Pro (110), changes the stable conformation to a straight hydrophobic helix that would span the membrane. Expression of the Cav-1 P110A mutant in HEK 293 cells followed by confocal immunofluorescence microscopy further confirmed the in silico data and the estimated topology [22].

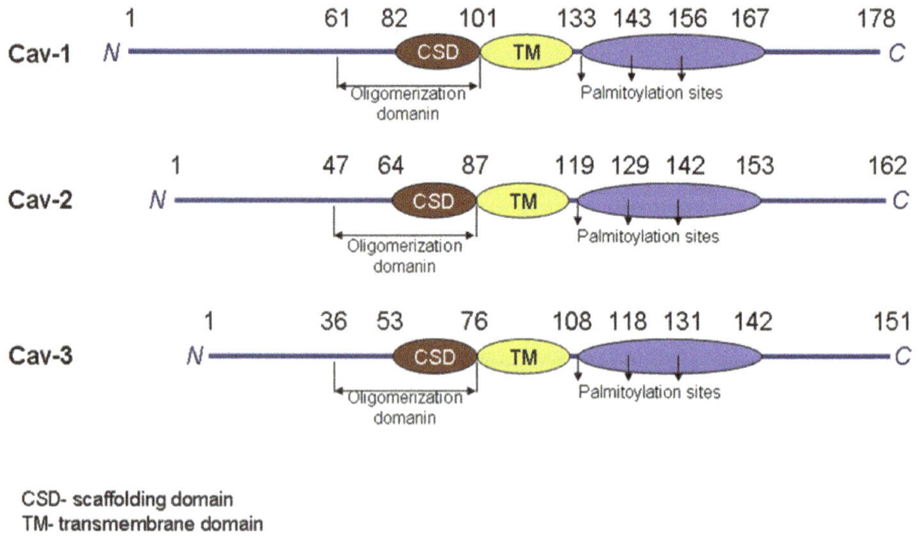

Figure 1. Schematic presentation of the structural similarities of the caveolin family of proteins.

Very recently, circular dichroism and NMR spectroscopy analysis have shown that the transmembrane domain of Cav-1 is primarily α-helical (57-65%). In addition, the helix–break–helix structure was suggested to be critical for the formation of the intra-membrane horseshoe conformation predicted for the protein. Interestingly, mutations of Ile (109) and Pro (110) to Ala dramatically altered the helix-break-helix structure. Moreover, it was shown that substitution of Pro (110) with any other residue results in disruption of the helix-break-helix structure, confirming the importance of the residue in the stability of the hydrophobic domain [23].

An important structural feature of Cav-1 is represented by its arrangement in high molecular mass oligomers of about 350 kDa, composed of 14-16 individual molecules [24,25]. Oligomerization is initiated in the endoplasmic reticulum (ER), where the Cav-1 monomer is co-translationally inserted into the membrane [19] and is rapidly assembled to form SDS-resistant, 8S complexes considered the building units of caveolae structure [24,25]. Intriguingly, blue native gel analysis evidenced only a few intermediate sized oligomers, suggesting that Cav-1 oligomerization is a highly cooperative process [26]. Oligomerization of the full-length protein requires the presence of the CSD and of the C-terminal domain [27] and appears to be stabilized by the palmitoylation of cysteine residues located at positions 133, 143 and 156 [18,28]. At this stage of their assembly, the complexes appear highly mobile in the ER membrane and rapidly concentrate at the ER exit sites, a process favored by the existence of a di-acidic export sequence located at the N-terminal domain. In the absence of this signal sequence, Cav-1 accumulates in lipid droplets [29,30]. This is an important observation, suggesting that ER exit and lipid droplets localization of caveolin complexes are competing processes, highlighting the role of the di-acidic motif in caveolin trafficking.

Interestingly, co-expression of Cav-1 and -2 results in assembly of mixt 8S complexes, where the Cav-1 to Cav-2 ratio may vary from 2:1 to 4:1 [31]; however, expression of Cav-2 alone does not result in oligomer formation.

The process continues in the trans-Golgi where the oligomers are exported in a COPII-dependent manner and self-associate into a large network of caveolin. However, formation of the 8S complex is not a prerequisite for Golgi transport, as expression of Cav- 2 alone, as well as that of an oligomerization-incompetent variant of Cav-1 does not result in their retention within the ER [32,33].

The Golgi oligomerization step is sensitive to BFA, clearly indicating that formation of large oligomer complexes is dependent on caveolins trafficking to this compartment, probably requiring a specific lipid composition of the membrane. The assembly process continues in a cholesterol-dependent manner, resulting in formation of 70S stable complexes, also evidenced by using a panel of anti-Cav-1 conformational antibodies. These complexes were assumed to correspond to the intact protein scaffold of the caveolae structure [34]. It is interesting that the caveolin assemblies colocalized with medial, rather than trans-Golgi markers, suggesting that, unlike other cargos transported through the secretory pathway, caveolar carrier vesicles are formed in the medial cisternae, being further exported to the plasma membrane in a dynamin-2 independent manner, similar to other raft-associated proteins [34]. This assembly process is schematically shown in Fig.2.

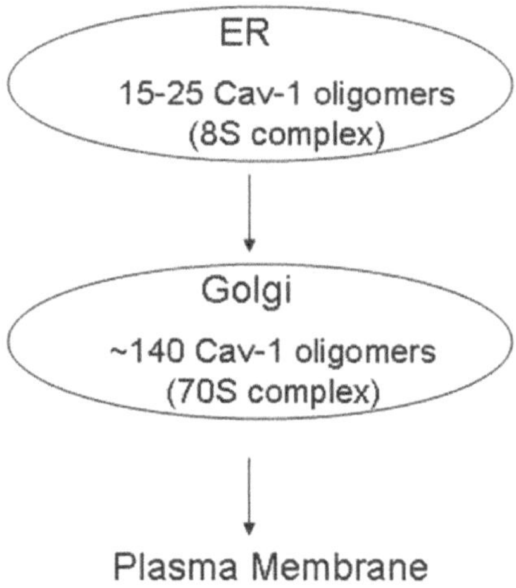

Figure 2. Caveolin oligomerization and assembly

It can be concluded from these observations that the tight regulation of Cav-1 trafficking along the secretory compartments (and the multiple check points) is totally justified by the complexity of the assembly process.

In addition to oligomerization, caveola formation involves association of the complexes with cholesterol-rich lipid-raft domains at plasma membrane. It is estimated that Cav-1 binds to 1–2 cholesterol molecules through the conserved basic and hydrophobic residues of the scaffolding domain [35]; thus, the relative amount of cholesterol concentrated in isolated caveolae can be as high as 20.000 molecules [36]. The relationship between cholesterol and caveolins is very complex. Treatment of cells with cholesterol binding or depleting agents results in caveolae with altered morphology and disrupted protein coat [7]. Moreover, cholesterol regulates Cav-1 expression at both, transcriptional and translational levels, through binding to either two steroid regulatory elements in the Cav-1 promoter, or the protein itself, thus modulating the level of Cav-1 mRNA or the protein stability [37,38].

Caveolae are enriched in glycosphingolipids (like GM1 and GM3) and sphingomyelin, the total lipid density being significantly higher than within the rest of the plasma membrane [36]. This is an important observation implying that certain lipids are recruited in caveolae, possibly to ensure their invagination-competent composition.

Recently, a crucial role in the last steps of caveolae biogenesis has been attributed to PTRF (Polymerase I and transcript release factor), originally described as an RNA Pol I transcription factor (also called Cav-P60 or cavin-1) [39,40]. Interestingly, cavin-1 is able to associate with plasma membrane caveolae but not with caveolins with other intracellular distribution (such as Golgi caveolins) [34]. Cavin-1 is recruited by Cav-1 to plasma membrane caveolar domains, where the two proteins are found to be in close proximity and an approximate ration of 1:1 [39]; however, whether or not they directly interact with each other is still a matter of debate. It was clearly demonstrated that cavins bind to phosphatidylserine and are phosphorylated at multiple sites, suggesting they may act as regulatory proteins of caveolae functions [41].

Based on sequence homology with cavin-1, three other proteins named cavin-2 to 4, sharing similar molecular organization, have been identified as part of the cavin family [42-44]. While cavin-1 expression is strictly associated to that of Cav-1 [39] contributing to the stability of the caveolae unit like a scaffolding protein, cavin-2 promotes recruitment of cavin-1 in caveolae and appears to have a role in the membrane-curvature [45].

The role of cavin-3 and -4 in caveolae biogenesis is less well understood. Cavin-3 was shown to regulate caveolae budding and Cav-1 trafficking, suggesting a function in coupling caveolae to the intracellular transport network [44]. Cavin-4 is co-expressed with Cav-3 in cardiac and muscle tissues where their function appears to be tightly correlated [42].

All members of the cavin family interact in a multimeric complex of about 60-80 cavins, in a Cav-1 independent manner. These complexes were detected both in the cytosol and plasma membrane fractions, suggesting they are the result of a succession of events, starting with cavin association into the cytosol and ending with the recruitment of the multimeric complexes to caveolae, during the final step of their biogenesis [42].

In contrast to caveolins, cavins are peripheral membrane proteins, and bind molecular components of the caveolar domain facing the cytosol. Given the high affinity of cavins for phosphatidylserine and the rapid dissociation from caveolae in the presence of nonionic detergents, it was suggested that binding to the lipid membrane, rather than to the protein scaffold, was highly probable.

The identification of cavins in caveolae opened new perspectives in understanding the complexity of caveolar structure. Although our knowledge on caveolae architecture and molecular composition has improved since their discovery, the main structural pillars defined at the time have not dramatically changed. Thus, today, caveolae are referred to as invaginations of the plasma membrane lipid bilayer, enriched in cholesterol and sphingolipids, embedding an integral membrane scaffold formed by caveolin oligomers assembled in a stable network, peripherally covered by a protein layer of cavin complexes. Once formed, this structure appears to remain stable also during endocytosis [46].

2. Caveolae signaling

The protein composition of caveolae has been addressed in a more systematic manner within the last years, using proteomic approaches [47-49]. A variety of signal transduction proteins were found to localize in caveolae, in tight connection with either the CSD or the lipid domains.

According to the caveolin signaling hypothesis, the role of caveolae is to trigger specific signal transduction by concentrating downstream effectors close to plasma membrane receptors, through direct interaction with the CSD [50,51]. Palmitoylation appears to play an important function in this process by facilitating the caveolar localization of proteins [52]. Of the signaling molecules identified, several have been more thoroughly investigated:

a. G proteins were abundant in caveolae; their binding to caveolin has a role in maintaining the Gα subunits in an inactive GDP-bound state [53]. Small GTP-binding proteins of the Ras superfamily also localized in caveolae, a process enhanced by palmitoylation of the C-terminal hypervariable region [54,55]. Binding of the H-Ras to the CSD results in inactivation of the protein, which is relevant in certain human cancers were H-Ras - caveolin interaction is prevented and the protein is maintained in an active state [56].
b. Src family kinases, such as c-Src, Fyn, Lyn [57] are nonreceptor tyrosine kinases, also enriched in caveolae. Their localization depends on the N-terminal myristoylation and subsequent interaction with the caveolins. Interestingly, Cav-1 palmitoylation is equally important for caveolae/c-Src interaction [52], which results in c-Src and Fyn inactivation [58]. In turn, the tyrosine phosphorylation of Cav-1 and -2 facilitates the recruitment of matrix metalloproteinases [58,59], and promotes caveolins localization to focal adhesions [60,61].
c. Several steroid hormone receptors were localized in caveolae, a process depending on both, palmitoylation and association with Cav-1 [62] and facilitating their activation [63].
d. Endothelial nitric oxide synthase (eNOS) is one of the most extensively studied Cav-1 interacting protein [64]. eNOS binds to the CSD of both Cav-1 and -3, which inhibit its enzymatic activity [65,66]. This observation lead to a novel concept of eNOS regulation, whereby, the interaction with caveolins is necessary to keep the enzyme inactive under basal conditions, while its concentration in caveolae will allow a quick response upon stimulation [67].
e. Many ion channels and pumps are targeted to caveolae and interact with caveolins, such as the calcium signaling molecules calmodulin, Ca^{2+}-ATPase, L-type Ca^{2+} channels [68,69]. Transient receptor potential (TRP) channels, and large-conductance Ca^{2+}-activated K^+ channels also localize in cholesterol-rich membrane areas, suggesting an important role of these domains in Ca^{2+} homeostasis [70,71]. Other transporters, like the Na/K-ATPase, involved in maintaining the Na^+ membrane gradient, are also found in caveolae, owing this localization to the existence of two caveolin-binding motifs in their amino acid sequence [72].
f. Protein kinases of different families were found in caveolae, due to their direct interaction with the Cav-1 CSD. For PKA, this interaction results in inhibition of the enzymatic activity [73], with consequences on regulation of other proteins, such as ATP-dependent K^+ channels or eNOS in muscle and endothelial cells, respectively [74,75]. Different isoforms of the PKC family of enzymes are also caveolae resident and appear to participate in regulation of caveolar proteins [76]. Caveolae interaction with PKC is more complex, leading to either activation (via ceramide interaction) [77], or inactivation, following endosomal delivery [78]. Caveolae also recruit the phosphatidylinositol-3-kinase (PI3K) through direct binding to Cav-1 [79] and the protein kinase B (PKB). Integration of this signaling pathway by caveolae plays a significant role in managing the cellular physiological stress, and regulating cell survival and death [80].
g. Phosphodiesterases (PDEs), involved in cyclic nucleotides (cAMP and cGMP) hydrolysis, have a preference for lipid rafts association (Abrahamsen H, 2004). For some

isoformes, such as PDE3B, a direct interaction with Cav-1, with a stabilizing effect on PDE3B, has been clearly confirmed by co-immunoprecipitation [81]. Other PDEs (PDE5, PDE4A4) appear to be recruited by caveoale through indirect mechanisms, possibly involving adapter proteins; [82]. Nevertheless, this association influences the establishment of cAMP/cGMP gradients and the downstream events [83].

3. Caveolae internalization and trafficking

Electron microscopy data show that caveolae are tightly connected to submembranous actin filaments [84,85], suggesting a function of the cytoskeleton in caveolae-mediated endocytosis. However, the exact role of the actin cytoskeleton is not clearly defined, as its disruption inhibits uptake of the caveolae ligand, alkaline phosphatase, on one hand [86] and promotes clustering of caveolae and internalization of Cav-1-labeled vesicles, on the other hand [6,87]. These observations fit into a model whereby actin would play a dual role in caveolae internalization: one is to keep the organization of caveolae and maintain their immobility at the plasma membrane, and the other to promote vesicle budding and release from the membrane.

Caveolae endocytosis relies heavily on dynamin, a multi-domain GTPase [88,89], which was shown to interact directly with Cav-1 [90]. Ligand binding initially disrupts the local actin cytoskeleton and promotes dynamin II recruitment to the site of internalization [91,92]. Dynamin oligomerization and the GTP-dependent conformational changes result in a structural collar around the neck of caveolae, directly mediating formation of free transport vesicles, following vesicular fission from the plasma membrane. It was shown that the protein regulates the actin tail formation [93,94], possibly through binding to cortactin [95-97] or intersectin, which promotes actin polymerization [98].

Another player in this complex molecular game was recently suggested, following the initial observation that Cav-1 binds to actin cross-linking proteins, filamin A and B, both *in vitro* and *in vivo* [99]. The main intracellular function of these proteins is to organize the actin cytoskeleton. The Cav-1 filamin A interaction was further confirmed in different cell types [100,101] and it was implicated in activation of the actin-folding and chaperone protein T-complex protein-1, [100] and inhibition of calpain-mediated cleavage of filamin A [102]. Thus, by providing the missing link between Cav-1 and the actin cytoskeleton, filamin is an important regulator of caveolae-mediated endocytosis and trafficking [103].

It was proposed that following budding, caveolae can fuse with either preformed vesicles, called "caveosome" at the time of their discovery, or early endosome, the latter process being dependent on Rab5 expression [46].

The caveosome was initially described as an immobile structure which did not co-localize with fluid phase markers or ligands of the clathrin-dependent pathway [104]. Moreover, the compartment was characterized by neutral pH and was unable to accumulate a lysosomal dye (lysotracker), reinforcing the notion of an independent organelle, clearly distinct from other endocytic compartments, which delivers its cargo to other cellular locations, such as the ER [104,105]. However, in a recent investigation of Cav-1 trafficking using pH-sensitive fluorophores, the existence of a neutral pH compartment was seriously doubted [106]. Rather, it was suggested that the caveosomes would correspond to modified late endosomes, where Cav-1 accumulates when over-expressed, undergoing ubiquitination and being further targeted to degradation [106]. Conversely, under physiological conditions, caveolae would bud from the plasma membrane transporting their viral cargo to early endosomes and eventually to the ER, in a microtubule-dependent manner [107], following a series of maturation events, which will be detailed below.

Clearly, more work is necessary to have the correct picture of the highly atypical caveolar trafficking, which appears to allow access of its ligands to intracellular destinations that are not reachable from other endocytic pathways. Despite the remaining uncertainties, the continuous development of the field has considerable advanced our knowledge of virus infection of host cells using caveolae endocytosis.

The current view of caveolae internalization and trafficking is depicted in Fig. 3.

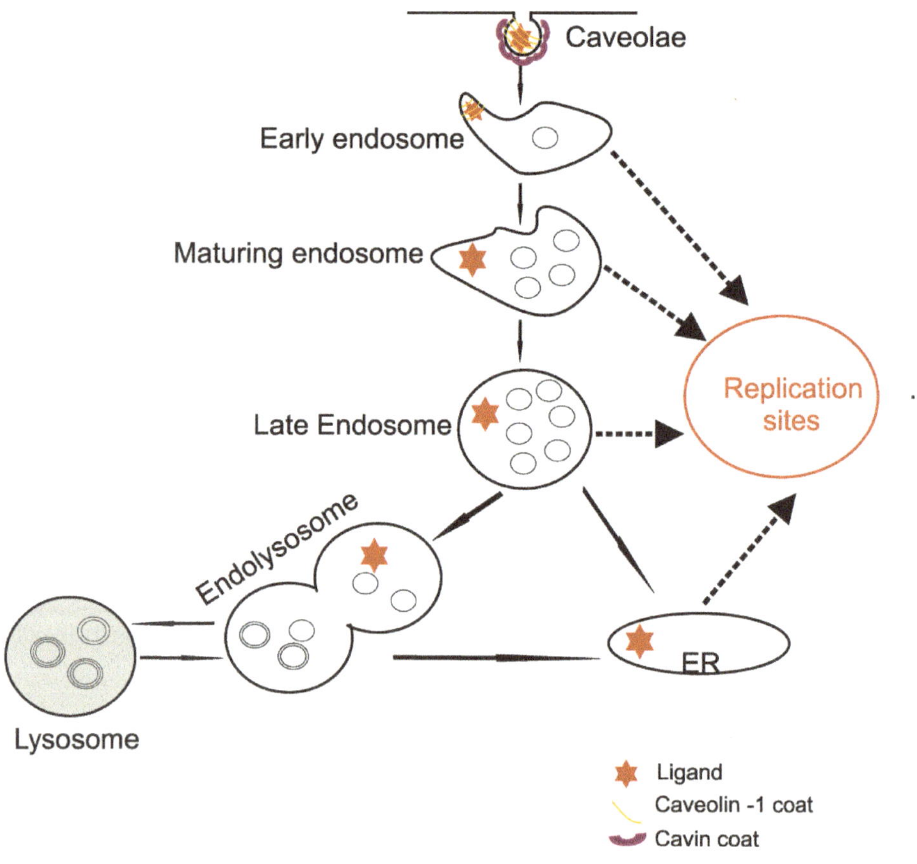

Figure 3. Ligand internalization and intracellular trafficking following caveolae-mediated endocytosis

4. Caveolae dependent viral infections

Owing to the vast amount of experimental data characterizing cell infection by the simian virus 40 (SV40), the pathogen has become the "star" of the caveolae-mediated entry pathway, being now extensively used in other studies as a marker of it [104,108]. Other well-characterized viruses using this entry pathway belong to the polyoma virus family, which has gained more interest recently, with the increasing number of human viruses identified. These include the KI polyoma virus, the WU polyoma virus and the Merkel cell polyoma virus [109-111], the latter being associated with an aggressive form of neuroendocrine skin cancer, the Merkel cell carcinoma. These are all non enveloped DNA viruses that replicate in the nucleus.

Viruses that use the same pathway to initiate a productive infection in target cells are listed in Table 1. Amongst them, Echovirus 1 (EV1), Human Hepatitis B virus (HBV) [112], Murine Leukemia Virus (MLV) [113], enteroviruses [114], have been more intensely investigated.

Virus	**Reference**
Simian virus 40	Stang, E., J. Kartenbeck, and R.G. Parton. 1997. Major histocompatibility complex class I molecules mediate association of SV40 with caveolae. *Mol Biol Cell*. 8:47-57. Pelkmans, L., J. Kartenbeck, and A. Helenius. 2001. Caveolar endocytosis of simian virus 40 reveals a new two-step vesicular-transport pathway to the ER. *Nat Cell Biol*. 3:473-83.
KI polyoma virus	Allander, T., K. Andreasson, S. Gupta, A. Bjerkner, G. Bogdanovic, M.A. Persson, T. Dalianis, T. Ramqvist, and B. Andersson. 2007. Identification of a third human polyomavirus. *J Virol*. 81:4130-6
WU polyoma virus	Gaynor, A.M., M.D. Nissen, D.M. Whiley, I.M. Mackay, S.B. Lambert, G. Wu, D.C. Brennan, G.A. Storch, T.P. Sloots, and D. Wang. 2007. Identification of a novel polyomavirus from patients with acute respiratory tract infections. *PLoS Pathog*. 3:e64
Merkel cell polyoma virus	Feng, H., M. Shuda, Y. Chang, and P.S. Moore. 2008. Clonal integration of a polyomavirus in human Merkel cell carcinoma. *Science*. 319:1096-100
Echovirus 1	Marjomaki, V., V. Pietiainen, H. Matilainen, P. Upla, J. Ivaska, L. Nissinen, H. Reunanen, P. Huttunen, T. Hyypia, and J. Heino. 2002. Internalization of echovirus 1 in caveolae. *J Virol*. 76:1856-65. Stuart, A.D., H.E. Eustace, T.A. McKee, and T.D. Brown. 2002. A novel cell entry pathway for a DAF-using human enterovirus is dependent on lipid rafts. *J Virol*. 76:9307-22
Human Hepatitis B virus	Macovei, A., C. Radulescu, C. Lazar, S. Petrescu, D. Durantel, R.A. Dwek, N. Zitzmann, and N.B. Nichita. 2010. Hepatitis B virus requires intact caveolin-1 function for productive infection in HepaRG cells. *J Virol*. 84:243-53.

Virus	Reference
Murine Leukemia Virus	Beer, C., D.S. Andersen, A. Rojek, and L. Pedersen. 2005. Caveola-dependent endocytic entry of amphotropic murine leukemia virus. *J Virol.* 79:10776-87
Tiger Frog Virus	Guo, C.J., D. Liu, Y.Y. Wu, X.B. Yang, L.S. Yang, S. Mi, Y.X. Huang, Y.W. Luo, K.T. Jia, Z.Y. Liu, W.J. Chen, S.P. Weng, X.Q. Yu, and J.G. He. 2011. Entry of tiger frog virus (an Iridovirus) into HepG2 cells via a pH-dependent, atypical, caveola-mediated endocytosis pathway. *J Virol.* 85:6416-26
Infectious spleen and kidney necrosis virus	Guo, C.J., Y.Y. Wu, L.S. Yang, X.B. Yang, J. He, S. Mi, K.T. Jia, S.P. Weng, X.Q. Yu, and J.G. He. 2011. Infectious spleen and kidney necrosis virus (a fish iridovirus) enters Mandarin fish fry cells via caveola-dependent endocytosis. *J Virol.* 86:2621-31.
Influenza viruses	Nunes-Correia, I., A. Eulalio, S. Nir, and M.C. Pedroso de Lima. 2004. Caveolae as an additional route for influenza virus endocytosis in MDCK cells. *Cell Mol Biol Lett.* 9:47-60.
Coronavirus 229E	Nomura, R., A. Kiyota, E. Suzaki, K. Kataoka, Y. Ohe, K. Miyamoto, T. Senda, and T. Fujimoto. 2004. Human coronavirus 229E binds to CD13 in rafts and enters the cell through caveolae. *J Virol.* 78:8701-8
Papillomavirus 31	Bousarghin, L., A. Touze, P.Y. Sizaret, and P. Coursaget. 2003. Human papillomavirus types 16, 31, and 58 use different endocytosis pathways to enter cells. *J Virol.* 77:3846-50
Respiratory syncytial virus	Brown G, Jeffree CE, McDonald T, Rixon HW, Aitken JD, et al. Analysis of the interaction between respiratory syncytial virus and lipid-rafts in Hep2 cells during infection. Virology 2004;327(2) 175-185. Werling D, Hope JC, Chaplin P, Collins RA, Taylor G, et al. Involvement of caveolae in the uptake of respiratory syncytial virus antigen by dendritic cells. J Leukoc Biol 1999;66(1) 50-58.
Newcastle disease virus	Cantin, C., J. Holguera, L. Ferreira, E. Villar, and I. Munoz-Barroso. 2007. Newcastle disease virus may enter cells by caveolae-mediated endocytosis. *J Gen Virol.* 88:559-69
Ebola virus	Empig, C.J., and M.A. Goldsmith. 2002. Association of the caveola vesicular system with cellular entry by filoviruses. *J Virol.* 76:5266-70.
BK polyoma virus	Eash, S., W. Querbes, and W.J. Atwood. 2004. Infection of vero cells by BK virus is dependent on caveolae. *J Virol.* 78:11583-90.
Marburg virus	Empig, C.J., and M.A. Goldsmith. 2002. Association of the caveola vesicular system with cellular entry by filoviruses. *J Virol.* 76:5266-70.

Table 1. Viruses that use caveolae-mediated endocytosis

After binding to the host cell, the virus particles are able to cluster the receptor molecules such as certain integrins (a2h1 in the case of EV1) [118] or glycosphingolipids (GM1 or GD1a in the case of SV40 or polyoma viruses), within the lipid rafts [119]. Accumulation of viral particles in caveolae induces a cascade of tyrosine phosphorylation reactions followed by rearrangements of the cortical actin cytoskeleton, as described above [91].

Several models addressing virus capturing into the plasma membrane invaginations have been proposed, the most recent relying on the observation that Cav-1 polymer remains intact once formed in the Golgi complex, during transport to the plasma membrane. According to this model, caveolae result from the fusion of a pre-existing caveolar vesicle with the plasma membrane [46]. As a consequence of virus binding to an increasing number of sphingolipids and/or integrins, the affinity of the pathogen for caveolar domains increases, facilitating entrapment in these areas. Another possibility is that virus particles bind and release the plasma membrane gangliosides in a transient manner, thus screening the whole cell surface. When reaching a caveolar region where multiple gangliosides interactions can occur simultaneously, binding becomes permanent and the virus particles are sequestered [104].

The intracellular trafficking of caveolae cargos has been recently re-evaluated using SV40 as a model and a series of complementary, state-of-the art techniques, including live-cells and electron microscopy, video recordings, pharmacological inhibitors and inhibition of expression of trafficking regulators [120]. It was shown that productive SV40 infection depends on the virus transport through a series of classical endocytic vesicles. Initially, the virus is found in Rab5-, EEA1-positive early endosomes and subsequently becomes associated with Rab7-positive domains, during endosome maturation. As this process proceeds, SV40 co-localizes with LAMP1-, Rab9-, and Rab7-positive late endosomes resembling multivesicular bodies and possibly endolysosomes. Endosome maturation also involves acidification of the compartments, as a consequence of vacuolar ATPase (v-ATPase) recruitment and activity. At this stage, acidification is required for SV40 subsequent transport steps and the initiation of productive infection. Interestingly, BK and JC viruses were also shown to enter the endosomes and depend on acidification for infection [121-123]. In the case of the mouse polyoma virus, the recycling, as well as late endosomes have been involved in infection [124,125].

From the late compartments of the endocytic pathway, SV40 appears to be directly transported to the ER, although an indirect ER targeting, via the Golgi complex, has not been completely excluded. Similarly, other polyoma viruses are transported to the ER lumen, before reaching the nucleus [121,125].

However, there are some notable exceptions of virus trafficking diverting from this pathway, despite being internalized through caveolae. Thus, cellular penetration of the EV1, a positive-stranded RNA human pathogen, depends on caveolins, dynamin II, and signaling events but does not require actin filaments or microtubules. The virus uptake was much faster than that of SV40 and was followed by rapid co-localization with Cav-1. Beyond this step, the virus failed to enter the Golgi complex, the ER, or the lysosomes, as none of the markers used to label these organelles co-localized with viral proteins. This observation raised the hypothesis that the virus particles remain sequestered in the Cav-1 positive endocytic vesicles until replication is initiated, further using them for cytoplasmic penetration and uncoating.

The lack of transport to the ER or Golgi was also confirmed by the absence of any inhibition of infection in the presence of nocodazole. Although more experimental data is needed to substantiate it, it is tempting to speculate that the different sorting pathway of the two viruses is related to their replication mechanism; thus, a DNA virus such as SV40 needs the nucleus for replication, which might be more accessible through the ER, whereas for a positive-stranded RNA virus the release of the genome into the cytoplasm is sufficient to initiate replication.

Either way, the precise molecular details characterizing this segment of the trafficking pathway are still to be defined, but understanding the factors involved in these unusual trafficking pathways is crucially important as other, yet uninvestigated viruses, may well use them when accessing the host cell via caveolae.

Importantly, there is accumulating evidence suggesting that several viruses take advantage of cross talk between endocytosis routes. For instance, JCV, bovine papillomavirus type 1 and human papillomavirus type 16 have been shown to access cells using the clathrin-dependent endocytosis, but intriguingly, they require Cav-1-mediated trafficking to initiate productive infection [123,126,127]. The internalized virions were trafficked to early endosomes before being transported to the caveolar pathway. From the Cav-1-positive vesicles, the viral cargo is further moved to the ER in a COPI-mediated, BFA-sensitive manner [127].

The newly described trafficking routes taken by these viruses may have an explanation in their absolute requirement to reach the ER compartment, a target that is not on the clathrin-mediated route.

The intriguing question as to why these viruses travel to the lumen of the ER, instead of using the endosomes for genome release, have received several interesting answers lately. One possibility is that the pathogens take advantage of the ER machinery of folding enzymes and chaperones, for uncoating and membrane penetration, being activated by lumenal thiol oxidoreductases before release into the cytosol/nucleoplasm [128].

Very recently, BiP and the ER-membrane protein BAP31 (both involved in ERAD) were shown to be essential factors for SV40 infection; thus incoming SV40 particles are structurally remodeled leading to exposure of the amino-terminal sequence of the minor viral protein VP2. These hydrophobic sequences anchor the virus to the ER membranes helping the particles release into the cytosol [129].

5. Investigation of the caveolae entry pathway

The molecular details of virus entry have been investigated through a variety of techniques, by perturbing endocytotic internalization with various inhibitors or interfering with the expression or function of key regulator proteins, using siRNA or dominant-negative mutants of the proteins, or by using transgenic animals. Because no single method to assess caveolae is perfect, the use of complementary techniques is crucial for such a task. These can employ cell fractionation, immunoprecipitation, protein and organelle labeling, immunofluorescence microscopy. Since caveolae are best characterized by their microscopic appearance, studies employing alteration of the intracellular level of caveolins, cholesterol, or different molecules enriched, but not exclusively present in caveolae, should ideally follow the impact of such changes to other internalization and trafficking pathways, by electron microscopy

5.1. Pharmacological approaches

A pharmacologic approach involves the treatment of cells with agents that deplete membranes of cholesterol, or inhibit various structural or signaling molecules involved in controlling the pathway. Having the advantage of being readily available and convenient to use, chemical inhibitors have been extensively employed to characterize different endocytic pathway; however, stringent controls must be included and results should be interpreted with care, because of the pleiotropic effects these drugs may have within the treated cell.

The usefulness as well as the pitfalls associated with the use of these agents will be detailed below.

Methyl-β-cyclodextrin (MβCD) – Cyclodextrins are cyclic oligomers of glucose that have the property to bind and extract lipophiles, including cholesterol, from their hydrophobic core [130]. Based on the tight dependency of caveolae stability and function on the amount of cholesterol present in the lipid rafts, the compound has been widely used to define the caveolar-mediated entry of many pathogens [6,117,131-135].

A major issue of MβCD treatment is its cellular toxicity, which was initially associated with longer incubation times (for example, the cell viability decreases form 90% during a 30 minutes incubation, to as low as 64%, if the drug is used for 12 h) [117]. A thorough study performed on multiple cell lines showed, however, that MβCD significantly decreased cellular viability, even after short treatment and at concentrations routinely used to inhibit endocytosis, a phenomenon which was cell line dependent [136]. Moreover, a low level of plasma membrane cholesterol was shown to interfere with other endocytic pathways, such as the clathrin-mediated endocytosis [137,138], or even with cholesterol independent endocytosis [139], demonstrating a rather poor specificity of the drug in inhibiting a distinct pathway.

Statins – are a group of drugs that lower the intracellular cholesterol level by competitively inhibiting the 3 hydroxy 3 methylglutaryl coenzyme A reductase involved in its biosynthesis. As a consequence, efficient depletion of membrane cholesterol and decreased formation of caveolae are observed [140]. Despite showing good toxicity profiles, statins also exert pleiotropic effects through a variety of mechanisms, which appear to be unrelated to their cholesterol-lowering activity. Thus, several immunosuppressive effects have been involved; amongst them well-documented are the prevention of activation of the transcription factor NF-kappaB or up-regulation of the pro-inflammatory cytokine production [141,142].

Filipin - is a macrolide pentene polyene with antibiotic properties relying on sterol binding with high affinity [143]. Filipin III has been employed to block caveolae entry since it complexes with membrane cholesterol, thus interfering with cholesterol-sensitive processes [144]. Treatment with filipin dispersed the receptors found in caveolae and promoted disassembly of these structures [145]. Similar to MβCD, filipin treatment is toxic and at least for certain cell lines, its inhibitory effect on endocytosis was exclusively due to cytotoxicity [146]. However, a narrow window of specific inhibitory function can be identified in most cell lines.

Genistein – inhibits several tyrosine kinases in mammalian cells and thus, caveolae internalization [86,147]. Genistein has been however shown to suppress the entry of several types of viruses that use different endocytic ways to gain access to the replication sites: SV40 [91], adenoviruses [148], human herpesvirus 8 (HHV-8) [149], HBV [112]. However, the general need of tyrosine kinases-mediated signaling of diverse families of viruses, both at early entry steps, or later in infection, makes it difficult to clearly ascertain the endocytic route used, only by using genistein.

U18666A - is an amphiphilic amine that arrests cholesterol transport and suppresses sterol biosynthesis. Treatment of cells with this inhibitor was shown to induce cholesterol accumulation in late endosomes/lysosomes and deplete cholesterol from the Golgi complex [150]. Interestingly, the mobility of Cav-1 significantly increased in the Golgi complex of U18666A-treated cells.

Phorbol 12-myristate 13-acetate (PMA) – is an activator of classical 2,3-diacylglycerol (DAG)-dependent protein kinase isoforms, owing this property to their high affinity for the DAG binding site. Interestingly, PMA treatment results in constitutive phosphorylation of caveolin [151] with significant inhibitory effects on caveola invagination from the intracellular face of the plasma membrane [86,152,153].

PMA has been shown to specifically inhibit the caveolae-mediated entry of Ebola and Marburg viruses, two negative-stranded RNA pathogens, members of the *Filoviridae* family [154]. PMA has a low citotoxicity even when used for longer incubation times (82% of the cell are still viable following incubation for 24 h, at concentrations required to inhibit caveolae endocytosis [117].

It is important to note that PMA can stimulate endocytosis of other ligands and interferes with the endocytic trafficking by stimulating a factor required for endosome fusion after Rab5 activation [155]. PMA treatment may have opposite effects on internalization of certain ligands (such as FITC-dextran) in polarized cells, increasing for instance, its basolateral, but not apical uptake.

Okadaic acid – is an inhibitor of phosphatases 1 and 2A, which are important in caveolae function [156]. Treatment with okadaic acid has been shown to promote removal of caveolar structures from the cell surface and stimulate endocytosis via these structures [6,86]. Importantly, the drug also inhibits the clathrin-mediated endocytosis [86] making the time of addition to the cells an important experimental factor in drawing the right conclusions on viral entry. Thus, pretreatment of cells with okadaic acid interferes with both, caveolae and clathrin pathways, while addition after virus binding to the target cells enhances infection, as it was shown for MLV [113].

Cytoskeleton inhibitors - the actin cytoskeleton localized near the plasma membrane appear to be a critical regulator of caveolae endocytosis [157].

Depolymerization of microtubules with colchicine or disruption of actin microfilaments with cytochalasin D resulted in a significant reduction of the amount of Cav-3 in plasma membrane fractions isolated from cardiac myocytes. Treatment with either drug also led to the exclusion of Cav-1 and -2 from similar fractions and the decrease of tyrosine-phosphorylated Cav-1 [101].

Other drugs interfering with actin polymerization are latrunculin A (actin monomer-sequestering drug) and jasplakinolide (an actin polymer-stabilizing compound), both were shown to reduced SV40 internalization by more than 60% [91]. Since function of many cellular processes such as trafficking and organelle movement are regulated by microtubules and the actin cytoskeleton, the biochemical assays should be combined with microscopy analysis to clearly define the role of the cytoskeleton in viral infection mediated by caveolae. Also, it is important to keep in mind that viruses induce cytoskeletal reorganization and reconfiguration to initiate, maintain and spread the infection. Therefore, the impact of the cytoskeleton perturbation on the outcome of infection highly depends on the stage of the viral life cycle the drug is acting upon [158].

5.2. Interference with expression and function of caveolae regulating proteins

As threshold levels of Cav-1 regulate caveolae formation [155, 159, 160], modulating the expression and/or function of this protein is one the most reliable approach to investigate caveolae entry. Cav-1 down-regulation using anti-sense, small interfering (si) or short hairpin (sh) RNA results in a significant decrease in the number of caveolae. For instance, the siRNA-mediated knockdown of Cav-1 expression was sufficient to inhibit albumin uptake in endothelial cells; however, intriguingly enough, the caveolae localization of signaling proteins, including eNOS, Rac, tyrosine kinase Src and insulin receptor was not altered. Using this technique, several viruses were shown to depend on caveolae for productive infection such as the Avian Reovirus [161], BK polyomavirus [133].

Cavin proteins are also important targets to study caveolae-mediated entry, since absence of cavin-1 results in lower expression level of caveolins and eventually, the loss of caveole. Down-regulation of cavin-1, using specific shRNA, increases mobility of caveolin-1, which is released from the cell surface and rapidly internalized and degraded [39]. Interfering with cavin-2 expression is also a valuable tool when assessing the role of caveolae in viral infection, since its down-regulation induces loss of cavin-1 and caveolin expression and therefore, it limits caveolae formation [43]. Similarly, suppression of cavin-3 biosynthesis uncouples caveolae from the intracellular transport machinery [44].

Caveolae budding from the plasma membrane and subsequent internalization strictly depends on dynamin II [89], thus silencing its expression is also often used in combination with caveolin inhibition.

An elegant alternative to silencing the expression of the proteins involved in caveolae architecture and function is over-expression of their mutant counterparts, which compete with the wild-type proteins for the same function. An important advantage of this technique is that, at any time during the experiment, the wild-type protein is still expressed ensuring the functioning of the pathway at a basal level and reducing toxicity. This approach was used to show the dependence of HBV internalization on functional Cav-1 and dynamin II (Fig. 4) [112].

A

B

C

D

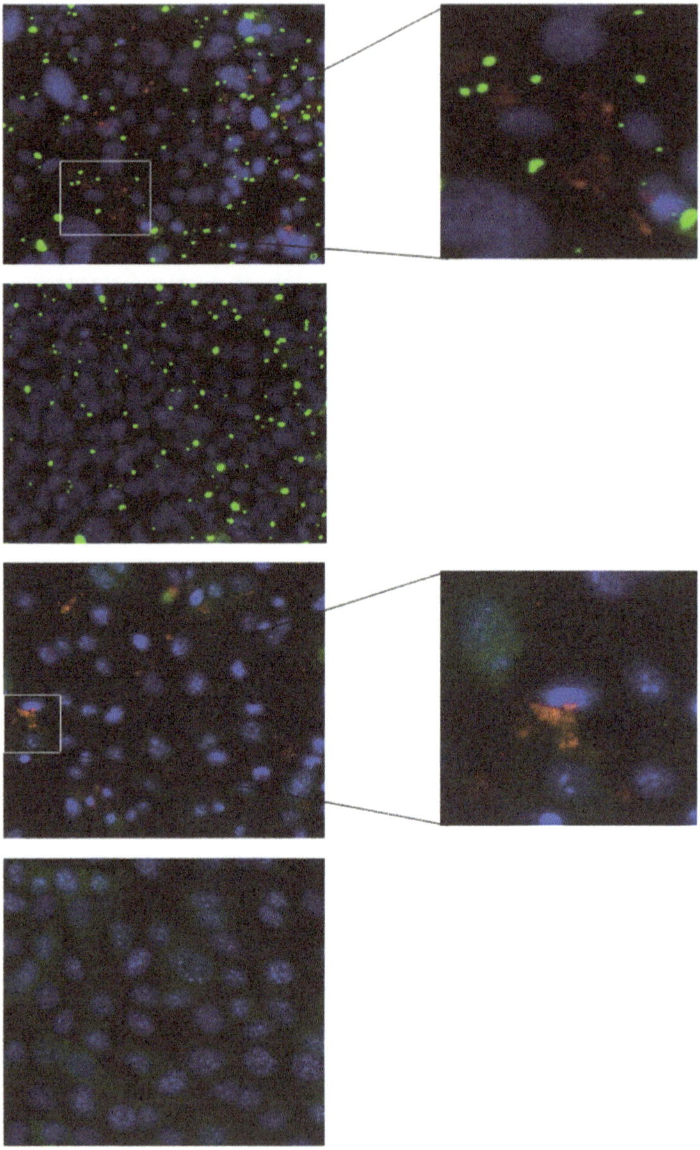

Figure 4. Hepatitis B Virus (HBV) infection of permissive HepaRG cells stably expressing dynamin II (A and B) and caveolin-1 (C and D) proteins with either wild-type (A and C) or dominant-negative (B and D) functions. HBV infection is evidenced by immunofluorescence microscopy using antibodies against the envelope proteins (in red). The dominant-negative dynamin II contains the K44A mutation which abolishes the GTP-ase activity (B). The dominant negative caveolin-1 contains a deletion of the 1-81 amino acid domain, at the N-terminal end (D). Expression of the wild-type and dominant-negative variants is evidenced through the Green Fluorescent Protein (GFP), which is either co-expressed from bicistronic GFP-caveolin DNA constructs (C and D) or expressed in fusion with dynamin II (A and B).

Cav -1 dominant negative proteins – disturb the formation of Cav-1-positive lipid rafts and cause the redistribution of endogenous caveolin to detergent soluble membrane fractions [162]. These are N-terminally truncated or N-terminally GFP-tagged caveolin constructs, which strongly inhibit SV40 entry [104,163] and were used to characterize the internalization pathway of many other viruses [134,164,165].

Other caveolin constructs containing the point mutations, Y14F and P132L were recently demonstrated to have dominant-negative activity [166]. Expression of the Cav-1 variant containing the P110A mutation was shown to determine a profound inhibition of caveolae endocytosis, cellular lipid accumulation and lipid droplet biogenesis. Moreover, this is a potent mutant to take into account when investigating the caveolae pathway, as it significantly reduces the Cav-1 localization into detergent-resistant domains of the plasma membrane and caveolae formation [22].

An interesting caveolin mutant is cav^{DGV}, a deleted Cav-3 form, which lacks the first 53 residues of the protein, but contains an intact scaffolding domain. The truncated protein acts as a dominant negative inhibitory mutant, causing the intracellular accumulation of free cholesterol in late endosomes, a reduction of surface cholesterol, efflux and synthesis [30].

Dynamin -II dominant negative proteins - as important regulators of clathrin, caveolae and other endocytic pathaways [88,167], dynamin II inhibition is often used in combination with modulation of other, more specific proteins involved in caveolae function (listed above). By far, the most used mutant dynamin is the K44A variant, defective in GTP hydrolysis, which was clearly shown to inhibit release of caveolae from plasma membranes in an *in vitro* assay [89].

A long term expression of a dominant negative protein may be toxic for the cells, determining changes of morphology. Also it is important to keep in mind that down-regulation of a certain pathway may promote up-regulation of other, compensatory entry mechanisms, if cells express dominant negative proteins for a long time. A solution to overcome these potential problems is the use controlled/inducible expression systems (such as the TetOn/Tet Off switch).

Generation of knockout (KO) mice – is a powerful approach for the study of caveolae *in vivo*. Caveolin-KO mice (Cav-1, -2, -3 and Cav-1/-3 double KO mice have already been generated and characterized. They displayed different phenotypes, but interestingly, were viable and fertile [168]. While Cav-2 KO mice retain normal expression of caveolae, Cav-1 KO mice are devoid of Cav-2 expression and caveolae in certain cell lines, and develop many cardiac and pulmonary diseases. More work is needed to understand whether or not these pathologies are directly correlated with the loss of expression of caveolins and caveolae and a good approach toward this aim would be to investigate each individual caveolins and the development of the corresponding phenotypes over a longer period of time.

6. Concluding remarks

The new experimental evidence emerged with the advance of the techniques used to investigate ligand internalization and intracellular trafficking, have consolidated the notion that endocytosis through caveolae is a true alternative to the clathrin-mediated pathway. By employing this route for entry into the target cell, viruses could benefit from the enormous advantage of being targeted directly to specific organelles that are essential for their replication; moreover, degradative compartments can be bypassed, which could enhance the efficiency of productive infection. Nevertheless, despite the tremendous development of the field in the last decade, many conceptual and mechanistic aspects are still to be clarified or reevaluated. Certainly, important issues regarding: a) the regulation of the crosstalk between different internalization pathways; b) the similarities between caveolae and other clathrin-independent entry routes; c) the exact mechanism of ligand sorting; d) the properties of the compartment(s) where it occurs; d) the preferential targeting of caveolae ligands to other intracellular compartments than the ER; are already under the scrutiny of many cell biologists and will find an answer in the near future.

Author details

Norica Branza-Nichita*, Alina Macovei and Catalin Lazar
Department of Viral Glycoproteins, Institute of Biochemistry Bucharest,
Romanian Academy, Romania

Acknowledgement

This work was supported by the Romanian Academy project 3 of the Institute of Biochemistry, the IDEI grant ID-84 and the POSDRU/89/1.5/S/60746 grant.

7. References

[1] Palade GE. An electron microscope study of the mitochondrial structure. J Histochem Cytochem 1953;1(4) 188-211.
[2] Yamada E. The fine structure of the renal glomerulus of the mouse. J Biophys Biochem Cytol 1955;1(6) 551-566.
[3] Stan RV, Roberts WG, Predescu D, Ihida K, Saucan L, et al. Immunoisolation and partial characterization of endothelial plasmalemmal vesicles (caveolae). Mol Biol Cell 1997;8(4) 595-605.
[4] Montesano R, Roth J, Robert A, Orci L. Non-coated membrane invaginations are involved in binding and internalization of cholera and tetanus toxins. Nature 1982;296(5858) 651-653.
[5] Parton RG, Way M, Zorzi N, Stang E. Caveolin-3 associates with developing T-tubules during muscle differentiation. J Cell Biol 1997;136(1) 137-154.
[6] Thomsen P, Roepstorff K, Stahlhut M, van Deurs B. Caveolae are highly immobile plasma membrane microdomains, which are not involved in constitutive endocytic trafficking. Mol Biol Cell 2002;13(1) 238-250.
[7] Rothberg KG, Heuser JE, Donzell WC, Ying YS, Glenney JR, et al. Caveolin, a protein component of caveolae membrane coats. Cell 1992;68(4) 673-682.

* Corresponding Author

[8] Tang Z, Okamoto T, Boontrakulpoontawee P, Katada T, Otsuka AJ, et al. Identification, sequence, and expression of an invertebrate caveolin gene family from the nematode Caenorhabditis elegans. Implications for the molecular evolution of mammalian caveolin genes. J Biol Chem 1997;272(4) 2437-2445.
[9] Scherer PE, Lewis RY, Volonte D, Engelman JA, Galbiati F, et al. Cell-type and tissue-specific expression of caveolin-2. Caveolins 1 and 2 co-localize and form a stable hetero-oligomeric complex in vivo. J Biol Chem 1997;272(46) 29337-29346.
[10] Song KS, Scherer PE, Tang Z, Okamoto T, Li S, et al. Expression of caveolin-3 in skeletal, cardiac, and smooth muscle cells. Caveolin-3 is a component of the sarcolemma and co-fractionates with dystrophin and dystrophin-associated glycoproteins. J Biol Chem 1996;271(25) 15160-15165.
[11] Tang Z, Scherer PE, Okamoto T, Song K, Chu C, et al. Molecular cloning of caveolin-3, a novel member of the caveolin gene family expressed predominantly in muscle. J Biol Chem 1996;271(4) 2255-2261.
[12] Way M, Parton RG. M-caveolin, a muscle-specific caveolin-related protein. FEBS Lett 1995;376(1-2) 108-112.
[13] Li S, Song KS, Lisanti MP. Expression and characterization of recombinant caveolin. Purification by polyhistidine tagging and cholesterol-dependent incorporation into defined lipid membranes. J Biol Chem 1996;271(1) 568-573.
[14] Fujimoto T, Kogo H, Nomura R, Une T. Isoforms of caveolin-1 and caveolar structure. J Cell Sci 2000;113 Pt 19 3509-3517.
[15] Lahtinen U, Honsho M, Parton RG, Simons K, Verkade P. Involvement of caveolin-2 in caveolar biogenesis in MDCK cells. FEBS Lett 2003;538(1-3) 85-88.
[16] Sowa G, Pypaert M, Fulton D, Sessa WC. The phosphorylation of caveolin-2 on serines 23 and 36 modulates caveolin-1-dependent caveolae formation. Proc Natl Acad Sci U S A 2003;100(11) 6511-6516.
[17] Galbiati F, Razani B, Lisanti MP. Caveolae and caveolin-3 in muscular dystrophy. Trends Mol Med 2001;7(10) 435-441.
[18] Dupree P, Parton RG, Raposo G, Kurzchalia TV, Simons K. Caveolae and sorting in the trans-Golgi network of epithelial cells. Embo J 1993;12(4) 1597-1605.
[19] Monier S, Parton RG, Vogel F, Behlke J, Henske A, et al. VIP21-caveolin, a membrane protein constituent of the caveolar coat, oligomerizes in vivo and in vitro. Mol Biol Cell 1995;6(7) 911-927.
[20] Schlegel A, Schwab RB, Scherer PE, Lisanti MP. A role for the caveolin scaffolding domain in mediating the membrane attachment of caveolin-1. The caveolin scaffolding domain is both necessary and sufficient for membrane binding in vitro. J Biol Chem 1999;274(32) 22660-22667.
[21] Arbuzova A, Wang L, Wang J, Hangyas-Mihalyne G, Murray D, et al. Membrane binding of peptides containing both basic and aromatic residues. Experimental studies with peptides corresponding to the scaffolding region of caveolin and the effector region of MARCKS. Biochemistry 2000;39(33) 10330-10339.
[22] Aoki S, Thomas A, Decaffmeyer M, Brasseur R, Epand RM. The role of proline in the membrane re-entrant helix of caveolin-1. J Biol Chem 2010;285(43) 33371-33380.

[23] Lee J, Glover KJ. The transmembrane domain of caveolin-1 exhibits a helix-break-helix structure. Biochim Biophys Acta 2012;1818(5) 1158-1164.

[24] Fernandez I, Ying Y, Albanesi J, Anderson RG. Mechanism of caveolin filament assembly. Proc Natl Acad Sci U S A 2002;99(17) 11193-11198.

[25] Sargiacomo M, Scherer PE, Tang Z, Kubler E, Song KS, et al. Oligomeric structure of caveolin: implications for caveolae membrane organization. Proc Natl Acad Sci U S A 1995;92(20) 9407-9411.

[26] Ren X, Ostermeyer AG, Ramcharan LT, Zeng Y, Lublin DM, et al. Conformational defects slow Golgi exit, block oligomerization, and reduce raft affinity of caveolin-1 mutant proteins. Mol Biol Cell 2004;15(10) 4556-4567.

[27] Song KS, Tang Z, Li S, Lisanti MP. Mutational analysis of the properties of caveolin-1. A novel role for the C-terminal domain in mediating homo-typic caveolin-caveolin interactions. J Biol Chem 1997;272(7) 4398-4403.

[28] Monier S, Dietzen DJ, Hastings WR, Lublin DM, Kurzchalia TV. Oligomerization of VIP21-caveolin in vitro is stabilized by long chain fatty acylation or cholesterol. FEBS Lett 1996;388(2-3) 143-149.

[29] Ostermeyer AG, Paci JM, Zeng Y, Lublin DM, Munro S, et al. Accumulation of caveolin in the endoplasmic reticulum redirects the protein to lipid storage droplets. J Cell Biol 2001;152(5) 1071-1078.

[30] Pol A, Luetterforst R, Lindsay M, Heino S, Ikonen E, et al. A caveolin dominant negative mutant associates with lipid bodies and induces intracellular cholesterol imbalance. J Cell Biol 2001;152(5) 1057-1070.

[31] Scheiffele P, Verkade P, Fra AM, Virta H, Simons K, et al. Caveolin-1 and -2 in the exocytic pathway of MDCK cells. J Cell Biol 1998;140(4) 795-806.

[32] Lee H, Park DS, Wang XB, Scherer PE, Schwartz PE, et al. Src-induced phosphorylation of caveolin-2 on tyrosine 19. Phospho-caveolin-2 (Tyr(P)19) is localized near focal adhesions, remains associated with lipid rafts/caveolae, but no longer forms a high molecular mass hetero-oligomer with caveolin-1. J Biol Chem 2002;277(37) 34556-34567.

[33] Mora R, Bonilha VL, Marmorstein A, Scherer PE, Brown D, et al. Caveolin-2 localizes to the golgi complex but redistributes to plasma membrane, caveolae, and rafts when co-expressed with caveolin-1. J Biol Chem 1999;274(36) 25708-25717.

[34] Hayer A, Stoeber M, Bissig C, Helenius A. Biogenesis of caveolae: stepwise assembly of large caveolin and cavin complexes. Traffic 2010;11(3) 361-382.

[35] Murata M, Peranen J, Schreiner R, Wieland F, Kurzchalia TV, et al. VIP21/caveolin is a cholesterol-binding protein. Proc Natl Acad Sci U S A 1995;92(22) 10339-10343.

[36] Ortegren U, Karlsson M, Blazic N, Blomqvist M, Nystrom FH, et al. Lipids and glycosphingolipids in caveolae and surrounding plasma membrane of primary rat adipocytes. Eur J Biochem 2004;271(10) 2028-2036.

[37] Fielding CJ, Bist A, Fielding PE. Caveolin mRNA levels are up-regulated by free cholesterol and down-regulated by oxysterols in fibroblast monolayers. Proc Natl Acad Sci U S A 1997;94(8) 3753-3758.

[38] Frank PG, Marcel YL, Connelly MA, Lublin DM, Franklin V, et al. Stabilization of caveolin-1 by cellular cholesterol and scavenger receptor class B type I. Biochemistry 2002;41(39) 11931-11940.

[39] Hill MM, Bastiani M, Luetterforst R, Kirkham M, Kirkham A, et al. PTRF-Cavin, a conserved cytoplasmic protein required for caveola formation and function. Cell 2008;132(1) 113-124.
[40] Liu L, Brown D, McKee M, Lebrasseur NK, Yang D, et al. Deletion of Cavin/PTRF causes global loss of caveolae, dyslipidemia, and glucose intolerance. Cell Metab 2008;8(4) 310-317.
[41] Aboulaich N, Vainonen JP, Stralfors P, Vener AV. Vectorial proteomics reveal targeting, phosphorylation and specific fragmentation of polymerase I and transcript release factor (PTRF) at the surface of caveolae in human adipocytes. Biochem J 2004;383(Pt 2) 237-248.
[42] Bastiani M, Liu L, Hill MM, Jedrychowski MP, Nixon SJ, et al. MURC/Cavin-4 and cavin family members form tissue-specific caveolar complexes. J Cell Biol 2009;185(7) 1259-1273.
[43] Hansen CG, Bright NA, Howard G, Nichols BJ. SDPR induces membrane curvature and functions in the formation of caveolae. Nat Cell Biol 2009;11(7) 807-814.
[44] McMahon KA, Zajicek H, Li WP, Peyton MJ, Minna JD, et al. SRBC/cavin-3 is a caveolin adapter protein that regulates caveolae function. Embo J 2009;28(8) 1001-1015.
[45] Zimmerberg J, Kozlov MM. How proteins produce cellular membrane curvature. Nat Rev Mol Cell Biol 2006;7(1) 9-19.
[46] Pelkmans L, Burli T, Zerial M, Helenius A. Caveolin-stabilized membrane domains as multifunctional transport and sorting devices in endocytic membrane traffic. Cell 2004;118(6) 767-780.
[47] Banfi C, Brioschi M, Wait R, Begum S, Gianazza E, et al. Proteomic analysis of membrane microdomains derived from both failing and non-failing human hearts. Proteomics 2006;6(6) 1976-1988.
[48] McMahon KA, Zhu M, Kwon SW, Liu P, Zhao Y, et al. Detergent-free caveolae proteome suggests an interaction with ER and mitochondria. Proteomics 2006;6(1) 143-152.
[49] Durr E, Yu J, Krasinska KM, Carver LA, Yates JR, et al. Direct proteomic mapping of the lung microvascular endothelial cell surface in vivo and in cell culture. Nat Biotechnol 2004;22(8) 985-992.
[50] Ostrom RS, Insel PA. The evolving role of lipid rafts and caveolae in G protein-coupled receptor signaling: implications for molecular pharmacology. Br J Pharmacol 2004;143(2) 235-245.
[51] Cohen AW, Razani B, Schubert W, Williams TM, Wang XB, et al. Role of caveolin-1 in the modulation of lipolysis and lipid droplet formation. Diabetes 2004;53(5) 1261-1270.
[52] Lee H, Woodman SE, Engelman JA, Volonte D, Galbiati F, et al. Palmitoylation of caveolin-1 at a single site (Cys-156) controls its coupling to the c-Src tyrosine kinase: targeting of dually acylated molecules (GPI-linked, transmembrane, or cytoplasmic) to caveolae effectively uncouples c-Src and caveolin-1 (TYR-14). J Biol Chem 2001;276(37) 35150-35158.
[53] Insel PA, Head BP, Ostrom RS, Patel HH, Swaney JS, et al. Caveolae and lipid rafts: G protein-coupled receptor signaling microdomains in cardiac myocytes. Ann N Y Acad Sci 2005;1047 166-172.

[54] Roy S, Plowman S, Rotblat B, Prior IA, Muncke C, et al. Individual palmitoyl residues serve distinct roles in H-ras trafficking, microlocalization, and signaling. Mol Cell Biol 2005;25(15) 6722-6733.
[55] Prior IA, Harding A, Yan J, Sluimer J, Parton RG, et al. GTP-dependent segregation of H-ras from lipid rafts is required for biological activity. Nat Cell Biol 2001;3(4) 368-375.
[56] Li S, Okamoto T, Chun M, Sargiacomo M, Casanova JE, et al. Evidence for a regulated interaction between heterotrimeric G proteins and caveolin. J Biol Chem 1995;270(26) 15693-15701.
[57] Sargiacomo M, Sudol M, Tang Z, Lisanti MP. Signal transducing molecules and glycosyl-phosphatidylinositol-linked proteins form a caveolin-rich insoluble complex in MDCK cells. J Cell Biol 1993;122(4) 789-807.
[58] Li S, Couet J, Lisanti MP. Src tyrosine kinases, Galpha subunits, and H-Ras share a common membrane-anchored scaffolding protein, caveolin. Caveolin binding negatively regulates the auto-activation of Src tyrosine kinases. J Biol Chem 1996;271(46) 29182-29190.
[59] Lee H, Volonte D, Galbiati F, Iyengar P, Lublin DM, et al. Constitutive and growth factor-regulated phosphorylation of caveolin-1 occurs at the same site (Tyr-14) in vivo: identification of a c-Src/Cav-1/Grb7 signaling cassette. Mol Endocrinol 2000;14(11) 1750-1775.
[60] Swaney JS, Patel HH, Yokoyama U, Head BP, Roth DM, et al. Focal adhesions in (myo)fibroblasts scaffold adenylyl cyclase with phosphorylated caveolin. J Biol Chem 2006;281(25) 17173-17179.
[61] Del Pozo MA, Schwartz MA. Rac, membrane heterogeneity, caveolin and regulation of growth by integrins. Trends Cell Biol 2007;17(5) 246-250.
[62] Acconcia F, Bocedi A, Ascenzi P, Marino M. Does palmitoylation target estrogen receptors to plasma membrane caveolae? IUBMB Life 2003;55(1) 33-35.
[63] Li L, Ren CH, Tahir SA, Ren C, Thompson TC. Caveolin-1 maintains activated Akt in prostate cancer cells through scaffolding domain binding site interactions with and inhibition of serine/threonine protein phosphatases PP1 and PP2A. Mol Cell Biol 2003;23(24) 9389-9404.
[64] Garcia-Cardena G, Fan R, Stern DF, Liu J, Sessa WC. Endothelial nitric oxide synthase is regulated by tyrosine phosphorylation and interacts with caveolin-1. J Biol Chem 1996;271(44) 27237-27240.
[65] Venema VJ, Ju H, Zou R, Venema RC. Interaction of neuronal nitric-oxide synthase with caveolin-3 in skeletal muscle. Identification of a novel caveolin scaffolding/inhibitory domain. J Biol Chem 1997;272(45) 28187-28190.
[66] Garcia-Cardena G, Martasek P, Masters BS, Skidd PM, Couet J, et al. Dissecting the interaction between nitric oxide synthase (NOS) and caveolin. Functional significance of the nos caveolin binding domain in vivo. J Biol Chem 1997;272(41) 25437-25440.
[67] Sbaa E, Frerart F, Feron O. The double regulation of endothelial nitric oxide synthase by caveolae and caveolin: a paradox solved through the study of angiogenesis. Trends Cardiovasc Med 2005;15(5) 157-162.
[68] Isshiki M, Anderson RG. Function of caveolae in Ca2+ entry and Ca2+-dependent signal transduction. Traffic 2003;4(11) 717-723.

[69] Bergdahl A, Sward K. Caveolae-associated signalling in smooth muscle. Can J Physiol Pharmacol 2004;82(5) 289-299.

[70] Wang XL, Ye D, Peterson TE, Cao S, Shah VH, et al. Caveolae targeting and regulation of large conductance Ca(2+)-activated K+ channels in vascular endothelial cells. J Biol Chem 2005;280(12) 11656-11664.

[71] Darby PJ, Kwan CY, Daniel EE. Caveolae from canine airway smooth muscle contain the necessary components for a role in Ca(2+) handling. Am J Physiol Lung Cell Mol Physiol 2000;279(6) L1226-1235.

[72] Liu L, Askari A. Beta-subunit of cardiac Na+-K+-ATPase dictates the concentration of the functional enzyme in caveolae. Am J Physiol Cell Physiol 2006;291(4) C569-578.

[73] Razani B, Lisanti MP. Two distinct caveolin-1 domains mediate the functional interaction of caveolin-1 with protein kinase A. Am J Physiol Cell Physiol 2001;281(4) C1241-1250.

[74] Sampson LJ, Hayabuchi Y, Standen NB, Dart C. Caveolae localize protein kinase A signaling to arterial ATP-sensitive potassium channels. Circ Res 2004;95(10) 1012-1018.

[75] Heijnen HF, Waaijenborg S, Crapo JD, Bowler RP, Akkerman JW, et al. Colocalization of eNOS and the catalytic subunit of PKA in endothelial cell junctions: a clue for regulated NO production. J Histochem Cytochem 2004;52(10) 1277-1285.

[76] Peart JN, Headrick JP. Adenosinergic cardioprotection: multiple receptors, multiple pathways. Pharmacol Ther 2007;114(2) 208-221.

[77] Fox TE, Houck KL, O'Neill SM, Nagarajan M, Stover TC, et al. Ceramide recruits and activates protein kinase C zeta (PKC zeta) within structured membrane microdomains. J Biol Chem 2007;282(17) 12450-12457.

[78] Prevostel C, Alice V, Joubert D, Parker PJ. Protein kinase C(alpha) actively downregulates through caveolae-dependent traffic to an endosomal compartment. J Cell Sci 2000;113 (Pt 14) 2575-2584.

[79] Krajewska WM, Maslowska I. Caveolins: structure and function in signal transduction. Cell Mol Biol Lett 2004;9(2) 195-220.

[80] Sedding DG, Hermsen J, Seay U, Eickelberg O, Kummer W, et al. Caveolin-1 facilitates mechanosensitive protein kinase B (Akt) signaling in vitro and in vivo. Circ Res 2005;96(6) 635-642.

[81] Nilsson R, Ahmad F, Sward K, Andersson U, Weston M, et al. Plasma membrane cyclic nucleotide phosphodiesterase 3B (PDE3B) is associated with caveolae in primary adipocytes. Cell Signal 2006;18(10) 1713-1721.

[82] Murray F, Insel PA, Yuan JX. Role of O(2)-sensitive K(+) and Ca(2+) channels in the regulation of the pulmonary circulation: potential role of caveolae and implications for high altitude pulmonary edema. Respir Physiol Neurobiol 2006;151(2-3) 192-208.

[83] Houslay MD, Baillie GS, Maurice DH. cAMP-Specific phosphodiesterase-4 enzymes in the cardiovascular system: a molecular toolbox for generating compartmentalized cAMP signaling. Circ Res 2007;100(7) 950-966.

[84] Morone N, Fujiwara T, Murase K, Kasai RS, Ike H, et al. Three-dimensional reconstruction of the membrane skeleton at the plasma membrane interface by electron tomography. J Cell Biol 2006;174(6) 851-862.

[85] Richter T, Floetenmeyer M, Ferguson C, Galea J, Goh J, et al. High-resolution 3D quantitative analysis of caveolar ultrastructure and caveola-cytoskeleton interactions. Traffic 2008;9(6) 893-909.

[86] Parton RG, Joggerst B, Simons K. Regulated internalization of caveolae. J Cell Biol 1994;127(5) 1199-1215.

[87] Mundy DI, Machleidt T, Ying YS, Anderson RG, Bloom GS. Dual control of caveolar membrane traffic by microtubules and the actin cytoskeleton. J Cell Sci 2002;115(Pt 22) 4327-4339.

[88] Henley JR, Krueger EW, Oswald BJ, McNiven MA. Dynamin-mediated internalization of caveolae. J Cell Biol 1998;141(1) 85-99.

[89] Oh P, McIntosh DP, Schnitzer JE. Dynamin at the neck of caveolae mediates their budding to form transport vesicles by GTP-driven fission from the plasma membrane of endothelium. J Cell Biol 1998;141(1) 101-114.

[90] Yao Q, Chen J, Cao H, Orth JD, McCaffery JM, et al. Caveolin-1 interacts directly with dynamin-2. J Mol Biol 2005;348(2) 491-501.

[91] Pelkmans L, Puntener D, Helenius A. Local actin polymerization and dynamin recruitment in SV40-induced internalization of caveolae. Science 2002;296(5567) 535-539.

[92] Sharma DK, Brown JC, Choudhury A, Peterson TE, Holicky E, et al. Selective stimulation of caveolar endocytosis by glycosphingolipids and cholesterol. Mol Biol Cell 2004;15(7) 3114-3122.

[93] Lee E, De Camilli P. Dynamin at actin tails. Proc Natl Acad Sci U S A 2002;99(1) 161-166.

[94] Orth JD, Krueger EW, Cao H, McNiven MA. The large GTPase dynamin regulates actin comet formation and movement in living cells. Proc Natl Acad Sci U S A 2002;99(1) 167-172.

[95] McNiven MA, Kim L, Krueger EW, Orth JD, Cao H, et al. Regulated interactions between dynamin and the actin-binding protein cortactin modulate cell shape. J Cell Biol 2000;151(1) 187-198.

[96] Schafer DA, Weed SA, Binns D, Karginov AV, Parsons JT, et al. Dynamin2 and cortactin regulate actin assembly and filament organization. Curr Biol 2002;12(21) 1852-1857.

[97] Krueger EW, Orth JD, Cao H, McNiven MA. A dynamin-cortactin-Arp2/3 complex mediates actin reorganization in growth factor-stimulated cells. Mol Biol Cell 2003;14(3) 1085-1096.

[98] Predescu SA, Predescu DN, Timblin BK, Stan RV, Malik AB. Intersectin regulates fission and internalization of caveolae in endothelial cells. Mol Biol Cell 2003;14(12) 4997-5010.

[99] Stahlhut M, van Deurs B. Identification of filamin as a novel ligand for caveolin-1: evidence for the organization of caveolin-1-associated membrane domains by the actin cytoskeleton. Mol Biol Cell 2000;11(1) 325-337.

[100] Doucey MA, Bender FC, Hess D, Hofsteenge J, Bron C. Caveolin-1 interacts with the chaperone complex TCP-1 and modulates its protein folding activity. Cell Mol Life Sci 2006;63(7-8) 939-948.

[101] Head BP, Patel HH, Roth DM, Murray F, Swaney JS, et al. Microtubules and actin microfilaments regulate lipid raft/caveolae localization of adenylyl cyclase signaling components. J Biol Chem 2006;281(36) 26391-26399.

[102] Ravid D, Chuderland D, Landsman L, Lavie Y, Reich R, et al. Filamin A is a novel caveolin-1-dependent target in IGF-I-stimulated cancer cell migration. Exp Cell Res 2008;314(15) 2762-2773.
[103] Sverdlov M, Shinin V, Place AT, Castellon M, Minshall RD. Filamin A regulates caveolae internalization and trafficking in endothelial cells. Mol Biol Cell 2009;20(21) 4531-4540.
[104] Pelkmans L, Kartenbeck J, Helenius A. Caveolar endocytosis of simian virus 40 reveals a new two-step vesicular-transport pathway to the ER. Nat Cell Biol 2001;3(5) 473-483.
[105] Le PU, Nabi IR. Distinct caveolae-mediated endocytic pathways target the Golgi apparatus and the endoplasmic reticulum. J Cell Sci 2003;116(Pt 6) 1059-1071.
[106] Hayer A, Stoeber M, Ritz D, Engel S, Meyer HH, et al. Caveolin-1 is ubiquitinated and targeted to intralumenal vesicles in endolysosomes for degradation. J Cell Biol 2010;191(3) 615-629.
[107] Gilbert J, Benjamin T. Uptake pathway of polyomavirus via ganglioside GD1a. J Virol 2004;78(22) 12259-12267.
[108] Stang E, Kartenbeck J, Parton RG. Major histocompatibility complex class I molecules mediate association of SV40 with caveolae. Mol Biol Cell 1997;8(1) 47-57.
[109] Allander T, Andreasson K, Gupta S, Bjerkner A, Bogdanovic G, et al. Identification of a third human polyomavirus. J Virol 2007;81(8) 4130-4136.
[110] Feng H, Shuda M, Chang Y, Moore PS. Clonal integration of a polyomavirus in human Merkel cell carcinoma. Science 2008;319(5866) 1096-1100.
[111] Gaynor AM, Nissen MD, Whiley DM, Mackay IM, Lambert SB, et al. Identification of a novel polyomavirus from patients with acute respiratory tract infections. PLoS Pathog 2007;3(5) e64.
[112] Macovei A, Radulescu C, Lazar C, Petrescu S, Durantel D, et al. Hepatitis B virus requires intact caveolin-1 function for productive infection in HepaRG cells. J Virol 2010;84(1) 243-253.
[113] Beer C, Andersen DS, Rojek A, Pedersen L. Caveola-dependent endocytic entry of amphotropic murine leukemia virus. J Virol 2005;79(16) 10776-10787.
[114] Stuart AD, Eustace HE, McKee TA, Brown TD. A novel cell entry pathway for a DAF-using human enterovirus is dependent on lipid rafts. J Virol 2002;76(18) 9307-9322.
[115] Brown G, Jeffree CE, McDonald T, Rixon HW, Aitken JD, et al. Analysis of the interaction between respiratory syncytial virus and lipid-rafts in Hep2 cells during infection. Virology 2004;327(2) 175-185.
[116] Werling D, Hope JC, Chaplin P, Collins RA, Taylor G, et al. Involvement of caveolae in the uptake of respiratory syncytial virus antigen by dendritic cells. J Leukoc Biol 1999;66(1) 50-58.
[117] Cantin C, Holguera J, Ferreira L, Villar E, Munoz-Barroso I. Newcastle disease virus may enter cells by caveolae-mediated endocytosis. J Gen Virol 2007;88(Pt 2) 559-569.
[118] Marjomaki V, Pietiainen V, Matilainen H, Upla P, Ivaska J, et al. Internalization of echovirus 1 in caveolae. J Virol 2002;76(4) 1856-1865.
[119] Tsai B, Gilbert JM, Stehle T, Lencer W, Benjamin TL, et al. Gangliosides are receptors for murine polyoma virus and SV40. Embo J 2003;22(17) 4346-4355.

[120] Engel S, Heger T, Mancini R, Herzog F, Kartenbeck J, et al. Role of endosomes in simian virus 40 entry and infection. J Virol 2011;85(9) 4198-4211.
[121] Jiang M, Abend JR, Tsai B, Imperiale MJ. Early events during BK virus entry and disassembly. J Virol 2009;83(3) 1350-1358.
[122] Liebl D, Difato F, Hornikova L, Mannova P, Stokrova J, et al. Mouse polyomavirus enters early endosomes, requires their acidic pH for productive infection, and meets transferrin cargo in Rab11-positive endosomes. J Virol 2006;80(9) 4610-4622.
[123] Querbes W, O'Hara BA, Williams G, Atwood WJ. Invasion of host cells by JC virus identifies a novel role for caveolae in endosomal sorting of noncaveolar ligands. J Virol 2006;80(19) 9402-9413.
[124] Mannova P, Forstova J. Mouse polyomavirus utilizes recycling endosomes for a traffic pathway independent of COPI vesicle transport. J Virol 2003;77(3) 1672-1681.
[125] Qian M, Cai D, Verhey KJ, Tsai B. A lipid receptor sorts polyomavirus from the endolysosome to the endoplasmic reticulum to cause infection. PLoS Pathog 2009;5(6) e1000465.
[126] Laniosz V, Holthusen KA, Meneses PI. Bovine papillomavirus type 1: from clathrin to caveolin. J Virol 2008;82(13) 6288-6298.
[127] Laniosz V, Dabydeen SA, Havens MA, Meneses PI. Human papillomavirus type 16 infection of human keratinocytes requires clathrin and caveolin-1 and is brefeldin a sensitive. J Virol 2009;83(16) 8221-8232.
[128] Schelhaas M, Malmstrom J, Pelkmans L, Haugstetter J, Ellgaard L, et al. Simian Virus 40 depends on ER protein folding and quality control factors for entry into host cells. Cell 2007;131(3) 516-529.
[129] Geiger R, Andritschke D, Friebe S, Herzog F, Luisoni S, et al. BAP31 and BiP are essential for dislocation of SV40 from the endoplasmic reticulum to the cytosol. Nat Cell Biol 2011;13(11) 1305-1314.
[130] Pitha J, Irie T, Sklar PB, Nye JS. Drug solubilizers to aid pharmacologists: amorphous cyclodextrin derivatives. Life Sci 1988;43(6) 493-502.
[131] Hussain KM, Leong KL, Ng MM, Chu JJ. The essential role of clathrin-mediated endocytosis in the infectious entry of human enterovirus 71. J Biol Chem 286(1) 309-321.
[132] Van Hamme E, Dewerchin HL, Cornelissen E, Verhasselt B, Nauwynck HJ. Clathrin- and caveolae-independent entry of feline infectious peritonitis virus in monocytes depends on dynamin. J Gen Virol 2008;89(Pt 9) 2147-2156.
[133] Moriyama T, Marquez JP, Wakatsuki T, Sorokin A. Caveolar endocytosis is critical for BK virus infection of human renal proximal tubular epithelial cells. J Virol 2007;81(16) 8552-8562.
[134] Macovei A, Radulescu C, Lazar C, Petrescu S, Durantel D, et al. Hepatitis B virus requires intact caveolin-1 function for productive infection in HepaRG cells. J Virol 84(1) 243-253.
[135] Spoden G, Freitag K, Husmann M, Boller K, Sapp M, et al. Clathrin- and caveolin-independent entry of human papillomavirus type 16--involvement of tetraspanin-enriched microdomains (TEMs). PLoS One 2008;3(10) e3313.

[136] Vercauteren D, Vandenbroucke RE, Jones AT, Rejman J, Demeester J, et al. The use of inhibitors to study endocytic pathways of gene carriers: optimization and pitfalls. Mol Ther 2010;18(3) 561-569.

[137] Rodal SK, Skretting G, Garred O, Vilhardt F, van Deurs B, et al. Extraction of cholesterol with methyl-beta-cyclodextrin perturbs formation of clathrin-coated endocytic vesicles. Mol Biol Cell 1999;10(4) 961-974.

[138] Krey T, Thiel HJ, Rumenapf T. Acid-resistant bovine pestivirus requires activation for pH-triggered fusion during entry. J Virol 2005;79(7) 4191-4200.

[139] Imelli N, Meier O, Boucke K, Hemmi S, Greber UF. Cholesterol is required for endocytosis and endosomal escape of adenovirus type 2. J Virol 2004;78(6) 3089-3098.

[140] Liao JK. Clinical implications for statin pleiotropy. Curr Opin Lipidol 2005;16(6) 624-629.

[141] Holschermann H, Schuster D, Parviz B, Haberbosch W, Tillmanns H, et al. Statins prevent NF-kappaB transactivation independently of the IKK-pathway in human endothelial cells. Atherosclerosis 2006;185(2) 240-245.

[142] Sun D, Fernandes G. Lovastatin inhibits bone marrow-derived dendritic cell maturation and upregulates proinflammatory cytokine production. Cell Immunol 2003;223(1) 52-62.

[143] Bolard J. How do the polyene macrolide antibiotics affect the cellular membrane properties? Biochim Biophys Acta 1986;864(3-4) 257-304.

[144] Orlandi PA, Fishman PH. Filipin-dependent inhibition of cholera toxin: evidence for toxin internalization and activation through caveolae-like domains. J Cell Biol 1998;141(4) 905-915.

[145] Schnitzer JE, Oh P, Pinney E, Allard J. Filipin-sensitive caveolae-mediated transport in endothelium: reduced transcytosis, scavenger endocytosis, and capillary permeability of select macromolecules. J Cell Biol 1994;127(5) 1217-1232.

[146] von Gersdorff K, Sanders NN, Vandenbroucke R, De Smedt SC, Wagner E, et al. The internalization route resulting in successful gene expression depends on both cell line and polyethylenimine polyplex type. Mol Ther 2006;14(5) 745-753.

[147] Dangoria NS, Breau WC, Anderson HA, Cishek DM, Norkin LC. Extracellular simian virus 40 induces an ERK/MAP kinase-independent signalling pathway that activates primary response genes and promotes virus entry. J Gen Virol 1996;77 (Pt 9) 2173-2182.

[148] Li E, Stupack DG, Brown SL, Klemke R, Schlaepfer DD, et al. Association of p130CAS with phosphatidylinositol-3-OH kinase mediates adenovirus cell entry. J Biol Chem 2000;275(19) 14729-14735.

[149] Sharma-Walia N, Naranatt PP, Krishnan HH, Zeng L, Chandran B. Kaposi's sarcoma-associated herpesvirus/human herpesvirus 8 envelope glycoprotein gB induces the integrin-dependent focal adhesion kinase-Src-phosphatidylinositol 3-kinase-rho GTPase signal pathways and cytoskeletal rearrangements. J Virol 2004;78(8) 4207-4223.

[150] Cubells L, Vila de Muga S, Tebar F, Wood P, Evans R, et al. Annexin A6-induced alterations in cholesterol transport and caveolin export from the Golgi complex. Traffic 2007;8(11) 1568-1589.

[151] Anderson RG. Caveolae: where incoming and outgoing messengers meet. Proc Natl Acad Sci U S A 1993;90(23) 10909-10913.

[152] Anderson HA, Chen Y, Norkin LC. Bound simian virus 40 translocates to caveolin-enriched membrane domains, and its entry is inhibited by drugs that selectively disrupt caveolae. Mol Biol Cell 1996;7(11) 1825-1834.

[153] Pho MT, Ashok A, Atwood WJ. JC virus enters human glial cells by clathrin-dependent receptor-mediated endocytosis. J Virol 2000;74(5) 2288-2292.

[154] Empig CJ, Goldsmith MA. Association of the caveola vesicular system with cellular entry by filoviruses. J Virol 2002;76(10) 5266-5270.

[155] Aballay A, Stahl PD, Mayorga LS. Phorbol ester promotes endocytosis by activating a factor involved in endosome fusion. J Cell Sci 1999;112 (Pt 15) 2549-2557.

[156] Cohen P, Holmes CF, Tsukitani Y. Okadaic acid: a new probe for the study of cellular regulation. Trends Biochem Sci 1990;15(3) 98-102.

[157] Lajoie P, Nabi IR. Regulation of raft-dependent endocytosis. J Cell Mol Med 2007;11(4) 644-653.

[158] Taylor MP, Koyuncu OO, Enquist LW. Subversion of the actin cytoskeleton during viral infection. Nat Rev Microbiol 2011;9(6) 427-439.

[159] Breuza L, Corby S, Arsanto JP, Delgrossi MH, Scheiffele P, et al. The scaffolding domain of caveolin 2 is responsible for its Golgi localization in Caco-2 cells. J Cell Sci 2002;115(Pt 23) 4457-4467.

[160] Hailstones D, Sleer LS, Parton RG, Stanley KK. Regulation of caveolin and caveolae by cholesterol in MDCK cells. J Lipid Res 1998;39(2) 369-379.

[161] Huang WR, Wang YC, Chi PI, Wang L, Wang CY, et al. Cell entry of avian reovirus follows a caveolin-1-mediated and dynamin-2-dependent endocytic pathway that requires activation of p38 mitogen-activated protein kinase (MAPK) and Src signaling pathways as well as microtubules and small GTPase Rab5 protein. J Biol Chem 2011;286(35) 30780-30794.

[162] Trouet D, Hermans D, Droogmans G, Nilius B, Eggermont J. Inhibition of volume-regulated anion channels by dominant-negative caveolin-1. Biochem Biophys Res Commun 2001;284(2) 461-465.

[163] Roy S, Luetterforst R, Harding A, Apolloni A, Etheridge M, et al. Dominant-negative caveolin inhibits H-Ras function by disrupting cholesterol-rich plasma membrane domains. Nat Cell Biol 1999;1(2) 98-105.

[164] Gianni T, Gatta V, Campadelli-Fiume G. {alpha}V{beta}3-integrin routes herpes simplex virus to an entry pathway dependent on cholesterol-rich lipid rafts and dynamin2. Proc Natl Acad Sci U S A 107(51) 22260-22265.

[165] Lakadamyali M, Rust MJ, Zhuang X. Endocytosis of influenza viruses. Microbes Infect 2004;6(10) 929-936.

[166] Shatz M, Lustig G, Reich R, Liscovitch M. Caveolin-1 mutants P132L and Y14F are dominant negative regulators of invasion, migration and aggregation in H1299 lung cancer cells. Exp Cell Res 2010;316(10) 1748-1762.

[167] Conner SD, Schmid SL. Regulated portals of entry into the cell. Nature 2003;422(6927) 37-44.

[168] Insel PA, Patel HH. Do studies in caveolin-knockouts teach us about physiology and pharmacology or instead, the ways mice compensate for 'lost proteins'? Br J Pharmacol 2007;150(3) 251-254.

Chapter 9

Endocytosis of Non-Enveloped DNA Viruses

Maude Boisvert and Peter Tijssen

Additional information is available at the end of the chapter

1. Introduction

Endocytosis is critical for the internalization and maturation of many DNA viruses. Since all non-enveloped DNA viruses replicate in the nucleus, they can also take advantage of endocytosis trafficking for transport from the cell surface to the nucleus. The internalization often involves clathrin-mediated endocytosis (ClaME), but also macropinycytosis, caveolae-mediated endocytosis (CavME) and other less characterized internalization mechanisms. Viruses strive to avoid evidence of cell entry, and thus of immune attack, and therefore as much as possible take advantage of existing systems. Consequently, virus entry has been extensively studied to understand endocytosis. In contrast to inert cargo, like dextran, well-defined virus particles can be mutated to change biological properties, e.g. receptor interaction, maturation steps or penetration of the endosomal membrane to the cytosol, and thus provide additional tools for the study of endocytosis. However, the use of viruses also poses major challenges since (i) the virus particle population may be heterogeneous; (ii), only one out of a thousand particles may be infectious; (iii), viruses may be using different pathways at the same time; (iv), viral entry pathways may depend on the cell types used; (v), some viruses, eg. papillomaviruses are difficult to reproduce in tissue culture; (vi) the mobility of the infecting particles may be heterogeneous; and (vii), viruses often developed mechanisms to avoid infected cells (e.g. viral neuraminidases). The use of purified viruses and cell cultures, although imparting useful models, may not reproduce the *in vivo* situation. Moreover, viruses may have adapted to *in vitro* cells and may use other pathways then *in vivo*. In this chapter, we focus on entry of non-enveloped DNA viruses, such as circoviruses, parvoviruses, polyomaviruses, papillomaviruses, adenoviruses and iridoviruses. Particularly, we will highlight the effects of binding of these viruses to the cell surface, the internalization and endocytosis processes, and escape from endosomes by breaching the endosomal barrier. The wide array of strategies employed by these viruses, even within the same virus family, will be highlighted. An understanding of these processes, which may result in a plethora of effects on cells, is essential for a wide variety of applications in basic research as well as providing blueprints for applied usages such as gene therapy.

2. Overview of non-enveloped DNA viruses and tools to study their entry

2.1. Non-enveloped DNA viruses

Viruses can be used as a tool to study endocytosis since their defined structure can be modified specifically by site-directed mutagenesis to study different parts of the endocytic pathways. Functions related to virus entry are embedded in the virus capsid. Therefore, we will summarize here the structures and properties of these viruses. All, except the iridoviruses, have been solved by X-ray crystallography and their properties are recapitulated in Table 1.

2.1.1. Circoviruses

Circoviruses are among the smallest DNA viruses; the diameter of the icosahedral particle is about 20 nm. The virus has a circular, 2 kb single-stranded DNA genome (ssDNA). The T=1 capsid consists of 60 subunits (1, 2). The best-known species infect pigs and birds. Soluble monomers of the porcine circovirus 2 (PCV2) capsid have been expressed in *E.coli* and could be assembled into virus-like particles (VLPs). X-ray crystallography (2) revealed two protrusions emanating from the icosahedral 2- and 5-axes. PEPSCAN analysis (3) and structural analysis (2) identified 4 surface epitopes, two epitopes at the subunit interface near the 3-fold axis and two at the interior surface. The interior epitope mapped to the N-terminus of the subunit, which, as for parvoviruses and some other animal viruses, may be transiently externalized. This domain may have membrane-disrupting activity required for infectivity. PCV2 binds to heparan sulfate receptors on cells (4). Clefts that bind sulfate ions, and are positively charged at neutral pH, on the exterior surface of the PCV2 structure surrounding the 2-fold axis are highly conserved. Heparan sulfate may thus bind to this region.

2.1.2. Parvoviruses

The icosahedral viral capsid of parvoviruses infecting vertebrates is made up of two or three proteins that have large common C-terminals part and different N-terminal extensions, due to different translation initiations or proteolysis. The rugged capsid structure is very resistant to acids, bases, solvents such as chloroform or butanol, and temperatures to beyond 50°C. The X-ray structures of many parvoviruses have been solved, but these structures lack the N-terminal extensions (5-10). Parvoviruses, such as the minute virus of mice (MVM), package their linear ssDNA genome into preformed capsids, in a process that is probably driven by a virus-encoded NS1 (helicase). Each of the twelve 5-fold vertices have a roughly cylindrically shaped pore created by the juxtaposition of 10 antiparallel beta-strands, two from each of the 5-fold-related capsid proteins. Mutant capsids that have their 5-fold channels blocked are unable to package DNA, strongly suggesting that the 5-fold pore is the packaging portal for genome entry (11). It may also be the DNA delivery site after the conformational changes occurring in these pores in the endosome. Virtually all parvoviruses also contain a phospholipase A2 activity (PLA2) in the N-terminal extension of

Family	Structure[2]	Capsid composition	Entry mode	Escape
Circoviridae[1] 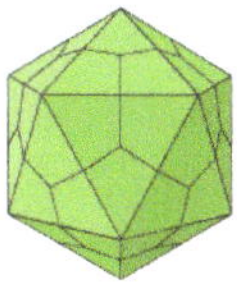	PDB : 3R0R 	T=1 60 proteins/subunits 20 nm	ClaME	Uncoating by serine proteases; Escape early (not at very low pH)
Parvoviridae[1] 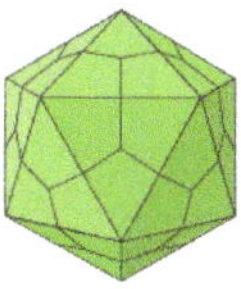	PDB : 2CAS 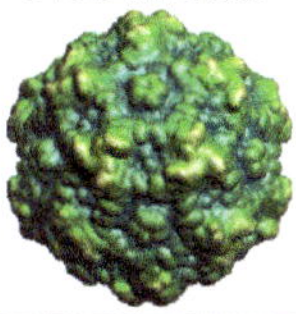	T=1 60 proteins/subunits 25 nm	Majority: ClaME Also: CavME Macropino-cytosis CLIC/GEEC	LE / lysosomes Viral PLA2 activity
Polyomaviridae[1] 	PDB : 1SVA 	T=7 360 VP1 proteins (72 capsomers) 45 nm	CavME	ER, Viral VP2 cellular proteins BAP31 and BiP
Papillomaviridae[1] 	PDB : 1L0T 	T=7 360 L1 proteins (72 capsomers) 55 nm	Majority : ClaME Other : CavME Macropino-cytosis	Cellular Nexin-17 viral capsid protein L2
Adenoviridae[1] 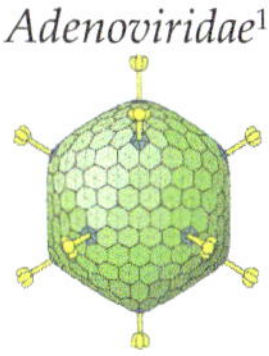	PDB : 1VSZ 	Pseudo-T=25 720 hexon proteins (240 subunits) 12 penton bases Cement proteins Fiber proteins 90 nm	Majority : ClaME Others : Macropino-cytosis	Sorting endosomes Penton base, L3/p23 protease, protein VI.
Iridoviridae 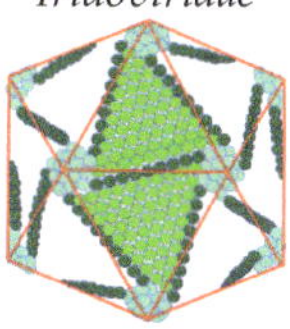	PDBe[3] : 1580 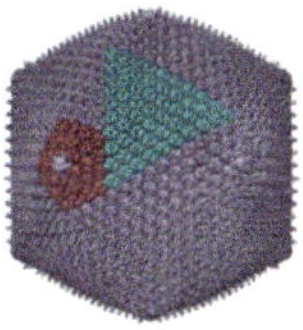	T=147 1460 hexon proteins (each forms trimer) 12 pentameric vertexes Fiber proteins (CIV) 185 nm	ClaME CavME	May involve partial uncoating and internal membrane

[1]Virus models, except of iridovirus, from Viral Zone, EXPASY (http://viralzone.expasy.org/); 2Crystal structures from VIPER data base (http://viperdb.scripps.edu/); 3CIV iridovirus cryo-EM structure is from http://www.ebi.ac.uk/pdbe/. In the iridovirus model, the trisymmetrons are depicted in green (quasi-equivalent capsomers in dark green) and pentasymmetrons in turquoise.

Table 1. Synopsis of nonenveloped DNA viruses and entry and exit of endosomes

the largest capsid protein (12, 13). This domain is located inside the capsid and needs to be externalized for it to be able to breach the endosomal membrane barrier (12). These multistep conformational changes occur in the endosomes.

2.1.3. *Polyomaviruses*

The 45 nm diameter T=7 icosahedral particles polyomaviruses virions (SV40, BK virus) are non-enveloped. The crystal structure of SV40 has been solved and found to be organized unlike other viruses (14-16). The 3 capsid proteins, VP1-3, form 72 pentameric capsomers, 60 hexagonally coordinated plus 12 pentamerically coordinated (at the vertices), i.e. the capsomers at the 5-fold axis have all 5 neighboring capsomers whereas the other 60 have 6 capsomer neighbors via different intercapsomer interactions. The 72 pentameric capsomers in each virus particle contain together 360 copies of VP1 plus 30-60 copies each of VP2 and VP3, i.e. ~1 copy in the tapering cavity inside of each pentamer. VP2 is important during cell entry. Each copy of VP1 has a sialic acid binding site on the surface and these are part of the receptor-binding site for the virus; hence the particles have haemagglutinating properties. The long C-terminal arms of VP1 stabilize the interpentameric contacts by invading neighboring capsomers. VP2 is myristylated at its NH2-terminus, and is also believed to be important in holding the particle together (17). Recently, it was also shown that the cysteine at position 9 in VP1 forms C9-C9 disulfide bonds and thus contributes to intercapsomeric crosslinks (18).

2.1.4. *Papillomaviruses*

Previously, the Papovaviridae family contained both the Papillomaviridae and the Polyomaviridae. Now, these two are separate families but their earlier co-classification indicates a certain degree of similarity (19, 20). The diameter of the papillomaviruses is about 20% larger than that of polyomaviruses, mainly due to the larger L1 as compared to VP1 (21, 22).The atomic structure resembles that of polyomaviruses (23), in particular the core capsomers, and the 55 nm diameter T=7 icosahedral particles also have 72 pentameric capsomers (20, 24, 25). However, the cysteine residues involved in intercapsomer disulfide bonds are C175 and C428 (23).

2.1.5. *Adenoviruses*

The over 50 serotypes of the *Adenoviridae* family can be categorized in about 6 subgroups. The icosahedral outer capsid of these viruses surrounds a (double stranded) dsDNA-protein core and has a diameter of about 90 nm (26). The most prevalent protein in the capsid are 720 subunits of hexon proteins arranged in 240 trimers (12 hexon trimers on each of the 20 facets); penton base proteins are found at each of the 12 vertices. A unique "spike" or trimeric fiber is associated to each of these penton bases. Moreover, the penton base contains an exposed arginine-glycine-aspartate (RGD) motif and both this motif and the fiber are involved in cell attachment. Minor capsid proteins (III, VI, VIII and IX) are also present to stabilize the capsid. The pseudo-T=25 capsid is the largest virus so far of which the near-atomic structure has been solved by X-ray crystallography (27, 28). The fibers were shortened in order to make crystallization possible. An unusual symmetry mismatch occurs between the fivefold symmetric penton base and the threefold symmetric fiber protein. This has a potential impact on cell receptor interactions and subsequent disassembly in the endosomal pathway. In this respect, the ability of the penton base to adapt large changes, resulting in very different central pore sizes, is striking. The large-pore, but not the small-pore, conformation would allow insertion of the fiber. These conformational changes seem to reflect early events in cell entry when the fiber is released from the penton base.

2.1.6. Iridoviruses

Iridoviruses have recently been reviewed in detail (29). Currently, the *Iriodviridae* family is divided into five genera: *Iridovirus, Chloriridovirus, Ranavirus, Lymphocystivirus*, and *Megalocytivirus*. Chilo Iridescent Virus (CIV) is the type specie of this family. Although, it is unlikely that the structures of these large icosahedral dsDNA viruses are amenable to X-ray crystallographic analysis, various details about their assembly and structure have been obtained by cryo-electron microscopy (cryo-EM) and 3D reconstruction (30, 31). Enveloped particles that acquire an envelope by budding from the plasma membrane seem also to exist for all iridoviruses and may be more infectious than naked particles (29). The CIV T=147 capsid has a diameter of about 185 nm. Its dsDNA-protein core is surrounded by a lipid bilayer, derived from the endoplasmic reticulum that follows the contour of the outer icosahedral capsid shells. The external capsid is essentially consistent with the classical quasiequivalent symmetry prediction. It has 1460 hexameric capsomers and 12 pentameric vertex complexes that are organized in 20 trisymmetrons and 12 pentasymmetrons. Each capsomer in CIV, but not in frog virus type 3 (FV3), has a central fiber that appears to be extended about 35 nm beyond the capsid surface. These flexible fibers are probably the first component that comes into contact with the host cell. The minor zip and finger proteins are suggested to stabilize the capsid by acting as intercapsomer cross-links and could be important during virus conformational changes during entry. Furthermore, the minor anchor protein appears to extend into the internal lipid membrane for further stabilization. Although the inner membrane is generally 4 nm thick, it is only 3 nm in regions just below the pentameric complex. Small icosahedral viruses are usually assembled from monomers having hexamers and pentamers, or only pentamers, in quasi-equivalent environments. The larger viruses, like adenovirus, iridoviruses and large algal and bacterial viruses, have significantly greater coding capacities that have allowed divergent evolution and specialized pentameric complexes. In contrast to adenovirus, CIV does not have fibers associated with its pentameric complexes but a complex mushroom-like structure that has the appearance of a five-blade propeller. This propeller could have an important role in assembly but a role in entry is not excluded.

2.2. Tools used to study endocytosis and trafficking of viruses

2.2.1. Imaging

Transmission electron microscopy (TEM) is still widely used, and enables visualization of virus in endocytic cups (size and shape, presence of coats). For instance, TEM imaging demonstrated that Adeno-associated virus 5 (AAV5) can enter cells in non-coated vesicles, pointing to a non-clathrin mechanism (32). Immunohistochemistry using gold-labeled antibody can be used in TEM to colocalize structures. Scanning electron microscopy (SEM) facilitates 3D imaging that can be helpful to see ruffles on cell surface, as shown in an AAV2 study (33).

Fluorescence, particularly confocal microscopy, is an excellent tool to visualize viruses and cellular structures. It is frequently used to confirm use of a pathway by co-localisation of the virus and pathway components. It can also be used in live imaging to see the movements of the virus along cellular structures, such as plasma membrane (filopodia) and actin or tubulin structural networks. One should be aware that conformational epitopes may change during the endosomal pathway and that some viruses have very high GCE/FFU ratios (genome copy equivalent/foci forming unit), so that an excess of non-infectious virus in the viral preparation could mask the real infectious pathway. For further details, see chapter by Aaron and Timlin.

2.2.2. Chemical inhibitors

Chemical inhibitors are widely used and provide an excellent basis for endocytosis studies of a particular cargo (34) (Table 2). These are usually not expensive, easy to work with, and have the major advantage of providing a uniform treatment of the cells. None of the extensively used pharmacological inhibitors have complete specificity, but, some parameters can be controlled to limit the possibility of artefacts. Dose-response curves to establish lowest concentrations and shortest exposition times possible to limit side effects on the cells are valuable, and limit upregulation of compensatory mechanisms. It is also easy to wash out the inhibitors to recover a normal state of the cells, in the case of reversible inhibition. This also remove virus that did not yet penetrate the cells, and allows continuing the infection with only endocytosed virus in normal conditions of the cells. Confirmation with another inhibitor, without the same side effects, is convenient. A combination of methods with siRNA or dominant negative/constitutive mutants can be exploited to identify the entry pathway of a specific cargo. Readers are referred to selected literature for further details on any particular inhibitor for comprehensive discussions (34, 35).

2.2.3. Dominant-negative/constitutive mutants

Dominant-negative mutants drive the expression of non-functional versions of the protein of interest (Table 3). The regular protein is still present but strong expression of the mutant form will result in an inactive pathway (35). The main challenges using dominant-negative mutants is to choose the appropriate protein and to achieve high efficiencies in the transfection of the constructs. Stable transfection, with selection of the transfected cells, is not a good choice, since long (constitutive) treatments that donwnregulate a pathway is prone to result in the upregulation of compensatory pathways that could be used by the cargo. Using dominant-negative mutants can also be more time-consuming and become more costly than chemical inhibitors. A constitutively active protein can block a pathway at a certain step, e.g. constitutively active Rab5 protein inhibits the passage from early endosomes and late endosomes (see section 4.2.2.). There is also a complex signalisation process involved in endocytosis, and any modification can have some side effect on the cells.

Pathway	Inhibitor	Mechanism of action	Side effect
Clathrin	Hypertonic sucrose	Dispersion of clathrin sub-units, preventing basket formation	Inhibits CavME and macropinocytosis (high doses); Induces important remodelling of the actin network
	Chlorpromazine	Clathrin and AP2 relocalized at the endosomes	Inhibits phagocytosis and formation of large vesicles
	Potassium depletion	Prevents clathrin basket formation	Reduces other non-clathrin endocytosis mechanisms; Alteration of actin cytoskeleton
	Cytosol acidification	Blocks budding of vesicles	Inhibition of macropinocytosis; Alters actin cytoskeleton
Caveolae	Nystatin	Induces cholesterol sequestration and aggregation	Can affect macropinocytosis at high concentrations
	Methyl-β-cyclodextrin	Disperse cholesterol, mislocalisation of caveolin-1	High concentration leads to inhibition of ClaME and fluid phase endocysotis; Affects actin structure
	Cholesterol oxydase	Relocalization of the caveolin-1 to the Golgi	Action of the enzyme is sensitive to several factors that may lead to false negatives
Macropino-cytosis	Amiloride	Inhibits Na^+/H^+-ATPase exchangers; Prevents membrane extension formation	Can affect ClaME; Can reduce lipid raft internalization; Alters actin structure
	Cytochalasin D	Prevents membrane projection by actin	Other mechanisms depend on actin and membrane modelling
	Wortmannin	Inhibition of PI3K, important for formation and fusion of vesicles	Not exclusive for macropinocytosis
Acidification of endosomes	Bafilomycin A	Block the V-ATPase proton pumps	No effect on recycling pathway, prevents transfer from EE to LE
	Weak base (NH_4Cl, chloroquine)	Gently raise the pH of the compartment;	Does not affect the transport of cargo to the lysosomes

Table 2. Chemical inhibitors to study endocytosis of viruses

Pathway / component	Protein	Mode of action
ClaME	EPS15	Deletion of EH domains results in loss of recruitment of clathrin and AP2 at endocytosis site (36)
	Clathrin	Overexpression of the HUB domain in the cells leads to the inhibition of the cage assembly at endocytosis site.
CavME	Caveolin-1	Overexpression of the WT protein stabilizes the caveolin structure located at the plasma membrane, thus prevent membrane curvature and/or vesicle budding
	Caveolin-3	Deletion of N-terminal, or fusion with EGFP in N-terminal inhibits endocytosis (37)
Dynamin	Dynamin-1/ Dynamin-2	K44A is widely used. Dynamin is involved in ClaME, CavME and other less described non-ClaME/ non-CavME mechanisms
Macropino-cytosis	Pak-1	Important in the regulation of vesicle cycling
	Rac-1	Important for membrane ruffling, inhibited with N17Rac1(33); Blocks cell cycle in G2/M phase
Other pathways	GRAF1	Key player in the newly described CLIC/GEEC pathway (33, 38)

Table 3. Dominant-negative / constitutive mutants

2.2.4. Gene silencing

Another way to target a particular protein, as opposed to chemical inhibitors and dominant-negative mutants, is gene silencing. Several endocytosis studies use transfection of small interfering RNA (siRNA) or small hairpin RNA (shRNA) to specifically knock down key proteins of different endocytosis pathways (see section 4.3). This method is relatively simple and selective, but depends on good delivery of the interfering RNA. A good knockdown of the target protein needs to be confirmed by Western blot analysis, and only then can endocytosis studies can be performed. However, since there are some entry pathways not fully described yet, it remains a possibility that some proteins that we believe specific to a certain entry mode will be demonstrated in the future to be involved with other pathways.

3. Docking of virus to cell surface and induced signalling pathways

3.1. Virus attachment to cells

The first interaction between a virus and a cell is crucial to promote endocytosis. In its simplest form, the virus binds to a cell surface receptor and triggers directly its endocytosis. However, several viruses have either multiple receptors, or they also bind abundant structure on the cell surface, e.g. as sialic acids, before binding to the "real" receptor. In its most sophisticated form, the virus needs to bind to a primary receptor to trigger conformational changes that will allow binding to a secondary receptor that will in turn trigger endocytosis. Binding to the cell surface is also the first element that can determine viral tropism, if the virus uses a specific receptor present only on certain cell type.

3.1.1. Cell attachment by circoviruses

Monocyte/macrophage lineage cells are target cells PCV2 replication. Misinzo et al. (4) used the porcine monocytic cell line 3D4/31, that supports PCV2 replication in vitro, and glycosaminoglycans (GAG), used by several viruses as receptors, for attachment studies. They observed that heparin, heparan sulfate (HS), chondroitin sulfate B (CS-B), but not CS-A, and keratan sulfate reduced PCV2 infection when these GAG were incubated with PCV2 prior to and during inoculation of 3D4/31 cells. Also, enzymatic removal of HS and CS-B prior to PCV2 inoculation of 3D4/31 cells significantly reduced PCV2 infection. Similarly, when PCV2 virus-like particles (VLP) were allowed to bind onto 3D4/31 cells in the presence of heparin and CS-B, attachment was strongly reduced. This was confirmed for the wild-type virus. Together, these results demonstrated that HS and CS-B are components of the attachment receptors for PCV2.

3.1.2. Cell attachment by parvoviruses

The *Parvoviridae* family is very large, and its members infect a wide range of hosts from invertebrates to vertebrates such as humans. Family members display different strategies to complete their replication cycle. Most parvovirus interactions with cells are neuraminidase-sensitive, indicating binding to abundant sialoglycoproteins, e.g. porcine parvovirus (PPV) (39), dependoviruses (AAVs) (40), minute virus of mice (MVM), canine (CPV) and feline (FPV) parvoviruses (41). These glycan-specific interactions occur usually in the depressions around the twofold symmetry axes. Specific receptors are not currently known for all parvoviruses. One of the best characterized interactions is between CPV/FPV and the transferrin receptor (TfR) (42). CPV and FPV is a well-documented evolution of virus characterised by species jump. Thus FPV, a virus that can infect only cats, acquired two mutations on the capsid protein, enabling the binding to the canine TfR and thus emerge as CPV2, a virus that could only infect dogs, and that caused serious health problems in dogs in the late 1970's. Shortly after, the CPV2a emerged and this virus was able to infect both species, because it could bind to both receptors. PPV, that can bind and enter many cell types, but that can complete the replication cycle only in few cell types (43).

3.1.3. Cell attachment by polyomaviruses

SV40 binds cell surfaces via major histocompatibility complex 1 (MHC-1) (37). This binding provokes the clustering of the virus in lipid rafts, and entry by CavME. Infection can be efficiently inhibited by pre-incubation of the cells with antibodies directed at the MHC-1. It was observed that initial binding (when endocytosis is inhibited by incubating the cells at 4°C) occurs outside of the lipid raft, and that at that point, there was no co-localisation with caveolin (37). However, when the cells are at 37°C, translocation to the caveolin-positive domain is fast and efficient.

3.1.4. Cell attachment by papillomaviruses

Uptake of papillomaviruses is very slow, and the virus can stay at the cell surface for several hours and the half-time of uptake can be as high as 14 hours. A major issue with this family of viruses that have slowed progress in their research was that, in the host, viruses only complete their replication cycle in terminally differentiated keratinocytes (44). Thus, it was only since the 2000's that virus could be efficiently produced, using "packaging cell lines" (45, 46). It is also hard to infect cells *in vitro*, since there are only a few permissive cell lines such as the HaCaT cells. There is still limited information concerning the uptake and nuclear delivery of those viruses. Concerning binding, growing evidence suggests that the extracellular matrix (ECM) can provide the first binding of the virus. Capsid protein L1 binds to the heparan sulfate proteoglycans (HSPG), or laminin 5 present in the ECM. Then the virus is transferred to the cellular filopodia, and slowly translocated to the body of the cell. Binding to this primary receptor will trigger conformational changes, lowering affinity for the primary receptor, and allowing binding to a still unidentified secondary receptor. The minor L2 capsid protein, usually not accessible at the virus surface will become exposed, and cleaved by cellular furin. This cleavage is absolutely required for successful infection and it is believed to be important later in the endosomal escape. Although *in vitro* binding to the surface receptor is critical, the *in vivo* situation is different in that the basement membrane is the primary site of binding (47).

3.1.5. Cell attachment by adenovirus

The first high-affinity interaction between the cell and the virus occurs between the globular knob of the C-terminal segment of the viral fiber protein and the cellular CAR (Coxackie and Adenovirus Receptor, belonging to the Ig superfamily) or CD46, for sub group C and B adenovirus respectively. The second interaction, of low affinity is between the RGD motive on the viral penton base and the cellular integrins. Together they will form a strong and irreversible connection between the cell and the virus, leading to endocytosis (48).

3.1.6. Cell attachment by iridovirus

Little is known about iridovirus receptors; although the central fibers of the capsomers have been proposed to be involved (31) not all iridoviruses have these fibers. Eaton et al. showed that anti-FV3 serum present at the time of FV3 (*Ranavirus*) infection enhances infectivity of the virus in non-immune teleost cell lines, but not in a non-immune mammalian cell line, suggesting a cell surface receptor specific to teleost cell lines (49). They observed that this antibody dependent enhancement (ADE) of FV3 was dependent on the Fc portion of anti-FV3 antibodies but not to complement.

3.2. Induced signalling pathways

Signalization from cell surface is an important process in coordination of different events of an organism. Readers are referred to two extensive reviews on the subject for further details (50,

51). Signalling also regulate endocytosis processes. Binding to a receptor can trigger signalling pathways that promote endocytosis by upregulating the quantity of clathrin at the cell surface. Src kinase was shown to increase new clathrin basket formation by phosphorylation of the clathrin, leading to assembly of the triskelions (52). Activation of p38 increases Rab5 recruitment to plasma membrane, promoting endocytosis (53). Raf kinase is implicated in recycling of the transferrin receptor to the cell surface (54). In the case of viruses, binding of the virus to the cell surface receptor can trigger other endocytosis mechanisms such as macropinocytosis which is not a constitutive process (55). Differential endocytosis can also be used by the cell to control the level of signalling. For example, activated EGFR can be endocytosed by ClaME or CavME. CavME is used when the level of activation is particularly high and lead to degradation of the receptor, while ClaME leads to recycling of the receptor.

4. Entry mechamisms

4.1. Initial steps of entry

Viruses have developed multiple ways to hijack cellular entry processes ideally without any remaining trace at the cell surface to prevent detection by the immune system. The three best understood mechanisms are clathrin-mediated endocytosis (ClaME), caveolae endocytosis (CavME) and macropinocytosis (Figure 1).

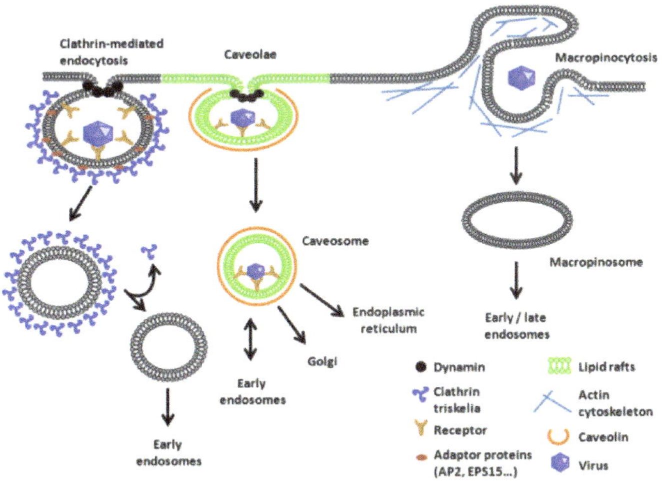

Figure 1. Major endocytosis pathways. Major components of 3 most described endocytosis mechanisms. Main intracellular targets are indicated (e.g. early endosome for most, but also others such as endoplasmic reticulum and Golgi for the CavME

ClaME is probably the best characterized cellular uptake mechanism. Cellular uptake was first observed using transmission electron microscopy (TEM) in the 1960's. One main characteristic is the assembly of clathrin proteins forming basket shape vesicles. As a result it was postulated that the size of material that can be taken via ClaME is limited in order to form the basket clathrin cage. ClaME proceeds in five steps (56). First, it requires specific receptor engagement to trigger intracellular signaling required for the beginning of membrane modeling, a process called "nucleation". The key element is the highly conserved adaptor protein AP2, recruited from the cytoplasm to the membrane. The next step is the "cargo selection". At this stage there will be several interactions between the cargo, its receptor and several adaptor and accessory proteins. The tetrameric AP2 adaptor protein, like clathrin, has disproportionally large numbers of interactors, such as dynamin and Eps15, and is a hub of protein networks. Accessory proteins may swap from the AP2 appendage hub to the clathrin hub during coated pit assembly. Clathrin protein will also be recruited at the membrane in the "clathrin coat assembly" step. Clathrin proteins are recruited as triskelions and will be polymerized into the basket shape forming pentagons and hexagons. When the polymerization reaches the point where only a neck shape remains between the forming coated vesicle and the plasma membrane the dynamin protein will be recruited to perform the "vesicle scission" step required for the budding. GTP hydrolysis will be necessary to complete the process. The final step will be the "uncoating and recycling" step. This uncoating is important before the nascent vesicle can move on and fuse with the endosomes. After this, there will be a sorting step deciding whether the vesicle is simply returned to the plasma membrane or pushed forward into the endosomal pathway.

CavME main characteristic is its origin in lipid rafts, small domains rich in sphingolipids and cholesterol (57). This endocytosis mechanism thus depends on cholesterol homeostasis of the cell. Caveolin is a membrane protein that binds to the cholesterol, as a result depletion of cholesterol leads to mis-localization of caveolin and inhibition of the pathway. Several components are involved such as caveolin-1 (caveolin-2 can be present but not required, caveolin-3 is present primarily in muscles), dynamin and actin (implicated in caveolin rigidity/motility). Vesicles acquire caveolin coats while forming; the released caveosomes will keep a neutral pH and can interact with the endosomal pathway, the Golgi system, or as in the case of SV40 endocytosis, transit to the smooth endoplasmic reticulum (37).

Macropinocytosis is an active uptake pathway that requires activation in most cells types (58). Activation of the pathway will lead to intense actin and microfilaments modulation and membrane ruffling. By definition pinocytosis is the uptake of fluids and membranes by the cell ("fluid phase endocytosis"). The main difference with other endocytosis mechanisms is that, once activated, uptake of material will not depend on receptor binding. In macropinocytosis, the membrane ruffles will or will not be closed and the size of the macropinosome is not directed by the cargo. The diameter will usually range between 0.5 to 10 μm, and thus is larger than in other endocytosis mechanisms. In contrast to phagocytosis, that is limited to few cell types, most cells are capable of macropinocytosis. The closure of the macropinosomes will require several cellular factors. Myosins will provide contractile activity, and fusion factors together with kinases, GTPases, will help the process. Once in the cytoplasm, macropinosomes will acidify and interact with the endosomal pathway.

Phagocytosis is a pathway limited to specialised immune cells such as macrophages, dendritic cells, neutrophils and monocytes. This pathway is mostly used by bacteria and fungi. Recent studies reported the use of this pathway also by herpes simplex virus (59) and mimivirus (60), both displaying membranes. The non-enveloped DNA viruses that fall in the scope of this chapter do not use it.

Novel pathways were revealed by investigating entry mechanism of viruses. They are often referred to as clathrin independent endocytosis. Since molecular components involved in these mechanisms depend not only on cargo but also on cell type, several groups suggested classifying them as dependent or independent of such components. For example, no matter which mechanism is involved the forming vesicle will need to be pinched of the cellular membrane. While ClaME, uses dynamin proteins some clathrin-independent endocytosis, were reported to be dependent on dynamin and some are independent.

4.2. Endosomal pathway

Regardless of the endocytosis pathways, virtually all the vesicles formed during entry will fuse or at least interact with the early endosomes, and acquire the Rab5 and EEA1 markers (61). After endocytosis, cargos enter the highly-coordinated endosomal pathways (62). General endocytosis is a highly active and dynamic process. In a mammalian cell up to 180%, of the cell surface will be endocytosed each hour (62). A simple model of the endosomal pathway is shown in Figure 2.

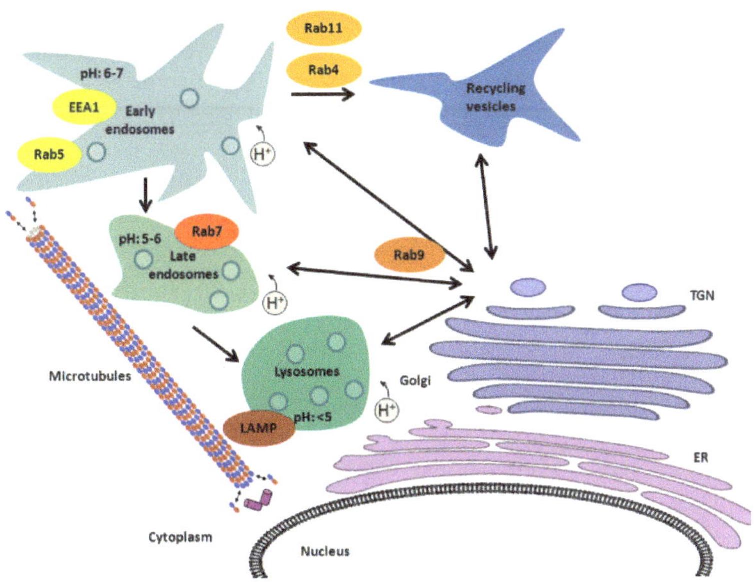

Figure 2. Endosomal pathway. Most endocytosis mechanisms lead to the early endosomes. Sorting will determine if the vesicle is recycled to the cell surface, caveosomes or continue in the degradation pathway towards the lysosomes. Acidification occurs throughout the endosomal pathway. Crosstalk between the *trans*-Golgi network (TGN) and the different compartments provides cellular components to the pathway. Transport of the vesicles on microtubules ensures steady traffic towards the nucleus. Rab proteins are regulators of the pathway and can be used as markers.

The first compartment encountered is the early endosome (EE) which is slightly acid (pH~6.5), highly pleomorphic, large and tubular shaped. The EEs are the main organelle where sorting will take place. The main markers of EEs are Rab5 and EEA1 proteins. The tubular projection contains microdomains for sorting, and will generate vesicles targeted to the recycling endosomes (with Rab4 and Rab11), or the *trans*-Golgi network (TGN). The TGN is constantly communicating with the endosomes at all stages of the pathway. It provides components necessary for the maturation of the endosomes. A marker for the shuttles between the TGN and the endosomes is Rab9. All endosomes contain internal vesicles that are used to isolate certain components such as receptors. While endosomes proceeds to the end of the pathway, these internal vesicles increase greatly in numbers. It has become clear for polyomaviruses and papillomaviruses that EE shuttle also to caveosomes (section 5). Maturation of the EEs into late endosomes (LE) includes replacement of the Rab5 proteins by Rab7 proteins (63). The maturation of EE in LE will create intermediate endosomes containing both Rab5 and Rab7 markers. The displacement of Rab5 by Rab7 begins by the recruitment of Rab7 by the activated form of Rab5. This recruitment will lead to the inactivation and dissociation of Rab5. This sequence is absolutely required for EE conversion into LE. The newly formed Rab7-positive endosomes will recruit specific proteins for the maturation of the LE.

LEs have an increased regular oval shape, their pH is more acidic (pH~5.5), and they contain many internal vesicles (62). LEs are generated close to the cell surface and will mature while moving towards the nuclear periphery. Fusion events will occur between LEs, and between LE and lysosomes. The lysosome is highly acid ($pH<5$) and contains various hydrolases. Acidification of the endosomes is provided by proton pumps within the membrane of the endosomes, the V-ATPases. The gradual acidification of the pathway is regulated by the concentration of these pumps, the isoforms present, and the presence of other complexes. Markers of the lysosomes are the LAMP proteins. In addition to transport of cargo to the lysosomes, there is also transport of cellular components to the lysosomes. The different hydrolases and membrane proteins are renewed to conserve the lysosome integrity, acidic pH and activities. Transport comes mostly from the TGN via the late endosomes. Mobility of the endosomes is dynamic, and involves both major transport on actin and the microtubule (Mt) networks. Also, transport is not unidirectional, but oscillating, since the vesicles interact both with the kinesins (transport toward cell periphery) and the dyneins (transport toward nucleus) on the Mts. As the maturation of the endosomes proceeds, the net movement will be toward the nucleus (64). Movement of the LEs toward the nucleus is important to reach the lysosomes (mostly located in the nuclear periphery). The actin network is also important for transport of the EEs to the LEs, and to generate small vesicles from the EEs that are targeted to the TGN or from LEs toward lysosomes. Actin is also important for fusion between the LEs and lysosomes.

There are more than 60 different Rab GTPases that have specific membrane localizations. Rab small GTPases are important regulators of the endosomal pathway (65) and thus they can be used as markers, or as inhibitors of specific parts of the pathway by expression of dominant-negative form. The regulation switch is the association of the Rab protein with either GDP or GTP, (66). Rab proteins are involved in multiple events. Rab5 is implicated in the recruitment of the clathrin subunits, with other regulator proteins, such as GAPGDI, will enable uncoating of the vesicle. Rab proteins were also shown to interact with actin (myosin) and microtubule (dynein and kinesin) motors. An example is the Rab11 family of proteins that link recycling vesicles to myosin (67). Also an effector of the Rab7 protein interacts with the dynein motor to promote minus-end traffic on the microtubules, transporting late ensosomes toward the lysosomes. Membrane fusion also involves several proteins, including the Rab protein, and SNAREs (soluble NSF attachment protein receptor) (68). In this case Rab protein could mediate targeting of the appropriate membrane and docking of the membranes, for SNAREs to achieve fusion event.

4.3. Pathways selected by various non-enveloped DNA viruses

Members of the same family or the same genus may use different pathways. There is growing evidence that certain viruses evolved the use of multiple entry pathways. This can be an advantage to infect different cell types. However, it can be difficult to determine whether multiple entry pathways are involved or if one complex pathway use different components. Often viruses can have multiple sequential binding to the cell surface, which can trigger different signaling pathways inside the cell, calling for components normally observed to drive different entry pathways.

4.3.1. Endocytosis of circoviruses

Porcine circovirus (PCV) infects a wide variety of cell types, including hepatocytes, cardiomyocytes, and macrophages via an unknown receptor. PCV utilizes ClaME to enter the cell, though other pathways may be involved (69, 70). In contrast, although a dynamin- and cholesterol-independent, but actin- and small GTPase-dependent pathway, allows PCV2 internalization in epithelial cells that leads to infection, a clathrin-mediated PCV2 internalization in epithelial cells is not followed by a full replication (69). Recent evidence suggests dendritic cells (DC) are involved through their particularly elevated endocytosis of the virus. PCV2 can accumulate to high levels both *in vitro* and *in vivo,* a phenomenon dependent on the virus capsid protein, inferring that the viral capsid or genome impedes DC endocytic degradation of the virus (71). However, PCV2 in DC does not interfere with processing of other antigens.

4.3.2. Endocytosis of parvoviruses

Most *Parvoviridae* family members were shown to enter cells by ClaME (39, 72, 73). When investigating early steps of infection of porcine parvovirus (PPV), we found that inhibition of ClaME affected PPV infection, but we could not achieve complete inhibition (39). Inhibition of fluid phase endocytosis components also reduced PPV infection. Interestingly combination of inhibition of both pathways reduced more the infection level, but still, some cells could be infected, suggesting a third pathway might be involved. It is important to consider the fact that inhibition of endocytosis pathways will result in upregulation of compensatory mechanisms that can influence viral infection. Laboratory experiments often use highly-purified viruses. We demonstrated, however, that after replication most PPV is present in aggregates which used preferentially fluid phase endocytosis, compared to the purified isolated particles that preferred ClaME (39). Purified viral preparations might not reflect the actual mode of infection inside the host. *In vivo,* aggregates are likely to form and be involved in further infection of other cells.

Adeno-associated viruses (AAVs) are members of the *Dependovirus* genus. They cannot complete their replication cycle without a helper virus, such as adenovirus. Many groups study AAVs entry and genome delivery in the nucleus, since they represent a good platform for gene therapy. However, the early steps of infection are not efficient (just like all parvoviruses) and a better understanding of the mechanism involved could lead to genetically modified viruses that are better suited for gene delivery. AAV5 can enter cells via ClaME (32), but it was shown to be able to use CavME too (32). To date, it is the only known parvovirus using this pathway. The authors observed the virus in non-coated vesicles using TEM. These vesicles contained caveolin-1. This pathway could not be found in all cell types, highlighting the fact that specific cargos can use different pathways depending on the cell type.

CLIC-GEEC (CLathrin-Independent Carrier/GPI-anchored protein-Enriched Endosomal Compartment) is a clathrin/caveolin independent entry pathway, identified in the past 10 years Figure 3. This pathway is constitutive, and thus attractive for pathogens. The size and shape of the vesicles can vary greatly, and could potentially accommodate large or multiple viruses. An important protein is GRAF1, serving as a marker or by inhibiting the pathway by gene silencing. Another interesting feature of the pathway, for viruses, is that resulting vesicles will acidify, and can provide proper environment for conformational changes.

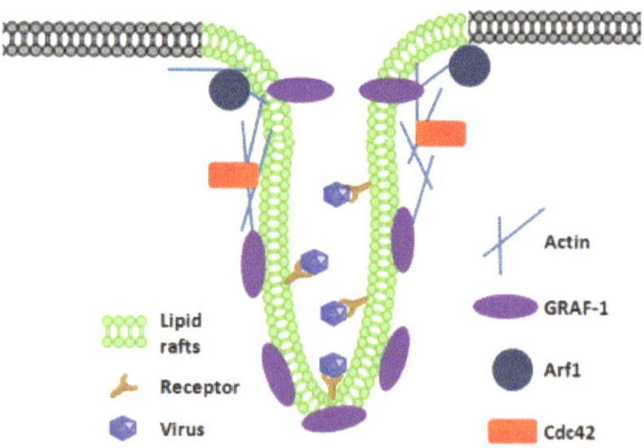

Figure 3. CLIC-GEEC entry mechanism. Characteristic features of this endocytosis mechanism include location in the lipid rafts, tubular shaped, large, pleomorphic invaginations, taking up large volume of membrane and fluids. Those required actin polymerisation involving Cdc42, and membrane curvature involving Arf1. GRAF-1 is a key player that can be used as marker or as target to inhibit this mechanism of uptake.

Adeno-associated viruses type 2 (AAV2) endocytosis was recently extensively studied, and revealed a complex mechanism (33). This endocytosis was independent of clathrin and caveolin. Also, only one known inhibitor (EIPA) of macropinocytosis could affect the entry of the virus. This inhibitor was shown to inhibit Rac1 and Cdc42 signaling, thus affecting other mechanisms as well (74). This endocytosis required actin cytoskeleton and cholesterol. Membrane and actin remodeling was observed by TEM and immunofluorescence, respectively. Virus bound to the cell surface was located in lipid rafts. These features led to the recently described CLIC/GEEC pathway (38). Dominant negative mutants revealed the critical role of Arf1 and Cdc42 for AAV2 endocytosis (33). GRAF1 (GTPase Regulator Associated with Focal adhesion kinase) was also demonstrated to be essential in CLIC-GEEC endocytosis, and postulated to act as a "coat" for that pathway (38). Expression of a truncated mutant of the protein, as well as the use of siRNA, both reduced the endocytosis of AAV2 (33). GRAF-1 is not implicated in ClaME or CavME, and thus appears specific to CLIC-GEEC endocytosis. Combination of GRAF1 and dynamin inhibition further diminished entry of AAV2, confirming that both proteins are important for endocytosis of AAV2. Interestingly, endosomes of the CLIC-GEEC pathway are known to acidify quickly and inhibition of acidification reduced transduction by AAVs (33). Then the virus is transported to the Golgi apparatus, and no co-localisation is observed with endosomal markers (Rab proteins).

4.3.3. Endocytosis of polyomaviruses

SV40 is probably the best described virus using CavME (75). Initial observations were made by electron microscopy (TEM) where the virus was not present in a coated vesicle, as opposed to ClaME. Entry is dependent of cholesterol, actin polymerization, dynamin and caveolin (37, 76). SV40 provides a good example of a particular traffic that does not involve classical endosomal pathway (37). Virus can be a great tool when used as a tracer for the endocytosis pathway, by using fluorescent viruses and either regular confocal microscopy or live imaging. Pelkmans et al. showed that SV40 traffic to the ER in two distinct steps (37). They saw, using live imaging that the virus once bound to its receptor will move on the surface of the cell until it reach the caveolae. Then the virus is endocytosed into very small vesicles. These will transport the virus to larger vesicle already present in the cytoplasm. These vesicle contained several virus particle and were caveolin-positive ("caveosomes"). Then the virus left the caveosomes in tubular-shaped small vesicles that moved along the microtubules toward the nucleus. The virus was then delivered to the smooth ER. No co-localisation with clathrin was observed, and inhibition of ClaME does not affect SV40 infection. SV40 was also reported to infect cells that do not express caveolin (77). In these cells, entry involved tyrosine kinases and cholesterol but again was not dependent of clathrin or dynamin (77). Entry could not be reduced when inhibiting ClaME or macropinocytosis. It was also faster than regular CavME, but led to the ER in a microtubule dependent manner similar to CavME. This is a good example of virus adaptation, enabling infection in a wide range of cells. In caveolae-expressing cells, both pathways are used. They share important characteristics such as dependence of cholesterol, tyrosine kinases and transit in neutral vesicles using the microtubule network leading to the ER (77).

BK virus is a polyomavirus that poses serious health problems (78). It causes diseases in immunosuppressed patients after renal transplants. Immunosuppression therapy can enable usually non-threatening viruses to replicate easily leading to diseases. More than 80% of the population has been infected often at young age and latent infection of the kidney is present in 50% of the population. Efforts are being made to understand infection by this virus. It was demonstrated that it hijacks CavME, as its relative virus SV40. Cholesterol depletion efficiently inhibited infection. Knock-down of caveolin-1 by siRNA inhibited the infection, while a knockout of clathrin had no effect. Immunofluorescence and co-localisation with confocal microscopy confirmed that BK virus use CavME in cell culture (78-80).

4.3.4. Endocytosis of papillomaviruses

Papillomaviruses are a good example of different family members using different entry pathways even after binding to similar receptors, thus co-receptors could shunt the virus to different pathways. BPV1 entered cells by ClaME (81), and HPV31 preferred CavME (82). To determine which pathway was important they used chemical inhibitors, dominant-negative mutants and TEM. Another publication showed that for HPV16 and HPV31 macropinocytosis was important in NK cells (83). In this case, membrane ruffling was observed when the cells were incubated with the virus-like particles (VLPs). Then, co-uptake of fluid phase was demonstrated with fluorescent dextran. Inhibition by chemical inhibitors prevented this fluid-phase uptake. In term of RhoGTPases, HPV VLPs induced activation of Cdc42 and inhibition of Rac1. Papillomaviruses are highly asynchronous and their pathways chosen are disputes in the literature.

4.3.5. Endocytosis of adenoviruses

Adenovirus could be observed in clathrin-coated vesicles by TEM (84). It is also present in non-coated vesicles. Dominant-negative constructs confirmed the ClaME (55). The involvement of integrins (85, 86) suggests that adenovirus endocytosis is highly regulated. These integrins activate PI(3)K which in turn induces actin polymerization and promotes adenovirus endocytosis (87). Another observation was that fluid uptake (macropinocytosis) increased with adenovirus infection (55). Amiloride was effective to inhibit infection. This additional infection pathway also required actin, small Rho GTPases and PKC.

4.3.6. Endocytosis of iridoviruses

The enveloped form of the iridovirus FV3 (genus *Ranavirus*) was discovered in early 1980s by TEM research, and shown to be internalized by ClaME whereas the naked particles were suggested to enter by fusion at the plasma membrane (29, 88). This model was questioned in recent reports with a closely related virus (TFV) and the ISKNV virus (89, 90) with both pointing to caveolae-dependent endocytosis. However, caution is warranted since infection of cells from non-physiological host, i.e. BHK-21, may cloud a direct comparison of TFV and FV3 entry. Similar studies coupled with TEM could resolve these issues. The TFV iridovirus, a ranavirus nearly identical to FV3, enters cells by an atypical, pH-dependent, CavME pathway. Experiments using chlorpromazine and over-expression of a dominant-negative form of Esp15 that inhibited assembly of clathrin-coated pits did not affect entry into HepG2 cells. Also, endocytosis of TFV was dependent on membrane cholesterol and was blocked by caveolin-1 scaffolding domain protein. Therefore, Guo et al. suggested that FV3, since its nearly identical in nucleotide sequence to TFV, may also enter via CavME. Later, these authors demonstrated that ISKNV, a fish iridovirus, enters fish cells also via CavME since inhibitors of ClaME had no effect on infection, in contrast to inhibitors of caveolin-1-involved signaling events (90). Moreover, ISKNV co-locates with caveolin-1 during virus infection and is dependent on dynamin and the microtubule cytoskeleton.

5. Virus maturation and endosomes escape

Non-enveloped viruses display a variety of strategies to gain access to the cytoplasm of the cell. Conformational changes are required in capsids to expose hydrophobic domains that anchor or disrupt the endosomal membrane during breaching. Parvoviruses have acquired an enzyme for membrane translocation. Virus trafficking is therefore highly temporally and spatially regulated and a plethora of tools have been developed to study these events.

5.1. Endosomal trafficking of circoviruses and escape to cytoplasm

PCV2 enters monocytic cells predominantly by ClaME and requires an acidic environment for infection. After endosomal escape, the virus aggregates in intracytoplasmic inclusion bodies (ICIs). Subsequently, PCV2 closely associates with mitochondria, completing a first cytoplasmic phase and enters the nucleus for replication (91). In epithelial cells, PCV2 is internalized via a clathrin-, caveolae-, and dynamin-independent small GTPase-regulated pathway (69).The latter leads to a more efficient PCV2 replication while the former seems to trap PCV2 leading to accumulation of the virus within epithelial cells. Inhibition of acidification of the endosomal/lysosomal system reduces PCV2 infection of the monocytic cells, indicating the requirement for a pH drop during replication. Surprisingly, inhibiting acidification highly increases PCV2 replication in epithelial cells, indicating that uncoating occurs at another pH. Serine proteases mediate PCV2 uncoating in both epithelial cells and cells of the monocytic cells but may have different optima in these cells (92). It is not clear how this uncoating process impacts on the endosomal escape.

5.2. Endosomal trafficking of parvoviruses and escape to cytoplasm

Parvoviruses have a unique enzyme domain, phospholipase A2, in their capsid (12, 13, 93). This protein domain with this motif is located inside the mature virion, and thus the virus needs to undergo a conformational change in order to expose and use this enzyme. The pentamer at the fivefold symmetry axis has a 0.8 nm central pore and is itself encircled by a 1.5 nm-deep depression ("canyon"). The hairpins of the central cylinder (see section 2.1.2) can thus move outward leading to a greater pore and allowing N-termini of VP2, during the transit in the endosome and acidification (94), to become exposed one by one and their N-termini proteolytically cleaved, yielding successively a larger pore. The internal unique part of VP1 (that resists proteolysis) is then able to externalize and expose its phospholipase A2 domain. Phospholipases catalyze hydrolysis of phospholipids, providing a means to breach the endosomal barrier. This motif is absolutely essential for parvoviruses infection and acts prior to the delivery of the genome to the nucleus (13). Millimolar calcium concentrations present in the endosome but not in the cytosol, are required for optimal activity of the enzyme (12). Since it was required after accumulation of the virus in the perinuclear region, but prior to DNA expression in the nucleus, it was hypothesized that its role could be the ensodomal escape. Co-uptake studies showed that small dextran particles, but not α-sarcin, could be released from the endosomes together with parvovirus escape. These observations suggested that parvovirus escape implicated small pore formation and not broad endosome disruption. Viruses harboring mutations in the PLA2 enzymatic domain remained trapped in endosomes in perinuclear areas, but could be complemented in *trans* in co-infection experiments with wild-type capsids (93). Moreover, induced endosomes disruption induced by treatment of the cells with polyethylenimine (PEI) was enough to rescue the infection, strongly suggesting that the main function of PLA2 is the escape of the virus from the ensosomal pathway.

5.3. Endosomal trafficking of polyomaviruses and escape to cytoplasm

SV40 enters via caveolin-1 containing vesicles that budded from the caveolae and that are transferred to caveosomes (lack markers for endosomes, lysosomes, ER and Golgi) (37). SV40 is then transferred from caveolin-free, tubular vesicles to the ER via an intermediate COP1 compartment that is brefeldin A-sensitive (95). Interestingly, polyomavirus is delivered to early endosomes and a crosstalk between ClaME and CavME has been reported for JCV (96). Nevertheless, after passage through early endosomes these are sorted to caveosomes. Escape occurs from the ER to the cytosol before reaching the nucleus (97). An elegant and elaborate study in an attempt to describe mechanisms involved in this escape was published in 2011 (97). TEM studies suggested profound conformational changes in the ER where the capsid diameter decreased from 45nm to 34nm. These conformational changes led to exposing structural protein VP2, which was not accessible in the mature virion that entered the cell. This VP2 protein contains hydrophobic structures that interact with the ER membrane. In the absence of VP2, the capsid can be targeted to the ER, but would not be able to transit back to the cytosol. There are several cellular proteins essential for SV40 escape to the cytosol. BAP31, highly-abundant membrane protein of the ER, is essential for SV40 infection. The chaperone BiP is also important for the escape, and will play its role after BAP31-induced conformational changes in the capsid suggesting that the ERAD-factors (isomerases, Derlin) assist is in SV4 ER- cytosol dislocation as used for misfolded proteins (97).

5.4. Endosomal trafficking of papillomaviruses and escape to cytoplasm

This subject is, due to the lack of suitable study models, still controversial. Conformational changes occur quickly after HPV entry and capsid protein L1 conformational epitope becomes inaccessible after endocytosis (44). Uncoating can be observed by L2 detection (normally buried and inaccessible in mature virions), or DNA detection. Papillomavirus escape from the endosomal pathway, at a relatively early stage, and only partial co-localisation between the virus and the lysosome can be seen (44, 98). Structural protein L2 is a major player in the escape. The protein has a hydrophobic C-terminal, adjacent to positively-charged residues, believed to destabilize the membrane structure and leading to disruption. It was recently described that the cellular protein nexin-17 was also important (98). The suggested role of this protein is to retain the virus in the late endosomes, preventing degradation in the lysosomes. When this protein is knocked down, strong co-localisation of the virus with the lysosomes was observed. Importantly, the binding sequence in L2 protein is conserved among *Papillomaviridae* family, both for low- and high-risk viruses. Recent studies suggested that BPV1 and human papillomavirus type 16 (HPV16, the most common etiological agent of cervical cancer) enter via ClaME, but that subsequent steps, as for JCV polyomavirus, require caveolin-1-mediated trafficking (81, 99). This novel trafficking may explain the requirement for the CavME pathway because ClaME typically does not lead to the ER.

5.5. Endosomal trafficking of adenoviruses and escape to cytoplasm

During entry, fibers are released, the penton base structures dissociated, the proteins connecting the DNA to the inside surface of the capsid degraded or shed, and the capsid-stabilizing minor proteins eliminated (100). The uncoating process starts immediately upon endocytic uptake with the loss of fibers and ends with the uptake of dissociated hexon proteins and DNA into the nucleus. For the adenoviruses, the escape will occur quickly, and is postulated to be as early as the sorting endosomes (48). An acidic pH is not sufficient for conformational changes in the virus structure to occur but a large number of factors, such as integrin, are involved (101). Integrin binding together with CAR-mediated drifts supported fiber shedding from adenovirus particles, leading to exposure of the membrane-lytic internal virion protein VI and enhanced viral escape from endosomes (102). The Ts1 mutant, which lacks the viral protease and cannot process the capsid or release the fibers during endosomal trafficking, will not be able to breach the endosomal membrane and ends up in lysosomes where it is degraded (103).

For subgroup B adenoviruses, Kalin et al. showed by confocal laser scanning microscopy, electron microscopy, and live cell imaging (104) that Ad35 colocalized with fluid-phase markers in large endocytic structures that were positive for CD46, alphanu integrins, and also CtBP1 (89).Their results extended observations with HAdV-3 (Ad3), using chemical inhibitors and dominant-negative mutants, that macropinocytosis is an infectious pathway for subgroup B human adenoviruses in epithelial and hematopoietic cells.

5.6. Endosomal trafficking of irdoviruses and escape to cytoplasm

Maturation of iridoviruses during entry has not been investigated in detail. The minor zip and finger proteins may play an essential role for capsid destabilization. Uncoating may occur at the nuclear membrane but, unlike herpesviruses, iridovirus genomes are not infectious, indicating that virion-associated proteins are required to initiate viral gene transcription (105). Early EM studies showed that FV3 ended up in the lysosome (88); however this may be a non-productive pathway. It has been suggested that ISKNV traffics via the caveosome to the ER, similar to SV40, and TFV via the caveosome to the *trans*-Golgi (89, 90). Chitnis et al. demonstrated that apoptosis induction by CIV iridovirus : (i) requires entry and endocytosis of virions or virion proteins, (ii) is inhibited under conditions permitting early viral expression, and (iii) requires the JNK signaling pathway (106). Apoptosis was inhibited by Z-IETD-FMK, an apical caspase inhibitor, indicating that CIV-induced apoptosis requires caspase activity. The JNK inhibitor SP600125 demonstrated drastic suppression of CVPE-induced apoptosis and showed that the JNK signaling pathway is significant for apoptosis in this system. Virus interaction with the cell surface was not sufficient for apoptosis since CIV(UV) particles bound to polysterene beads failed to induce apoptosis. Furthermore, blocking viral DNA replication with aphidicolin or phosphonoacetic acid suppressed apoptosis and Cf-caspase-i activity, indicating that early viral expression is necessary for inhibition of apoptosis, and de novo synthesis of viral proteins is not required for induction.

6. Conclusions

Endocytosis is a wide subject that has been studied for a very long time. Viruses are definitely great tool to study endocytosis since they are masters in hijacking every single possibility of the cell to achieve infection. Elucidating steps in viral replication often leads to hitherto unknown mechanisms that expand our knowledge of normal cellular processes. It is also interesting to realise that significant differences are found inside a family of closely related viruses. Co-evolution with different hosts, in different conditions pushed viruses to adapt in different direction and provide a great variety of tools. However, in the case of life threatening or economically-important viruses, it is sometimes very hard to find a way to prevent infection without impeding with normal cellular processes. This growing field of research will surely provide new exiting insights to previously unknown cellular process, and will continue to provide good targets for antiviral drugs.

Author details

Maude Boisvert and Peter Tijssen*
INRS-Institut Armand-Frappier, Institut Pasteur International Network, Université du Québec, QC, Canada

Acknowledgement

Authors wish to thank Natural Sciences and Engineering Research Council of Canada for grant to support P.T. and for the scholarship to M.B.

* Corresponding Author

7. References

[1] Crowther RA, Berriman JA, Curran WL, Allan GM, Todd D. Comparison of the structures of three circoviruses: chicken anemia virus, porcine circovirus type 2, and beak and feather disease virus. J Virol. 2003;77(24):13036-41. Epub 2003/12/04.

[2] Khayat R, Brunn N, Speir JA, Hardham JM, Ankenbauer RG, Schneemann A, et al. The 2.3-angstrom structure of porcine circovirus 2. J Virol. 2011;85(15):7856-62. Epub 2011/06/03.

[3] Mahe D, Blanchard P, Truong C, Arnauld C, Le Cann P, Cariolet R, et al. Differential recognition of ORF2 protein from type 1 and type 2 porcine circoviruses and identification of immunorelevant epitopes. J Gen Virol. 2000;81(Pt 7):1815-24. Epub 2000/06/22.

[4] Misinzo G, Delputte PL, Meerts P, Lefebvre DJ, Nauwynck HJ. Porcine circovirus 2 uses heparan sulfate and chondroitin sulfate B glycosaminoglycans as receptors for its attachment to host cells. J Virol. 2006;80(7):3487-94. Epub 2006/03/16.

[5] Tsao J, Chapman MS, Agbandje M, Keller W, Smith K, Wu H, et al. The three-dimensional structure of canine parvovirus and its functional implications. Science. 1991;251(5000):1456-64. Epub 1991/03/22.

[6] Kaufmann B, El-Far M, Plevka P, Bowman VD, Li Y, Tijssen P, et al. Structure of Bombyx mori densovirus 1, a silkworm pathogen. J Virol. 2011;85(10):4691-7. Epub 2011/03/04.

[7] Kaufmann B, Bowman VD, Li Y, Szelei J, Waddell PJ, Tijssen P, et al. Structure of Penaeus stylirostris densovirus, a shrimp pathogen. J Virol. 2010;84(21):11289-96. Epub 2010/08/13.

[8] Simpson AA, Hebert B, Sullivan GM, Parrish CR, Zadori Z, Tijssen P, et al. The structure of porcine parvovirus: comparison with related viruses. J Mol Biol. 2002;315(5):1189-98. Epub 2002/02/06.

[9] Simpson AA, Chipman PR, Baker TS, Tijssen P, Rossmann MG. The structure of an insect parvovirus (Galleria mellonella densovirus) at 3.7 A resolution. Structure. 1998;6(11):1355-67. Epub 1998/11/18.

[10] Agbandje-McKenna M, Llamas-Saiz AL, Wang F, Tattersall P, Rossmann MG. Functional implications of the structure of the murine parvovirus, minute virus of mice. Structure. 1998;6(11):1369-81. Epub 1998/11/18.

[11] Plevka P, Hafenstein S, Li L, D'Abrgamo A, Jr., Cotmore SF, Rossmann MG, et al. Structure of a packaging-defective mutant of minute virus of mice indicates that the genome is packaged via a pore at a 5-fold axis. J Virol. 2011;85(10):4822-7. Epub 2011/03/04.

[12] Canaan S, Zadori Z, Ghomashchi F, Bollinger J, Sadilek M, Moreau ME, et al. Interfacial enzymology of parvovirus phospholipases A2. J Biol Chem. 2004;279(15):14502-8. Epub 2004/01/17.

[13] Zadori Z, Szelei J, Lacoste MC, Li Y, Gariepy S, Raymond P, et al. A viral phospholipase A2 is required for parvovirus infectivity. Dev Cell. 2001;1(2):291-302. Epub 2001/11/13.

[14] Yan Y, Stehle T, Liddington RC, Zhao H, Harrison SC. Structure determination of simian virus 40 and murine polyomavirus by a combination of 30-fold and 5-fold electron-density averaging. Structure. 1996;4(2):157-64. Epub 1996/02/15.
[15] Liddington RC, Yan Y, Moulai J, Sahli R, Benjamin TL, Harrison SC. Structure of simian virus 40 at 3.8-A resolution. Nature. 1991;354(6351):278-84. Epub 1991/11/28.
[16] Baker TS, Drak J, Bina M. The capsid of small papova viruses contains 72 pentameric capsomeres: direct evidence from cryo-electron-microscopy of simian virus 40. Biophys J. 1989;55(2):243-53. Epub 1989/02/01.
[17] Streuli CH, Griffin BE. Myristic acid is coupled to a structural protein of polyoma virus and SV40. Nature. 1987;326(6113):619-22. Epub 1987/04/09.
[18] Schelhaas M, Malmstrom J, Pelkmans L, Haugstetter J, Ellgaard L, Grunewald K, et al. Simian Virus 40 depends on ER protein folding and quality control factors for entry into host cells. Cell. 2007;131(3):516-29. Epub 2007/11/06.
[19] Sapp M, Day PM. Structure, attachment and entry of polyoma- and papillomaviruses. Virology. 2009;384(2):400-9. Epub 2009/01/23.
[20] Belnap DM, Olson NH, Cladel NM, Newcomb WW, Brown JC, Kreider JW, et al. Conserved features in papillomavirus and polyomavirus capsids. J Mol Biol. 1996;259(2):249-63. Epub 1996/06/07.
[21] Chen XS, Casini G, Harrison SC, Garcea RL. Papillomavirus capsid protein expression in Escherichia coli: purification and assembly of HPV11 and HPV16 L1. J Mol Biol. 2001;307(1):173-82. Epub 2001/03/13.
[22] Chen XS, Garcea RL, Goldberg I, Casini G, Harrison SC. Structure of small virus-like particles assembled from the L1 protein of human papillomavirus 16. Mol Cell. 2000;5(3):557-67. Epub 2000/07/06.
[23] Modis Y, Trus BL, Harrison SC. Atomic model of the papillomavirus capsid. EMBO J. 2002;21(18):4754-62. Epub 2002/09/18.
[24] Hagensee ME, Olson NH, Baker TS, Galloway DA. Three-dimensional structure of vaccinia virus-produced human papillomavirus type 1 capsids. J Virol. 1994;68(7):4503-5. Epub 1994/07/01.
[25] Baker TS, Newcomb WW, Olson NH, Cowsert LM, Olson C, Brown JC. Structures of bovine and human papillomaviruses. Analysis by cryoelectron microscopy and three-dimensional image reconstruction. Biophys J. 1991;60(6):1445-56. Epub 1991/12/01.
[26] Chailertvanitkul VA, Pouton CW. Adenovirus: a blueprint for non-viral gene delivery. Curr Op Biotech. 2010;21(5):627-32. Epub 2010/07/20.
[27] Reddy VS, Natchiar SK, Stewart PL, Nemerow GR. Crystal structure of human adenovirus at 3.5 A resolution. Science. 2010;329(5995):1071-5. Epub 2010/08/28.
[28] Reddy VS, Natchiar SK, Gritton L, Mullen TM, Stewart PL, Nemerow GR. Crystallization and preliminary X-ray diffraction analysis of human adenovirus. Virology. 2010;402(1):209-14. Epub 2010/04/17.

[29] Chinchar VG, Yu KH, Jancovich JK. The molecular biology of frog virus 3 and other iridoviruses infecting cold-blooded vertebrates. Viruses. 2011;3(10):1959-85. Epub 2011/11/10.
[30] Yan X, Olson NH, Van Etten JL, Bergoin M, Rossmann MG, Baker TS. Structure and assembly of large lipid-containing dsDNA viruses. Nature Struct Biol. 2000;7(2):101-3. Epub 2000/02/03.
[31] Yan X, Yu Z, Zhang P, Battisti AJ, Holdaway HA, Chipman PR, et al. The capsid proteins of a large, icosahedral dsDNA virus. J Mol Biol. 2009;385(4):1287-99. Epub 2008/11/26.
[32] Bantel-Schaal U, Braspenning-Wesch I, Kartenbeck J. Adeno-associated virus type 5 exploits two different entry pathways in human embryo fibroblasts. J Gen Virol. 2009;90(Pt 2):317-22. Epub 2009/01/15.
[33] Nonnenmacher M, Weber T. Adeno-associated virus 2 infection requires endocytosis through the CLIC/GEEC pathway. Cell Host Microbe. 2011;10(6):563-76. Epub 2011/12/20.
[34] Ivanov AI. Pharmacological inhibition of endocytic pathways: is it specific enough to be useful? Methods Mol Biol. 2008;440:15-33. Epub 2008/03/29.
[35] Sieczkarski SB, Whittaker GR. Dissecting virus entry via endocytosis. J Gen Virol. 2002;83(Pt 7):1535-45. Epub 2002/06/21.
[36] Benmerah A, Bayrou M, Cerf-Bensussan N, Dautry-Varsat A. Inhibition of clathrin-coated pit assembly by an Eps15 mutant. J Cell Sci. 1999;112 (Pt 9):1303-11. Epub 1999/04/09.
[37] Pelkmans L, Kartenbeck J, Helenius A. Caveolar endocytosis of simian virus 40 reveals a new two-step vesicular-transport pathway to the ER. Nat Cell Biol. 2001;3(5):473-83. Epub 2001/05/02.
[38] Lundmark R, Doherty GJ, Howes MT, Cortese K, Vallis Y, Parton RG, et al. The GTPase-activating protein GRAF1 regulates the CLIC/GEEC endocytic pathway. Curr Biol. 2008;18(22):1802-8. Epub 2008/11/28.
[39] Boisvert M, Fernandes S, Tijssen P. Multiple pathways involved in porcine parvovirus cellular entry and trafficking toward the nucleus. J Virol. 2010;84(15):7782-92. Epub 2010/05/21.
[40] Opie SR, Warrington KH, Jr., Agbandje-McKenna M, Zolotukhin S, Muzyczka N. Identification of amino acid residues in the capsid proteins of adeno-associated virus type 2 that contribute to heparan sulfate proteoglycan binding. J Virol. 2003;77(12):6995-7006. Epub 2003/05/28.
[41] Hueffer K, Govindasamy L, Agbandje-McKenna M, Parrish CR. Combinations of two capsid regions controlling canine host range determine canine transferrin receptor binding by canine and feline parvoviruses. J Virol. 2003;77(18):10099-105. Epub 2003/08/28.

[42] Parker JS, Murphy WJ, Wang D, O'Brien SJ, Parrish CR. Canine and feline parvoviruses can use human or feline transferrin receptors to bind, enter, and infect cells. J Virol. 2001;75(8):3896-902. Epub 2001/03/27.
[43] Oraveerakul K, Choi CS, Molitor TW. Restriction of porcine parvovirus replication in nonpermissive cells. J Virol. 1992;66(2):715-22. Epub 1992/02/01.
[44] Sapp M, Bienkowska-Haba M. Viral entry mechanisms: human papillomavirus and a long journey from extracellular matrix to the nucleus. FEBS J. 2009;276(24):7206-16. Epub 2009/11/03.
[45] Buck CB, Pastrana DV, Lowy DR, Schiller JT. Efficient intracellular assembly of papillomaviral vectors. J Virol. 2004;78(2):751-7. Epub 2003/12/25.
[46] Pyeon D, Lambert PF, Ahlquist P. Production of infectious human papillomavirus independently of viral replication and epithelial cell differentiation. Proc Natl Acad Sci U S A. 2005;102(26):9311-6. Epub 2005/06/17.
[47] Roberts JN, Buck CB, Thompson CD, Kines R, Bernardo M, Choyke PL, et al. Genital transmission of HPV in a mouse model is potentiated by nonoxynol-9 and inhibited by carrageenan. Nature Med. 2007;13(7):857-61. Epub 2007/07/03.
[48] Leopold PL, Crystal RG. Intracellular trafficking of adenovirus: many means to many ends. Adv Drug Del Rev. 2007;59(8):810-21. Epub 2007/08/21.
[49] Eaton HE, Penny E, Brunetti CR. Antibody dependent enhancement of frog virus 3 infection. Virol J. 2010;7:41. Epub 2010/02/20.
[50] Polo S, Di Fiore PP. Endocytosis conducts the cell signaling orchestra. Cell. 2006;124(5):897-900. Epub 2006/03/15.
[51] Sorkin A, von Zastrow M. Endocytosis and signalling: intertwining molecular networks. Nat Rev Mol Cell Biol. 2009;10(9):609-22. Epub 2009/08/22.
[52] Wilde A, Beattie EC, Lem L, Riethof DA, Liu SH, Mobley WC, et al. EGF receptor signaling stimulates SRC kinase phosphorylation of clathrin, influencing clathrin redistribution and EGF uptake. Cell. 1999;96(5):677-87. Epub 1999/03/25.
[53] Cavalli V, Vilbois F, Corti M, Marcote MJ, Tamura K, Karin M, et al. The stress-induced MAP kinase p38 regulates endocytic trafficking via the GDI:Rab5 complex. Mol Cell. 2001;7(2):421-32. Epub 2001/03/10.
[54] Nekhoroshkova E, Albert S, Becker M, Rapp UR. A-RAF kinase functions in ARF6 regulated endocytic membrane traffic. PloS one. 2009;4(2):e4647. Epub 2009/02/28.
[55] Meier O, Boucke K, Hammer SV, Keller S, Stidwill RP, Hemmi S, et al. Adenovirus triggers macropinocytosis and endosomal leakage together with its clathrin-mediated uptake. J Cell Biol. 2002;158(6):1119-31. Epub 2002/09/11.
[56] McMahon HT, Boucrot E. Molecular mechanism and physiological functions of clathrin-mediated endocytosis. Nat Rev Mol Cell Biol.12(8):517-33. Epub 2011/07/23.
[57] Lajoie P, Nabi IR. Regulation of raft-dependent endocytosis. J Cell Mol Med. 2007;11(4):644-53. Epub 2007/09/01.
[58] Mercer J, Helenius A. Virus entry by macropinocytosis. Nat Cell Biol. 2009;11(5):510-20. Epub 2009/05/01.

[59] Clement C, Tiwari V, Scanlan PM, Valyi-Nagy T, Yue BY, Shukla D. A novel role for phagocytosis-like uptake in herpes simplex virus entry. J Cell Biol. 2006;174(7):1009-21. Epub 2006/09/27.
[60] Ghigo E, Kartenbeck J, Lien P, Pelkmans L, Capo C, Mege JL, et al. Ameobal pathogen mimivirus infects macrophages through phagocytosis. PLoS Pathog. 2008;4(6):e1000087. Epub 2008/06/14.
[61] Mayor S, Pagano RE. Pathways of clathrin-independent endocytosis. Nat Rev Mol Cell Biol. 2007;8(8):603-12. Epub 2007/07/05.
[62] Huotari J, Helenius A. Endosome maturation. EMBO J. 2011;30(17):3481-500. Epub 2011/09/01.
[63] Rink J, Ghigo E, Kalaidzidis Y, Zerial M. Rab conversion as a mechanism of progression from early to late endosomes. Cell. 2005;122(5):735-49. Epub 2005/09/07.
[64] Driskell OJ, Mironov A, Allan VJ, Woodman PG. Dynein is required for receptor sorting and the morphogenesis of early endosomes. Nat Cell Biol. 2007;9(1):113-20. Epub 2006/12/19.
[65] Stenmark H. Rab GTPases as coordinators of vesicle traffic. Nat Rev Mol Cell Biol. 2009;10(8):513-25. Epub 2009/07/16.
[66] Pfeffer SR. Structural clues to Rab GTPase functional diversity. J Biol Chem. 2005;280(16):15485-8. Epub 2005/03/05.
[67] Seabra MC, Coudrier E. Rab GTPases and myosin motors in organelle motility. Traffic. 2004;5(6):393-9. Epub 2004/05/01.
[68] Stenmark H, Parton RG, Steele-Mortimer O, Lutcke A, Gruenberg J, Zerial M. Inhibition of rab5 GTPase activity stimulates membrane fusion in endocytosis. EMBO J. 1994;13(6):1287-96. Epub 1994/03/15.
[69] Misinzo G, Delputte PL, Lefebvre DJ, Nauwynck HJ. Porcine circovirus 2 infection of epithelial cells is clathrin-, caveolae- and dynamin-independent, actin and Rho-GTPase-mediated, and enhanced by cholesterol depletion. Virus Res. 2009;139(1):1-9. Epub 2008/10/28.
[70] Misinzo G, Meerts P, Bublot M, Mast J, Weingartl HM, Nauwynck HJ. Binding and entry characteristics of porcine circovirus 2 in cells of the porcine monocytic line 3D4/31. J Gen Virol. 2005;86(Pt 7):2057-68. Epub 2005/06/17.
[71] McCullough KC, Ruggli N, Summerfield A. Dendritic cells--at the front-line of pathogen attack. Vet Immun Immunopathol. 2009;128(1-3):7-15. Epub 2008/11/28.
[72] Douar AM, Poulard K, Stockholm D, Danos O. Intracellular trafficking of adeno-associated virus vectors: routing to the late endosomal compartment and proteasome degradation. J Virol. 2001;75(4):1824-33. Epub 2001/02/13.
[73] Parker JS, Parrish CR. Cellular uptake and infection by canine parvovirus involves rapid dynamin-regulated clathrin-mediated endocytosis, followed by slower intracellular trafficking. J Virol. 2000;74(4):1919-30. Epub 2000/01/22.

[74] Koivusalo M, Welch C, Hayashi H, Scott CC, Kim M, Alexander T, et al. Amiloride inhibits macropinocytosis by lowering submembranous pH and preventing Rac1 and Cdc42 signaling. J Cell Biol.188(4):547-63. Epub 2010/02/17.

[75] Norkin LC. Simian virus 40 infection via MHC class I molecules and caveolae. Immunol Rev. 1999;168:13-22. Epub 1999/07/10.

[76] Pelkmans L, Puntener D, Helenius A. Local actin polymerization and dynamin recruitment in SV40-induced internalization of caveolae. Science. 2002;296(5567):535-9. Epub 2002/04/20.

[77] Damm EM, Pelkmans L, Kartenbeck J, Mezzacasa A, Kurzchalia T, Helenius A. Clathrin- and caveolin-1-independent endocytosis: entry of simian virus 40 into cells devoid of caveolae. J Cell Biol. 2005;168(3):477-88. Epub 2005/01/26.

[78] Moriyama T, Marquez JP, Wakatsuki T, Sorokin A. Caveolar endocytosis is critical for BK virus infection of human renal proximal tubular epithelial cells. J Virol. 2007;81(16):8552-62. Epub 2007/06/08.

[79] Dugan AS, Eash S, Atwood WJ. Update on BK virus entry and intracellular trafficking. Transplant Infect Dis. 2006;8(2):62-7. Epub 2006/06/01.

[80] Moriyama T, Sorokin A. Repression of BK virus infection of human renal proximal tubular epithelial cells by pravastatin. Transplantation. 2008;85(9):1311-7. Epub 2008/05/14.

[81] Laniosz V, Holthusen KA, Meneses PI. Bovine papillomavirus type 1: from clathrin to caveolin. J Virol. 2008;82(13):6288-98. Epub 2008/04/18.

[82] Bousarghin L, Touze A, Sizaret PY, Coursaget P. Human papillomavirus types 16, 31, and 58 use different endocytosis pathways to enter cells. J Virol. 2003;77(6):3846-50. Epub 2003/03/01.

[83] Renoux VM, Bisig B, Langers I, Dortu E, Clemenceau B, Thiry M, et al. Human papillomavirus entry into NK cells requires CD16 expression and triggers cytotoxic activity and cytokine secretion. Eur J Immunol. 2011;41(11):3240-52. Epub 2011/08/11.

[84] Brown DT, Burlingham BT. Penetration of host cell membranes by adenovirus 2. J Virol. 1973;12(2):386-96. Epub 1973/08/01.

[85] Suomalainen M, Nakano MY, Boucke K, Keller S, Greber UF. Adenovirus-activated PKA and p38/MAPK pathways boost microtubule-mediated nuclear targeting of virus. EMBO J. 2001;20(6):1310-9. Epub 2001/03/17.

[86] Wickham TJ, Mathias P, Cheresh DA, Nemerow GR. Integrins alpha v beta 3 and alpha v beta 5 promote adenovirus internalization but not virus attachment. Cell. 1993;73(2):309-19. Epub 1993/04/23.

[87] Li E, Stupack D, Klemke R, Cheresh DA, Nemerow GR. Adenovirus endocytosis via alpha(v) integrins requires phosphoinositide-3-OH kinase. J Virol. 1998;72(3):2055-61. Epub 1998/03/14.

[88] Braunwald J, Nonnenmacher H, Tripier-Darcy F. Ultrastructural and biochemical study of frog virus 3 uptake by BHK-21 cells. J Gen Virol. 1985;66 (Pt 2):283-93. Epub 1985/02/01.

[89] Guo CJ, Liu D, Wu YY, Yang XB, Yang LS, Mi S, et al. Entry of tiger frog virus (an Iridovirus) into HepG2 cells via a pH-dependent, atypical, caveola-mediated endocytosis pathway. J Virol. 2011;85(13):6416-26. Epub 2011/05/06.
[90] Guo CJ, Wu YY, Yang LS, Yang XB, He J, Mi S, et al. Infectious spleen and kidney necrosis virus (a fish iridovirus) enters Mandarin fish fry cells via caveola-dependent endocytosis. J Virol. 2012;86(5):2621-31. Epub 2011/12/16.
[91] Rodriguez-Carino C, Duffy C, Sanchez-Chardi A, McNeilly F, Allan GM, Segales J. Porcine circovirus type 2 morphogenesis in a clone derived from the l35 lymphoblastoid cell line. J Comp Pathol. 2011;144(2-3):91-102. Epub 2010/08/31.
[92] Nauwynck HJ, Sanchez R, Meerts P, Lefebvre DJ, Saha D, Huang L, et al. Cell tropism and entry of porcine circovirus 2. Virus Res. 2012;164(1-2):43-5. Epub 2011/11/22.
[93] Farr GA, Zhang LG, Tattersall P. Parvoviral virions deploy a capsid-tethered lipolytic enzyme to breach the endosomal membrane during cell entry. Proc Natl Acad Sci U S A. 2005;102(47):17148-53. Epub 2005/11/15.
[94] Mani B, Baltzer C, Valle N, Almendral JM, Kempf C, Ros C. Low pH-dependent endosomal processing of the incoming parvovirus minute virus of mice virion leads to externalization of the VP1 N-terminal sequence (N-VP1), N-VP2 cleavage, and uncoating of the full-length genome. J Virol. 2006;80(2):1015-24. Epub 2005/12/28.
[95] Norkin LC, Anderson HA, Wolfrom SA, Oppenheim A. Caveolar endocytosis of simian virus 40 is followed by brefeldin A-sensitive transport to the endoplasmic reticulum, where the virus disassembles. J Virol. 2002;76(10):5156-66. Epub 2002/04/23.
[96] Querbes W, O'Hara BA, Williams G, Atwood WJ. Invasion of host cells by JC virus identifies a novel role for caveolae in endosomal sorting of noncaveolar ligands. J Virol. 2006;80(19):9402-13. Epub 2006/09/16.
[97] Geiger R, Andritschke D, Friebe S, Herzog F, Luisoni S, Heger T, et al. BAP31 and BiP are essential for dislocation of SV40 from the endoplasmic reticulum to the cytosol. Nat Cell Biol. 2011;13(11):1305-14. Epub 2011/09/29.
[98] Bergant Marusic M, Ozbun MA, Campos SK, Myers MP, Banks L. Human papillomavirus L2 facilitates viral escape from late endosomes via sorting nexin 17. Traffic. 2012;13(3):455-67. Epub 2011/12/14.
[99] Laniosz V, Dabydeen SA, Havens MA, Meneses PI. Human papillomavirus type 16 infection of human keratinocytes requires clathrin and caveolin-1 and is brefeldin a sensitive. J Virol. 2009;83(16):8221-32. Epub 2009/06/06.
[100] Greber UF, Willetts M, Webster P, Helenius A. Stepwise dismantling of adenovirus 2 during entry into cells. Cell. 1993;75(3):477-86. Epub 1993/11/05.
[101] Meier O, Greber UF. Adenovirus endocytosis. J Gene Med. 2004;6 Suppl 1:S152-63. Epub 2004/02/24.
[102] Burckhardt CJ, Suomalainen M, Schoenenberger P, Boucke K, Hemmi S, Greber UF. Drifting motions of the adenovirus receptor CAR and immobile integrins initiate virus uncoating and membrane lytic protein exposure. Cell Host Microbe. 2011;10(2):105-17. Epub 2011/08/17.

[103] Greber UF, Webster P, Weber J, Helenius A. The role of the adenovirus protease on virus entry into cells. The EMBO J. 1996;15(8):1766-77. Epub 1996/04/15.

[104] Kalin S, Amstutz B, Gastaldelli M, Wolfrum N, Boucke K, Havenga M, et al. Macropinocytotic uptake and infection of human epithelial cells with species B2 adenovirus type 35. J Virol. 2010;84(10):5336-50. Epub 2010/03/20.

[105] Willis DB, Goorha R, Granoff A. Nongenetic reactivation of frog virus 3 DNA. Virology. 1979;98(2):476-9. Epub 1979/10/30.

[106] Chitnis NS, D'Costa SM, Paul ER, Bilimoria SL. Modulation of iridovirus-induced apoptosis by endocytosis, early expression, JNK, and apical caspase. Virology. 2008;370(2):333-42. Epub 2007/10/19.

Section 4

Pathogens and Toxins

Chapter 10

Pathogen and Toxin Entry – How Pathogens and Toxins Induce and Harness Endocytotic Mechanisms

Thorsten Eierhoff, Bahne Stechmann and Winfried Römer

Additional information is available at the end of the chapter

1. Introduction

Humans have been exposed to a plethora of pathogens (bacteria, viruses) ever since. Infectious diseases are among the leading causes of death worldwide. For example, in 2011, 1.34 million people died of tuberculosis, which is caused by an infection with *Mycobacterium tuberculosis.* Even more died of an infection by the human immunodeficiency virus (HIV; 1.78 million) or lower respiratory tract infection (3.46 million) [1]. In addition, recurring pandemic outbreaks of the influenza A virus, as in 2009, or an epidemic outbreak of enterohemorrhagic *E. coli* (EHEC) in Germany in 2011, show quite plainly that pathogens in the 21th century still are a severe health problem, not only in developing countries.

During evolution, defence mechanisms have been developed by the host to counter pathogens, which in turn needed to respond with new strategies to gain entry into host cells. As a consequence, a wide variety of invasion mechanisms have evolved, of which only a few have been characterised in molecular detail to date.

In this chapter, we describe the different invasion strategies of bacteria, viruses and toxins by illustrating the mechanisms using prominent examples. Rather than relying passively on cellular mechanisms of their hosts, diverse pathogens and toxins actively induce the first steps of their uptake into a wide range of target cells. In most cases, the pathogen plays a key role in subverting the cellular machinery to stimulate actin re-arrangements, which facilitates the invasion process. As an example, recent progress in our understanding of the molecular mechanism of lipid-mediated endocytosis of carbohydrate-binding viruses and toxins is presented. In particular, we highlight the critical role of lipid species underlying these processes.

Bacteria invade cells as a means to escape host immune responses. Once inside a hijacked cell, the pathogen is protected against active factors of the immune system (e. g. complement factors, antibodies) and is conveniently provided with nutrients. In addition, viruses critically depend on the cellular machinery of host cells for their replication.

The initial step in the cellular uptake process of diverse pathogens and toxins is characterised by the binding to carbohydrate moieties exposed by a lipid or a protein in the plasma membrane of target cells. For example: cholera toxin binds with its B-subunit to the ganglioside GM1 in intestinal cells, the opportunistic human pathogen *Pseudomonas aeruginosa* attaches to respiratory cells by binding to asialo-GM1 and asialo-GM2 through type IV pili [2, 3], and the influenza A virus initiates its uptake by binding to sialic acids in the host cell membrane [4]. Conventionally considered as adhesion receptors for toxins, viruses and bacteria, recent data indicate that glycosphingolipids are also crucial parameters for the self-induced endocytosis of toxins and viruses [5, 6].

Glycosphingolipids, such as Gb3 or GM1, are enriched in the external leaflet of the plasma membrane and comprise a glycan and a ceramide lipid moiety of sphingosine (a long-chain amino alcohol) linked to a fatty acid [7]. The structure of the ceramide moiety of glycosphingolipids is highly divers and varies in length, saturation degree and hydroxylation. However, glycosphingolipids are traditionally classified by the structure of their glycans.

A second important type of lipids, which is critically important for the uptake of pathogens, is phosphatidylinositol-phosphate (PIP). PIPs are essential components of cell membranes implicated in a variety of signalling events. They are glycero-phospholipids with a negatively charged myo-inositol head group, which can be phosphorylated at different OH-positions of the inositol ring (D1-D5) [8]. More than 50 enzymes have been identified to combinational phosphorylate and dephosphorylate the inositol ring [9].

PIPs are signalling molecules rather than structural components of the plasma membrane, considered to be involved in dynamic cellular processes like (plasma) membrane dynamics, vesicle trafficking and actin polymerisation [8]. Probably this is the reason why many invasive bacteria as well as viral pathogens hijack these lipids to manipulate the plasma membrane in order to ensure their proper uptake into host cells.

One strategy by which invasive bacteria manipulate the PIP metabolism is the translocation of effector proteins, which act as phosphatidylinositol phosphatases (e.g. IpgD of *Shigella* and SopB/SigD of *Salmonella,* discussed below). A second option to interfere with the PIP metabolism is the engagement of specific host cell receptors.

The invasion process of pathogens is highly complex because it involves a specific spatiotemporally regulated interplay of different subsets of host cell and pathogenic factors. In addition, the composition and architecture of biological membranes is extensive. To understand how individual factors contribute to the entry process, less complex and easy-to-manipulate synthetic systems are needed. Artificial membrane systems gain more and more in importance as simpler and controllable systems to reconstitute and study endocytotic processes (see "EXCURSUS" box).

EXCURSUS: Artificial Membrane systems – powerful tools to study endocytotic membrane processes

Liposomes (also called vesicles) are more and more used as simple synthetic models for biological membranes. They represent a sphere of a cytosol-free unilamellar lipid bilayer and consist of a defined lipid composition resembling e.g. that of the plasma membrane. Manipulations at the outside can be conducted with ease. Giant unilamellar liposomes (GUVs) with diameters between 1 and 50 μm can be obtained by swelling of a phospholipid bilayer in water within an electrical field [10] (Figure 1). Because of its simplicity compared to native cells, this type of liposomes has been used recently to identify the initial steps of the cellular uptake of Shiga toxin [5, 11].

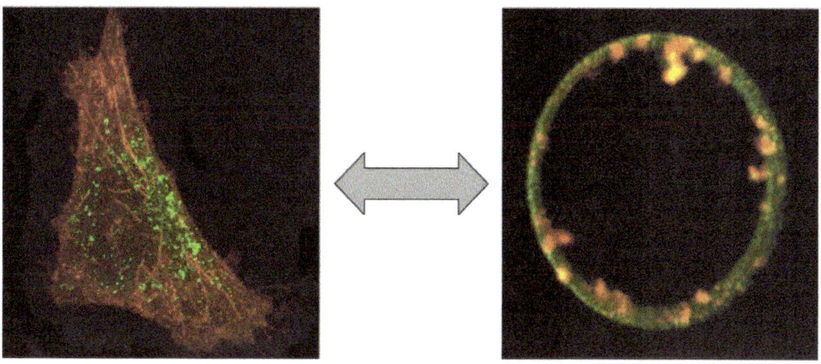

Figure 1. Giant unilamellar vesicles (GUVs) as a minimal membrane system to investigate endocytotic processes. Shiga toxin-induced tubular membrane invaginations on a HeLa cell (left image; in red colour) can be reconstituted on GUVs in the complete absence of cellular energy and cytosolic proteins (right image; red colour). Systems of different complexity are helpful and complementary to identify the molecular mechanisms of cellular processes.

The electroformation technique to produce liposomes is simple and rapid, but less suitable for embedding cytosolic proteins into the lumen of the lipid bilayer. For instance, the examination of the scission process of Shiga toxin-induced membrane invaginations requires the addition of the protein machinery at the internal (cytosolic) side of the membrane. For this, proteins could be microinjected into GUVs, or GUVs could be grown in the presence of protein machinery in the buffer solution. While the former method is very time-consuming and may easily provoke vesicle rupture, the latter is inefficient, not really well controlled, and proteins might be denatured (due to the application of an electric field during liposome formation). Moreover, in this setup, it is practically impossible to apply acute changes in buffer conditions (e.g. addition of ATP).

A model membrane system that better responds to the challenge of manipulating the protein machinery at both sides of the membrane are liposomes produced by the inverse emulsion technique [12] which allows for the use of two different buffers inside and outside. This technique has been used, for example, to reconstitute nucleation and assembly of cortical actin at the inner side of a liposome. In principle, an inverted phospholipid micelle containing the protein(s) of interest is sedimented through a phospholipid monolayer at the phase boundary of oil and an aqueous buffer solution. By the addition of a pore forming α-hemolysin, ATP and ions can pass the liposome membrane to induce actin polymerization within the liposome [13]. In addition, this technique has already been successfully used to study the actin-driven scission of endocytotic membrane invaginations [11].

The inverse emulsion technique has several advantages compared to the now classical electroformation technique, apart from the preservation of protein functionality. First, it allows for the incorporation of proteins in very small volumes (a few microliters). Second, it allows for the use of two different buffers inside and outside of the vesicle (providing that the osmotic pressure is kept constant). Third, since the two membrane monolayers are prepared independently of each other, lipid asymmetry within the lipid bilayer can be taken into account.

However, the inverse emulsion technique still has some major limitations. Most notably, the inclusion of proteins within the liposomal lumen remains tedious as liposomes must be generated in the presence of the protein mix. Furthermore, the high diffusion mobility of small liposomes makes observation by microscopy cumbersome and prevents from following the same object over periods longer than a few seconds.

Already established for impedance spectroscopy measurements and single ion channel recordings [14, 15], pore-suspending membranes based on highly ordered pore arrays might represent marvellous tools to study endocytotic processes. This hybrid membrane system combines the advantages of freestanding and solid supported lipid membranes. While part of the lipid bilayer is anchored to the surface of the porous matrix and resembles a solid supported membrane, the pore suspending parts can be viewed as freestanding lipid membranes. Highly ordered alumina and silicon pore arrays with pore sizes in the nano- and micrometer range can be chosen as supports for lipid bilayer immobilization.

The porous support can either be covered by a synthetic lipid bilayer or native plasma membrane sheets, depending on the application. The advantage of this technique is that both sides of the membrane are freely accessible. This is particularly interesting, when native plasma membrane sheets are analyzed, considering their asymmetric composition. In general, two different methods exist to produce native plasma membrane sheets: the rip off and the lysis-squirting technique. In both cases adherent cells are first grown on the porous membranes. For the rip off technique, a chip with a nanostructured, porous surface is pressed on top of cultured, adherent cells to form direct molecular contacts with the cells. The chip is then removed, thereby peeling off the upper plasma membranes of the adhering cells, which are now located on the chip as supported membranes [16]. With the lysis-squirting technique, adherent cells are incubated in hypotonic buffer for a few minutes before being squirted with the same buffer. The basal part of the membrane remains attached to the support and can be studied further [17]. The pore-suspending membranes can be analyzed by e.g. atomic force microscopy, fluorescence microscopy or electrophysiological measurements.

Besides already established concepts of pathogen and toxin uptake, we will discuss the novel concept of glycosphingolipid-driven uptake starting at a low complexity level with toxins. With an increase in size and complexity of the objects of interest, we will continue our reflection about viruses and bacteria.

2. Toxins

The pathology of infections caused by different species of *Shigella*, enterohemorrhagic strains of *E. coli* (EHEC) and *Vibrio cholerae* is closely linked to the action of their glycosphingolipid-binding toxins: Shiga toxin (Stx), Shiga-like toxins (SLTs) and cholera toxin (Ctx), respectively. Commonly these toxins lead to severe diarrhoea, accompanied by hemorrhagic lesions in the intestine in the case of infections by EHEC and some strains of *Shigella*.

Because of their structural organisation with a monomeric A-subunit and a pentameric B-subunit, Stx, Ctx and SLTs belong to a group of toxins referred to as AB_5 toxins. The A-subunit consists of the enzymatic part of the toxin, which modifies intracellular targets: The RNA N-glycosidase activity of Stx, for example, targets cellular 28S rRNA, rendering ribosomes inactive for protein synthesis [18] and the ADP-ribosyltranferase activity of Ctx targets heterotrimeric G-proteins, thereby activating adenylate cyclases in mucosal epithelial cells of the small intestine [19]. The A-subunit is non-covalently linked to the B-subunit, which binds to host cell glycosphingolipids [20]. To exert their catalytic functions in the cytosol, the toxins have to be endocytosed and the A-subunits translocated into the cytosol. We will also introduce the plant toxin ricin, which gained notoriety due to its misuse as a bioterrorism weapon.

2.1. Shiga toxins and other bacterial toxins

Shigella dysenteriae and certain other *Shigella* strains secrete two types of enterotoxins: the so-called Shiga toxins I and II (Stx1 and Stx2; or verotoxins). These toxins are functionally and structurally related to the Shiga-like toxins I and II (SLTs), which are produced by enterohemorrhagic *Escherichia coli* (EHEC) strains [21]. In humans, these toxins cause serious complications in the gastrointestinal tract, including haemolytic colitis, which may (especially in children and elderly people) further progress to hemolytic-uremic syndrome (HUS) and severe complications of the central nervous system [22, 23]. Most recently, more than 50 patients in Germany died of EHEC infections [24-26].

While the sequence homology between Stx1 and Stx2 is only modest, *S. dysenteriae*'s Shiga toxin I and *E. coli*'s Shiga-like toxin I are 99% identical [27]. Shiga toxins are composed of a catalytic active A-subunit (StxA) of 32 kDa, which is non-covalently associated to the receptor-binding B-subunit pentamer (StxB; molecular mass of 5 x 7.7 kDa) [28, 29]. Despite the modest sequence homology (only 56% of the amino acid sequence), the B-subunits of Stx1 and Stx2 form a similarly structured homo-pentamer and bind to the same cellular receptor, the neutral glycosphingolipid globotriaosylceramide (Gb3, also known as CD77 or P^k blood group antigen) [30].

Other bacterial toxins, which use glycosphingolipids as their cellular receptors, include the GM1-binding cholera toxin (Ctx) of *Vibrio cholerae,* which is the causative agent of cholera [31], the respectively GM1 and GD1a-binding Heat-labile enterotoxins 1 and IIb of certain *E.coli* strains [32], Tetanus neurotoxin of *Clostridium tetani* and Botulinum toxin of *Clostridium botulinum.*

Crystal structures of Shiga toxins, cholera toxin, and Heat-labile enterotoxins revealed that the B-subunits of all these toxins fold into a doughnut-shaped pentamer and are of remarkable resemblance, although no amino acid sequence homology exists. Even more striking, the GM1-binding simian virus 40 capsid protein VP1, which we will present in more detail later, shares a structurally very closely related pentameric structure with each binding pocket arranged some 30 Å apart.

The binding affinity of the B-subunit of Shiga toxin (StxB) and cholera toxin (CtxB) to individual Gb3 and GM1 molecules, respectively, is very low (in the mM range) [33, 34], but the cooperative binding of multiple lipid molecules (up to 15 Gb3 molecules in the case of StxB) markedly increases the apparent affinity of the toxin to its receptor (in the nM range) [35-37]. Mutations of individual binding-pockets in the B-subunits of Stx and Ctx dramatically decrease the ability of the toxins to strongly associate to its receptor, and consequently to efficiently infect cells [38-40].

After receptor binding, Stx is internalized by clathrin-dependent as well as clathrin-independent endocytosis [41-44]. Even though Ctx has been found to be associated with caveolae, Ctx is efficiently endocytosed into cells devoid of caveolin-1 (a critical structural component of caveolae) [45], arguing that the caveolae-mediated endocytosis is not the major internalization pathway for Ctx in certain cells.

Studies on artificial membrane systems and energy-depleted cells (i.e. under conditions where the functionality of the cytosolic machinery is efficiently impaired) showed that StxB and CtxB are able to strongly cluster their glycosphingolipid receptors in the outer membrane leaflet, provoking the inward-bending of the plasma membrane and the generation of deep tubular membrane invaginations (Figure 1 and 2) [5, 6], suggesting that the toxin is able to trigger its own internalization into host cells, independently of host cell factors. These studies have uncovered a previously unknown mechanism for generating negative membrane curvature, and they have created a new paradigm that allows the conceptualization of why endocytotic coats are not detected at many sites of clathrin-independent endocytosis.

After binding of the toxins to glycosphingolipids, the invaginated membrane remains connected to the extracellular space as long as scission does not occur. The scission process requires cellular energy in contrast to the tubule formation. As a key factor in membrane scission, the GTPase dynamin has been described [46, 47]. However, dynamin-independent and cholesterol-dependent scission can also be observed, e.g. for the clathrin-independent endocytosis of clustered glycosylphosphatidylinositol (GPI)-anchored proteins when they laterally associate with proximal transmembrane proteins

[48]. Interestingly, the dynamin-independent scission involves glycosphingolipids. In line with this, it was shown that Stx-induced membrane tubules that undergo dynamin-independent and cholesterol-dependent scission by an Arp2-based reorganization of the cortical actin covering the tubule, ultimately leading to membrane constrictions [11]. How actin is linked to the plasma membrane is not clear, but binding of Stx to Gb3 leads to a redistribution of different proteins that are involved in the regulation of the cytoskeletal organisation including ezrin, CD44, vimentin, cytokeratin, paxillin, focal adhesion kinase (FAK), alpha- and gamma-tubulins beneath the plasma membrane [49]. Interestingly, among the redistributed proteins, ezrin (one of many proteins that links the actin cytoskeleton to the plasma membrane) [50] was shown to be phosphorylated in response to the binding of Stx to Gb3, in a process that is dependent on cholesterol, phosphoinositide 3-kinase (PI3K), Src family kinases and Rho associated kinase 1 [49]. These studies hint to ezrin as a possible linker between the plasma membrane and actin on Stx-induced membrane tubules. However, how the different identified kinases regulate the Stx-induced phosphorylation of ezrin and if ezrin phosphorylation is important for tubule scission has to be investigated.

On Hela cells and other toxin-sensitive cells, Stx- and Ctx-containing plasma membrane derived vesicles localise to early endosomes from which they are transported along the retrograde pathway via the Golgi apparatus to the ER. There, the A-subunit is retro-translocated into the cytosol to inhibit protein biosynthesis [51]. The escape of Stx from the early endocytotic pathway to enter the retrograde pathway critically depends on clathrin [42], the phosphatidylinositol (4)-phosphate-binding clathrin adaptor epsinR [44] and the curvature- and cargo-recognizing retromer complex [52]. Following a model proposed by Popoff et al. [53], clathrin, which is recruited to early endosomal membranes by the PI(4)P-binding protein epsinR [44], induces membrane curvatures on early endosomes to form retrograde tubules that are processed by retromer-dependent scission [21]. The subsequent transport from early endosomes to the Golgi complex is specifically regulated by the delta isoform of the protein kinase C (PKCδ), which is activated by binding of Stx to Gb3. Inhibiton of PKCδ results in the accumulation of Stx in early endosomes, which fails to reach the Golgi complex [54]. In addition, Stx also activates spleen tyrosine kinase (Syk) upon binding to Gb3, which causes a rapid phosphorylation of the clathrin heavy chain (CHC). Prevention of CHC-phosphorylation results in an ineffective transport of Stx from the early endosome to the Golgi complex. These findings again establish clathrin as a critical regulator of the endosome-to-Golgi transport of Stx. It needs to be addressed by additional studies how PKC and Syk interact on this part of the retrograde transport of Stx.

The retrograde transport is the main route for intoxication of Stx (and other AB_5-toxins), and highly specific, protective small-molecule inhibitors of intracellular toxin transport have recently been identified [55]. Indeed, human monocyte-derived macrophages and dendritic cells are resistant to Stx intoxication, probably because StxB fails to associate with membrane microdomains and does not detectably enter the retrograde route [56].

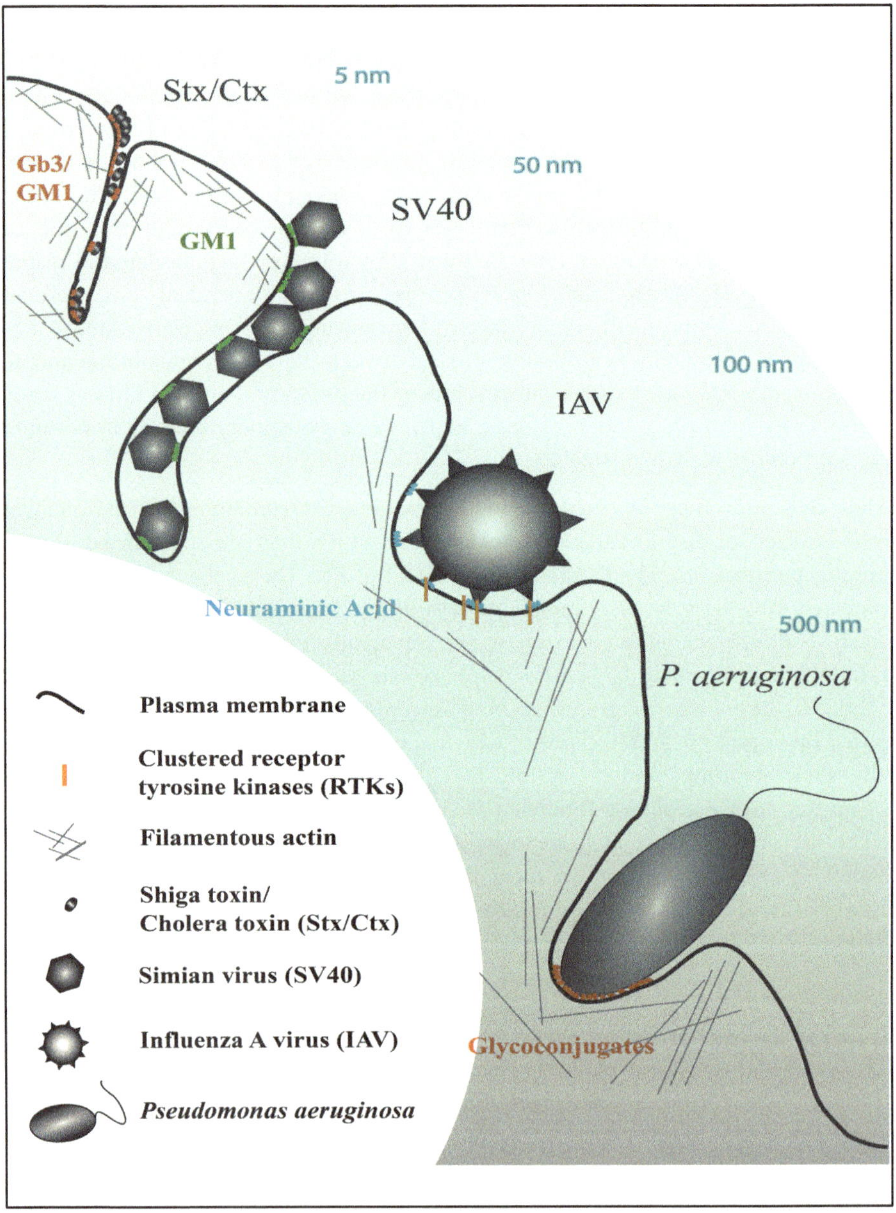

Figure 2. Schematic representation of entry strategies of some prominent pathogens and pathogenic products (toxins) and their (proposed) glycolipid receptors. As a common feature, clustering of glycolipids or glycosylated receptors has been proposed to trigger the endocytotic uptake of the toxin or the pathogen. The different elements in the scheme are not drawn to scale.

Lipid-mediated endocytosis represents a unique opportunity for several bacterial toxins to actively impose their invasion into host cells. This concept of a self-induced, lipid-mediated cellular uptake, seems not to be restricted to toxins alone. We will see in the section on viruses that a member of the *polyomaviridae* family, simian virus 40, follows this strategy in order to ensure efficient endocytosis into the host cell.

2.2. Ricin

Other lipid-binding toxins are the plant toxins ricin and the less-well characterized abrin, which are found in the seeds of castor beans (*Ricinus communis*) and of rosary pea (*Abrus precatorius*), respectively. These thermally stable proteins can be purified with ease in larger quantities, which prompted the Centers for Disease Control and Prevention (CDC) to categorise them as "Category B" agents (second highest priority) [57]. On the flipside, abrin and ricin were instrumental in Paul Ehrlich´s seminal work on the induction of immunity in mice, which have been fed with small amounts of ricin and later become immune against otherwise lethal toxin doses [58].

Despite being produced by unrelated plants and sharing only limited sequence homology, abrin and ricin exhibit a similar overall molecular structure [59, 60]: they are composed of a catalytic (toxic) A-subunit, which is covalently linked via a disulfide bond to a receptor-binding B-subunit.

The A-subunit possesses an N-glycosidase activity (identical to Shiga toxins and Shiga-like toxins) and strongly inhibits protein biosynthesis by removing a specific adenine residue from the 28S ribosomal RNA (A4324 in rat ribosomes), which prevents the binding of elongation factors [61]. These enzymes are highly efficient and a single toxin molecule suffices to kill a HeLa cell [62]. It is a remarkable observation that toxins produced by certain plants (ricin, abrin) and bacteria (Shiga toxins, Shiga-like toxins) exert their deleterious effect by an identical catalytic mechanism.

The B-subunit of ricin is a lectin with two functional binding pockets [63, 64] for ß1→4 linked galactose and N-acetylgalactoseamine residues of glycoproteins and glycolipids [65]. Whether these two binding sites operate independently from each other or cooperativity exists, remains a matter of debate. However, it is unlikely that an active lipid clustering effect, by which the toxin imposes its own uptake into cells as it has been proposed for Shiga toxin and simian virus 40, applies to ricin. Rather, ricin opportunistically binds to any glycoprotein or glycolipid with a terminal galactose at the cell surface and is passively taken up piggyback along with its binding partner. Due to its promiscuous binding, ricin enters cells by multiple endocytosis pathways (clathrin-dependent as well as clathrin-independent) [66] and its retrograde transport from the plasma membrane to the ER is highly inefficient: only an estimated 5% of cell-bound ricin is transported via the trans-Golgi network to the ER, from which it translocates into the cytosol, while the vast majority of toxin is either being recycled back to the cell surface or degraded in late endosomes and lysosomes [67]. This indiscriminate binding of ricin also explains why it is so difficult to identify distinct molecular players involved in the intracellular trafficking of ricin, while the retrograde transport of the Shiga toxin, which specifically uses the glycosphingolipid Gb3 as cellular receptor, is currently much better characterised [21, 68]. Recently, a genome-wide RNAi screen shed some light on the molecular requirements of ricin intoxication [69]. This study corroborates earlier observations that only a subset of molecular factors that are required for ricin trafficking is also involved in the retrograde transport of other toxins, such as Shiga toxin or *Pseudomonas* exotoxin, and that several intertwined retrograde transport pathways exist in parallel. Ablating specifically the retrograde transport of ricin by small-molecule inhibitors in vivo protects mice against an otherwise lethal dosis of ricin [55].

3. Viruses

As opposed to bacteria (discussed in the next section), viruses require the cellular machinery of the host for their replication and therefore must deliver their genome into their eukaryotic target cells. In contrast to viruses that directly fuse with the plasma membrane, for example retroviruses, herpes viruses and HIV, most of the enveloped viruses, including Influenza A Virus, hijack endocytotic pathways for their cellular entry [70-72], thereby taking advantage of the endosomal sorting machinery to reach a defined cellular compartment for uncoating. In the following section, we exemplify the entry of non-enveloped as well as enveloped viruses through the simian virus 40 and the Influenza A Virus, respectively.

3.1. Simian virus 40

Several non-enveloped viruses bind with their capsid to glycosphingolipids on the host cell and use them as viral receptors for efficient endocytosis and infection. The best characterized of these viruses are two members of the *polyomaviridae* family, the simian virus 40 (SV40) and the mouse polyoma virus (mPy), which bind to the gylcosphingolipids GM1, GD1a and GT1b, respectively. Other lipid-binding, non-enveloped viruses include the BK virus, Merkel cell polyomavirus and murine norovirus [73, 74].

Interestingly, the binding to glycosphingolipids at the plasma membrane pre-determines the uptake mechanism and intracellular trafficking route of viruses: instead of being degraded in the late endocytotic pathway or recycled back to the plasma membrane after endocytosis, these viruses are transported from the plasma membrane to the ER [75], from which they translocate into the cytosol [76]. Once they reach the nucleus, viruses subvert the cellular machinery and replicate.

Natural hosts for SV40 are Asian macaques, where it induces persistent infections in the kidneys. However, it was also shown that SV40 is significantly associated with human brain tumours and bone cancer [77], indicating its cell transforming properties.

The viral capsid is mainly composed of 72 VP1 protein pentamers in an icosahedral organisation [78]. The VP1 of SV40 folds into a doughnut conformation and bears five highly specific GM1-binding sites [79, 80]. Minor differences in the carbohydrate moiety of GM1, which is exposed into the extracellular space, strongly affect the binding affinity of the virus [80].

The multivalent binding of the VP1 pentamer to GM1 enables the tight association of the virus to the host cell despite the otherwise low affinity of individual binding sites of SV40 to GM1 [79]. In addition, a recent study on cellular and artificial membranes revealed that by virtue of this multivalent binding of GM1, SV40 induces the reorganization of membrane lipids and the segregation of specific lipids into membrane nanoscale domains, and thereby actively promotes its own uptake into the host cell [6]. This process critically depends on the lipid structure of GM1 and is essential for efficient infection by SV40 (Figure 2).

The precise physiological function of caveolae – uncoated, flask-like pits, enriched in cholesterol and glycosphingolipids (e.g. GM1, GM3) – still remains debated. A recent study supports the notion that caveolae act as a membrane reservoir to counter mechanical stress [81]. The role of caveolae in clathrin-independent endocytosis is equally a matter of much debate [82, 83]. FRAP (fluorescence recovery after photobleaching)-experiments on cells, which express GFP-tagged caveolin-1 (the major protein component of caveolae in epithelial cells) show that caveolae are rather immobile structures, a finding that argues against a major role of caveolae in constitutive endocytosis [84]. Though earlier studies suggest that the uptake of SV40 occurs via caveolae [85-88], recent work shows that the majority of SV40 does not partition into caveolae and that SV40 efficiently infects cells devoid of caveolin-1 [6, 89], corroborating the idea that the caveolin-independent, lipid-induced pathway represents the major route for efficient SV40-infection.

3.2. Influenza A Virus

Influenza A Virus (IAV) is the causative agent of flu, which is an infectious disease, primary affecting the deep respiratory tract. IAV is an enveloped virus, which possesses a single stranded RNA genome in a negative orientation. Infectious particles of influenza viruses are pleomorphic, filamentous or spherically shaped particles with a mean diameter of 120 nm [90]. IAV particles attach to their host cells by binding with their trimeric haemagglutinin (HA) to terminal α2,6 or α2,3 glycosidic-bound N-acetylneuraminic acids (sialic acids) on the surface of the host cells [4]. Following receptor binding, virions undergo endocytosis and become uncoated in a pH-dependent manner [91]. The low pH of late endosomes induces a conformational change in the HA, resulting in the fusion of HA with the endosomal membrane [92, 93] and the release of the RNA into the cytosol of the infected cell.

Electron microscopy-based studies revealed that plasma membrane-derived vesicles containing IAV are surrounded by clathrin, indicating that clathrin is involved in the uptake of IAV [93]. Since IAV particles have also been observed in smooth, non-coated vesicles, it was speculated that IAV also enter host cells by clathrin- and caveolin-independent endocytosis. This notion was supported by the observation that cells expressing dominant negative Eps15 (a clathrin adaptor) and caveolin-1 were still infected by IAV. Interestingly, subsequent studies showed that IAV actively induce the *de novo* formation of clathrin-coated pits by binding to the host cell surface [72]. The mechanism that triggers the recruitment of clathrin is unknown. It was speculated that IAV by binding to the host cell surface induces negative membrane curvatures that are sensed by BAR-domain containing proteins, which in turn recruit clathrin. However, the membrane-bending properties of IAV have not yet been shown. In this regard it is interesting that IAV, SV40 VP1 and Shiga toxin all bind to glycosphingolipids to induce endocytotic processes. Although experimental data are missing to this end, it is conceivable that IAV is able to bend membranes through clustering of sialic acid receptors for entry (Figure 2). The process of membrane bending and receptor clustering would likely be more complex than for Shiga toxin and SV40 VP1, considering that two different membranes (i.e. plasma membrane of the host cell and the viral envelope) are involved in this process.

Although more specific receptors than sialic acids are not yet identified for IAV, the virus activates specific cellular kinases for its efficient uptake, for example PI3K. PI3K is activated during the first 60 min of infection and was demonstrated to be required for efficient uptake [94]. The precise function of PI3K for the entry of IAV is not completely understood. Interestingly, IAV-activated PI3K seems to regulate an entry step, which precedes endosomal sorting [94]. In the context of bacterial invasion, PI3K activation is often associated with dramatic actin re-arrangements leading to macropinocytosis of the bacterium. Interestingly, this pathway has been reported recently as an alternative entry pathway for IAV that is dependent on the kinases Rac1 and Src, but independent of dynamin [95]. It was speculated that the virus activates this pathway by interacting with receptor tyrosine kinases (RTK) in the plasma membrane of the host cell. A study published at the same time reports that EGF receptor, a RTK, is activated by sialic acid-dependent IAV binding to ensure the efficient uptake of IAV [96]. This study hints to RTK, such as the EGF receptor, as entry receptors that promote the efficient uptake of the virus in a sialic acid dependent manner. It was hypothesized that sialic acids containing signalling receptors and/or glycosphingolipids become clustered upon binding by viral HA, leading to the activation of these signalling receptors and subsequent induction of PI3K signalling required for the cellular uptake [96]. So far, PI3K activation has not been linked to the re-arrangements of the cytoskeleton by actin polymerisation, although it was shown for polarized epithelial cells; actin dynamics and the motor protein myosin IV are apparently indispensable for the internalisation of IAV [97].

Despite the vast number of reports analysing the entry processes of IAV at the plasma membrane, additional studies are required to understand the exact mechanistic role of sialic acids during the entry mechanism. In particular, less is known about the trans-bilayer signalling of sialic acids on the outer membrane leaflet towards the cytosolic machinery in the context of IAV entry.

So far, we have addressed the endocytotic mechanisms of toxins and viruses. In the following, we will review the internalization strategies of invasive bacteria. The most significant difference to toxin molecules and viruses regarding the initial entry steps, is that bacteria sense environmental changes (e.g. Ca^{2+} levels, temperatures, surfaces) and dynamically *respond* to them in a more complex manner than toxins and viruses can (e.g. by the expression of a secretion system for their uptake [98], as discussed below). By doing so, these pathogens can manipulate their local microenvironment to a certain level, which makes the invasion process more complex as compared to toxins and viruses.

4. Non-invasive bacteria and the role of phagocytosis

A limited number of immune cells, such as macrophages, monocytes, dendritic cells, and neutrophils, are able to incorporate large particles in an actin-dependent process called phagocytosis and thus eliminate cellular debris, apoptotic bodies and pathogens [99]. During phagocytosis, the tight interaction between the particle and cell surface receptors of the host cell (e.g. Fc or complement receptors) induces a transient reorganization of the actin cytoskeleton and the generation of local membrane protrusions that engulf the particle.

Several pathogens, such as *Mycobacteria* (including *M. tuberculosis* and *M. leprae*) and *Brucella,* have exploited this mechanism for their uptake into host cells. During the invasion by means of phagocytosis, the pathogen is passively taken up into the cell together with extracellular fluid. After internalization, pathogens alter the cellular machinery (e.g. prevention of the fusion of phagosomes and lysosomes) or are equipped to counter the phagocytic attacks (e.g. certain components in the outer bacterial membrane protect the pathogen against lysosomal enzymes; secreted enzymes neutralize toxic oxygen species) in order to survive inside the phagocytic cell, where they can replicate. The uptake of *M. tuberculosis* nicely illustrates the role for host cell PIPs in the invasion of pathogenic bacteria. Under normal conditions PI(4,5)P_2 and PI(3,4,5)P_3 are mainly localized and formed at the plasma membrane and recruit proteins important for phagocytosis. During the maturation of the phagosomes the small GTPase Rab5, the most abundant protein on pre-mature phagocytic vacuoles, recruits the PI3K hVps34 to generate PI(3)P. PI(3)P is now the dominating PIP species on the phagosomal membrane and attracts PI(3)P-binding proteins. These include the early endosomal antigen 1 (EEA1), which is critical for the further maturation of pre-mature phagocytic vesicles into phagolysosomes [100]. *M. tuberculosis* has evolved a mechanism to prevent the fusion of the bacterium-containing phagosome with early endosomes. By secreting the phosphatidylinositol analogon lipoarabinomannan (LAM), *M. tuberculosis* inhibits an increase of the cytosolic calcium in infected cells, thereby blocking Ca^{2+}/calmodulin kinase II, which is required for the activation of the PI3K hVps34 and the generation of PI(3)P [101]. Phosphatidylinositol mannoside (PIM), which is another mycobacterium-secreted phosphoinositide, stimulates early endosome fusion and consequently blocks phagosomal maturation [102]. PI(3)P is also directly dephosphorylated by the mycobacterium-secreted PI phosphatase SapM, which additionally contributes to the arrest of phagosomal maturation. [103]. These examples illustrate the powerful defence of *M. tuberculosis* to prevent its digestion in lysosomes by interfering with the host cell PIP metabolism.

5. Invasive bacteria: trigger versus zipper mechanism

As only a subset of cells phagocytose, so-called "invasive bacteria" have developed strategies to actively induce their own uptake into non-phagocytic cells (e.g. intestinal epithelial cells). These invasive bacteria are categorized by their entry mechanism into two groups: "triggering" and "zippering" bacteria. Swanson and Baer were the first who proposed these mechanisms for particle phagocytosis in 1995 [99].

Principally, "triggering" bacteria secret effector proteins into their target cells. These cells respond with a re-arrangement of the cytoskeleton that promotes the entry of the bacterium. In contrast, bacteria that enter cells via the "zipper" mechanism engage specific surface receptors of the target cell, leading to just moderate actin re-modelling concomitant by less dramatic alterations of the host cell surface. As a result of both strategies, these bacteria are tightly engulfed by the host cell plasma membrane [104] (Figure 3). In the following, we discuss these invasion strategies by using the examples of the best-characterised "triggering" and "zippering" bacteria.

5.1. "Triggering" bacteria

Many Gram-negative bacteria, such as *Salmonella enterica, Shigella flexneri,* and *Pseudomonas aeruginosa,* invade cells through a "trigger" mechanism.

Salmonella, as well as *Shigella,* (described in the toxin section above), *Listeria* and *Yersinia* (described below), are foodborne pathogens, which cause gastritic infections by ingestion of contaminated food or water. Typical symptoms of an infection by *Salmonella* species, e.g. *Salmonella enterica,* are diarrhoea, abdominal cramps and fever.

By virtue of a type III secretion system (T3SS), which serves as a translocation pore, *Salmonella, Shigella* and *P. aeruginosa* inject virulence factors directly into the host cell cytosol during infection. The T3SS spans the bacterial membrane and is then inserted into the host cell membrane. It is assembled in the different species by SipB and SipC in *Salmonella*; IpaB and IpaC in *Shigella*; and EspB and EspD in enteropathogenic *Escherichia coli* [105, 106].

For an efficient invasion, a specific membrane microenvironment is critical. Indeed, it has been shown for *Salmonella, Shigella,* FimH-expressing *E. coli* and *P. aeruginosa* that specialized lipid membrane microdomains, which are enriched in cholesterol and sphingolipids, are required for efficient binding of bacteria to target cells, the activation of their T3SS, the translocation of effectors into the host cell cytosol and for the activation of cellular signalling pathways essential for bacterial invasion [107-110].

The translocated bacterial virulence factors subvert various cellular activities of the host cell, which leads to a massive polymerization of actin and enables the internalization of the pathogen into the target cell [111, 112]. In the case of *Shigella,* VirA binds to αβ-tubulin heterooligomers and induces a local destabilization and depolymerisation of microtubules, which triggers the activation of the kinase Rac1 and promotes membrane ruffling [113]. Other virulence factors, such as SipC, SipA, SopE and SopE2 in *Salmonella* and IpaB and IpaC in *Shigella,* induce the polymerization of actin directly (SipC) or in a Rac1- and Cdc42-dependent manner via activation of the Arp2/3 complex (IpaC and SopE/SopE2) [114-116]. A further stimulus of the actin polymerization is mediated by phosphatidylinositol phosphatases (IpgD in *Shigella* and SopB/SigD in *Salmonella*), which hydrolyse $PI(4,5)P_2$ into PI(5)P, causing the disconnection of cortical actin from the plasma membrane and enhancing actin dynamics at the bacterial entry site [117, 118]. The resulting membrane protrusions engulf the pathogen, which is then (i.e. upon actin depolymerisation) internalized into the host cell.

Pseudomonas aeruginosa is a Gram-negative opportunistic pathogen leading to acute infections of the respiratory tract, the urinary tract and the skin. Like other Gram-negative bacteria mentioned above, *P. aeruginosa* also injects effector proteins into host cells by virtue of a T3SS. Among the different strains presented so far, those that produce the T3SS effector proteins ExoT and ExoS are more efficiently internalized than those that do not produce these proteins [119]. Both, ExoS and ExoT are bifunctional proteins containing a C-terminal ADP-ribosylase and a N-terminal GTPase activating protein (GAP)-activity. Both functions redundantly disrupt the actin cytoskeleton [120]. Surprisingly, ExoS and ExoT act as anti-internalization factors [121], probably by interfering with components of the Abl pathway [122]: Rac1, Cdc42 and Crk were demonstrated to be activated by, and necessary for, the cellular uptake of *P. aeruginosa* [123]. Interestingly, additional experiments performed with ΔExoS and ΔExoT deletion mutants of *P. aeruginosa* revealed that ExoT abrogated Rac1 and Cdc42 activation, whereas ExoS activates these GTPases in order to promote efficient uptake [123]. However, the role of ExoS/T in the *P. aeruginosa* entry seems to be complex, in particular considering that ExoS/T production is characteristic for invasive strains.

Early studies demonstrate that *P. aeruginosa* binds to asialo-gangliosides [2]. Among these, asialo-GM1 seems to be important for the attachment of the bacteria to target cells in the respiratory tract by type IV pili [3]. Surprisingly, studies applying the small-molecule inhibitor PPMP, which inhibits the glucosylceramide synthase and consequently the biosynthesis of glycosphingolipids [124], revealed that glycosphingolipids are rather important for the internalization instead of the adhesion to target cells [125]. Based on these observations, one can speculate that those bacterial lectins, which bind to glycosphingolipids, might also promote the cellular invasion of *P. aeruginosa* into host cells. In this scenario, such lectins bend the plasma membrane by multivalent binding to their glycosphingolipid receptors in a similar manner as StxB (see section: Toxins), which might at least facilitate the initial steps of bacterial uptake. However, additional studies are required to proof this concept.

On the cytoplasmic side of the plasma membrane, PIPs were found to be important for the entry of *P. aeruginosa*. It was shown that invasive *P. aeruginosa* activate, and depend on, PI3K activity for efficient internalization [126]. Interestingly, PI(3,4,5)P_3 was found to accumulate at the bacteria entry site. However, it is not clear, which bacterial factor activates PI3K signalling and how PI3K leads to the internalization of *P. aeruginosa*. It was speculated that PI3K-regulated actin dynamics in the context of macropinocytosis leads to the internalization of the bacterium. This idea is supported by the fact that actin is required for the internalization of *P. aeruginosa* into host cells [127]. In subsequent studies using polarized epithelial cells, *P. aeruginosa* activates and recruits PI3K to the bacterial attachment site at the apical membrane of the cells. These processes were accompanied by the induction of membrane protrusions, enriched in filamentous actin and PI(3)P [128]. It needs to be determined how PI(3)P induces actin dynamics at the plasma membrane in order to promote the internalization of *P. aeruginosa*.

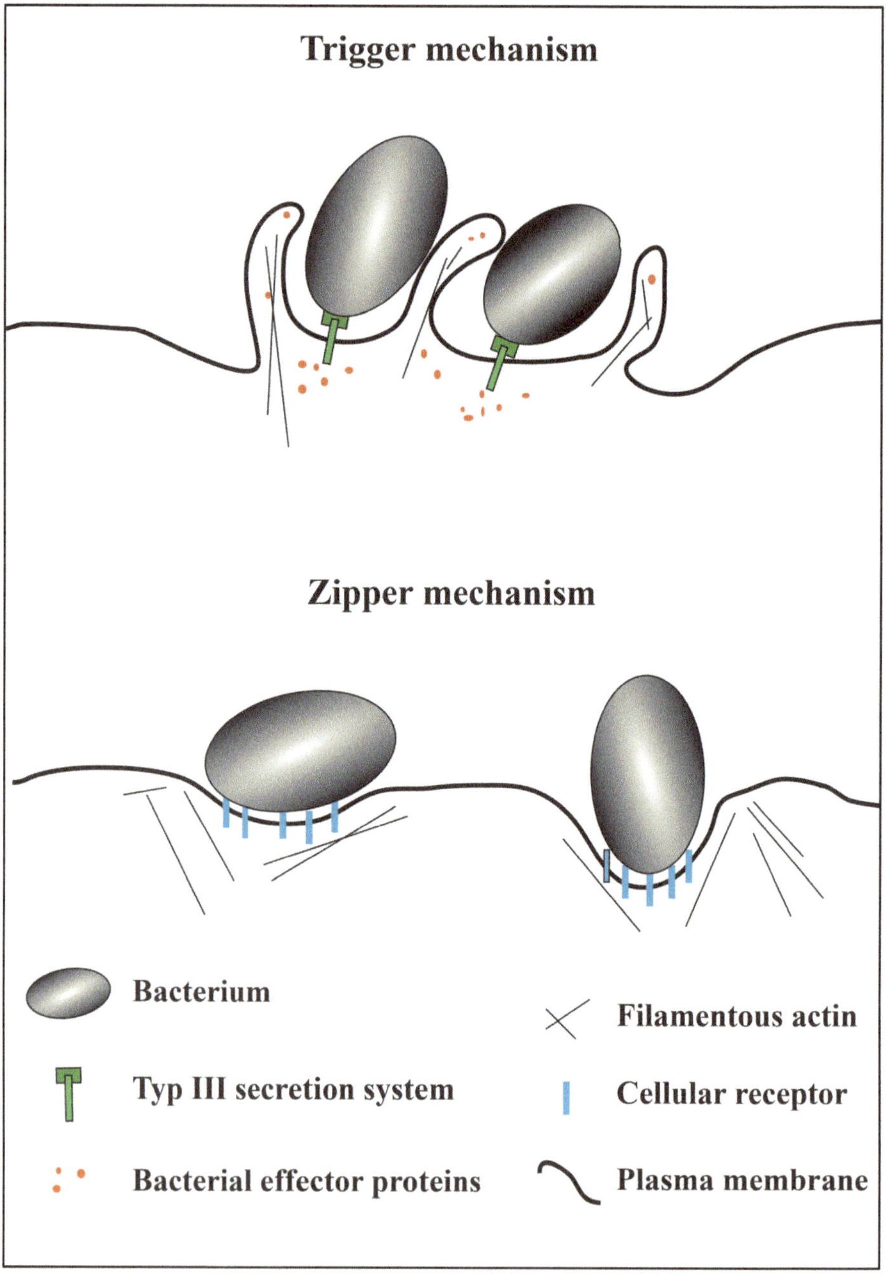

Figure 3. The two principle invasion mechanisms for invasive bacteria: the trigger and the zipper mechanism.

In general, less is known about the PIP metabolism in the context of *P. aeruginosa* uptake. It would be interesting to analyse whether *P. aeruginosa* also affects other PIPs and whether the T3SS effectors ExoS and ExoT also interfere with the PIP metabolism to induce the cellular uptake of the bacterium. These open questions necessitate additional studies to obtain more detailed insights into the molecular invasion mechanism of *P. aeruginosa.*

5.2. "Zippering" bacteria

The zipper mechanism is best characterized for the cellular invasion of *Listeria monocytogenes* and *Yersinia pseudotuberculosis*. By interacting with host cell receptors, these pathogens induce signalling events in the host cell to stimulate modest cytoskeletal rearrangements and membrane extensions for efficient invasion. *L. monocytogenes* is a foodborne pathogen, which crosses several host barriers leading to listeriosis, gastroenteritis and central nervous system infections [129]. The foodborne *Y. pseudotuberculosis* causes gastrointestinal and extra-intestinal infections, which can be accompanied by an abscess-forming mesenteric lymphadenitis [130].

Internalin (InlA) of *L. monocytogenes* binds to the cellular adhesion glycoprotein E-cadherin [131, 132], which is critically involved in the formation and integrity of adherens junctions in epithelial cells. Actin reorganization required for the entry of *L. monocytogenes* is triggered by the cytoplasmic domain of E-cadherin, which binds transiently to the actin cytoskeleton via interactions with α- and β-catenins and in concert with ARHGAP10 (the guanosine-activating protein (GAP) for RhoA and Cdc42), and the GTPase Arf6 [133, 134]. E-cadherin-mediated actin remodelling is further stimulated by the phosphorylation of cortactin [135] and the activation of the Arp2/3 complex in a Rac-dependent manner [136]. The efficient entry of *L. monocytogenes* also requires caveolin for the clustering of E-cadherin. In this scenario, caveolin is a prerequisite for the InlA-induced clustering of E-cadherin, which is localized around the bacterium. The clustered E-cadherin becomes tyrosine phosphorylated by Src, recruiting the ubiquitin ligase Hakai, which mediates the ubiquitination of E-cadherin. Finally, the ubiquitinated E-cadherin triggers the recruitment of clathrin to the entry site, which leads to the internalization of *Listeria* [137]. Furthermore, myosin VIIA and its ligand vezatin have been implicated in the endocytosis of *L. monocytogenes* [138].

The second invasion protein of *L. monocytogenes*, Internalin B (InlB), interacts with the ubiquitously expressed HGF (hepatocyte growth factor)-receptor Met [139]. Upon binding, the tyrosine-kinase Met dimerizes and autophosphorylates, which leads to the recruitment of the adaptor proteins Cbl, Gab1, Shc and Crk2 [140]. The subsequent activation of PI3K, which phosphorylates $PI(4,5)P_2$ into $PI(3,4,5)P_3$, promotes actin polymerization through the stimulation of the Arp2/3 complex in a Rac1-dependent manner [141]. Interestingly, depletion of membrane cholesterol with methyl-ß-cyclodextrin (MβCD) diminishes the activation of Rac1, but not of PI3K, indicating a possible need for the repartitioning of $PI(3,4,5)P_3$ into cholesterol-enriched membrane microdomains [142]. Surprisingly, clathrin, dynamin and several other components of the endocytotic machinery co-localize with the bacterial entry site and are essential for invasion of *L. monocytogenes* and *Y. pseudotuberculosis* [143].

While clathrin is crucial for the internalization of these "zippering" bacteria, it is not required for the entry of "triggering" bacteria (*Salmonella, Shigella*) [144]. More studies are needed to elucidate the functional interplay between clathrin and cholesterol-enriched membrane domains.

Recently, another class of GTPases – Septins - have been identified to be required for the efficient, InlB-dependent entry of *L. monocytogenes* [145]. Septins regulate actin organisation [146] and phagosome formation in macrophages [147]. The mechanism of septins during the uptake of *L. monocytogenes* is not yet clear. Interestingly, septin and actin colocalize at the same bacterial entry site. However, based on the ring-like assembly of septins, which is different from the actin architecture at the bacterial entry site, it is suggested that actin and septins fulfil distinct or complementary roles during the internalization process [145].

Upon internalisation, *L. monocytogenes* is located within a vacuole, from which it eventually escapes into the cytosol by the synergistic action of the pore forming toxin listeriolysin O and the bacterial encoded, PIP-specific PLC [140].

Similar to *L. monocytogenes*, enteropathogenic *Yersinia* species activate PI kinases by interaction of the *Yersinia* outer membrane protein invasine to the heterodimeric β1 integrin receptor [148] and subsequent activation of the Rac1 pathway. The activation of the Rac1 pathway leads to the local enrichment of $PI(4,5)P_2$ through the recruitment and activation of PI5K [148], which is a lipid kinase that selectively phosphorylates the inositol ring at D5 position of PIs. However, the final activation of PI5K seems to be induced by Arf6, a GTP binding protein that normally regulates the production of $PI(4,5)P_2$ [149]. It is assumed that Arf6-driven actin dynamics mediate the formation of the phagocytic cup surrounding *Yersinia* [150]. At this stage of entry, *Yersinia* is located in an intermediate compartment, termed prevacuole, which is still connected to the plasma membrane. It could be shown that $PI(4,5)P_2$ needs to be hydrolyzed in order to proceed the maturation of the prevacuole into a separate, sealed compartment. This step is mediated by the inositol-5-phosphatases OCRL and Inpp5b, which are recruited to the *Yersinia*-containing prevacuole. As a prerequisite for this recruitment, the GTPase Rab5 must associate with the prevacuole and it could indeed be shown that PI3K seems to mediate this step [151].

This is an intriguing example that illustrates how invasive pathogens dynamically regulate host cell PIPs to complete their internalization.

6. Conclusion

The example of SV40 shows how pathogens are able to initiate their uptake by engaging glycosphingolipids on the surface of host cells. It is intriguing to see that this concept of endocytotic uptake resembles those found for Stx uptake. This suggests that glycosphingolipid-driven endocytosis is not restricted to specific pathogens or toxins, but rather seems to be a general concept for initiating endocytotic processes at biological membranes. Considering that invasive bacteria bind carbohydrate receptors in the plasma membrane for proper attachment to host cells, an interesting question is whether such adhesion receptors are also engaged as internalization receptors allowing invasive pathogens to gain access to host cells. In such a scenario, the invasive bacterium binds via some of its lectins to host cell receptors, which subsequently become clustered, thereby creating asymmetrical stress in the lipid bilayer and leading to membrane invaginations that facilitate the bacterial uptake. In particular, it would be interesting to know if the putative lectin-induced membrane invagination modulates host cell PIP metabolism at the cytoplasmic side of the plasma membrane for the efficient entry of the pathogen. Additionally, it needs be clarified whether the lectin-bended plasma membrane is sensed by cellular effector proteins, for example by BAR domains, that could stabilise and/or further assist in the invagination of the plasma membrane to accomplish the endocytosis process.

As most pathogens, in particular invasive bacteria, depend on the rearrangement of actin for their cellular uptake, it remains to be determined to which extent actin dynamics, as observed for Stx-induced membrane tubules, also contribute to the scission of the bacteria-containing vacuoles.

Glycolipids are often found co-clustered with protein receptors in the plasma membrane. It remains to be identified, if specific glycolipids, which are used as pathogen or toxin receptors, preferentially interact with other proteins in the plasma membrane to specify the endocytotic route of a toxin or an invasive pathogen.

As yet, less is known about these specific issues of the initial steps of pathogen internalization, and many more studies need to be carried out to examine the exact functional and mechanistic role of glycosphingolipids and phosphoinositides in pathogen invasion.

Author details

Thorsten Eierhoff
Albert-Ludwigs-University Freiburg,
Institute of Biology II and BIOSS Centre for Biological Signalling Studies, Freiburg, Germany

Bahne Stechmann
Leibniz-Institut für Molekulare Pharmakologie (FMP), Berlin, Germany

Winfried Römer
Albert-Ludwigs-University Freiburg,
Institute of Biology II and BIOSS Centre for Biological Signalling Studies, Freiburg, Germany

7. Acknowledgement

This work was supported by the German Excellence Initiative of the Deutsche Forschungsgemeinschaft (EXC 294) and by a starting grant of the European Research Council to W.R. (Programme "Ideas" - call identifier: ERC-2011-StG 282105-lec&lip2invade).

8. References

[1] WHO, The top 10 causes of death. WHO Fact sheet 2011. N° 310. Available: http://www.who.int/mediacentre/factsheets/fs310/en/index.html

[2] Krivan, H.C., V. Ginsburg, and D.D. Roberts, Pseudomonas aeruginosa and Pseudomonas cepacia isolated from cystic fibrosis patients bind specifically to gangliotetraosylceramide (asialo GM1) and gangliotriaosylceramide (asialo GM2). Archives of biochemistry and biophysics, 1988. 260(1): p. 493-6.

[3] Saiman, L. and A. Prince, Pseudomonas aeruginosa pili bind to asialoGM1 which is increased on the surface of cystic fibrosis epithelial cells. The Journal of clinical investigation, 1993. 92(4): p. 1875-80.

[4] Skehel, J.J. and D.C. Wiley, Receptor binding and membrane fusion in virus entry: the influenza hemagglutinin. Annual review of biochemistry, 2000. 69: p. 531-69.

[5] Romer, W., et al., Shiga toxin induces tubular membrane invaginations for its uptake into cells. Nature, 2007. 450(7170): p. 670-5.

[6] Ewers, H., et al., GM1 structure determines SV40-induced membrane invagination and infection. Nature cell biology, 2010. 12(1): p. 11-8; sup pp 1-12.

[7] Ajit Varki (2009) Essentials of Glycobiology. Cold Spring Harbor (NY).

[8] Falkenburger, B.H., et al., Phosphoinositides: lipid regulators of membrane proteins. The Journal of physiology, 2010. 588(Pt 17): p. 3179-85.

[9] Sasaki, T., et al., The physiology of phosphoinositides. Biological & pharmaceutical bulletin, 2007. 30(9): p. 1599-604.

[10] Mathivet, L., S. Cribier, and P.F. Devaux, Shape change and physical properties of giant phospholipid vesicles prepared in the presence of an AC electric field. Biophysical journal, 1996. 70(3): p. 1112-21.

[11] Romer, W., et al., Actin dynamics drive membrane reorganization and scission in clathrin-independent endocytosis. Cell, 2010. 140(4): p. 540-53.

[12] Pautot, S., B.J. Frisken, and D.A. Weitz, Engineering asymmetric vesicles. Proceedings of the National Academy of Sciences of the United States of America, 2003. 100(19): p. 10718-21.

[13] Pontani, L.L., et al., Reconstitution of an actin cortex inside a liposome. Biophysical journal, 2009. 96(1): p. 192-8.

[14] Romer, W., et al., Channel activity of a viral transmembrane peptide in micro-BLMs: Vpu(1-32) from HIV-1. Journal of the American Chemical Society, 2004. 126(49): p. 16267-74.

[15] Romer, W. and C. Steinem, Impedance analysis and single-channel recordings on nano-black lipid membranes based on porous alumina. Biophysical journal, 2004. 86(2): p. 955-65.

[16] Danelon, C., et al., Cell membranes suspended across nanoaperture arrays. Langmuir : the ACS journal of surfaces and colloids, 2006. 22(1): p. 22-5.

[17] Ziegler, U., et al., Preparation of basal cell membranes for scanning probe microscopy. FEBS letters, 1998. 436(2): p. 179-84.

[18] Endo, Y., et al., Site of action of a Vero toxin (VT2) from Escherichia coli O157:H7 and of Shiga toxin on eukaryotic ribosomes. RNA N-glycosidase activity of the toxins. European journal of biochemistry / FEBS, 1988. 171(1-2): p. 45-50.
[19] Sharp, G.W. and S. Hynie, Stimulation of intestinal adenyl cyclase by cholera toxin. Nature, 1971. 229(5282): p. 266-9.
[20] Falnes, P.O. and K. Sandvig, Penetration of protein toxins into cells. Current opinion in cell biology, 2000. 12(4): p. 407-13.
[21] Johannes, L. and W. Romer, Shiga toxins--from cell biology to biomedical applications. Nature reviews. Microbiology, 2010. 8(2): p. 105-16.
[22] Karmali, M.A., et al., Sporadic cases of haemolytic-uraemic syndrome associated with faecal cytotoxin and cytotoxin-producing Escherichia coli in stools. Lancet, 1983. 1(8325): p. 619-20.
[23] Bale, J.F., Jr., C. Brasher, and R.L. Siegler, CNS manifestations of the hemolytic-uremic syndrome. Relationship to metabolic alterations and prognosis. American journal of diseases of children, 1980. 134(9): p. 869-72.
[24] Frank, C., et al., Epidemic profile of Shiga-toxin-producing Escherichia coli O104:H4 outbreak in Germany. The New England journal of medicine, 2011. 365(19): p. 1771-80.
[25] Beutin, L. and A. Martin, Outbreak of Shiga toxin-producing Escherichia coli (STEC) O104:H4 infection in Germany causes a paradigm shift with regard to human pathogenicity of STEC strains. Journal of food protection, 2012. 75(2): p. 408-18.
[26] Bielaszewska, M., et al., Characterisation of the Escherichia coli strain associated with an outbreak of haemolytic uraemic syndrome in Germany, 2011: a microbiological study. The Lancet infectious diseases, 2011. 11(9): p. 671-6.
[27] Jackson, M.P., et al., Nucleotide sequence analysis of the structural genes for Shiga-like toxin I encoded by bacteriophage 933J from Escherichia coli. Microbial pathogenesis, 1987. 2(2): p. 147-53.
[28] Stein, P.E., et al., Crystal structure of the cell-binding B oligomer of verotoxin-1 from E. coli. Nature, 1992. 355(6362): p. 748-50.
[29] Fraser, M.E., et al., Crystal structure of the holotoxin from Shigella dysenteriae at 2.5 A resolution. Nature structural biology, 1994. 1(1): p. 59-64.
[30] Lindberg, A.A., et al., Identification of the carbohydrate receptor for Shiga toxin produced by Shigella dysenteriae type 1. The Journal of biological chemistry, 1987. 262(4): p. 1779-85.
[31] Heyningen, S.V., Cholera toxin: interaction of subunits with ganglioside GM1. Science, 1974. 183(4125): p. 656-7.
[32] Lencer, W.I. and D. Saslowsky, Raft trafficking of AB5 subunit bacterial toxins. Biochimica et biophysica acta, 2005. 1746(3): p. 314-21.
[33] St Hilaire, P.M., M.K. Boyd, and E.J. Toone, Interaction of the Shiga-like toxin type 1 B-subunit with its carbohydrate receptor. Biochemistry, 1994. 33(48): p. 14452-63.
[34] Schon, A. and E. Freire, Thermodynamics of intersubunit interactions in cholera toxin upon binding to the oligosaccharide portion of its cell surface receptor, ganglioside GM1. Biochemistry, 1989. 28(12): p. 5019-24.

[35] Fuchs, G., et al., Pathogenesis of Shigella diarrhea: rabbit intestinal cell microvillus membrane binding site for Shigella toxin. Infection and immunity, 1986. 53(2): p. 372-7.
[36] Kuziemko, G.M., M. Stroh, and R.C. Stevens, Cholera toxin binding affinity and specificity for gangliosides determined by surface plasmon resonance. Biochemistry, 1996. 35(20): p. 6375-84.
[37] Turnbull, W.B., B.L. Precious, and S.W. Homans, Dissecting the cholera toxin-ganglioside GM1 interaction by isothermal titration calorimetry. Journal of the American Chemical Society, 2004. 126(4): p. 1047-54.
[38] Bast, D.J., et al., The identification of three biologically relevant globotriaosyl ceramide receptor binding sites on the Verotoxin 1 B subunit. Molecular microbiology, 1999. 32(5): p. 953-60.
[39] Soltyk, A.M., et al., A mutational analysis of the globotriaosylceramide-binding sites of verotoxin VT1. The Journal of biological chemistry, 2002. 277(7): p. 5351-9.
[40] Wolf, A.A., et al., Attenuated endocytosis and toxicity of a mutant cholera toxin with decreased ability to cluster ganglioside GM1 molecules. Infection and immunity, 2008. 76(4): p. 1476-84.
[41] Sandvig, K., et al., Endocytosis from coated pits of Shiga toxin: a glycolipid-binding protein from Shigella dysenteriae 1. The Journal of cell biology, 1989. 108(4): p. 1331-43.
[42] Lauvrak, S.U., M.L. Torgersen, and K. Sandvig, Efficient endosome-to-Golgi transport of Shiga toxin is dependent on dynamin and clathrin. Journal of cell science, 2004. 117(Pt 11): p. 2321-31.
[43] Nichols, B.J., et al., Rapid cycling of lipid raft markers between the cell surface and Golgi complex. The Journal of cell biology, 2001. 153(3): p. 529-41.
[44] Saint-Pol, A., et al., Clathrin adaptor epsinR is required for retrograde sorting on early endosomal membranes. Developmental cell, 2004. 6(4): p. 525-38.
[45] Kirkham, M. and R.G. Parton, Clathrin-independent endocytosis: new insights into caveolae and non-caveolar lipid raft carriers. Biochimica et biophysica acta, 2005. 1746(3): p. 349-63.
[46] Ferguson, S.M. and P. De Camilli, Dynamin, a membrane-remodelling GTPase. Nature reviews. Molecular cell biology, 2012. 13(2): p. 75-88.
[47] Schmid, S.L. and V.A. Frolov, Dynamin: functional design of a membrane fission catalyst. Annual review of cell and developmental biology, 2011. 27: p. 79-105.
[48] Mayor, S. and R.E. Pagano, Pathways of clathrin-independent endocytosis. Nature reviews. Molecular cell biology, 2007. 8(8): p. 603-12.
[49] Takenouchi, H., et al., Shiga toxin binding to globotriaosyl ceramide induces intracellular signals that mediate cytoskeleton remodeling in human renal carcinoma-derived cells. Journal of cell science, 2004. 117(Pt 17): p. 3911-22.
[50] Vaheri, A., et al., The ezrin protein family: membrane-cytoskeleton interactions and disease associations. Current opinion in cell biology, 1997. 9(5): p. 659-66.
[51] Sandvig, K., et al., Retrograde transport of endocytosed Shiga toxin to the endoplasmic reticulum. Nature, 1992. 358(6386): p. 510-2.

[52] Bujny, M.V., et al., The retromer component sorting nexin-1 is required for efficient retrograde transport of Shiga toxin from early endosome to the trans Golgi network. Journal of cell science, 2007. 120(Pt 12): p. 2010-21.
[53] Popoff, V., et al., The retromer complex and clathrin define an early endosomal retrograde exit site. Journal of cell science, 2007. 120(Pt 12): p. 2022-31.
[54] Torgersen, M.L., et al., Protein kinase Cdelta is activated by Shiga toxin and regulates its transport. The Journal of biological chemistry, 2007. 282(22): p. 16317-28.
[55] Stechmann, B., et al., Inhibition of retrograde transport protects mice from lethal ricin challenge. Cell, 2010. 141(2): p. 231-42.
[56] Falguieres, T., et al., Targeting of Shiga toxin B-subunit to retrograde transport route in association with detergent-resistant membranes. Molecular biology of the cell, 2001. 12(8): p. 2453-68.
[57] Roxas-Duncan VI, Smith LA, Of Beans and Beads: Ricin and Abrin in Bioterrorism and Biocrime. J Bioterr Biodef, 2012. S7:002. doi:10.4172/2157-2526.S7- 002
[58] Arthur M Silverstein (2002) Paul Ehrlich's receptor immunology: The magnificent obsession. Academic Press.
[59] Montfort, W., et al., The three-dimensional structure of ricin at 2.8 A. The Journal of biological chemistry, 1987. 262(11): p. 5398-403.
[60] Tahirov, T.H., et al., Crystal structure of abrin-a at 2.14 A. Journal of molecular biology, 1995. 250(3): p. 354-67.
[61] Endo, Y. and K. Tsurugi, RNA N-glycosidase activity of ricin A-chain. Mechanism of action of the toxic lectin ricin on eukaryotic ribosomes. The Journal of biological chemistry, 1987. 262(17): p. 8128-30.
[62] Eiklid, K., S. Olsnes, and A. Pihl, Entry of lethal doses of abrin, ricin and modeccin into the cytosol of HeLa cells. Experimental cell research, 1980. 126(2): p. 321-6.
[63] Sphyris, N., et al., Mutational analysis of the Ricinus lectin B-chains. Galactose-binding ability of the 2 gamma subdomain of Ricinus communis agglutinin B-chain. The Journal of biological chemistry, 1995. 270(35): p. 20292-7.
[64] Wales, R., et al., Mutational analysis of the galactose binding ability of recombinant ricin B chain. The Journal of biological chemistry, 1991. 266(29): p. 19172-9.
[65] Baenziger, J.U. and D. Fiete, Structural determinants of Ricinus communis agglutinin and toxin specificity for oligosaccharides. The Journal of biological chemistry, 1979. 254(19): p. 9795-9.
[66] Lord, M.J., et al., Ricin. Mechanisms of cytotoxicity. Toxicological reviews, 2003. 22(1): p. 53-64.
[67] van Deurs, B., et al., Estimation of the amount of internalized ricin that reaches the trans-Golgi network. The Journal of cell biology, 1988. 106(2): p. 253-67.
[68] Johannes, L. and V. Popoff, Tracing the retrograde route in protein trafficking. Cell, 2008. 135(7): p. 1175-87.
[69] Moreau, D., et al., Genome-wide RNAi screens identify genes required for Ricin and PE intoxications. Developmental cell, 2011. 21(2): p. 231-44.
[70] Pelkmans, L. and A. Helenius, Insider information: what viruses tell us about endocytosis. Current opinion in cell biology, 2003. 15(4): p. 414-22.

[71] Sieczkarski, S.B. and G.R. Whittaker, Dissecting virus entry via endocytosis. The Journal of general virology, 2002. 83(Pt 7): p. 1535-45.

[72] Rust, M.J., et al., Assembly of endocytic machinery around individual influenza viruses during viral entry. Nature structural & molecular biology, 2004. 11(6): p. 567-73.

[73] Ewers, H. and A. Helenius, Lipid-mediated endocytosis. Cold Spring Harbor perspectives in biology, 2011. 3(8): p. a004721.

[74] Rydell, G.E., et al., QCM-D studies of human norovirus VLPs binding to glycosphingolipids in supported lipid bilayers reveal strain-specific characteristics. Glycobiology, 2009. 19(11): p. 1176-84.

[75] Kartenbeck, J., H. Stukenbrok, and A. Helenius, Endocytosis of simian virus 40 into the endoplasmic reticulum. The Journal of cell biology, 1989. 109(6 Pt 1): p. 2721-9.

[76] Schelhaas, M., et al., Simian Virus 40 depends on ER protein folding and quality control factors for entry into host cells. Cell, 2007. 131(3): p. 516-29.

[77] Vilchez, R.A., et al., Simian virus 40 in human cancers. The American journal of medicine, 2003. 114(8): p. 675-84.

[78] Liddington, R.C., et al., Structure of simian virus 40 at 3.8-A resolution. Nature, 1991. 354(6351): p. 278-84.

[79] Neu, U., et al., Structural basis of GM1 ganglioside recognition by simian virus 40. Proceedings of the National Academy of Sciences of the United States of America, 2008. 105(13): p. 5219-24.

[80] Campanero-Rhodes, M.A., et al., N-glycolyl GM1 ganglioside as a receptor for simian virus 40. Journal of virology, 2007. 81(23): p. 12846-58.

[81] Sinha, B., et al., Cells respond to mechanical stress by rapid disassembly of caveolae. Cell, 2011. 144(3): p. 402-13.

[82] Rothberg, K.G., et al., Caveolin, a protein component of caveolae membrane coats. Cell, 1992. 68(4): p. 673-82.

[83] Kurzchalia, T.V., et al., VIP21, a 21-kD membrane protein is an integral component of trans-Golgi-network-derived transport vesicles. The Journal of cell biology, 1992. 118(5): p. 1003-14.

[84] Thomsen, P., et al., Caveolae are highly immobile plasma membrane microdomains, which are not involved in constitutive endocytic trafficking. Molecular biology of the cell, 2002. 13(1): p. 238-50.

[85] Anderson, H.A., Y. Chen, and L.C. Norkin, Bound simian virus 40 translocates to caveolin-enriched membrane domains, and its entry is inhibited by drugs that selectively disrupt caveolae. Molecular biology of the cell, 1996. 7(11): p. 1825-34.

[86] Pelkmans, L., J. Kartenbeck, and A. Helenius, Caveolar endocytosis of simian virus 40 reveals a new two-step vesicular-transport pathway to the ER. Nature cell biology, 2001. 3(5): p. 473-83.

[87] Pelkmans, L., et al., Genome-wide analysis of human kinases in clathrin- and caveolae/raft-mediated endocytosis. Nature, 2005. 436(7047): p. 78-86.

[88] Stang, E., J. Kartenbeck, and R.G. Parton, Major histocompatibility complex class I molecules mediate association of SV40 with caveolae. Molecular biology of the cell, 1997. 8(1): p. 47-57.

[89] Damm, E.M., et al., Clathrin- and caveolin-1-independent endocytosis: entry of simian virus 40 into cells devoid of caveolae. The Journal of cell biology, 2005. 168(3): p. 477-88.
[90] Harris, A., et al., Influenza virus pleiomorphy characterized by cryoelectron tomography. Proceedings of the National Academy of Sciences of the United States of America, 2006. 103(50): p. 19123-7.
[91] Lakadamyali, M., M.J. Rust, and X. Zhuang, Endocytosis of influenza viruses. Microbes and infection / Institut Pasteur, 2004. 6(10): p. 929-36.
[92] Maeda, T. and S. Ohnishi, Activation of influenza virus by acidic media causes hemolysis and fusion of erythrocytes. FEBS letters, 1980. 122(2): p. 283-7.
[93] Matlin, K.S., et al., Infectious entry pathway of influenza virus in a canine kidney cell line. The Journal of cell biology, 1981. 91(3 Pt 1): p. 601-13.
[94] Ehrhardt, C., et al., Bivalent role of the phosphatidylinositol-3-kinase (PI3K) during influenza virus infection and host cell defence. Cellular microbiology, 2006. 8(8): p. 1336-48.
[95] de Vries, E., et al., Dissection of the influenza A virus endocytic routes reveals macropinocytosis as an alternative entry pathway. PLoS pathogens, 2011. 7(3): p. e1001329.
[96] Eierhoff, T., et al., The epidermal growth factor receptor (EGFR) promotes uptake of influenza A viruses (IAV) into host cells. PLoS pathogens, 2010. 6(9): p. e1001099.
[97] Sun, X. and G.R. Whittaker, Role of the actin cytoskeleton during influenza virus internalization into polarized epithelial cells. Cellular microbiology, 2007. 9(7): p. 1672-82.
[98] Hueck, Type III Protein Secretion Systems in Bacterial Pathogens of Animals and Plants. Microbiol. Mol. Biol. Rev. , 1998. 62(2).
[99] Swanson, J.A. and S.C. Baer, Phagocytosis by zippers and triggers. Trends in cell biology, 1995. 5(3): p. 89-93.
[100] Kinchen, J.M. and K.S. Ravichandran, Phagosome maturation: going through the acid test. Nature reviews. Molecular cell biology, 2008. 9(10): p. 781-95.
[101] Vergne, I., J. Chua, and V. Deretic, Tuberculosis toxin blocking phagosome maturation inhibits a novel Ca2+/calmodulin-PI3K hVPS34 cascade. The Journal of experimental medicine, 2003. 198(4): p. 653-9.
[102] Vergne, I., et al., Mycobacterium tuberculosis phagosome maturation arrest: mycobacterial phosphatidylinositol analog phosphatidylinositol mannoside stimulates early endosomal fusion. Molecular biology of the cell, 2004. 15(2): p. 751-60.
[103] Vergne, I., et al., Mechanism of phagolysosome biogenesis block by viable Mycobacterium tuberculosis. Proceedings of the National Academy of Sciences of the United States of America, 2005. 102(11): p. 4033-8.
[104] Cossart, P. and P.J. Sansonetti, Bacterial invasion: the paradigms of enteroinvasive pathogens. Science, 2004. 304(5668): p. 242-8.
[105] Menard, R., et al., Extracellular association and cytoplasmic partitioning of the IpaB and IpaC invasins of S. flexneri. Cell, 1994. 79(3): p. 515-25.

[106] Ide, T., et al., Characterization of translocation pores inserted into plasma membranes by type III-secreted Esp proteins of enteropathogenic Escherichia coli. Cellular microbiology, 2001. 3(10): p. 669-79.
[107] Lafont, F., et al., Initial steps of Shigella infection depend on the cholesterol/sphingolipid raft-mediated CD44-IpaB interaction. The EMBO journal, 2002. 21(17): p. 4449-57.
[108] van der Goot, F.G., et al., Rafts can trigger contact-mediated secretion of bacterial effectors via a lipid-based mechanism. The Journal of biological chemistry, 2004. 279(46): p. 47792-8.
[109] Schoehn, G., et al., Oligomerization of type III secretion proteins PopB and PopD precedes pore formation in Pseudomonas. The EMBO journal, 2003. 22(19): p. 4957-67.
[110] Shin, J.S., Z. Gao, and S.N. Abraham, Involvement of cellular caveolae in bacterial entry into mast cells. Science, 2000. 289(5480): p. 785-8.
[111] Ogawa, M., et al., The versatility of Shigella effectors. Nature reviews. Microbiology, 2008. 6(1): p. 11-6.
[112] McGhie, E.J., et al., Salmonella takes control: effector-driven manipulation of the host. Current opinion in microbiology, 2009. 12(1): p. 117-24.
[113] Yoshida, S., et al., Shigella deliver an effector protein to trigger host microtubule destabilization, which promotes Rac1 activity and efficient bacterial internalization. The EMBO journal, 2002. 21(12): p. 2923-35.
[114] Tran Van Nhieu, G., et al., IpaC induces actin polymerization and filopodia formation during Shigella entry into epithelial cells. The EMBO journal, 1999. 18(12): p. 3249-62.
[115] Hayward, R.D. and V. Koronakis, Direct nucleation and bundling of actin by the SipC protein of invasive Salmonella. The EMBO journal, 1999. 18(18): p. 4926-34.
[116] Srikanth, C.V., et al., Salmonella effector proteins and host-cell responses. Cellular and molecular life sciences : CMLS, 2011. 68(22): p. 3687-97.
[117] Norris, F.A., et al., SopB, a protein required for virulence of Salmonella dublin, is an inositol phosphate phosphatase. Proceedings of the National Academy of Sciences of the United States of America, 1998. 95(24): p. 14057-9.
[118] Niebuhr, K., et al., Conversion of PtdIns(4,5)P(2) into PtdIns(5)P by the S.flexneri effector IpgD reorganizes host cell morphology. The EMBO journal, 2002. 21(19): p. 5069-78.
[119] Engel, J. and P. Balachandran, Role of Pseudomonas aeruginosa type III effectors in disease. Current opinion in microbiology, 2009. 12(1): p. 61-6.
[120] Barbieri, J.T. and J. Sun, Pseudomonas aeruginosa ExoS and ExoT. Reviews of physiology, biochemistry and pharmacology, 2004. 152: p. 79-92.
[121] Ha, U. and S. Jin, Growth phase-dependent invasion of Pseudomonas aeruginosa and its survival within HeLa cells. Infection and immunity, 2001. 69(7): p. 4398-406.
[122] Pendergast, A.M., The Abl family kinases: mechanisms of regulation and signaling. Advances in cancer research, 2002. 85: p. 51-100.
[123] Pielage, J.F., et al., RNAi screen reveals an Abl kinase-dependent host cell pathway involved in Pseudomonas aeruginosa internalization. PLoS pathogens, 2008. 4(3): p. e1000031.

[124] Abe, A., et al., Improved inhibitors of glucosylceramide synthase. Journal of biochemistry, 1992. 111(2): p. 191-6.
[125] Emam, A., W.G. Carter, and C. Lingwood, Glycolipid-Dependent, Protease Sensitive Internalization of Pseudomonas aeruginosa Into Cultured Human Respiratory Epithelial Cells. The open microbiology journal, 2010. 4: p. 106-15.
[126] Kierbel, A., et al., The phosphoinositol-3-kinase-protein kinase B/Akt pathway is critical for Pseudomonas aeruginosa strain PAK internalization. Molecular biology of the cell, 2005. 16(5): p. 2577-85.
[127] Evans, D.J., et al., Pseudomonas aeruginosa invasion and cytotoxicity are independent events, both of which involve protein tyrosine kinase activity. Infection and immunity, 1998. 66(4): p. 1453-9.
[128] Kierbel, A., et al., Pseudomonas aeruginosa exploits a PIP3-dependent pathway to transform apical into basolateral membrane. The Journal of cell biology, 2007. 177(1): p. 21-7.
[129] Lecuit, M., Understanding how Listeria monocytogenes targets and crosses host barriers. Clinical microbiology and infection : the official publication of the European Society of Clinical Microbiology and Infectious Diseases, 2005. 11(6): p. 430-6.
[130] Galindo, C.L., et al., Pathogenesis of Y. enterocolitica and Y. pseudotuberculosis in Human Yersiniosis. Journal of Pathogens, 2011. 2011: p. 1-16.
[131] Mengaud, J., et al., E-cadherin is the receptor for internalin, a surface protein required for entry of L. monocytogenes into epithelial cells. Cell, 1996. 84(6): p. 923-32.
[132] Schubert, W.D., et al., Structure of internalin, a major invasion protein of Listeria monocytogenes, in complex with its human receptor E-cadherin. Cell, 2002. 111(6): p. 825-36.
[133] Lecuit, M., et al., A role for alpha-and beta-catenins in bacterial uptake. Proceedings of the National Academy of Sciences of the United States of America, 2000. 97(18): p. 10008-13.
[134] Sousa, S., et al., ARHGAP10 is necessary for alpha-catenin recruitment at adherens junctions and for Listeria invasion. Nature cell biology, 2005. 7(10): p. 954-60.
[135] Helwani, F.M., et al., Cortactin is necessary for E-cadherin-mediated contact formation and actin reorganization. The Journal of cell biology, 2004. 164(6): p. 899-910.
[136] Kovacs, E.M., et al., E-cadherin homophilic ligation directly signals through Rac and phosphatidylinositol 3-kinase to regulate adhesive contacts. The Journal of biological chemistry, 2002. 277(8): p. 6708-18.
[137] Bonazzi, M., et al., Successive post-translational modifications of E-cadherin are required for InlA-mediated internalization of Listeria monocytogenes. Cellular microbiology, 2008. 10(11): p. 2208-22.
[138] Sousa, S., et al., Unconventional myosin VIIa and vezatin, two proteins crucial for Listeria entry into epithelial cells. Journal of cell science, 2004. 117(Pt 10): p. 2121-30.
[139] Shen, Y., et al., InlB-dependent internalization of Listeria is mediated by the Met receptor tyrosine kinase. Cell, 2000. 103(3): p. 501-10.

[140] Pizarro-Cerda, J. and P. Cossart, Listeria monocytogenes membrane trafficking and lifestyle: the exception or the rule? Annual review of cell and developmental biology, 2009. 25: p. 649-70.

[141] Ireton, K., et al., A role for phosphoinositide 3-kinase in bacterial invasion. Science, 1996. 274(5288): p. 780-2.

[142] Seveau, S., et al., A FRET analysis to unravel the role of cholesterol in Rac1 and PI 3-kinase activation in the InlB/Met signalling pathway. Cellular microbiology, 2007. 9(3): p. 790-803.

[143] Veiga, E. and P. Cossart, Listeria hijacks the clathrin-dependent endocytic machinery to invade mammalian cells. Nature cell biology, 2005. 7(9): p. 894-900.

[144] Veiga, E., et al., Invasive and adherent bacterial pathogens co-Opt host clathrin for infection. Cell host & microbe, 2007. 2(5): p. 340-51.

[145] Mostowy, S., et al., Septins regulate bacterial entry into host cells. PloS one, 2009. 4(1): p. e4196.

[146] Kremer, B.E., L.A. Adang, and I.G. Macara, Septins regulate actin organization and cell-cycle arrest through nuclear accumulation of NCK mediated by SOCS7. Cell, 2007. 130(5): p. 837-50.

[147] Huang, Y.W., et al., Mammalian septins are required for phagosome formation. Molecular biology of the cell, 2008. 19(4): p. 1717-26.

[148] Wong, K.W. and R.R. Isberg, Arf6 and phosphoinositol-4-phosphate-5-kinase activities permit bypass of the Rac1 requirement for beta1 integrin-mediated bacterial uptake. The Journal of experimental medicine, 2003. 198(4): p. 603-14.

[149] Brown, F.D., et al., Phosphatidylinositol 4,5-bisphosphate and Arf6-regulated membrane traffic. The Journal of cell biology, 2001. 154(5): p. 1007-17.

[150] Pizarro-Cerda, J. and P. Cossart, Subversion of phosphoinositide metabolism by intracellular bacterial pathogens. Nature cell biology, 2004. 6(11): p. 1026-33.

[151] Sarantis, H., et al., Yersinia Entry into Host Cells Requires Rab5-Dependent Dephosphorylation of PI(4,5)P(2) and Membrane Scission. Cell host & microbe, 2012. 11(2): p. 117-28.

Chapter 11

The Unique Endosomal/Lysosomal System of *Giardia lamblia*

Maria C. Touz

Additional information is available at the end of the chapter

1. Introduction

Endocytosis, which is important for the internalization of nutrients from the plasma membrane as well as extracellular fluids, has been extensively described in mammalian cells and yeast. The protozoan parasite *Giardia lamblia* (syn. *Giardia intestinalis, Giardia duodenalis*) is an early branching eukaryote that possesses a reduced membrane network with highly polarized vesicles, denominated peripheral vacuoles (PVs), neighboring the plasma membrane (reviewed by Adam, 2001; Faso and Hehl, 2011; Touz, 2011). This is an important zone of interaction between the parasite and its environment (the host intestine *in vivo* or the culture medium *in vitro*), and was shown to be the place involved in fluid-phase and receptor-mediated endocytosis (Gaechter *et al.*, 2008; Rivero *et al.*, 2010). Another characteristic of these PVs is that they contain hydrolytic activity resembling the function performed by lysosomes. These vacuoles have a high lytic capacity and a low luminal pH, both properties of mature lysosomes (Ward *et al.*, 1997; Touz *et al.*, 2002b).

More than twenty years after the first morphological description of the PVs, there is a consensus that these vesicles are rather an unusual combination of endosomal and lysosomal compartments. Conserved markers and mechanisms that govern trafficking to the PVs have been found, but there are particularities that show *Giardia* as a simplified organism compared with higher eukaryotes. This makes *Giardia* a unique biological cell model for investigating the minimal machinery employed by a eukaryote to regulate endocytosis and degradation. Here, we will discuss emerging data that are beginning to shed light on the endosomal-lysosomal system in *Giardia* and the molecules involved in this selective trafficking.

2. The parasite

Giardia lamblia is a flagellated protozoan that inhabits the upper small intestine of its vertebrate hosts and is the most common cause of defined waterborne diarrhea worldwide.

Clinical manifestations of giardiasis vary from asymptomatic infection to acute or chronic disease associated with diarrhea and malabsorption (Adam, 2001). It is the most common cause of diarrheal disease in the United States (Barwick *et al.*, 2000). In developing countries, there is a very high prevalence and incidence of infections, and data suggest that long-term growth retardation can result from chronic giardiasis (Fraser *et al.*, 2000).

Giardia was initially described by van Leeuwenhoek in 1681 from examining his own diarrheal stools under the microscope (Dobell, 1950), but it was not until 1981 that the World Health Organization classified *Giardia* as a human pathogen. Infections initiate with the ingestion of the cyst forms, which excyst in the upper small intestine of the host. The trophozoites replicate and colonize the intestinal surface and some of them encyst in the lower small intestine after sensing the stimulus for encystation (Lujan *et al.*, 1997; Adam, 2001). Interestingly, during differentiation (encystation/excystation), trophozoites undergo important biochemical and morphological modifications involving the secretory machinery of the cell. Recent studies about these changes have provided new insights into the mechanisms of secretion in this organism, but the molecular events leading to intracellular protein trafficking and secretion in *Giardia* remain poorly understood or controversial (Lujan and Touz, 2003; Faso and Hehl, 2011).

The trophozoite is between 10 and 15 microns long and 5 microns wide, pear-shaped and cut along the longitudinal axis (pyriform morphology) (Figure 1A). It presents bilateral symmetry, has two diploid nuclei with nuclear membranes, four pairs of flagella and two media bodies consisting of microtubules (Adam, 2001), suggested to be the storage reservoir of microtubules of the cell (Piva and Benchimol, 2004). The trophozoite also possesses a complex cytoskeleton and endomembrane system including the endoplasmic reticulum (ER), which extends symmetrically throughout the cell body, and the PVs located underneath the plasma membrane. In the front half, on the ventral surface, is the adhesive disc which is used to bind to the intestinal epithelium of the host (Elmendorf *et al.*, 2003). The oval cyst size is between 9 and 12 microns and contains four nuclei, the axostyle (structure at the base of the flagella) and remnants of flagella (Figure 1B). It is characterized by a rigid outer wall glycoprotein, composed of proteins and carbohydrate (Jarroll *et al.*, 1989; Manning *et al.*, 1992; Gerwig *et al.*, 2002). The construction of the extracellular cyst wall (CW; Cyst Wall) is of paramount importance because it allows the parasite to persist in fresh water, survive even the action of disinfectants and resist stomach acid in its new host, and then start infection in the gut.

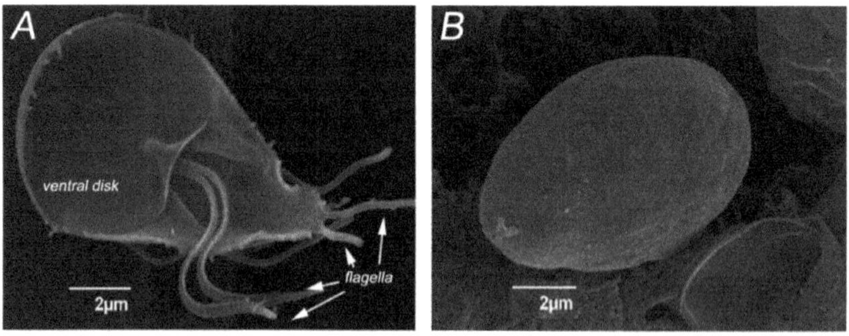

Figure 1. *Giardia lamblia* stages. (A) *Giardia* trophozoite, the ventral disk structure and the flagella are highlighted. (B) *Giardia* cyst.

3. *Giardia* secretory pathway

Eukaryotic cells have to deal with the fact that, after translation at the ribosomes, most proteins must be specifically targeted from the cytoplasm to the organelle in which they must function. As cellular components became more complex and abundant during evolution, subcellular compartmentalization developed into an essential feature to prevent the inappropriate meeting of certain intracellular components, as well as facilitating efficient ordered reactions (Munro, 2004). To maintain these compartments, cells have evolved mechanisms to ensure that specific proteins are delivered to specific organelles. In most eukaryotes, the Golgi complex serves as a major sorting point in the secretory pathway, selectively targeting proteins and lipids to different organelles (Gu *et al.*, 2001). *Giardia* possesses a distinctive endomembrane system involving the nuclear membrane, the ER, and lysosome-like PVs (Figure 2).

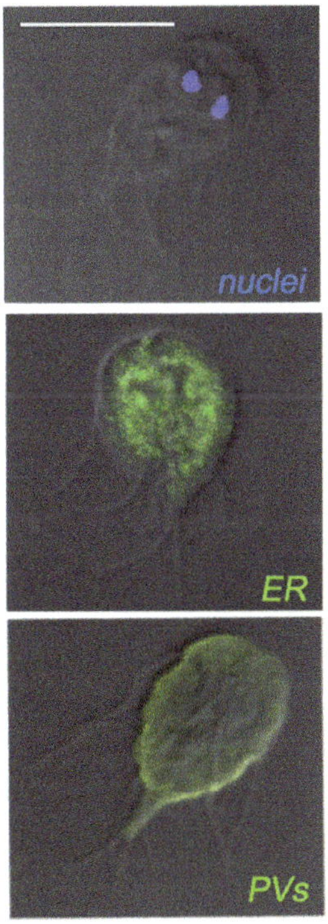

Figure 2. *Giardia* organelles. Confocal microscopy showing the nuclei, endoplasmic reticulum (ER) and PVs labeled with DAPI (4',6-diamidino-2-phenylindole), anti-BiP mAb, and anti-AP2 mAb, respectively. Bar, 10 μm.

On the other hand, it lacks other organelles characteristic of higher eukaryotes such as canonical endosomes, lysosomes, mitochondria, peroxisomes, and Golgi apparatus (reviewed by Lujan and Touz, 2003; Hehl and Marti, 2004). In most eukaryotic cells, the Golgi apparatus consists of a series of flattened cisternal membranes forming a stack

(Ladinsky *et al.*, 2002). The architecture of this organelle is remarkably conserved throughout eukaryotic evolution (Mellman and Simons, 1992); however, a typical Golgi complex with organized and parallel cisternae is not apparent in vegetative *Giardia* trophozoites. Several pieces of evidence suggest that *Giardia* trophozoites may possess organelle(s) in which typical Golgi functions take place, even though they do not have a Golgi-like appearance. The fact that both constitutive and regulated mechanisms for protein transport exist in *Giardia* is an example of Golgi functions, since the sorting and selection process generally occurs in the *trans*-Golgi network in more complex cells (Gu *et al.*, 2001). The constitutive pathway in *Giardia* is represented by continuous expression and trafficking to the plasma membrane of the transmembrane-anchored variant-specific surface proteins (VSPs) (Nash, 2002; Marti *et al.*, 2003a; Touz *et al.*, 2005). The regulated pathway takes place only during encystation and associates with the appearance of encystation-specific vesicles (ESVs), which transport cyst wall components to the plasma membrane of the encysting cell and release their content to the cell exterior during cyst wall formation (Reiner *et al.*, 1989; McCaffery and Gillin, 1994; Lujan *et al.*, 1995; Sun *et al.*, 2003). A selective pathway sorting proteins to the endosomal/lysosome membrane system has been recently demonstrated by our group, although *Giardia* do not possess distinctive endosomes or lysosomes (Touz *et al.*, 2003; Touz *et al.*, 2004; Rivero *et al.*, 2010; Rivero *et al.*, 2011).

Analyses of genes and proteins used for phylogenetic classification indicate that *Giardia* is in fact one of the earliest branching eukaryotes (Sogin *et al.*, 1989; Hashimoto *et al.*, 1998), but some of the particular cellular characteristics of this organism are probably a result of the secondary loss of complex cell structures, as a consequence of its parasitic life style, rather than the primitive simplicity supposed for early diverging protists (Dacks and Doolittle, 2002; Lujan and Touz, 2003).

4. The endosomal/lysosomal system of *Giardia*

Most eukaryotes have a system of endosomes and lysosomes that mediate the internalization, recycling, transport and breakdown of cellular and extracellular components and facilitate dissociation of receptors from their ligands. Early endosomes (EE) internalize endocytosed proteins to allow for their subsequent return to the cell membrane. Later, conversion of the EEs to late endosomes (LE) takes place, undergoing homotypic fusion reactions, growing in size, and acquiring more intraluminal vesicles. What follows is the fusion of an endosome with a lysosome and maturation of the subsequent endolysosome into lysosome, which constitutes a storage organelle for lysosomal hydrolases at acidic pH, and membrane components. Although this is a highly dynamic system, discrete compartments can be distinguished (Huotari and Helenius, 2011).

In contrast to most eukaryotes, *Giardia* has highly polarized vacuoles, located underneath the plasma membrane of the dorsal side, which combine some of the characteristics of endosomes and some of lysosomes (Lindmark, 1988; Lanfredi-Rangel *et al.*, 1998; Touz *et al.*, 2004). These PVs, distinguished by their localization, are about 150 nm in diameter with variable oval shapes and contain a core of low electron density (Figure 3). They are acidic, as shown by the uptake of acridine orange and the lysosomal markers Lyso-sensor and Lyso-tracker (Lanfredi-Rangel *et al.*, 1998; Touz *et al.*, 2002b; Touz *et al.*, 2003). The first description of hydrolase activity in the PVs came from studies in which acid phosphatase activity was tested, showing a cytochemical localization in these vacuoles as well as in the ER and nuclear envelope cisternae (Feely and Dyer, 1987). The presence of hydrolase activities in the PVs was also proved for cysteine proteases and RNases, demonstrating their lysosomal characteristics (Lindmark, 1988; Ward *et al.*, 1997; Touz *et al.*, 2002b). In addition, their potential role in endocytosis was demonstrated by the uptake of exogenous ferritin and Lucifer yellow (Bockman and Winborn, 1968; Lanfredi-Rangel *et al.*, 1998). Pulse-chase experiments with horseradish peroxidase and fluorescent dextran showed an early and persistent labeling of the PVs, suggesting that there is no distinction between early and late endocytic vesicles in *Giardia*, in contrast to what occurs in higher eukaryotes (Lanfredi-Rangel *et al.*, 1998; Gaechter *et al.*, 2008).

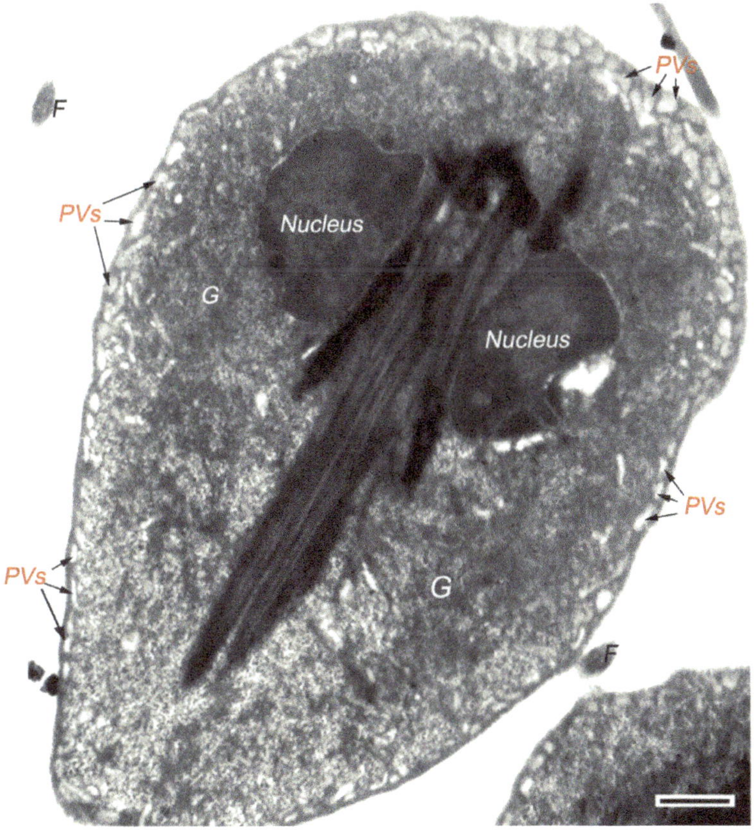

Figure 3. *Giardia* trophozoite ultrastructure. Electromicrograph of a growing *Giardia* trophozoite, showing the peripheral vacuoles (PVs) located underneath the plasma membrane (arrows). Nuclei are also denoted. G: electron-dense glycogen deposits. Bar, 0.5 μm. From Rivero *et al.*, 2010.

Analysis of fluid-phase endocytosis demonstrated that there is no lateral exchange of fluid phase markers between individual PVs (Gaechter et al., 2008). It was observed that, after internalization, some fluid-phase markers translocated rapidly to the ER or to an associated membrane compartment termed the tubulo-vesicular network (TVN) (Abodeely et al., 2009). Moreover, the presence of protease functions within the TVN, plus 3-D reconstruction and electron microscopy tomography of trophozoites stained for acid phosphatase and glucose-6-phosphatase, suggest that there might be a connection between some vesicles and profiles of the ER (Lanfredi-Rangel *et al.*, 1998; Abodeely *et al.*, 2009a). However, a recent work showed, by immunofluorescence and 3-D reconstruction, that the ER membranes are found throughout the cytoplasm, but do not permeate the space occupied by PVs (Faso and Hehl, 2011). Recently, it was shown that a mechanism of receptor-mediated endocytosis occurs in this organism, with specific molecules selectively directed to the PVs through a classical endocytic mechanism (Rivero *et al.*, 2010; Rivero *et al.*, 2011). These data suggest that uptake of soluble material into PVs is not selective at this step but is still capable of redirecting specific molecules to the TVN (Hernandez *et al.*, 2007; Abodeely *et al.*, 2009b). In terms of receptor-mediated endocytosis, movement between vesicles could be observed (Figure 4), which suggests, not only that *Giardia* possesses a refined and conserved mechanism of endocytosis, but also that the PVs population might not be as homogeneous as was thought but rather organized depending on their functions (Rivero & Touz, unpublished).

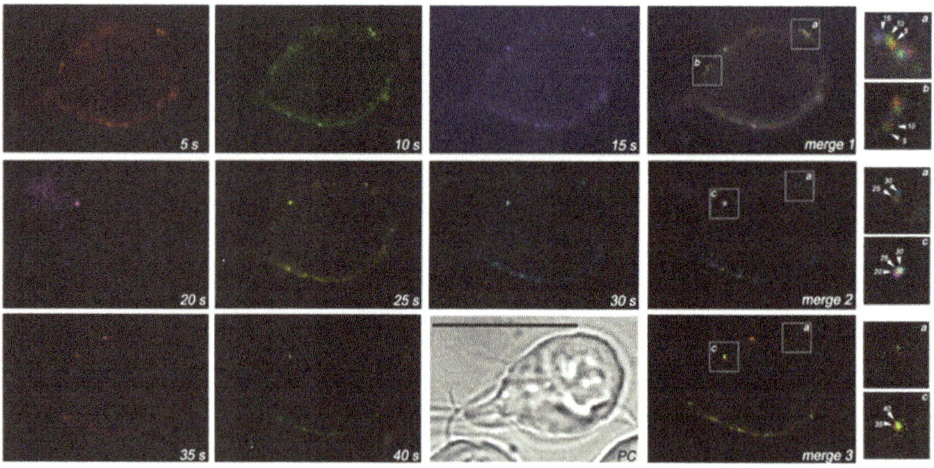

Figure 4. Epifluorescence microscopy shows the internalization and delivery of LDL to the PVs. Eight frames from 5 s to 40 s were artificially colored and combined (merge 1-3) to determine the movement of endocytosed BODIPY-LDL. The lateral movement of the LDL between the PVs in living cells is observed following the sequence red, green, blue (5 s, 10 s, 15 s) for merge 1; magenta, yellow, cyan (20 s, 25 s, 30 s) for merge 2; and red and green (35 s and 40 s) for merge 3. In the insertions a, b and c, movement of BODIPY-LDL is observed in detail. PC: phase contrast. Bars, 10 μm. From Rivero & Touz, unpublished.

During evolution, eukaryotic cells differentiated, adapting to their increasingly complex environment by acquiring new abilities for nutrient uptake, internalization of growth factors, and entry of pathogens, among others. A complex array of endosomal compartments are preserved as modules that are able to perform specific functions in modern eukaryotic cells. In the case of less-evolved eukaryotes, the variety and simplicity of these systems is only starting to become apparent. Thus, *Giardia* possesses the endosomal-lysosomal system concentrated in the PVs, which may represent an ancient organellar system that later subdivided into compartments as early and late endosomes and lysosomes.

5. Endosomal-lysosomal protein trafficking

5.1. Clathrin

Clathrin-mediated endocytosis (CME) regulates many cell physiological processes, such as the internalization of growth factors and receptors, entry of pathogens, and synaptic transmission. Within the endocytic network, clathrin functions as a central organizing platform for coated pit assembly and dissociation via its terminal domain. As isolated from coated vesicles, clathrin is a trimer of 190-kDa heavy chains, each with an associated 25-kDa light chain forming a spiderlike molecule, the 'triskelion' (Kirchhausen, 2000). The unusual geometry of the triskelion allows it to assemble into regular polyhedral structures, the 'clathrin coats', which eventually give rise to clathrin-coated vesicles (CCVs). The dense protein coat of the CCV and its bristle-like morphology were first described by Roth & Porter (1964), who noted the involvement of these vesicles in RME of yolk proteins in mosquito oocytes (Roth and Porter, 1964). The formation of CCVs occurs at the plasma membrane, *trans*-Golgi network and endosomes (Kirchhausen and Toyoda, 1993), and follows a sequence of coordinated steps, in which membrane invagination is coupled to growth of the clathrin lattice, leading to lattice closure and vesicle budding (Kirchhausen, 2000).

The *Giardia* genome encodes an ortholog of the clathrin heavy chain (CHC) (Morrison *et al.*, 2007), and has a molecular weight of about 200 kDa, with three C-terminal clathrin repeats and one N-terminal propeller, according to a protein family database (Finn *et al.*, 2008). Analysis of GlCHC expression showed that clathrin is expressed almost equally in both stages of the parasite and is located in close association with the PVs in trophozoites, and in the ESVs in immature cysts (Marti *et al.*, 2003a; Marti *et al.*, 2003b; Hehl and Marti, 2004; Gaechter *et al.*, 2008). On the basis of these observations, it was suggested that recruitment of clathrin to late ESVs could serve to disperse large ESVs into smaller transport vesicles in response to the secretion signal (see below). The identification of a clathrin light chain (CL) ortholog in the GDB has so far been unsuccessful, probably because the sequences of CLs are not uniformly conserved among species. The differential expression of the CLs (e.g. one from yeast, two mammalian tissues with the presence of isoforms in mammalian neurons) might be associated with the high degree of specialization involved in clathrin vesicle trafficking.

Several groups have presented evidence for a role of clathrin in endocytosis in *Giardia* (Touz *et al.*, 2004; Hernandez *et al.*, 2007; Gaechter *et al.*, 2008; Rivero *et al.*, 2010). However, neither typical membrane-associated clathrin lattices nor emerging clathrin-coated pits have been observed in this parasite. Instead, uncharacteristic coated pits were seen in close association with the PVs (Lanfredi-Rangel *et al.*, 1998), suggesting that a distinct arrangement of clathrin might occur in this parasite. It is then possible that, as was observed *in vitro* (Zhang *et al.*, 2007), clathrin may be organized in a hexagonal array, forming tubes instead of vesicles. Indeed, a different type of clathrin-coated transport carriers (TCs), consisting of larger tubular/vesicular structures having one or more clathrin-coated buds, have been identified (Polishchuk *et al.*, 2006). These TCs travel long distances from the juxtanuclear area of the cell until they fuse with peripheral endosomes. The function of the TCs might be to mediate long-range distribution of mannose 6-phosphate receptors (MPR) and their cargo hydrolases to the peripheral cytoplasm (Puertollano *et al.*, 2003). Similar to TCs in HeLa cells (Polishchuk *et al.*, 2006), the tubules might not break down into CCVs en route to PVs in *Giardia*. In mammalian cells, it has been shown that TCs contain mannose-phosphate receptors, clathrin, Golgi-localizing Gamma-ear containing ARF-binding proteins (GGAs), and/or adaptor protein 1 (AP-1), and it was suggested that these might be uncoated during the TC-endosome fusion or could become integrated into the endosome membrane (Polishchuk *et al.*, 2006). Possibly supporting this hypothesis, giardial clathrin and AP-1 were observed not only on ER-exist sites but also in PVs (Marti *et al.*, 2003a; Marti *et al.*, 2003b; Touz *et al.*, 2004; Gaechter *et al.*, 2008). As we said, the ER tubular-vesicular network apparently extends to and contacts the PVs in the periphery of the cell (Abodeely *et al.*, 2009b) but it was recently reported that no ER membranes invade the space occupied by PVs (Faso and Hehl, 2011). An explanation that reconciles these observations might be that at least some of the clathrin-dependent trafficking in *Giardia* involves tubular carriers that extend from the ER-exit sites to the peripheral cytoplasm until they meet with distally located PVs.

5.2. Adaptor proteins

The classic model for clathrin-dependent sorting comprises the participation of cargo receptors, adaptor heterotetramers and clathrin triskelia. Because clathrin has no affinity for biological membranes, its recruitment to membranes and capture of transmembrane cargo requires the action of clathrin-associated adaptor proteins (AP), which bind to clathrin through the amino-terminal domain of the CLH (Bonifacino and Traub, 2003). Among these adaptors are AP-1, AP-2 and AP-3, which comprise two large chains (one each of $\gamma/\alpha/\delta$ and β1-3, respectively), one medium-sized chain (μ1-3), and one small chain (σ1-3) (Boehm and Bonifacino, 2002). These complexes are localized to different subcellular compartments, where they function in cargo selection (Boehm and Bonifacino, 2002). At least one of the large subunits in each AP complex ($\gamma/\alpha/\delta$) mediates binding to the target membrane. The other large subunit, β1-3, recruits clathrin through a 'clathrin-box' motif (Boehm and Bonifacino, 2001; Brodsky *et al.*, 2001). The μ1-3 subunits are involved in the recognition of tyrosine-based, YXXØ signals (where X represents any amino acid and Ø indicates a residue with a bulky hydrophobic side chain), and combinations of $\alpha\sigma2$, $\gamma\sigma1$ and $\delta\sigma3$ recognize dileucine-based, [DE]XXXl[LI] signals (Collins *et al.*, 2002; Bonifacino and Traub, 2003). A fourth AP complex, AP-4, is thought to be a component of a non-clathrin coat and to recognize a different type of signal (Burgos *et al.*). Besides the putative GlCHC, orthologs of two large, one medium, and one small subunit of each AP-1 and AP-2 are present in the *Giardia* genome. The colocalization of AP-1 with lysosomal proteins, its interaction with the GlCHC, together with the observation that lysosomal protein trafficking is altered in μ1-depleted trophozoites, support the participation of this complex in the forward transport of proteins towards the PVs in *Giardia* (Touz *et al.*, 2004). AP-1 also plays a central role during parasite differentiation, since μ1 depletion impairs encystation (Touz *et al.*, 2004). On the other hand, AP-2 is localized to the PVs and plasma membrane in trophozoites and also neighboring the ESV in encysting cells (Rivero *et al.*, 2010). AP-2 participates in RME and is crucial in the internalization of lipoproteins (Rivero *et al.*, 2010). Although the β1–2 and μ2 mRNA transcripts change little during the completion of the cell cycle (Marti *et al.*, 2003b; Rivero *et al.*, 2010), the role of the corresponding AP complexes appears essential for the adaptation of the parasite. AP-1 is not critical for *Giardia* trophozoite survival and multiplication, but it is necessary for cyst formation, acting indirectly in this process by transporting a transmembrane protein to the PVs (Touz *et al.*, 2002b; Touz *et al.*, 2003; Touz *et al.*, 2004). In contrast, AP-2 is essential for *Giardia* growth and survival, being involved in the endocytosis of essential molecules (*e.g.*, exogenous lipids) (Rivero *et al.*, 2010) and in the fragmentation of ESVs into small transport vesicles containing cyst wall proteins during encystation (Rivero & Touz, unpublished). The fast secretion and deposition of cyst wall material has been reported to involve clathrin- and dynamin-dependent breakup of ESVs into small vesicles targeted for the plasma membrane (Hehl and Marti, 2004; Gaechter *et al.*, 2008). It is possible that this parasite requires the concerted action of clathrin and adaptors as well as accessory proteins at the time of cyst wall formation.

Taken together, these results support the hypothesis that *Giardia* possesses molecular mechanisms for lysosomal protein trafficking involving adaptor proteins similar to those of other eukaryotes. AP-1 and AP-2 appear to be the only two adaptors involved in lysosomal protein trafficking in *Giardia,* since there is no evidence of the participation of other adaptor proteins such as AP-3, AP-4, and monomeric adaptors (i.e., the GGAs). It has been suggested that the two prototypic *Giardia* AP complexes predict the point of separation of *Giardia* after the first coordinated round of gene duplications, resulting in an AP-3 and an AP-1/2/4 ancestor (Marti *et al.*, 2003b). Phylogenetic reconstruction from comparative genomics has shown that all four AP complexes were present in the Last Common Eukaryotic Ancestor (LCEA), as was the F-COP subcomplex (Boehm and Bonifacino, 2001). However, the GGAs, which also exhibit homology to the ear region of the AP-1 γ protein, are restricted to animal and fungal lineages (Field *et al.*, 2007). Therefore, individuality of the species lineage and secondary loss are common characteristics in the evolutionary history of the adaptins. Secondary losses of adaptors can be observed in *Drosophila melanogaster, Caeanorhabditis elegans, Saccharomyces cerevisiae* and *Schizosaccharomyces pombe,* which lack the AP-4 complex (Field *et al.*, 2007). In addition, comparative genomic and phylogenetic analyses of protozoan parasites have shown loss of the AP-3 complex in the species *Theileria, Cryptosporidium parvum,* and *Babesia bovis,* while *Trypanosome brucei* and *Leishmania major* lack AP-2 and AP-4, respectively (Nevin and Dacks, 2009). Thus, examination of the role of the reduced set of AP complexes in protozoa provides insight into the depths of our cellular history and highlights the importance of essential cell biology adaptations of the ancestral cellular organization.

5.3. Lysosomal proteins

Lysosomal integral membrane proteins (e.g., LAMP/LIMP family proteins) are transported to lysosomes by binding of their cytosolic motifs to AP complexes (Bonifacino and Lippincott-Schwartz, 2003). The carboxy-terminal β-sandwich domain of the μ subunits of AP-1, AP-2, and AP-3 binds directly to YXXØ-type sequences, while the γσ1, ασ2 and δσ3 hemicomplexes bind to the [DE]XXXL[LI] sequences (Janvier *et al.*, 2003; Doray *et al.*, 2007). Although lysosomal integral membrane glycoproteins have not been identified in *Giardia,* it was reported that a cysteine protease termed ESCP (encystation-specific cysteine protease) is transported to the PVs through a tyrosine-based motif. This enzyme is homologous to cathepsin C enzymes of higher eukaryotes and possesses a transmembrane domain and a YRPI motif within the cytoplasmic tail. ESCP localizes to the PVs in growing trophozoites and also to the plasma membrane in encysting cells (Touz *et al.*, 2003). Deletion of the YRPI motif or suppression of μ1 mislocalizes this protein to the plasma membrane or to the ER-exit sites, respectively (Touz *et al.*, 2003; Touz *et al.*, 2004).

Soluble acid hydrolases, on the other hand, are synthesized in the ER and transported to the Golgi complex, where their carbohydrate chains are modified by resident enzymes before delivery to lysosomes. In mammalian cells, the hydrolases are modified with mannose 6-phosphate residues that function as recognition signals for MPRs in late Golgi compartments. In yeast, the vacuolar hydrolases lack mannose 6-phosphate, and the sorting receptor is the product of the VPS10 (Vacuolar Protein Sorting 10) gene. In both cases, however, sorting signals present in the cytosolic tails of the receptors interact with clathrin adaptors and direct packaging of the hydrolase-receptor complexes within CCVs or clathrin-coated TCs (Ohno *et al.*, 1995; Ohno *et al.*, 1996). Recently, multi-ligand type-1 receptors Sortilin, SorCS1, SorCS2, SorCS3, and SorLA were discovered, containing an N-terminal Vps10p domain (Rezgaoui *et al.*, 2001; Hermey, 2009). These are transmembrane proteins that convey Golgi-endosome transport and bind a number of unrelated ligands.

In *Giardia,* high hydrolase activity in the PVs has been implicated in protein degradation during growth (Lujan and Touz, 2003), encystation (Touz *et al.*, 2002a; Touz *et al.*, 2003) and excystation (Slavin *et al.*, 2002). A family of three cysteine protease genes (CP1, CP2, and CP3) has been shown to encode members of the cathepsin B subgroup of the peptidase family C1 (Ward *et al.*, 1997), and soluble CP2 has been found in PVs and ER of trophozoites (Ward *et al.*, 1997; Abodeely *et al.*, 2009a). Also, AcPh activity has been examined cytochemically, revealing communication of the PVs with the ER (Feely and Dyer, 1987; Lanfredi-Rangel *et al.*, 1998; de Souza *et al.*, 2004). Unlike AcPh in other eukaryotes, including protozoa, *Giardia* AcPh is a soluble protein that is transported from the ER-exit site to the PVs via AP-1 (Touz *et al.*, 2004). It is thus possible that a specific receptor, possessing a function similar to the MPR or Vps10p, is involved in the trafficking of soluble hydrolases toward the PVs. Recent studies have identified a type-I membrane protein that interacts with AcPh and contains an YQII (YXXØ-type) motif in its cytosolic tail (Rivero & Touz, unpublished). *In silico* analysis revealed that this protein (GDB: GL50803_28954) might be orthologous to the Vps10p receptor. Further biochemical studies on this putative receptor in both vegetative and encysting trophozoites and its participation in hydrolase delivery are necessary to elucidate the exact function of this protein.

Comparative analysis of lysosomal proteins present in *Giardia* and other cells reveals some intriguing differences. For instance, AcPh is soluble in *Giardia* but exists as a type-I membrane protein containing a YXXØ-type internalization sequence in cells as different as *Leishmania* and humans (Gottlieb and Dwyer, 1981; Waheed *et al.*, 1988; Shakarian *et al.*, 2002), with transport to the lysosome occurring through several cycles of plasma membrane internalization and recycling. In the lysosome of mammalian cells, the luminal domain of AcPh is processed and released in soluble form (Peters *et al.*, 1990). Moreover, while the AcPh tail interacts with AP-2 in these cells, the lysosomal traffic of *Giardia* AcPh depends on AP-1 (Touz *et al.*, 2004). Because much of the machinery involved in lysosomal trafficking is derived from a few protein families (where the various family members perform the same basic mechanistic function), the analysis of the similarities and differences between organisms might provide further insight into eukaryotic cell evolution.

As mentioned above, recent studies have shown that AP-2 participates in endocytosis of the *Giardia* Low-density lipoprotein Receptor-related Protein or LRP (Rivero *et al.*, 2010). *Giardia* LRP is a type-I membrane protein, which shares the substrate-N-terminal binding domain and a FXNPXY-type endocytic motif with human LRP1. This receptor localizes predominantly to the ER but is also found in the PVs and plasma membrane in *Giardia*, and internalizes both low density lipoproteins (LDL) and chylomicrons as shown by *in vitro* studies. The FXNPXY motif of LRP was shown to bind directly to the μ2 subunit of AP-2, with this interaction being necessary for its proper localization, processing, and function.

One common characteristic of LDLR family members such as the LRPs is that they have at least one copy of the FXNPXY-like sequence in their cytosolic tail, which serves as the signal for endocytosis or as a binding element for adaptor proteins involved in signal transduction (Harris-White and Frautschy, 2005). In other eukaryotes, FXNPXY signals are recognized by the adaptor proteins Disabled homolog 2 (Dab2) and Autosomal Recessive Hypercholesterolemia (ARH), which contain a phosphotyrosine-binding (PTB) domain (Traub, 2009). However, no PTB-containing proteins such as Dab2 or ARH are encoded in the *Giardia* genome, supporting the idea that AP-2 might be the key endocytic adaptor in this parasite. Indeed, it has been shown by surface plasmon resonance and photoaffinity labeling that the FXNPXY-like motif binds to μ2 purified from bovine-brain-coated vesicles (Boll et al., 2002). Interestingly, in spite of the strong interaction found between ARH and

LRP1 in an in vitro binding assay, the subcellular localization of LRP1 was not affected in the liver of ARH-deficient mice, whereas LDLR was found to be redistributed from intracellular localizations to the cell membrane (Jones et al., 2003). Thus, the importance of the availability of intracellular adaptor proteins might determine the specific cellular function of lipoprotein receptors. Since *Giardia* trophozoites do not have the capacity of *de novo* synthesis of cholesterol, its acquisition may depend on the internalization of chylomicrons from the host intestine by LRP. Moreover, because the trophozoites normally thrive in an environment where they never come in contact with LDL, it is possible that the binding of LDL to LRP represents a result of exaptation, thereby shifting the function of this protein allowing growth of parasites in culture medium.

5.4. Accessory proteins

5.4.1. Dynamin

Dynamins are large GTPases that belong to a protein superfamily, which are involved in many processes including clathrin-mediated endocytosis, clathrin-independent endocytosis, budding of transport vesicles, organelle division, cytokinesis and pathogen resistance (McNiven *et al.*, 2000; Ochoa *et al.*, 2000; Cao *et al.*, 2003; Krueger *et al.*, 2003). Substantial evidence indicates that dynamin oligomerization around the necks of endocytosing vesicles and subsequent dynamin-catalyzed GTP hydrolysis are responsible for membrane fission (Sweitzer and Hinshaw, 1998). Mammalian dynamins 1, 2 and 3 are the founder members of the dynamin family that, along with other large GTPases, possesses five identifiable domains: GTPase domain, middle domain, a lipid binding Pleckstrin-homology (PH) domain, GTPase Effector domain (GED) and C-terminal proline-arginine rich domain (PRD) (Hinshaw, 2000).

Recent studies indicate that *Giardia* possesses a single dynamin homolog (GlDRP, dynamin-related protein), with the predicted protein containing the N-terminal GTPase domain (33–219), the middle domain (230–523) and a C-terminal GED (628–719) (Marti *et al.*, 2003b). The giardial dynamin (GlDRP, dynamin-related protein), has a PRD of 70 amino acids (538–608) (Gaechter *et al.*, 2008). While the PRDs of dynamins are normally localized at the C-terminus, after the GED, in other eukaryotes, the PDR of the giardial dynamin is inserted between the middle domain and the GED. Interestingly, a typical PH domain that could mediate direct interaction with membrane lipids is missing.

GlDRP partially colocalizes with clathrin at the PVs and is necessary for endocytosis of plasma membrane proteins but not for fluid-phase endocytosis in *Giardia* trophozoites. Moreover, the expression of a mutant GlDRP, with reduced affinity for GTP and GDP, impaired endocytosis and resulted in enlarged PVs, indicative of blocked vesicular fission in these organelles. Also in these cells, GlDRP is detected at the ER, with only a minor proportion being present as a cytoplasmic pool. During encystations, however, both clathrin and GlDRP localize in part in the ESVs containing cyst wall material (Marti *et al.*, 2003b; Hehl and Marti, 2004; Gaechter *et al.*, 2008). Interestingly, matching the function observed in depleted-μ2 encysting trophozoites (Rivero *et al.*, 2010), expression of a dominant-negative GlDRP affects the formation of small vesicles containing cyst wall proteins (CWPs) and blocks its exocytosis to form the cyst wall. Because close contact of the ESVs and PVs has been frequently reported (Marti *et al.*, 2003b; Touz *et al.*, 2003; Hehl and Marti, 2004; Gaechter *et al.*, 2008), exchange of material between these two structures may occur, with the PVs finally acting as sorting organelles, probably by delivering the CWPs to the plasma membrane and/or returning other proteins to the ER-exit site.

With the characterization of dynamic-like proteins from *Homo sapiens, Drosophila melanogaster, Caenorhabditis elegans*, yeast species, *Arabidopsis thaliana*, and *Giardia,* it may now be possible to reveal the many varied functions of the members of the dynamin superfamily in different organisms.

5.4.2. *Snares and rabs*

The basic steps underlying vesicle-mediated transport between the secretory and endocytic pathways are vesicle formation from a donor compartment, translocation of transport intermediates to a target organelle, tethering of transport intermediates with the target compartment, and, finally, the docking and fusion of vesicles with the membrane target. SNAREs (soluble N-ethylmaleimide-sensitive factor-attachment protein receptors) function in the final event of docking of vesicles with the target membrane, catalyzing the fusion of the opposing membranes. Compared to other organisms, *Giardia* has a relatively small number of SNARE proteins (Morrison *et al.*, 2007; Elias *et al.*, 2008). Seventeen putative SNAREs have been identified and partially characterized, with five representing Qa-SNAREs, five Qb-SNAREs, four Qc-SNAREs and three R-SNAREs. Although some of these SNAREs localize to the PV area, their function has not been investigated and has rather been inferred from the participation of their orthologs in other cells (Elias *et al.*, 2008). For example, different gSNAREs are present in the PVs/plasma membrane area, Qa1, gQa3, gQa5, gQb2, gQb4, gQb5, and gR3, with the presence of three different gQa SNAREs suggested to be involved in distinct pathways such as exocytosis, endocytosis, and PV-PV fusion (Elias *et al.*, 2008).

The Rab family GTPases regulates many steps of membrane traffic, including vesicle formation, vesicle movement along actin and tubulin networks, and membrane fusion. So far, over 70 members of the Rab family have been identified in mammalian cells, and each seems to have a characteristic intracellular localization and function. For instance, Rab5 plays roles in endocytosis, early endosome fusion, and caveolar vesicle targeting to early endosomes (Barbieri *et al.*, 2000; Pelkmans *et al.*, 2004), while Rab11 mediates slow endocytic recycling through recycling endosomes, and Rab4 mediates fast endocytic recycling directly from early endosomes (van der Sluijs *et al.*, 1992; Ullrich *et al.*, 1996). There is also a coordinated action of Rab5 and Rab7 in the sorting of cargo receptors by the retromer complex (Rojas *et al.*, 2008).

Similar to the SNAREs, there is a remarkable reduction in the members of this family, with only seven Rab proteins being present in the *Giardia* genome (Morrison *et al.*, 2007). The giardial Rab11 has been localized to the PVs and cytoplasm but relocalizes to the ESVs during encystation (Morrison *et al.*, 2007; Abodeely *et al.*, 2009b). A *Giardia* Rab1 that localizes to ER-exit sites and PVs is also associated with the ESVs during encystation (Langford *et al.*, 2002).

The question of whether Rabs and SNAREs participate in delivery of lysosomal proteins to the PVs remains unanswered and could be addressed in greater detail by functional analysis. Since each member of the Rab and SNARE family retains analogous biological functions in almost all the species analyzed, it will be interesting to determine whether, in *Giardia*, selective pressures might have been operating on distinguishing aspects of the lysosomal trafficking pathway, adapting the specificities of these proteins to accomplish their function.

6. Conclusion

Over the past few years, our understanding of the cell biological processes underlying the function of the PVs during *Giardia* growth and differentiation has advanced considerably. This is largely because of the complete sequencing and annotation of the *Giardia* genome, the development of transfection systems, highly sophisticated morphological analyses, and the identification of new parasite proteins that participate in endosomal/lysosomal trafficking. Endosomal/lysosomal trafficking pathways exhibit significant complexity and diversity in terms of morphology, function, and mechanisms among different organisms and cell types. As shown by several studies, part of the *Giardia* transport machinery is fairly well conserved. The existence in this organism of the GlCHC and dynamin, the presence of tetrameric adaptor proteins, and endosomal-lysosomal sorting motifs within cargo proteins, support an early acquisition of genes necessary for endosomal/lysosomal trafficking during eukaryotic evolution. Nevertheless, this parasite has experienced considerable diversification (Figure 5). The constraints of living under parasitic conditions have probably been the major driver for the reductive evolution of lysosome/endosome and Golgi compartments to maintain only those components that are essential for specific compartmentalization needs. For example, this parasite possesses only two of the four AP complexes, AP-1 and AP-2, which are involved in sorting signal recognition. No monomeric adaptor proteins have been identified so far. Also, reduced members of the Rab family are present and, although Rab11 has been associated with the PVs, further analysis will be necessary to assess Rab participation in membrane tethering and fusion to preserve the PVs identity. Similarly, investigation on the SNARE proteins closely associated with PVs will shed light on the mechanism of vesicle-vacuole fusion. Despite the progress in the field, it is clear that our molecular understanding of this complex situation remains far from complete. In particular, the current view that this parasite needs a reduced set of organelles and machinery to control nutrient uptake as well as degradation of intracellular proteins and endocytosed macromolecules is without precedent and raises several conceptual questions, some of which have been addressed in this chapter. In the last decade, PVs are emerging as sorting stations were molecules can be sorted and selected for plasma membrane or ER delivery. The future challenge will be to complete the pieces of this important puzzle to understand and unravel functions of the PVs and to throw light on the fundamental organizational principles of endosome/lysosome biogenesis in all eukaryotes.

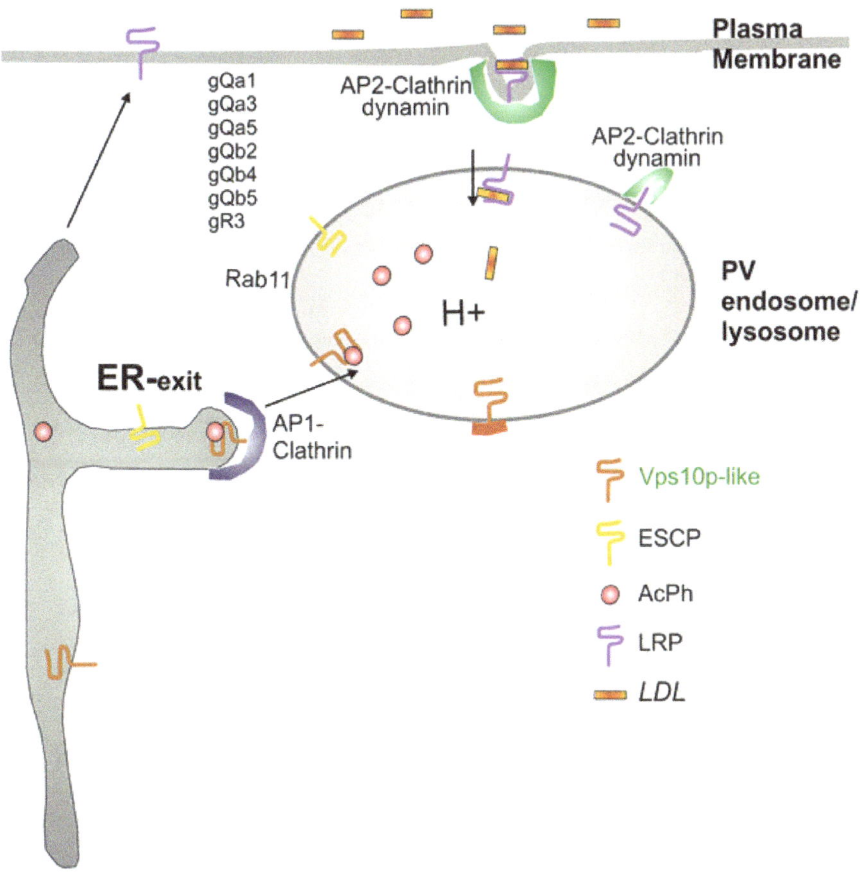

Figure 5. Schematic representation of lysosomal protein trafficking in growing *Giardia* trophozoites. From the ER-exit sorting site, the membrane protease ESCP is directed to the lysosome-like PVs in AP-1 and clathrin-coated vesicles. By the same pathway, the hydrolase AcPh is probably associated with the membrane receptor Vps10p and AP-1. AP-2 is involved in LDL/LRP endocytosis and PV delivery. The cytosolic proteins, clathrin and dynamin, are localized in the PVs. Rab11, and the SNAREs Qa1, gQa3, gQa5, gQb2, gQb4, gQb5, and gR3 may participate in vesicle trafficking to and/or from the PVs. H+ represents the acidic pH of the PV lumen. Unconfirmed protein participation is depicted in green. Modified from (Touz, 2011).

Author details

Maria C. Touz
Instituto de Investigación Médica Mercedes y Martín Ferreyra, INIMEC – CONICET, National University of Cordoba, Córdoba, Argentina

Acknowledgement

I thank numerous colleagues for helpful suggestions. I am grateful to Nahuel Zamponi and Dr. Adriana Lanfredi-Rangel for the scanning microscopy images of *Giardia* trophozoite and cyst.

7. References

Abodeely, M., DuBois, K.N., Hehl, A., Stefanic, S., Sajid, M., DeSouza, W., Attias, M., Engel, J.C., Hsieh, I., Fetter, R.D., and McKerrow, J.H. (2009a). A contiguous compartment functions as endoplasmic reticulum and endosome/lysosome in Giardia lamblia. Eukaryot Cell *8*, 1665-1676.

Abodeely, M., Dubois, K.N., Hehl, A., Stefanic, S., Sajid, M., Desouza, W., Attias, M., Engel, J.C., Hsieh, I., Fetter, R.D., and McKerrow, J.H. (2009b). A Contiguous Compartment Functions as Er and Endosome/Lysosome in Giardia Lamblia. Eukaryot Cell.

Adam, R.D. (2001). Biology of Giardia lamblia. Clin Microbiol Rev *14*, 447-475.

Barbieri, M.A., Roberts, R.L., Gumusboga, A., Highfield, H., Alvarez-Dominguez, C., Wells, A., and Stahl, P.D. (2000). Epidermal growth factor and membrane trafficking. EGF receptor activation of endocytosis requires Rab5a. J Cell Biol *151*, 539-550.

Barwick, R.S., Levy, D.A., Craun, G.F., Beach, M.J., and Calderon, R.L. (2000). Surveillance for waterborne-disease outbreaks--United States, 1997-1998. MMWR CDC Surveill Summ *49*, 1-21.

Bockman, D.E., and Winborn, W.B. (1968). Electron microscopic localization of exogenous ferritin within vacuoles of Giardia muris. J Protozool *15*, 26-30.

Boehm, M., and Bonifacino, J.S. (2001). Adaptins: the final recount. Mol Biol Cell *12*, 2907-2920.

Boehm, M., and Bonifacino, J.S. (2002). Genetic analyses of adaptin function from yeast to mammals. Gene *286*, 175-186.

Boll, W., Rapoport, I., Brunner, C., Modis, Y., Prehn, S., and Kirchhausen, T. (2002). The mu2 subunit of the clathrin adaptor AP-2 binds to FDNPVY and YppO sorting signals at distinct sites. Traffic *3*, 590-600.

Bonifacino, J.S., and Lippincott-Schwartz, J. (2003). Coat proteins: shaping membrane transport. Nat Rev Mol Cell Biol *4*, 409-414.

Bonifacino, J.S., and Traub, L.M. (2003). Signals for sorting of transmembrane proteins to endosomes and lysosomes. Annu Rev Biochem *72*, 395-447.

Brodsky, F.M., Chen, C.Y., Knuehl, C., Towler, M.C., and Wakeham, D.E. (2001). Biological basket weaving: formation and function of clathrin-coated vesicles. Annu Rev Cell Dev Biol *17*, 517-568.

Burgos, P.V., Mardones, G.A., Rojas, A.L., daSilva, L.L., Prabhu, Y., Hurley, J.H., and Bonifacino, J.S. Sorting of the Alzheimer's disease amyloid precursor protein mediated by the AP-4 complex. Dev Cell *18*, 425-436.

Cao, S., Yao, J., and Shah, V. (2003). The proline-rich domain of dynamin-2 is responsible for dynamin-dependent in vitro potentiation of endothelial nitric-oxide synthase activity via selective effects on reductase domain function. J Biol Chem *278*, 5894-5901.

Collins, B.M., McCoy, A.J., Kent, H.M., Evans, P.R., and Owen, D.J. (2002). Molecular architecture and functional model of the endocytic AP2 complex. Cell *109*, 523-535.

Dacks, J.B., and Doolittle, W.F. (2002). Novel syntaxin gene sequences from Giardia, Trypanosoma and algae: implications for the ancient evolution of the eukaryotic endomembrane system. J Cell Sci *115*, 1635-1642.

de Souza, W., Lanfredi-Rangel, A., and Campanati, L. (2004). Contribution of microscopy to a better knowledge of the biology of Giardia lamblia. Microsc Microanal *10*, 513-527.

Dobell, C. (1950). Samuel Hoole, translator of Leeuwenhoek's Select Works; with notes on that publication. Isis *41*, 171-180.

Doray, B., Lee, I., Knisely, J., Bu, G., and Kornfeld, S. (2007). The gamma/sigma1 and alpha/sigma2 hemicomplexes of clathrin adaptors AP-1 and AP-2 harbor the dileucine recognition site. Mol Biol Cell *18*, 1887-1896.

Elias, E.V., Quiroga, R., Gottig, N., Nakanishi, H., Nash, T.E., Neiman, A., and Lujan, H.D. (2008). Characterization of SNAREs determines the absence of a typical Golgi apparatus in the ancient eukaryote Giardia lamblia. J Biol Chem.

Elmendorf, H.G., Dawson, S.C., and McCaffery, J.M. (2003). The cytoskeleton of Giardia lamblia. Int J Parasitol *33*, 3-28.

Faso, C., and Hehl, A.B. (2011). Membrane trafficking and organelle biogenesis in Giardia lamblia: Use it or lose it. Int J Parasitol.

Feely, D.E., and Dyer, J.K. (1987). Localization of acid phosphatase activity in Giardia lamblia and Giardia muris trophozoites. J Protozool *34*, 80-83.

Field, M.C., Gabernet-Castello, C., and Dacks, J.B. (2007). Reconstructing the evolution of the endocytic system: insights from genomics and molecular cell biology. Adv Exp Med Biol *607*, 84-96.

Finn, R.D., Tate, J., Mistry, J., Coggill, P.C., Sammut, S.J., Hotz, H.R., Ceric, G., Forslund, K., Eddy, S.R., Sonnhammer, E.L., and Bateman, A. (2008). The Pfam protein families database. Nucleic Acids Res *36*, D281-288.

Fraser, D., Bilenko, N., Deckelbaum, R.J., Dagan, R., El-On, J., and Naggan, L. (2000). Giardia lamblia carriage in Israeli Bedouin infants: risk factors and consequences. Clin Infect Dis *30*, 419-424.

Gaechter, V., Schraner, E., Wild, P., and Hehl, A.B. (2008). The single dynamin family protein in the primitive protozoan Giardia lamblia is essential for stage conversion and endocytic transport. Traffic *9*, 57-71.

Gerwig, G.J., van Kuik, J.A., Leeflang, B.R., Kamerling, J.P., Vliegenthart, J.F., Karr, C.D., and Jarroll, E.L. (2002). The Giardia intestinalis filamentous cyst wall contains a novel beta(1-3)-N-acetyl-D-galactosamine polymer: a structural and conformational study. Glycobiology *12*, 499-505.

Gottlieb, M., and Dwyer, D.M. (1981). Leishmania donovani: surface membrane acid phosphatase activity of promastigotes. Exp Parasitol *52*, 117-128.

Gu, F., Crump, C.M., and Thomas, G. (2001). Trans-Golgi network sorting. Cell Mol Life Sci *58*, 1067-1084.

Harris-White, M.E., and Frautschy, S.A. (2005). Low density lipoprotein receptor-related proteins (LRPs), Alzheimer's and cognition. Curr Drug Targets CNS Neurol Disord *4*, 469-480.

Hashimoto, T., Sanchez, L.B., Shirakura, T., Muller, M., and Hasegawa, M. (1998). Secondary absence of mitochondria in Giardia lamblia and Trichomonas vaginalis revealed by valyl-tRNA synthetase phylogeny. Proc Natl Acad Sci U S A *95*, 6860-6865.

Hehl, A.B., and Marti, M. (2004). Secretory protein trafficking in Giardia intestinalis. Mol Microbiol *53*, 19-28.

Hermey, G. (2009). The Vps10p-domain receptor family. Cell Mol Life Sci *66*, 2677-2689.

Hernandez, Y., Castillo, C., Roychowdhury, S., Hehl, A., Aley, S.B., and Das, S. (2007). Clathrin-dependent pathways and the cytoskeleton network are involved in ceramide endocytosis by a parasitic protozoan, Giardia lamblia. Int J Parasitol *37*, 21-32.

Hinshaw, J.E. (2000). Dynamin and its role in membrane fission. Annu Rev Cell Dev Biol *16*, 483-519.

Huotari, J., and Helenius, A. (2011). Endosome maturation. EMBO J *30*, 3481-3500.

Janvier, K., Kato, Y., Boehm, M., Rose, J.R., Martina, J.A., Kim, B.Y., Venkatesan, S., and Bonifacino, J.S. (2003). Recognition of dileucine-based sorting signals from HIV-1 Nef and LIMP-II by the AP-1 gamma-sigma1 and AP-3 delta-sigma3 hemicomplexes. J Cell Biol *163*, 1281-1290.

Jarroll, E.L., Manning, P., Lindmark, D.G., Coggins, J.R., and Erlandsen, S.L. (1989). Giardia cyst wall-specific carbohydrate: evidence for the presence of galactosamine. Mol Biochem Parasitol *32*, 121-131.

Jones, C., Hammer, R.E., Li, W.P., Cohen, J.C., Hobbs, H.H., and Herz, J. (2003). Normal sorting but defective endocytosis of the low density lipoprotein receptor in mice with autosomal recessive hypercholesterolemia. J Biol Chem *278*, 29024-29030.

Kirchhausen, T. (2000). Clathrin. Annu Rev Biochem *69*, 699-727.

Kirchhausen, T., and Toyoda, T. (1993). Immunoelectron microscopic evidence for the extended conformation of light chains in clathrin trimers. J Biol Chem *268*, 10268-10273.

Krueger, E.W., Orth, J.D., Cao, H., and McNiven, M.A. (2003). A dynamin-cortactin-Arp2/3 complex mediates actin reorganization in growth factor-stimulated cells. Mol Biol Cell *14*, 1085-1096.

Ladinsky, M.S., Wu, C.C., McIntosh, S., McIntosh, J.R., and Howell, K.E. (2002). Structure of the Golgi and distribution of reporter molecules at 20 degrees C reveals the complexity of the exit compartments. Mol Biol Cell *13*, 2810-2825.

Lanfredi-Rangel, A., Attias, M., de Carvalho, T.M., Kattenbach, W.M., and De Souza, W. (1998). The peripheral vesicles of trophozoites of the primitive protozoan Giardia lamblia may correspond to early and late endosomes and to lysosomes. J Struct Biol *123*, 225-235.

Langford, T.D., Silberman, J.D., Weiland, M.E., Svard, S.G., McCaffery, J.M., Sogin, M.L., and Gillin, F.D. (2002). Giardia lamblia: identification and characterization of Rab and GDI proteins in a genome survey of the ER to Golgi endomembrane system. Exp Parasitol *101*, 13-24.

Lindmark, D.G. (1988). Giardia lamblia: localization of hydrolase activities in lysosome-like organelles of trophozoites. Exp Parasitol *65*, 141-147.

Lujan, H.D., Mowatt, M.R., Conrad, J.T., Bowers, B., and Nash, T.E. (1995). Identification of a novel Giardia lamblia cyst wall protein with leucine-rich repeats. Implications for secretory granule formation and protein assembly into the cyst wall. J Biol Chem *270*, 29307-29313.

Lujan, H.D., Mowatt, M.R., and Nash, T.E. (1997). Mechanisms of Giardia lamblia differentiation into cysts. Microbiol Mol Biol Rev *61*, 294-304.

Lujan, H.D., and Touz, M.C. (2003). Protein trafficking in Giardia lamblia. Cell Microbiol *5*, 427-434.

Manning, P., Erlandsen, S.L., and Jarroll, E.L. (1992). Carbohydrate and amino acid analyses of Giardia muris cysts. J Protozool *39*, 290-296.

Marti, M., Li, Y., Schraner, E.M., Wild, P., Kohler, P., and Hehl, A.B. (2003a). The secretory apparatus of an ancient eukaryote: protein sorting to separate export pathways occurs before formation of transient Golgi-like compartments. Mol Biol Cell *14*, 1433-1447.

Marti, M., Regos, A., Li, Y., Schraner, E.M., Wild, P., Muller, N., Knopf, L.G., and Hehl, A.B. (2003b). An ancestral secretory apparatus in the protozoan parasite Giardia intestinalis. J Biol Chem *278*, 24837-24848.

McCaffery, J.M., and Gillin, F.D. (1994). Giardia lamblia: ultrastructural basis of protein transport during growth and encystation. Exp Parasitol *79*, 220-235.

McNiven, M.A., Cao, H., Pitts, K.R., and Yoon, Y. (2000). The dynamin family of mechanoenzymes: pinching in new places. Trends Biochem Sci *25*, 115-120.

Mellman, I., and Simons, K. (1992). The Golgi complex: in vitro veritas? Cell *68*, 829-840.

Morrison, H.G., McArthur, A.G., Gillin, F.D., Aley, S.B., Adam, R.D., Olsen, G.J., Best, A.A., Cande, W.Z., Chen, F., Cipriano, M.J., Davids, B.J., Dawson, S.C., Elmendorf, H.G., Hehl, A.B., Holder, M.E., Huse, S.M., Kim, U.U., Lasek-Nesselquist, E., Manning, G., Nigam, A., Nixon, J.E., Palm, D., Passamaneck, N.E., Prabhu, A., Reich, C.I., Reiner, D.S., Samuelson, J., Svard, S.G., and Sogin, M.L. (2007). Genomic minimalism in the early diverging intestinal parasite Giardia lamblia. Science *317*, 1921-1926.

Munro, S. (2004). Organelle identity and the organization of membrane traffic. Nat Cell Biol *6*, 469-472.

Nash, T.E. (2002). Surface antigenic variation in Giardia lamblia. Mol Microbiol *45*, 585-590.

Nevin, W.D., and Dacks, J.B. (2009). Repeated secondary loss of adaptin complex genes in the Apicomplexa. Parasitol Int *58*, 86-94.

Ochoa, G.C., Slepnev, V.I., Neff, L., Ringstad, N., Takei, K., Daniell, L., Kim, W., Cao, H., McNiven, M., Baron, R., and De Camilli, P. (2000). A functional link between dynamin and the actin cytoskeleton at podosomes. J Cell Biol *150*, 377-389.

Ohno, H., Fournier, M.C., Poy, G., and Bonifacino, J.S. (1996). Structural determinants of interaction of tyrosine-based sorting signals with the adaptor medium chains. J Biol Chem *271*, 29009-29015.

Ohno, H., Stewart, J., Fournier, M.C., Bosshart, H., Rhee, I., Miyatake, S., Saito, T., Gallusser, A., Kirchhausen, T., and Bonifacino, J.S. (1995). Interaction of tyrosine-based sorting signals with clathrin-associated proteins. Science *269*, 1872-1875.

Pelkmans, L., Burli, T., Zerial, M., and Helenius, A. (2004). Caveolin-stabilized membrane domains as multifunctional transport and sorting devices in endocytic membrane traffic. Cell *118*, 767-780.

Peters, C., Braun, M., Weber, B., Wendland, M., Schmidt, B., Pohlmann, R., Waheed, A., and von Figura, K. (1990). Targeting of a lysosomal membrane protein: a tyrosine-containing endocytosis signal in the cytoplasmic tail of lysosomal acid phosphatase is necessary and sufficient for targeting to lysosomes. Embo J *9*, 3497-3506.

Piva, B., and Benchimol, M. (2004). The median body of Giardia lamblia: an ultrastructural study. Biol Cell *96*, 735-746.

Polishchuk, R.S., San Pietro, E., Di Pentima, A., Tete, S., and Bonifacino, J.S. (2006). Ultrastructure of long-range transport carriers moving from the trans Golgi network to peripheral endosomes. Traffic *7*, 1092-1103.

Puertollano, R., van der Wel, N.N., Greene, L.E., Eisenberg, E., Peters, P.J., and Bonifacino, J.S. (2003). Morphology and dynamics of clathrin/GGA1-coated carriers budding from the trans-Golgi network. Mol Biol Cell *14*, 1545-1557.

Reiner, D.S., Douglas, H., and Gillin, F.D. (1989). Identification and localization of cyst-specific antigens of Giardia lamblia. Infect Immun *57*, 963-968.

Rezgaoui, M., Hermey, G., Riedel, I.B., Hampe, W., Schaller, H.C., and Hermans-Borgmeyer, I. (2001). Identification of SorCS2, a novel member of the VPS10 domain containing receptor family, prominently expressed in the developing mouse brain. Mech Dev *100*, 335-338.

Rivero, M.R., Miras, S.L., Quiroga, R., Ropolo, A.S., and Touz, M.C. (2011). Giardia lamblia low-density lipoprotein receptor-related protein is involved in selective lipoprotein endocytosis and parasite replication. Mol Microbiol *79*, 1204-1219.

Rivero, M.R., Vranych, C.V., Bisbal, M., Maletto, B.A., Ropolo, A.S., and Touz, M.C. (2010). Adaptor Protein 2 Regulates Receptor-Mediated Endocytosis and Cyst Formation in Giardia lamblia. Biochem J.

Rojas, R., van Vlijmen, T., Mardones, G.A., Prabhu, Y., Rojas, A.L., Mohammed, S., Heck, A.J., Raposo, G., van der Sluijs, P., and Bonifacino, J.S. (2008). Regulation of retromer recruitment to endosomes by sequential action of Rab5 and Rab7. J Cell Biol *183*, 513-526.

Roth, T.F., and Porter, K.R. (1964). Yolk Protein Uptake in the Oocyte of the Mosquito Aedes Aegypti. L. J Cell Biol *20*, 313-332.

Shakarian, A.M., Joshi, M.B., Ghedin, E., and Dwyer, D.M. (2002). Molecular dissection of the functional domains of a unique, tartrate-resistant, surface membrane acid phosphatase in the primitive human pathogen Leishmania donovani. J Biol Chem *277*, 17994-18001.

Slavin, I., Saura, A., Carranza, P.G., Touz, M.C., Nores, M.J., and Lujan, H.D. (2002). Dephosphorylation of cyst wall proteins by a secreted lysosomal acid phosphatase is essential for excystation of Giardia lamblia. Mol Biochem Parasitol *122*, 95-98.

Sogin, M.L., Gunderson, J.H., Elwood, H.J., Alonso, R.A., and Peattie, D.A. (1989). Phylogenetic meaning of the kingdom concept: an unusual ribosomal RNA from Giardia lamblia. Science *243*, 75-77.

Sun, C.H., McCaffery, J.M., Reiner, D.S., and Gillin, F.D. (2003). Mining the Giardia lamblia genome for new cyst wall proteins. J Biol Chem *278*, 21701-21708.

Sweitzer, S.M., and Hinshaw, J.E. (1998). Dynamin undergoes a GTP-dependent conformational change causing vesiculation. Cell *93*, 1021-1029.

Touz, M.C., Conrad, J.T., and Nash, T.E. (2005). A novel palmitoyl acyl transferase controls surface protein palmitoylation and cytotoxicity in Giardia lamblia. Mol Microbiol *58*, 999-1011.

Touz, M.C., Gottig, N., Nash, T.E., and Lujan, H.D. (2002a). Identification and characterization of a novel secretory granule calcium-binding protein from the early branching eukaryote Giardia lamblia. J Biol Chem *277*, 50557-50563.

Touz, M.C., Kulakova, L., and Nash, T.E. (2004). Adaptor protein complex 1 mediates the transport of lysosomal proteins from a Golgi-like organelle to peripheral vacuoles in the primitive eukaryote Giardia lamblia. Mol Biol Cell *15*, 3053-3060.

Touz, M.C., Lujan, H.D., Hayes, S.F., and Nash, T.E. (2003). Sorting of encystation-specific cysteine protease to lysosome-like peripheral vacuoles in Giardia lamblia requires a conserved tyrosine-based motif. J Biol Chem *278*, 6420-6426.

Touz, M.C., Nores, M.J., Slavin, I., Carmona, C., Conrad, J.T., Mowatt, M.R., Nash, T.E., Coronel, C.E., and Lujan, H.D. (2002b). The activity of a developmentally regulated cysteine proteinase is required for cyst wall formation in the primitive eukaryote Giardia lamblia. J Biol Chem *277*, 8474-8481.

Touz, M.C., Rivero, M. R., Miras, S. L. and Bonifacino, J. S. (2011). Lysosomal Protein Trafficking in *Giardia lamblia*: Common and Distinct Features Front Biosci (Elite Ed). *4*, 1898-1909.

Traub, L.M. (2009). Tickets to ride: selecting cargo for clathrin-regulated internalization. Nat Rev Mol Cell Biol *10*, 583-596.

Ullrich, O., Reinsch, S., Urbe, S., Zerial, M., and Parton, R.G. (1996). Rab11 regulates recycling through the pericentriolar recycling endosome. J Cell Biol *135*, 913-924.

van der Sluijs, P., Hull, M., Webster, P., Male, P., Goud, B., and Mellman, I. (1992). The small GTP-binding protein rab4 controls an early sorting event on the endocytic pathway. Cell *70*, 729-740.

Waheed, A., Gottschalk, S., Hille, A., Krentler, C., Pohlmann, R., Braulke, T., Hauser, H., Geuze, H., and von Figura, K. (1988). Human lysosomal acid phosphatase is transported as a transmembrane protein to lysosomes in transfected baby hamster kidney cells. Embo J *7*, 2351-2358.

Ward, W., Alvarado, L., Rawlings, N.D., Engel, J.C., Franklin, C., and McKerrow, J.H. (1997). A primitive enzyme for a primitive cell: the protease required for excystation of Giardia. Cell *89*, 437-444.

Zhang, F., Yim, Y.I., Scarselletta, S., Norton, M., Eisenberg, E., and Greene, L.E. (2007). Clathrin adaptor GGA1 polymerizes clathrin into tubules. J Biol Chem *282*, 13282-13289.

Section 5

Cell Surface Receptors

Chapter 12

Endocytosis in Notch Signaling Activation

Elisa Sala, Luca Ruggiero, Giuseppina Di Giacomo and Ottavio Cremona

Additional information is available at the end of the chapter

1. Introduction

In mechanistic terms, endocytosis is the process by which plasma membrane (PM) components, together with extracellular solutes, macromolecules and particles, are internalized in the cell. Once the endocytic vesicle (or vacuole) is formed by fission of the PM, it is generally delivered to a specialized membrane compartment – the endosome – for recycling, degradation or re-routing.

In cell-physiological terms, endocytosis exerts multiple functions, which are only partially known and characterized. At a minimum, it maintains PM homeostasis by counterbalancing the apposition of new membrane (due to exocytosis) and by renewing PM components. More extensively, endocytosis constantly modulates PM composition and takes an active part in a variety of normal and pathological cell processes, including cell nutrition, cell motility, mitosis, neurotransmission, immune response, and microorganism entry (reviewed in [1-8]).

1.1. Endocytosis and signaling

In recent years, much of the effort to investigate this extensive endocytic activity has been focused upon unveiling the reciprocal interplay between endocytosis and cell signaling. In this introductory section, we provide a quick overview of the key concepts in the field to explain the endocytic function in Notch signaling. We refer those readers who wish to explore the relationship between endocytosis and cell signaling in details to other papers in this volume, and to recent reviews in the field [3, 9-11].

Originally, endocytosis was linked to the termination of PM-generated signals by reducing the availability of membrane receptors for ligand binding, and by degrading the ligand-receptor complex (reviewed in [10, 12-14]). Hence, blockage or dysregulation of endocytosis closely correlates with increased cell proliferation by the activation of receptor-tyrosine kinase pathways and cancer promotion (reviewed in [10, 15, 16]). More recently, new strategies for the endocytosis-mediated regulation of signaling have been uncovered: (i) endocytosis can activate/modulate some PM-generated signals either directly (e.g. by controlling ligand availability, as in the case of the cell-to-cell Eph/Ephrin signaling pathway [17]), or indirectly (e.g. by regulating the composition of specific signaling platforms, as in the case of phospholipase C and PI3Kinase signaling activated via EGF receptor [18]); (ii) endocytosis can propagate signals to intracellular compartments, especially the endosomal compartment, where these signals are sustained, specified, spread over long distances or rerouted (reviewed in [9, 19-22]); (iii) endocytosis can ensure spatial restriction to signaling responses emanating from the PM and/or from the endosome (e.g. the endocytosis/recycling function in the spatial restriction of signaling controlling migratory programs (reviewed in [23], or in determining the timing, levels, and localization of guidance receptors, thus determining the outcome of guidance decisions [24]). On the other hand, signaling can modulate endocytosis: (i) activation of specific signaling pathways can upregulate or downregulate endocytosis, thus modulating other PM- and/or endosome-generated signals (e.g. EGF receptor activation increases SRC kinase-mediated phosphorylation of the clathrin heavy chain, which redistributes to the cell periphery, potentiating endocytosis [25]); (ii) actin dynamics/signaling takes an active part in the endocytic reaction by helping membrane invagination [26, 27], vesicle transportation [28, 29], and endosomal microdomain organization [30]. We will see that many of these endocytic strategies to control signaling are exploited in Notch signaling.

1.2. Types of endocytosis and their regulation

In order to promote its many functions, endocytosis relies on a variety of specialized mechanisms and accessory factors to guarantee selectivity, vectoriality and plasticity.

Regarding these mechanisms, there are multiple forms of endocytosis that act concomitantly in the cell (reviewed in [4, 31-33]). The best studied, and perhaps the most common, forms are clathrin-based. Their central paradigm is the recruitment and assembly of clathrin to the PM, triggered by a variety of adaptor proteins, which bind (and sometimes bend) the PM by means of lipid- and protein-interacting domains (reviewed in [34]). Invagination of the PM to form a bud depends on a concerted action of the clathrin lattice rearrangement (which shapes the high curvature profile of the bud [35]), polymerization of bending proteins (which shape the neck of the bud (reviewed in [36])), and actin polymerization (which helps the extension and constriction of the neck of the bud [37, 38]). Constriction of the bud and its fission to form free clathrin-coated vesicles requires the pinchase action of the GTPase dynamin(s) (reviewed in [39-41]). Free vesicles are then stripped of their coat by an uncoating complex, composed of the ATPase heat shock cognate 70 (HSC70) [42] and the J-domain-containing co-chaperone auxilin [43]. Structural requirements for the uncoating reaction are reviewed in [44, 45]. An essential function is also carried out by phosphoinositides, and more specifically by the PtdIns(4,5)P_2 present in the membrane of coated vesicles, which has to be hydrolyzed by synaptojanin for efficient release of endocytic adaptors, which precedes clathrin disassembly [46, 47].

Clathrin-mediated endocytosis (CME) is not the only type of endocytosis. It is now many years since evidence for clathrin-independent endocytosis (CIE) has been accumulated, although the mechanism behind this process is as yet poorly characterized. The development of specialized techniques, reagents and markers to trace endocytosis has unveiled a whole new world of internalization routes that persist after inhibition of the clathrin function (reviewed in [32, 48-50]). A common finding is the exquisite sensitivity of CIE to cholesterol depletion, although CME is also somewhat sensitive to cholesterol levels, and some forms of CIE can still occur without membrane cholesterol [51]. On the PM, cholesterol is transiently enriched in microdomains, commonly known as lipid rafts [52]. This fact, together with the absence of rafts in clathrin intermediates and with the observation that most of the raft components are endocytosed by non-clathrin-dependent pathways, has led to the idea that most CIE occurs in these lipid microdomains. At rafts, signaling events are subcompartmentalized in specific nanoplatforms, whose composition and, therefore, activity is continuously changing [52]. The wealth of proteins that participate in raft signaling events also gives rise to a number of different CIE pathways which differ for (i) fission machinery (i.e. dynamin-dependence), (ii) coat composition and (iii) Rab effector specificity (reviewed in [4, 32, 50]). Both CME and CIE forms participate in Notch signaling activation and regulation.

Regarding accessory factors, tens of molecules, both proteins and lipids, interact with endocytic machinery at various stages. Several recognition modules have been identified, including protein-lipid (e.g. PH domain) and protein-protein interaction modules (e.g. BAR, SH3-, proline rich-, EH-, coiled-coil domains, ubiquitin interacting motifs (UIM)). It is conceivable that this array of interactions may help regulate both CME and CIE by: (i) assisting coat assembly/disassembly, (ii) regulating membrane shaping/sculpting and fission, and (iii) mediating interaction of the coat with signaling molecules and the cytoskeleton (reviewed in [4, 9, 34, 53-55].

In this review, we will focus on the molecular details at the basis of the endocytic control of Notch activation. Specific emphasis will be made on genetic data in mammals and invertebrates that support or validate *in vitro* interaction and/or models. Since dysregulation of Notch signaling contributes to the multi-step progression of a variety of cancers by inducing uncontrolled proliferation, the possible therapeutic value of this information can be clearly envisaged.

2. Overview of the Notch signaling pathway

The first Notch gene was identified in *Drosophila melanogaster* by J.S. Dexter, in T.H. Morgan's laboratory about a century ago, as a dominant mutant with a peculiar toothed/notched wing margin in heterozygosity [56]; this phenotype was later associated with additional defects, including thickened wing veins, and bristle abnormalities [57]. In hemizygosity, or in homozygous females in flies, Notch loss-of-function mutations are embryonic lethal with neuralization of important parts of the ectoderm, leading to hypertrophy of the central nervous system and corresponding hypotrophy of the epidermis of the fully developed embryo [58, 59]. As anticipated by these data, in higher metazoan the Notch pathway is one of a handful of signaling pathways (including Wnt/wingless,

BMP/TGF-beta, Sonic Hedgehog, receptor tyrosine kinases, nuclear receptors, JAK/STAT) that act reiteratively in cell fate decision and determination in tissues that derive from all three germ layers (reviewed in [60, 61]). After development, Notch signaling is required for the homeostasis of tissues and stem cells, as underscored by the high number of tumors associated with Notch signaling dysregulation, which was an early finding in Notch research in mammals.

The Notch signaling is a cell-to-cell communication pathway that is activated when Notch ligands (**D**elta/**S**errate/**L**ag2-DSL in invertebrates and Delta-like/Jagged, in mammals) on the sending cell bind to Notch receptor(s) on the receiving cell. This triggers a sequence of proteolytic cleavages, which starts with an ADAM-mediated cleavage at **s**ite **2** (S2) [62]. While ADAM17/TACE seems to be the main metalloprotease able to cleave Notch receptors *in vitro* [63], animal models point to ADAM10/Kuzbanian metalloprotease for this essential function in vivo [62, 64-68]. ADAM proteases leaves a short-lived fragment anchored to the PM, called NEXT (for **N**otch **ex**tracellular **t**runcation, see Fig.1), which becomes a substrate for the aspartyl-protease presenilin(s), a component of the γ-secretase complex [69, 70]. This protease complex (which includes four core proteins, i.e. presenilin 1 or 2, **a**nterior **ph**arynx defective 1 (APH1), nicastrin, and **p**resenilin **en**hancer 2 (PEN2) [71]) operates an intramembrane cleavage at site 3 (S3), which releases the **N**otch **i**ntra**c**ellular **d**omain (NICD). Then, NICD translocates to the nucleus and turns a transcription factor of the CSL family (**C**p-**b**inding **f**actor 1 (CBF-1)/**r**ecombination signal sequence-**b**inding **p**rotein Jk (RBP-Jk) in mammals, Su(H) (***S****uppressor of* ***H****airless)* in *Drosophila*, and LAG-1 in nematodes) from a repressor [72-74] to a transcriptional activator. Although Notch signaling has such a broad impact in a variety of cellular functions, only a limited number of Notch primary targets have so far been identified, of which the best characterized are the helix-loop-helix transcription factors of the **H**airy/**e**nhancer of **s**plit (Hes) and **Hes**-**r**elated (Hesr, also known as Hey/HRT, CHF and gridlock) families (reviewed in [75-78]). CSL binding sites have also been identified in the promoter region of other genes, including c-myc, cyclinD1, p21/Waf1, NFk B2, glial fibrillary acidic protein *(GFAP)*, Nodal, GATA3, bcl-2 and CD25 (alpha chain of the IL-2 receptor), although the role of these genes as direct Notch targets has still not been unambiguously shown (reviewed in [78, 79]).

2.1. Domain structure of Notch components

Let us now briefly analyze the architecture of Notch receptors and ligands to highlight those structural features that are key factors in endocytosis-mediated signaling activation (reviewed in [80, 81]).

While *Drosophila* has a single Notch receptor gene, C. Elegans has two (Glp-1 and Lin-12 [82], which are highly redundant [83]), and mammals have four paralogues (Notch1-4) with only partially superimposable functions. The Notch receptor is a type-I transmembrane protein which is cleaved in mammals by a furin-like convertase at an external site close to the PM (the site 1 (S1)). Proteolytic cleavage occurs in the trans-Golgi network to generate a heterodimer at the cell surface composed of two non-covalently associated fragments: the

Notch extracellular domain (NECD) and the Notch transmembrane domain (NTMD); NTMD contains a small portion of the extracellular region, the transmembrane region, and the intracellular domain [84, 85] (see Fig.1). The impact of cleavage on signaling activation is a matter of open discussion: while it was seen by some laboratories to be a prerequisite for delivery of receptors to the cell surface [84, 85], other groups have shown that Notch receptors which are defective for S1 cleavage are normally exposed to the cell surface, but fail in ligand-mediated activation of canonical Notch signaling [86, 87]; these latter data support the hypothesis that dissociation of NECD from NTMD (by endocytosis, see later) may be a prerequisite for S2 proteolysis. Notably, S1 cleavage-defective Notch receptors exhibit little change in their crystal and NMR structure in comparison with wild-type receptors [88]; this is in contrast to what happens with some viruses (including avian influenza virus, HIV-1, measles and papilloma virus [89-92]) in which furin cleavage induces major conformational changes leading to protein activation (reviewed in [93]).

2.1.1. Notch receptor architecture

Notch receptors are multidomain proteins, which have been conserved from invertebrates to man. Going from the N- to the C-terminus, mammalian Notch receptors contain five regions (Fig.1): (i) a variable number of **E**pidermal **g**rowth **f**actor (EGF)-like domains (spanning from 29 to 36 domains in mammals), many of which contain **c**alcium-**b**inding sites (cbEGF-like domains); (ii) three Lin-12-Notch repeats (LNR); (iii) a hydrophobic region for receptor **h**etero**d**imerization (HD domain), which is cleaved at the S1 site (at 70 amino acids from the transmembrane domain), and which contains the S2 site (at 12 amino acids from the transmembrane domain); (iv) the transmembrane domain, which bears the S3 cleavage site, a substrate for **r**egulated **i**ntramembrane **p**roteolysis (RIP) by the presenilin/γ-secretase complex to liberate the intracellular domain; (v) the NICD, which contains a RAM domain (for Notch binding to the transcription factor CSL/CBF1/Suppressor of Hairless/Lag-1)), seven **ank**yrin repeats (ANK), a **t**ranscription **a**ctivation **d**omain (TAD) and a PEST domain (which is implicated in NICD degradation by proteolysis and whose mutation leads to increased receptor stability , a condition that closely correlates with cancer, including some T-cell leukemias) (reviewed in [80]). LNRs, plus the heterodimerization domain, form the so-called **n**egative **r**egulatory **r**egion (NRR), which folds onto the S2 cleavage site by means of extensive interdomain interactions [94-98]. As analyzed in greater details in section 3.1.1, this conformation makes the S2 site inaccessible to ADAM metalloproteases, thus protecting the Notch receptor from ligand-independent activation; the key importance of this region is underlined by the fact that mutations of the NRR, which activates the Notch receptor, closely correlates with T-cell acute lymphoblastic leukemia lymphoma (T-ALL) [95]. The TAD region is found in Notch-1/-2, but it is not present in Notch-3/-4 in mammals.

2.1.2. Notch ligand architecture

As for Notch receptors, Notch ligands are genes that have been conserved throughout evolution. *Drosophila* has two such ligand genes, Delta and Serrate, while there are five mammalian ligands, three belonging to the **Del**ta-**l**ike family (Dll-1, -3 and -4) and two to the Jagged (Serrate homologous) family (Jagged-1 and -2). Starting from the N-terminus, the domain structure of the Notch ligands can be outlined as follows (Fig.1): (i) a **m**odule at the **N**-terminus of **N**otch **l**igands (also known as the MNNL domain) of unknown structure but functionally relevant since Jag-1 mutations in this region are present in a subset of patients with the Alagille syndrome [99, 100]; (ii) a DSL domain; (iii) a number of EGF-like repeats (ranging in number from 16 in the Jagged family to 5–9 in the Delta-like family); (iv) a **c**ysteine-**r**ich **d**omain (CRD) present in Jagged but not in Dll ligands; (v) a transmembrane domain and (vi) an intracellular domain, highly divergent among Notch ligands, but with a conserved PDZ-binding domain in the mammalian Jagged-1, Delta-like-1 and -4. The function of this latter domain is unknown, although there is some evidence that its interaction with PDZ-containing adherens-junction proteins inhibits cell motility and favors epithelial cell assembly [101-103]. Similar to Notch receptors, Notch ligands undergo ectodomain shedding by ADAM metalloproteases [104, 105] and RIP) by the presenilin/γ-secretase complex [104] with the release of a C-terminal intracellular fragment (CTIF). As for many other γ-secretase products, including Notch and **a**myloid **p**recursor **p**rotein (APP) intracellular domains, CTIF translocates to the nucleus where it may help transcriptional activities [104]. In particular, Jagged1 CTIF selectively stimulates the expression of reporter genes driven by the AP-1 response element, but not by other broad-spectrum enhancer elements [104]; these data point to a possible role of CTIF as a transcriptional co-activator.

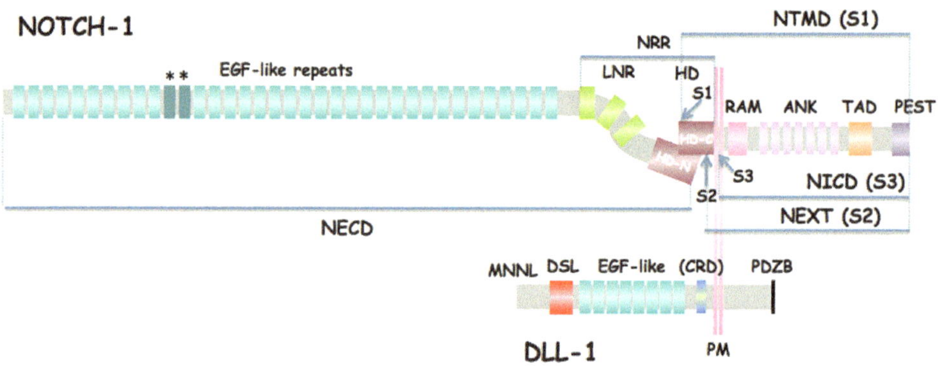

Figure 1. The domain architecture of mammalian Notch receptors (i.e. Notch-1) and Notch ligands/DSLs (i.e. Dll-1) is schematized in this drawing. Asterisks (*) indicate the EGF-like 11 and 12 of Notch receptors, which are keys for ligand binding. PM=Plasma Membrane; for other abbreviations, please refer to the text.

2.2. Notch ligand-receptor interaction

In the last few years, major advances have been made in clarifying the structural details of Notch-DSL interaction. This information is highly relevant to understand the effect of internalization and membrane trafficking on Notch signaling activation.

2.2.1. *Structural requirements*

By screening *Drosophila* Notch deletion mutants for their ability to promote aggregation of S2 cells [106], the EGF-like repeats 11-12 of Notch has been identified as the major interacting site for DSLs [107, 108]. Subsequent analyses have confirmed this initial observation, but have also shown that optimal binding between Notch and its ligands requires many of the 36 EGF-like repeats: a minimal Notch interacting fragment composed only of the EGF domains 10–13 has a 45-fold lower binding ability to Delta-expressing cells in comparison with full-length Notch receptors [109]. The relatively large size of EGF domains, their need for extensive disulphide bonding and, possibly, glycosylation for proper folding, have hampered the possibility to obtain structural data for Notch ligand-receptor interaction for many years. However, when an unglycosylated fragment of Notch-1 encompassing the EGF domains 11-13, was *in vitro* redox-refolded and demonstrated to be able to bind Notch ligands in a calcium-dependent manner, the way to get structural data for EGF repeats was discovered: the NMR structure of this fragment was readily solved, showing a well-defined, rod-like orientation of EGF-like 11-12, rigidified by calcium [110]. In the meantime, other studies have identified the DSL domain of Jagged-1 as the minimum binding site for Notch-2, both *in vivo* and *in vitro*; interestingly, the EGF-like repeats of Jagged-1 immediately downstream of the DSL domain, in particular the first and second EGFs, had been shown to considerably improve this interaction with their stabilizing action [111]. The convergence of previous information has made it possible to better define the structure of the Notch ligand-receptor interaction: the crystal structure of the minimal DSL binding site (i.e. the Jagged1 fragment comprising DSL plus EGF 1-3 ($Jagged1_{DSL\text{-}EGF3}$)), and that of its Notch receptor counterpart (i.e. the Notch-1 EGF 11-13 fragment ($Notch\text{-}1_{11\text{-}13}$)), were separately obtained [112]. Although it was not possible to make a co-crystal of the interaction, *in silico* docking of $Jagged1_{DSL\text{-}EGF3}$ and $Notch\text{-}1_{11\text{-}13}$ structures using restraints from parallel NMR binding data gave precious information: a single DSL surface is responsible for both *cis*-inhibiting and *trans*-activating complex of ligand and receptor [112]. A parallel study from the same group has also demonstrated that selective mutagenesis of the calcium binding site in the EGF-like repeat 12 abrogates ligand binding, thus strongly supporting the idea that this EGF repeat is actually the major DSL-binding site [113].

2.2.2. *Glycosylation function*

Post-translational modifications play a key role in modulating Notch activation. NECD is heavily glycosylated and many studies have tried to address the impact of these modifications on Notch signaling.

A **p**rotein **O-fu**cosyl**t**ransferase (Ofut1 in *Drosophila* and Pofut1 in mammals [114]) binds fucose to specific serine and threonine residues of the EGF-like repeats (reviewed in [115]); afterwards, acetylglucosaminyltransferases of the Fringe family (Fringe in *Drosophila*, Lunatic fringe (Lfng), Radical fringe (Rfng) and Manic fringe (Mfng) in mammals) can elongate the sugar chain by adding *N*-acetylglucosamine residues [116].

In *Drosophila*, Ofut1 downregulation by RNAi or gene inactivation results in classic Notch loss-of-function phenotypes, including neurogenic defects [115, 117]. The requirement of fucosylation for Notch activation is even more evident in mammals, where constitutive inactivation of the Pofut1 gene produces developmental defects that are undistinguishable from the most aggressive Notch mutants [118]; notably, defects due to Pofut1 absence can be fully rescued by expressing a constitutively active form of Notch1, at least in the hematopoietic compartment [119]. Taken together, these data show that fucosylation is required for proper Notch signaling activation in all species.

The precise role played by Pofut1 in this process was partially addressed by creating a mouse mutant bearing a Notch1 allele which was deficient for the fucosylation in a critical EGF-like repeat for DSL binding, i.e. EGF-12: trans-heterozygous mice carrying the $Notch1^{12f}$ allele and a Notch1 null allele exhibit embryonic lethality, and defects similar to Notch1 knockouts [120]. However, homozygous $Notch1^{12f}$ mice are viable, but with defects in T cell specification and functions and, notably, a sharply decreased binding capacity to Delta1-expressing cells [120], thus pointing to a key function of fucosylation in regulating the affinity of Notch receptors for Notch ligands.

Quite recently, experimental evidence has accumulated for other key roles of Ofut1 besides fucosylation. In *Drosophila*, depletion of Ofut1 determines Notch accumulation in the endoplasmic reticulum, where the fucosyltransferase is resident; since transfection of the mouse Pofut1 rescues this accumulation defect, it was proposed that Ofut1 might have additional chaperone activity for the trafficking of Notch out of the endoplasmic reticulum [121].

Gene knockdown of the *Drosophila* **G**DP-4,6-**m**annose-**d**eshydratase (GMD, a cytosolic enzyme that converts GDP-mannose in GDP-4-keto-6-deoxymannose, an intermediate in the synthesis of GDP-fucose [122]) generates a loss-of-function Notch phenotype with increased Notch degradation but no accumulation in the ER; instead, co-silencing of Ofut1 and GMD restored ER accumulation, thus further supporting to the idea that an additional Ofut1 activity (independent from *O*-fucosylation) is required for proper exiting of Notch from the ER [123]. An Ofut1 chaperone function was suggested to be related to a quality control mechanism that scrutinizes Notch receptors for inappropriate inter- or intramolecular bonds between EGF-like repeats [124-127]; evidence of such a conserved function in mammals is lacking, at the moment.

Regarding the extension of *O*-fucosylated residues, Fringe deletion in flies results in a partial Notch phenotype with dorsal-ventral boundary defects during wing development [128, 129], thus formally proving the relevance of this additional glycosylation step at least for some aspects of Notch signaling activation (reviewed in [130, 131]). In mammals, evidence for a similar conserved function is accumulating. Although the three Fringe homologues have very similar enzymatic activity, substrate specificities and tissue distribution [132], only Lfng inactivation results in Notch-related defects, including (i) impairment of T cell maturation (since a developmental stage–specific expression of Lfng is required for the access of T cell progenitors to intrathymic niches that support Notch1-dependent T cell development [133]), and (ii) subversion of somitogenesis in some body districts, with major alterations in vertebral and rib cage morphogenesis [134, 135]). Conversely, Rfng and Mfng knockouts display no obvious phenotypic defects. Furthermore, no synergistic defects were observed in mice lacking all fringe genes, thus questioning function redundancy in this gene family [136-138]. Notably, *in vitro* glycosylation and ligand binding studies and *in vivo* genetic data have established that the addition of N-acetylglucosamine onto O-fucose indeed modulates the affinity of Notch ligand-receptor interaction in an opposite manner, i.e. by enhancing Notch binding to Delta and inhibiting Notch binding to Serrate [139-141].

In addition to fucosylation, Notch receptors are also modified by *O*-glucosylation. Genetic studies in flies and mammals have shown that inactivation of the only enzyme responsible for the addition of O-glucose to EGF-like repeats, i.e. **p**rotein **O-glu**cosyl**t**ransferase (Poglut/ Rumi), results in severe Notch phenotypes [142, 143]. Notably, Rumi activity is required in the signal-sending cell, where it has neither chaperone-like activity [142, 143] nor a function in ligand binding, since Notch in fly rumi knockdown cells binds Delta as efficiently as in control cells [142]. The concentration of NICD is dramatically reduced in several tissues of Rumi mutants [142], thus pointing to a function of *O*-glucosylation in Notch proteolysis, rather than in ligand binding. The structural basis of this function is unknown, but it is plausible that O-glucosylation may affect the structure of the NECD so that an initial constraint for S2 cleavage is removed [131, 144].

Therefore, genetic and in vitro data on Notch glycosylation indicate (i) that *O*-fucosylation and *O*-glucosylation play a general role in Notch signaling activation (albeit with different mechanisms), while *N*-acetylglucosamine addition is required for more specific aspects of this signaling activation, and (ii) that *O*-fucosylation and its acetylglucosamine extension mainly acts by regulating the affinity of DSL-Notch interaction.

3. Endocytosis in notch signaling activation

An absolute requirement for endocytosis was an early finding in Notch studies. Notably, the *shibire* mutant (i.e. a temperature-sensitive mutant of the endocytic fission protein dynamin in flies [145]), results in a developmental phenotype with an excess of neural cells when raised to restrictive temperature, i.e. with defects which closely phenocopy the Notch mutant [146, 147]. This phenotype is in sharp contrast to other signaling pathways, which are not severely disrupted in the *shibire* mutant (e.g. wingless [148], although even this signaling is affected by endocytic defects [149-151], as for most other signaling pathways (reviewed in [152])). The staminal observation of genetic interaction between dynamin and Notch prompted investigation of the requirement for dynamin function (i.e. endocytosis) for Notch signaling during the segregation of sensory bristles of the fly [153]. Overexpression of activated Notch isoforms (either membrane tethered or soluble) suppresses the *shibire* phenotype, thus indicating that endocytosis main action is upstream of the signal transduction promoted by Notch activation [153]. Notably, when wild-type Notch has to be activated by its ligand Delta, dynamin is required in both signaling and receiving cells, as shown by mosaic analysis in which bristles along the border can be either wild-type or mutant [153].

Localization studies of Notch and Delta during fly development showed that both Notch and Notch ligands were in a dynamic equilibrium between a PM pool and an intracellular vesicle pool, with a transition to internalized pool upon interaction of adjacent cells [154]. Delta is detected at both the PM and in vesicles only at some stages of specific developmental systems, while it is mostly internalized in others, including all stages and cell types of retinal development [154-158]. Morphological analyses of Delta subcellular localization in this latter development system have clarified that most, if not all, Delta-containing vesicles have an endocytic origin [159, 160]: Delta is re-localized to the PM in the full endocytic mutants, *hook* and *shibire*, thus supporting the idea that Delta is initially transported to the cell surface, but then it is taken up very quickly and efficiently by endocytosis to be delivered to the endocytic compartment [159, 160]. These and other observations [106, 161] suggested that an endocytic event could precede Notch activation. Direct evidence in support of this hypothesis came from (i) antibody uptake assays in living *Drosophila* tissue and in mammalian cells, which showed that DSLs are rapidly and efficiently internalized upon antibody binding and clustering [162-164], (ii) transfection assays with endocytosis-defective DSLs, which provided direct evidence that Notch ligand internalization is required to activate Notch signaling [164, 165], and (iii) uptake assays of recombinant forms of DSL (Delta1-Fc chimeric protein) [166] and of Notch-1 (N1Fc chimeric protein) [167], which showed that even soluble fragments of Notch ligands and Notch receptors, upon clustering, could potently promote the internalization of their cognate partners. Under these conditions, the Delta1-Fc chimera was also proved to be able to fully activate canonical Notch signaling [166].

Thus, the paradigm in the field is that Notch signaling critically depends on DSL endocytosis for its activation and modulation. We shall now analyze the molecular machinery involved in this process. At present, as the reader will see, only partial information is available, with (many) puzzling and (some) conflicting results.

3.1. Notch ligand endocytosis

The molecular characterization of DSL endocytosis began to attract great interest when it was published an in-depth morphological analysis of the effect of the *shibire* mutation on the localization of Delta and Notch in retinal development. This seminal work prompted a wealth of new studies on the relationship between endocytosis and Notch signaling [165]. In this paper, it was shown with stunning morphological data that NECD detaches from the Notch receptor on the signal-receiving cells (i.e. on the latticework cells) and is internalized or, more specifically, *trans*-endocytosed in the signal-sending cells (i.e. in the cone cells) in a complex with Delta. Notably, this process tightly correlates with the Notch signaling activation that underlies the cell fate specification of the retinal latticework, thus supporting the idea that NECD *trans*-endocytosis is requested for Notch activation. As expected, Notch receptor dissociation and its *trans*-endocytosis were severely hampered in the *shibire* mutant (i.e. in a condition in which endocytosis is blocked at the fission reaction of the clathrin-coated pit from the PM - see above), as well as when endocytosis-defective mutants of Delta were expressed in cultured cells [165]. Furthermore, this trans-endocytic mechanism was also found to be active in another developmental system under strict Notch control, i.e. wing vein development, thus suggesting its universal use in Notch activation [165]. Besides further supporting a role of endocytosis in Notch activation, these data suggest a possible mechanism of how DSL endocytosis might control Notch signaling: the dissociation of NECD, which is triggered by DSL endocytosis, is the event that activates Notch by possibly giving access to its cleavable sites. In partial support of this hypothesis, previous studies indicated that Delta proteins lacking the intracellular domain (i.e. lacking the binding site for endocytic adaptors so that DSL endocytosis cannot occur) acted as dominant-negative proteins for Notch signaling in *Drosophila* and in vertebrates [168-170].

However, direct evidence that NECD shedding by endocytosis is required to trigger the Notch proteolytic cascade was lacking in the *Drosophila* analyses [165]. This issue was addressed by a follow-up study on mammalian cultured cells by another group [87]. In this latter paper, it was demonstrated that an NTMD construct (Fig.1) transfected in mammalian cells was constitutively active in a reporter assay for Notch activation, and that treatment with BB94 (a metalloprotease inhibitor) and/or DAPT (a γ-secretase inhibitor) reduced the level of this signaling. These data indicate that, when the NTMD fragment is generated, it will be constitutively processed by proteolytic cleavage to free up the NICD. These findings support a two-step model in which (1st step) ligand endocytosis non-enzymatically dissociates and internalizes NECD in ligand cells, and then (2nd step) the membrane-bound NTMD undergoes constitutive cleavage by ADAM metalloproteases to produce NEXT, followed by γ-secretase cleavage to produce NICD (see Fig.1 for fragments description).

3.1.1. Mechanistic models

Following the staminal observation by Parks et al. [165], many laboratories investigated the machinery and the types of endocytosis that are at the basis of DSL-mediated Notch activation. A large collection of data has been produced which points at two distinct, but not mutually exclusive, models (see Fig.2): (i) according to the "pulling force" model (that derives directly from the observation of NECD shedding [165]), DSL internalization can exert a mechanical stretching, or detaching, action on the NECD to unmask the cleavage sites (especially S2) of Notch receptors; (ii) alternatively ("Notch ligand/DSL trafficking model"), or in combination with the previous model, an inactive DSL is activated either by trafficking through a recycling compartment ("ligand maturation" or "recycling" model), or by transcytosis to a membrane domain where interaction with Notch receptor occurs with increased frequency ("highly polarized cells" model).

It is plausible that, if (and/or when) the two models combine, they will act sequentially: DSL activation by intracellular trafficking should precede the mechanical shedding of the Notch receptor by DSL endocytosis, making the two mechanistic models not just compatible, but even synergistic.

The initial hint for the existence of a trafficking event that could activate the Notch ligand in order to make it competent for Notch activation came from the analysis of the fly mutant *liquid facet* (*lqf*), whose gene encodes the *Drosophila* epsin ([171] and see later). In this work, epsin was found to be implicated in a subset of DSL endocytic events which were able to activate Notch, while the bulk of DSL endocytosis (i.e. the constitutive endocytosis of Delta) was neither related to epsin function nor to Notch signaling [171]. Rescue of epsin absence was achieved by expressing a chimeric DSL, in which the intracellular tail was replaced by a short internalization signal of the LDL receptor, which was known to mediate the internalization and recycling of many proteins through the endosome [172]. Further studies in *Drosophila* and in mammalian cells have substantiated this initial observation of the existence of a possible trafficking step for the maturation of Notch ligands: (i) a defect in Delta trafficking through the recycling endosome was proposed to cause the aberrant cell fate transformation in sec15 mutant sensory lineages (see next paragraph for links of this developmental pathway to Notch signaling) [173]; (ii) expression of a dominant negative Rab11 (a small GTPase which regulates trafficking from the recycling endosome to the PM) was associated with DSL accumulation in endosomes and Notch signaling failure in a mammalian co-culture system [174]; (iii) an ubiquitylation-defective mutant of Dll1 can be efficiently endocytosed, but in contrast to the wild-type isoform is unable to recycle back to the cell surface and, possibly as a consequence of this trafficking defect, to efficiently bind Notch1 in a mammalian cell system [175].

An important question on the "ligand maturation" model regards the nature of the DSL activation process. In the *lqf* paper [171], a proteolytic step for *lqf* was identified, which was absent in epsin mutants but present in wild type-cells; it was speculated that this processing could indeed be Delta's activation step. Other Authors have looked for DSL processing in another system in which Notch ligand trafficking is essential, i.e. the **s**ensory **o**rgan **p**recursors (SOP) system (see next paragraph), but they failed to detect any evidence of DSL pre-cleavage [176], thus leaving unsolved the question of which molecular action eventually makes DSLs competent for Notch activation.

Intracellular trafficking can also activate DSLs with another mechanism, i.e. by re-localizing DSL from a membrane domain where it cannot interact with Notch to a membrane domain where this interaction can efficiently occur. This "highly-polarized cell" model is supported by at least two key sets of experiments undertaken in the *Drosophila* SOP system. This system is related to the development of the sensory organs (i.e. the mechanosensory bristles) located along the cuticle of the adult *Drosophila*, and is critically dependent on Notch: during a program of three rounds of asymmetric cell division [177, 178], each division generates one daughter cell that assumes the signal-sending role and uses DSLs to activate Notch in its sibling, which acts as a signal-receiving cell [179]. In the signal-sending cells of SOP, Delta localizes both at the apical and at the basolateral membrane, while Notch accumulates apically [180]. By using a pulse chase antibody uptake assay coupled to confocal microscopy sectioning, it was demonstrated that the basolateral pool of Delta is continuously endocytosed and delivered to the apical PM, where the interaction with Notch is most likely to occur [180]. This observation was extended by the same Authors to highly-polarized mammalian cells: by using a compartmentalized antibody uptake assay, they showed that Dll1 is similarly internalized from the basolateral membrane of Madin-Darby canine kidney cells and then transcytosed to the apical plasma membrane where Notch1 accumulates

[180]. In a second set of experiments on SOP the function of two primary regulators of actin dynamics was explored, i.e. the Arp2/3 complex and the **W**iskott-**A**ldrich **s**yndrome **p**rotein (Wasp) (reviewed in [181-183]). It was found that Arp2/3 and WASp were responsible for the nucleation of a filamentous **a**ctin-**r**ich **s**tructure (termed ARS) underneath the PM of pIIb (signal-sending) and pIIa (signal-receiving) cells at two different locations: (i) the apical domain, where it is responsible for microvilli elongation, and (ii) the lateral cell-cell contacts between pIIa and pIIb cells following SOP division [176]. Besides giving rise to this specialized actin cytomatrix, Arp2/3 and WASp also regulated the trafficking of Delta along the ARS in quite a unique manner: Delta was internalized from that part of the apical cell surface where microvilli were not clustered, and travelled basally, where it was then re-localized apically to the microvilli-rich portion, exactly where the contact with Notch usually occurs [176]. By inactivating Arp2/3 function, the ARS architecture was perturbed while Delta was still internalized. However, Delta failed to be delivered to the apical microvillar portion of the PM, being stopped in the basal portion of pIIb cells. Collectively, these data support a fundamental function of actin cytoskeleton in Delta trafficking, which is requested in the SOP system in order to localize DSL where interaction with Notch can occur. Conversely, a role of actin in DSL internalization is not requested in this developmental organ system [176, 184], while it is essential in other systems, cells and organisms (reviewed in [185-188].

A universal requirement for DSL trafficking does not seem exist for all tissues or developmental systems. Rab11 function, which was found to be essential for Delta trafficking and activation in the SOP system, is not required for *Drosophila* eye development [189], nor for germinal cell signaling [190]. In both these Notch-dependent events, endocytosis was required and a specific need for epsin, but not recycling, was evident at least in eye development [189].

More recently, the emerging structural findings described in section 2.2.1 have steered and re-focused the attention on the "pulling force" model. Direct evidence supports this model. As already discussed, using classical techniques to study membrane trafficking events it was possible to demonstrate that DSL-mediated receptor dissociation precedes and permits the proteolytic activation of Notch both in flies and in mammalian cells [87, 165]. A better structural appreciation of this event was acquired by using a material science technique to study surface morphology at the atomic level. Atomic force microscopy, applied to protein (or other molecules) interactions, can quite precisely measure the force applied to make contact between two interacting surfaces (the contact force) and the force applied to detach them after contact (the detachment force). This technique was then adapted to measure cell-cell adhesion [191] and was used to characterize Notch interaction [192]. A specific setup was engineered to mount a single S2-Delta-expressing *Drosophila* cell on the "tip-less" cantilevers, while immobilizing a S2-Notch-expressing cell in a plate well, and adhesion forces derived from this cell-cell interaction were measured [192]. The results of this elegant experiment showed that (i) expression of full-length Notch is required to produce maximal adhesion force (in the order of ~14 nN, comparable to a cell-cell adhesion contact) and signaling with S2-Delta cells; (ii) upon contact, this considerable adhesion force is lost within minutes (~10min.), as a direct result of the proteolytic cleavage of Notch, then signaling starts to rise very quickly (this timing is compatible with the time-course of Notch signaling in lateral inhibition models where the signaling event is accomplished in less than 20 min [193, 194]); (iii) Ofut1 RNAi in S2-Notch cells abolish the detachment force from S2-Delta cells, consistent with a reduced binding capacity of unfucosylated Notch to Delta [120, 195] (iv) if pulling is applied to S2-Delta cells by cantilever retraction, the detachment force from the S2-Notch cell drops with an increased kinetics, suggesting that a stretching action accelerates proteolytic cleavage of Notch. Taken together, these observations support the notion that Notch proteolytic cleavage depends on the strength of Delta binding [192], thus providing key structural data to support the "pulling force model" of Notch activation.

Since Notch proteolysis proceeds constitutively after exposing the S2 cleavage site (either by shedding of the Notch ectodomain or through its stretching), this unmasking reaction has to be considered the true rate limiting step of Notch signaling activation [196]. What are the structural constraints that keep S2 in an inactive silent state, preventing unwanted activation before ligand interaction? How are these constraints lifted/eased/modified during Delta-mediated NECD pulling? As anticipated in section 2.1.1, experimental evidence points to the NRR region of NECD for this key inhibitory action of Notch cleavage. Receptors that lack EGF-like repeats cannot undergo constitutive proteolytic cleavage and are functionally inert [70, 94, 108, 197-199]; conversely, an NTMD construct undergoes constitutive cleavage to release NICD [87]. Taken together, these data indicate that the restraints on ligand-independent activation of Notch receptors reside in a region downstream of EGF-like repeats but upstream of NTMD. This region corresponds to the three LNR repeats plus the HD domain, i.e. the NRR (see Fig.1). Key evidence to support this idea came from the isolation of Notch gain-of-function or loss-of-function phenotypes directly related to the NRR. (i) Antibodies raised against the NRR region did not compete with ligand binding to the receptor, but strongly inhibited Notch activation [96, 200]. Notably, those inhibitory antibodies recognized a conformational epitope lying on a face where the first LNR repeat (LNR-A) approaches the β-sheets of the HD (HD2), supporting the idea that autoinhibition is due to the clamp of LNR1 and HD2 together (see later for structural considerations) [96]. Conversely, an anti-Notch-1 antibody that recognized a linear epitope in the LNR1 domain only was activating Notch signaling, possibly by inducing a conformational change of the LNR1 that opened the access to the S2 site [96]. (ii) Mutations in the NRR of Notch receptors produced gain-of-function phenotypes in various biological contexts, including invertebrate developments. An activating mutation of the *glp-1* Notch receptor in *C. Elegans* was located in the LNR1 of the molecule [201], while a bunch of activation mutations of the other Notch receptor in worms (i.e. *lin-12*) were found to be spread in both the LNR and HD region, with a preference for the latter domain [202]. (iii) A subset of patients, who develop T-cell acute lymphoblastic leukemia, have gain-of-function mutations of Notch-1, which clusters in two regions: the HD domain and the C-terminus (including the TAD and PEST domains). While the C-terminal mutations were found to increase NICD stability, the HD mutations increase Notch-1 proteolysis, as suggested by the blocking of their stimulatory activity by γ-secretase inhibitors [203].

Many structural data have been collected in recent years that have helped to clarify the mechanistic details of NRR function, in particular (i) the NRR role in protecting S2 from constitutive cleavage and (ii) the kinetics of S2 autoinhibition. At present, the crystal structure of the NRR of human Notch-1 [88], of human Notch-2 [98], and the co-crystal of inhibiting antibodies, together with their target NRR epitopes [204], have been solved at high resolution. The NRR of Notch-2, the first NRR analyzed, was seen to form a very compact structure with overall dimensions of 60Åx45Åx25Å [98]. The three irregularly folded LNRs wrap around the HD domain forming "a cauliflower-like shape, in which the LNRs 'florets' cover and protect the HD domain 'stem'" [98]. The two halves of the HD domain, (i.e. the HD-N before the S1 site and HD-C after this site, see Fig.1) form an intimately intertwined α/β sandwich containing three α-helices and five β-strands connected by several conserved loops [98]. The inner, concave face of the HD domain has hydrophobic residues pointing toward its center. The S2 site is on the $\beta5$-strand of the HD-C and it is actually buried in a small pocket that prevents protease accessibility; the pocket is formed by the hydrophobic residues of α-HD -C and of the LNR-AB linker. In particular, it is thought that a leucine residue (L1457) extends from the LNR-AB linker toward a critical valine residue (V1666) at the C-terminus of the S2, thus obliterating the access to the ADAM cleavage site [98]. The $\alpha3$-helix above the S2 site is stabilized by hydrophobic interactions with residues in the LNR-B and in the LNR-AB linker plus a conserved hydrogen bond from LNR-A [98]. Consistent with previous structural data, deletion of LNR-A, the LNR-AB linker and LNR-B makes a constitutively activated Notch-2 [98]. The Notch-1 NRR structure is similar, although not identical, to that of the human Notch-2 NRR, with the classic conformation of the LNR-AB linker providing a key leucine residue that packs tightly against the C-terminal valine of the S2 site. As for NRR2, the folding of the HD domain has a rather stiff structure that is stabilized by extensive interaction between helices and strands. These data confirm a common autoinhibition strategy that is implemented among Notch family members [95].

Additional and fundamental structural data on Notch NRR function and dynamics came from the field of Notch immunotherapy and from the application of unconventional structural techniques. In an effort to overcome problems generated by the clinical use of presenilin inhibitors to silence the Notch pathway (in particular, the lack of selectivity for this pathway with a consequent broad toxicity), phage display technology was used to generate highly specific antibodies that could selectively antagonize a single Notch paralog (i.e. able to distinguish between Notch-1 and Notch-2) [204]. A co-crystal of this interaction shows that inhibitory, anti-NRR1 Fab-fragments bridge the LNR and HD domains, thus locking the NRR in a clamped conformation, which makes the S2 site unreachable for metalloproteases [204]. Further key data for the understanding of NRR-dependent S2 activation came from the application of hydrogen exchange mass spectrometry, a technique that monitors the exchange of deuterium between the solvent and the backbone amides during conformational changes [205, 206]. More specifically, when a surface of a protein is exposed it is rapidly deuterated, while when it is masked the exchange of hydrogen for deuterium is slow, or it does not happen at all. This technique was used to monitor the accessibility of the S2 cleavage site in a condition which should mimic ligand-dependent Notch activation, i.e. by chelation of Ca2+, a condition which causes the dissociation of the Notch receptor and triggers its signaling [207] (although widely used, Ca2+ chelation cannot

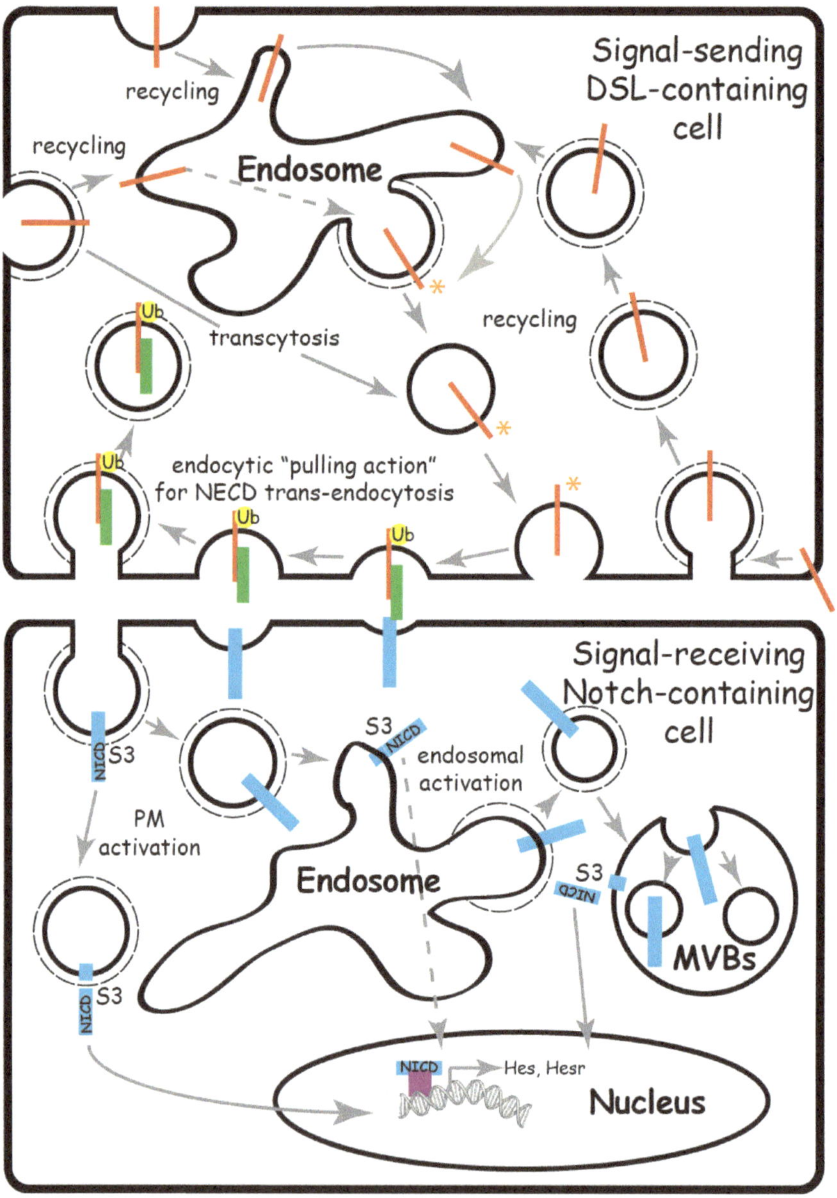

Figure 2. The different recycling and activation pathways of Notch ligands (DSLs) and of Notch receptors are outlined. In Red=Notch ligands/DSLs, in green=NECD, in blue=NTDM, in violet=the Notch transcriptional complex, PM=plasma membrane, S3=the Notch S3 site, which is cleaved by the γ-secretase complex to release NICD (where indicated, it represents the location of its presumable action); the orange (*) asterisk indicates a putative, activated state of DSL, after its recycling. For abbreviations, please refer to the text and to Fig.1.

be considered a surrogate of DSL action on the Notch receptor but obvious experimental constraints prevented the use of a more physiological condition). The results of these experiments showed that (i) upon Ca2+ chelation, LNR-A unfolding was the first event to occur, followed by the unfolding of LNR-B; (ii) after unfolding of the first two LNRs, the S2 site became accessible to the external environment, thus confirming previous results with deletion mutants in which removal of the LNR-A and LNR-B regions was sufficient to obtain a constitutively activated receptor [98]; (iii) Ca2+ is fundamental in stabilizing the secondary structure of LNR repeats [98]; (iv) HD-N and HD-C do not separate when S2 is exposed, and the HD domain maintains its folding for a very long time after Ca2+ chelation (i.e. well beyond the proteolytic cleavage of the receptor is terminated). This latter observation may indicate that ectodomain shedding is not an absolute prerequisite for the activation of the Notch proteolytic cascade [97]. To summarize, these structural data suggest that LNR-A and –B repeats are the fundamental gatekeepers of Notch activation as they control access to the Notch S2 cleavage site. Interestingly, in a recent paper, topology-based coarse-grained and physics-based atomistic molecular dynamics simulations were used to predict the conformational changes that occurred in the NRR by intrinsic and force-induced mechanisms [208]. These computer simulations showed that LNR unfolding is not sufficient to unmask the S2 site, but the continuous application of an external stretching/pulling force is needed to unfold the HD domain and, in particular, its β-5 strand [208]. Notably, the extension force required to unfold the β5 strand should be much lower than the force needed for heterodimer dissociation [208], suggesting that dissociation of Notch receptor is not needed for its activation, since an intermediate state with exposed S2 site might persist for a significant period of time before global unfolding and heterodimer disassociation occur. These predictions provide new and unforeseen roles for HD in Notch activation that definitely need experimental support.

3.1.2. Specialized endocytic machinery

Genetic evidence in invertebrates and mammals points to ubiquitylation (also referred to as ubiquitination) as the master regulatory mechanism controlling the endocytosis implicated in Notch signaling activation (reviewed in [209-211]).

Ubiquitylation, i.e. the conjugation of ubiquitin to proteins, is a rather common post-translational modification that regulates protein stability, localization, and activity (reviewed in [9, 11, 212-215]). Ubiquitin is a small conserved protein, whose C-terminal glycine (Gly76) can be engaged in a covalent isopeptide bond with the ε-amino group of lysine residues in substrate proteins. Ubiquitin can serve as an acceptor to form a polyubiquitin chain via one of its seven lysine residues (K6, K11, K27, K29, K33, K48 and K63). A hierarchical set of three enzymes acts in a sequential process to operate ubiquitin modification: (i) ubiquitin-activating (E1), (ii) -conjugating (E2), and (iii) -ligating (E3) enzymes. The large numbers of the latter enzymes (of which, the best studies are the **R**eally **I**nteresting **N**ew **G**enes (RING)-type and **H**omologous to the **E**6-AP **C**arboxyl **T**erminus (HECT)-type E3s) provide specificity to this post-translational modification in determining which substrate proteins will be modified. Ubiquitin can be attached in different amounts which have an impact on protein stability, localization and activity: (i) as a single molecule (monoubiquitylation), (ii) as multiple monomers linked to different lysine residues of a protein (multiubiquitylation), and/or (iii) as chains of ubiquitin molecules of various lengths and linkages (polyubiquitylation). Based on the linkage type, polyubiquitylation can be homotypic or heterotypic: it is homotypic when the same lysine residue is used for the sequential conjugation of ubiquitin moieties, while it is heterotypic (or mixed-linkage) when different ubiquitin's lysines are used to add monomers to the growing chain [216]. Furthermore, polyubiquitin chains can be linear (usually when homotypically built) or ramified (when the heterotypical linkage is used). Ubiquitin moieties are recognized and non-covalently bound by specific modular elements, collectively called **u**biquitin-**b**inding **d**omains (UBDs), which are now classified in different families, according to their structural homology [217]. Ubiquitylation is requested for many cellular processes, including proteasomal targeting and degradation of proteins, cell division, apoptosis, immune response, cytoskeleton dynamics, DNA transcription and repair, signal transduction, quality control and, last but not least, membrane trafficking, of which endocytosis and endosomal sorting are the best characterized ubiquitin-regulated events (reviewed in [6, 11, 212, 214, 218-224]}. Ubiquitylation can be reversed by multiple deubiquitinating enzymes (DUBs), the study of which constitutes a fast growing field of research (reviewed in [225-230]).

The first hint that ubiquitylation might be a necessary step for Notch activation came from the correlation between two sets of data, obtained almost twenty years apart: (i) a mutation screening in *Drosophila* identified *neuralized* (*neur*) among various genes phenocopying Notch neurogenic defects [59], thus indicating genetic interaction between that protein and Notch signaling; (ii) *neur* was seen to encode for a RING-type ubiquitin E3 ligase [231, 232]}, whose mutations in the catalytic domain were not able to rescue neur mutant embryos, thus formally proving that ubiquitylation is essential for Notch signaling *in vivo* [232]. In the same year, Xenopus neuralized was seen to carry out the same functional and biochemical activities as for the fly homologue [233].

Since then, an impressive number of experiments has been carried out on Neuralized activity and action in invertebrates. Key advances can be summarized as follows. (i) The RING domain was found to be critically required for Delta endocytosis: as expected, when the mutant *neur* is expressed, Delta stays mainly on the PM but re-localizes to internal vesicles upon (over) expression of the wild-type gene [163, 231, 234, 235]. The activity of Neur was firmly localized in the Notch-sending/DSL-bearing cell following cell-transplantation experiments [234]. As a collateral observation, it was documented that fly cells overexpressing Neur had a reduced level of Delta due to increased proteasomal activity, secondary to massive polyubiquitylation [231, 234]. (ii) Two critical lysine residues for Neur-mediated ubiquitylation (K688, K742R) were identified in a screening of Delta mutants for aberrant subcellular trafficking (i.e. mutants with a stable PM localization) [236]. Similar results were seen in a study on Serrate to uncover motifs leading to its internalization: two highly conserved lysines (K1272, K1290) were identified which are conserved between *Drosophila* Serrate and mammalian Jagged, and whose mutation resulted in blockage of DSL endocytosis and Notch activation [237]. However, the sites and types of ubiquitylation for endogenous DSLs are not yet known [237]. (iii) Neuralized-binding motifs, independent of ubiquitylated lysines, were identified on Delta and Serrate in the form of an NXXN sequence conserved among species, [238, 239]. (iv) Since a function of Neur in *cis*-inhibition was supposed due to its activity in Delta degradation [231, 233], it was demonstrated that overexpression of Neur indeed causes *cis*-inhibition, but Neur activity is not requested for this fundamental function during development [237].

In mammals, two **neur**alized-**l**ike genes, Neurl1 and Neurl2, are present. Quite surprisingly, inactivation of these genes does not result in major phenotypic defects [240-242]. Only subtle defects were scored in Neurl1$^{-/-}$ mice: (i) male mutants are sterile due to a defect in the axonemal organization of spermatozoa that leads to immotile sperm [242]; (ii) female KOs are defective in the final stages of mammary gland maturation during pregnancy [242]; (iii) Neurl1$^{-/-}$ mice are hypersensitive to ethanol effects on motor coordination and exhibit a defect in olfactory discrimination [241]. Only these latter defects can be putatively connected to an impairment of some subtle (yet to be defined) function of Notch in mammalian neurons, but no classical Notch signaling defects are identifiable in these mutants. Clearly, a compensation by the remaining Neurl2 gene was suspected, but, surprisingly and unexpectedly, inactivation of both Neurl1 and 2 did not result in any overt Notch defect in mice [240].

Neurls are not the only Notch-ligand specific E3 ligases present in vertebrate genomes. Another family, named after its first member *mind bomb* (*mib*), was identified in Zebrafish in a screening for neurogenic phenotypes (in which several Notch signaling components were also isolated [164]). Mib encodes for another RING-type E3 ligase, whose loss-of-function mutants cause major Notch developmental defects in the *Danio R.* [164]. Mib and Neuralized show complementary functions: (i) as for Neur, Mib(s) act(s) in the signal-sending cell [164] by promoting endocytosis of various DSLs, including Xenopus Delta [243] and Zebrafish Delta [164]; (ii) two mib genes are present in *Drosophila* with tissue distribution that complements that of Neur: inactivation of Mib indeed caused Notch defects in flies, but only in those tissues in which Neur was not expressed, while in tissues in which both Neur and Mib(s) were expressed, Notch phenotypes arose only upon co-inactivation of all E3 ligases [162, 243-245]. (iii) Mib1 cannot rescue *Drosophila neur* mutants [162], and, conversely, Neur and Mib1 cannot compensate for *mib2* defects in myoblast fusion and muscle homeostasis [246], thus showing that Mib(s) and Neur probably have other functions besides the ubiquitylation of DSL substrates.

Inactivation of Mib1 in mice finally results in a pure Notch phenotype, which recapitulates the most severe mammalian mutants of this signaling pathway [240]. Surprisingly, triple Neurl1/Neurl2/Mib2 knockout mice do not show major phenotypic defects, suggesting that Mib1 is the only essential E3 ligase for Notch activation. In support of these genetic data, knockdown of mib1 expression by siRNA dramatically reduces Notch activation in mammalian co-culture experiments [247, 248].

Activation of DSL internalization by ubiquitin moieties requires UBDs recognition and functional binding. Genetic experiments in mammals and invertebrates point to epsin family members as the principal actors in linking endocytosis, ubiquitylation and Notch activation. Epsins are highly conserved genes with two homologues in yeast (Ent1, Ent2) [249], one in Drosophila (Lqf) [250], and three epsin genes (Epn1, 2 and 3) in mammals. While epsins 1 and 2 are expressed in all tissues [251, 252], epsin 3 is restricted to surface epithelia [253, 254]. Epsins have a characteristic, highly conserved, three domain structure: (i) a **E**psin **N**-**T**erminal **H**omology (ENTH)-domain for phosphoinositides binding, in particular PtdIns(4,5)P2 [255, 256]; (ii) a central region which interacts with clathrin and its adaptor AP2; (iii) a C-terminal domain with multiple NPF motifs for the recognition of **E**ps15-**h**omology (EH)-domain-containing proteins, including the endocytic adaptors Eps15(R) and intersectin1/2 [251, 257]; (iv) multiple **U**biquitin **I**nteracting **M**otifs (UIMs) between the ENTH domain and the central domain for mono/polyubiquitin binding and for epsin (mono)ubiquitylation [258]. Epsin was initially characterized to be at the center of a highly regulated network of ubiquitinating and deubiquitinating enzymes: (i) Drosophila epsin (*lqf*) is the substrate of *fat facets* (*faf*), a deubiquitinating enzyme whose mutation is embryonically lethal in the fly [250]; (ii) RPM1/Highwire/Hiw, an E3 ligase of the RING type, regulates synaptic morphology (in flies and nematodes [259]), where a *lqf* function was also demonstrated [260]; (iii) although Lqf is not a substrate for Highwire [259], Hiw and Fat facets interact genetically and act as mutually antagonistic regulators of presynaptic growth [261]; (iv) epsin in neurons undergoes cycles of multi(mono)ubiquitylation/deubiquitylation, that change epsin affinities for interactors [262]. Based on this interaction, epsins were classified as housekeeping **c**lathrin-**as**sociated **s**orting **p**roteins (CLASPs) with specificity for ubiquitylated cargos (e.g. the EGF receptor upon ligand binding [251, 258, 263-267]), with the additional function of promoting membrane curvature [256, 268]. Genetic studies in yeast and *Dictyostelium* have also shown an additional role of epsin hortologues in the actin dynamics, which correlates with the endocytic function [269, 270]. However, at least in yeast, endocytosis has different requirements, being actin- but not clathrin-dependent, as in multicellular organisms (reviewed in [187, 188]).

Genetic studies in invertebrates have shown that the only epsin gene present in these species is required for the activation of Notch signaling [171, 245, 271, 272], and that this function is closely related to DSL ubiquitylation [245]. Genetic experiments in mammals have confirmed those studies and firmly established the essential role of epsin1 and 2 in Notch activation in vertebrates [273]: (i) the absence of epsin1/2 expression during mouse development correlates with embryonic lethality at midgestation, with multiorgan defects highly reminiscent of the most severe Notch mutants; (ii) accordingly, expression of Notch primary target genes is severely reduced in epsin1/2 double knockout embryos. Surprisingly, housekeeping forms of clathrin-mediated endocytosis were not impaired in cells deriving from those embryos [273].

A very recent study has provided evidence that epsins might have a previously unforeseen role in membrane fission [274]. In particular, predictions based on biophysical models support the idea that amphipathic helices (as those present in the epsin ENTH domain) could create a higher energy state due to their limited insertion into the polar head region, but not into the hydrocarbon region of the PM. This accumulated energy, when released, will crucially favor the fission reaction. This hypothesis was carefully tested *in vitro* by cell-free vesiculation assays and some correlative morphological tests in cultured cells: results confirm the prediction, thus sustaining a role for epsins that could parallel to or substitute that of dynamin [274]. Furthermore, it was found that simultaneous depletion of epsin1/2/3 by knockdown experiments results in the impairment of all current paradigms of clathrin-dependent endocytosis, thus suggesting a general role of epsin in the core machinery of this endocytic pathway [274]. Interestingly, single epsin KDs, or any combination of two of them, have little effect on endocytosis [274].

Taken together, these experiments suggest that epsins are the best candidates to explain the molecular action of ubiquitylation in DSL endocytosis, although the machinery behind this function has still to be fully uncovered. Triple epsin knockout mice could be the key to shed light on this molecular network.

Regarding the types of endocytosis, most of the evidence cited in section 3.1 strongly supports a clathrin-dependent pathway for DSL uptake. However, in invertebrates and, more specifically, in their oogenesis, Delta endocytosis could occur in an AP-2- and clathrin-independent way, as assayed by Notch activation of surrounding follicular cells triggered by germline clones bearing mutations of clathrin and AP-2 adaptor subunits, but not dynamin [190]. In the same system, it was also analyzed the dependence of Notch activation on endosomal trafficking in signal-sending cells: germline clones mutant for small GTPases that critically regulate the endosomal compartment, including Rab5 and Rab11, normally activate Notch in follicular cells. Taken together, these data support the absolute requirement for dynamin in DSL uptake. Conversely, neither CME nor endosomal entry of DSLs are universally required for Notch activation [190] (and, see section 3.1.1).

3.2. Notch receptor endocytosis

As discussed at the beginning of section 3, a strict requirement for endocytosis in the signal-receiving cell is supported by *Drosophila* studies on the *shibire* mutation in the sensory bristle development [153]. However, the mechanistic and molecular information available for Notch receptor endocytosis is very poor (and, sometimes, contradictory) in comparison with the large amount of data available for DSL internalization and trafficking.

3.2.1. Notch receptor internalization and PM-emanating signals

Some recent results seem to question the requirement of Notch receptor internalization for the activation of its signaling. In mammalian HeLa cells, overexpression of a dominant negative form of dynamin (the K44A mutation) does not prevent the processing of a chimeric NEXT to generate the NICD, which then translocates to the nucleus and activates signaling [275]. Blockage of the internalization step increases γ-secretase-mediated Notch processing and downstream signaling, suggesting that Notch receptor endocytosis might tame the Notch signaling emanating from the PM, as observed for other signaling pathways (see section 1).

This puzzling result is supported by other observations both *in vivo* and *in vitro.* (i) Presenilin can cleave any single-pass transmembrane protein provided that its extracellular domain is sufficiently small (<300 amino acids) [276]. Such presenilin substrates can normally be processed in the initial rounds of neuroblast segregation in *shibire*^{TS} embryos, suggesting that presenilin-dependent cleavage is not inherently dependent on Notch receptor endocytosis (at least during the first few hours upon temperature shifting, while experiments in [153] were scored after more than 6 hours) [276]. (ii) Proteolysis at S3 does not occur at a unique site but at multiple sites of NEXT, both in HEK293T cells and in a cell-free system. NICD fragments showed different stability and, therefore, signaling intensity, according to the proteasome N-end rule, where N-terminal valine provided maximal stability and signaling. Notably, PM-derived NICDs contain preferentially N-val, i.e. the most stable NICD, while endosome-generated NICD showed the lowest stability [277].

In the same set of experiments on HeLa cells, the machinery responsible for Notch internalization was also partially characterized. It was found that Notch uptake is strictly dependent on clathrin, since it is suppressed by knockdown of this latter gene and of its adaptor AP-2, while it is attenuated in the absence of epsin1 [275]. Notably, epsin1 interaction with Notch was ubiquitin-dependent, and the HECT domain-containing E3 ligase Nedd4 was found to participate in that action [275]. In the *Drosophila* system, Nedd4 is a negative regulator of Notch signaling by targeting Notch and Deltex (see later) to endocytosis and degradation, possibly protecting unstimulated cells from sporadic activation of Notch signaling [278].

To summarize, these data suggest that, in specific cell systems, PM emanating signals (from Notch receptors) can be (down)regulated by endocytosis, which uses the same machinery of the Notch signal-sending cell, i.e. clathrin-mediated endocytosis triggered by ubiquitylation with a role of epsin in coat formation and membrane invagination (and perhaps fission). The suppressive action of endocytosis on Notch activation can have many functions, including the termination of Notch signaling and the cell-fate determination of the Notch signal-sending cell, as Numb function seems to suggest (see next section).

3.2.2. Notch receptor trafficking and endosomal-emanating signals

In elegant morphological experiments, Notch receptor localization, processing, and signaling output in subsequent steps of its endocytic route were monitored by analyzing imaginal discs in *Drosophila* bearing homozygous mutations for key endocytic factors [279]. In the *shibire* and *Rab5* mutations, Notch accumulated at or below the plasma membrane, respectively, with no signaling effect in either case as scored by activation of a transcriptional reporter of Notch signaling. These data confirm the role of dynamin in Notch activation (in sharp contrast with that the role of endocytosis reported in the previous section), but, more importantly, they identify a new membrane compartment that is required for Notch activation, i.e. the endosome, in which Rab5 regulates the entry of endocytic cargos (reviewed in [280]). As expected, wing discs that express a constitutively active Rab5 show strong up-regulation of signaling, but similar results were also obtained with overexpression of Hrs, which regulates entry into multivescicular bodies (MVBs). Taken together, these experiments suggest that the transport of Notch receptor to endosomes and to MVBs potently stimulates Notch signaling possibly making the endosome the preferential station for the full activation of Notch receptor.

In a search for factors that regulate Notch activation in endosomes, it was found that mutations of the vacuolar proton pump (V-ATPase) produce defects in the processing of the internalized Notch receptor and its signaling [281, 282]. These results, together with the observation that presenilin works optimally in an acidic environment such as that present in the endosome/lysosome [283], support the idea that endosomal sorting of Notch is required for best activation of its S3 cleavage. However, unrestricted access of Notch receptors to the endosome should be prevented, since the acidic pH could dissociate the NECD, thus triggering ligand-independent Notch activation [210].

Another somewhat newer protagonist in Notch activation from endosomes is Deltex, whose mutation results in a lethal phenotype when associated to a gene dosage defect of one of the DSLs or Notch. Deltex encodes for a highly conserved gene endowed with three domains (reviewed in [284]): (i) a N-terminal WWE domain which binds the ANK repeats of Notch, (ii) a central proline-rich region for the binding to yet unknown SH3 domain-containing proteins, and (iii) a C-terminal RING-domain which has the signature of an E3 ligase, yet formal evidence of a Deltex direct ubiquitylation of Notch is lacking [284]. All domains are necessary for Deltex function, whose action has been studied intensively in recent years.

Data support a Deltex action both in Notch internalization and activation. Evidence for these functions can be summarized as follows: (i) in the *Deltex*-null *Drosophila* mutant, Notch accumulates on the cell surface and in some unknown endosomal compartment, but failes to be efficiently incorporated into internalized vesicles from the PM and in transport vesicles from early endosomes to lysosomes [285]; (ii) Deltex overexpression promotes Notch accumulation in late endosomes, where its signaling activity is potently stimulated [286]; (iii) Deltex makes a functional complex with critical regulators of late endosome formation or maturation [287], i.e. AP-3 (which selects cargos for late endosomes and lysosomes [288]) and HOPS (which participates in late endosome maturation in lysosomes [289]). To summarize, Deltex regulates Notch activation by stably localizing Notch in the late endosomal compartment, thus avoiding its delivery to MVBs where signaling is suppressed (since internalization of Notch in MVBs would prevent NICD release in the cytosol, see Fig.2).

However, the positive or negative outcome of endosomal sorting on Notch activation depends on other regulatory factors that control or antagonize the action of Deltex. (i) A member of the Nedd4 family of E3 ligases, *Suppressor of Deltex (Su(dx)),* permits the exit of Notch from the late endosomal compartment to incorporation into MVBs, thus terminating Notch signaling by avoiding the cytoplasmic release of NICD [287]. It was hypothesized that this negative regulatory function might be favored by a direct ubiquitylation of Notch by Su(dx), which, however, has not yet been detected [284]. In contrast, other members of Nedd4 family promote Notch ubiquitylation and degradation both in Drosophila and mammals (see section 3.2.1). (ii) A binding partner of Deltex is Kurtz (Krz), the homologue of mammalian non-visual β-arrestins. This protein family is involved in the desensitization and endocytosis of G-coupled receptors [290], TGF β [291] and Frizzled 4 [292]. Deltex, Krz and Notch form a complex in endocytic vesicles [293]. *Krz* mutants show upregulation of Notch signaling without altering Deltex levels, thus suggesting that the trimeric complex is important in the ubiquitin/proteasome-mediated degradation of Notch and subsequent signaling termination [293]. (iii) A critical component of the ESCRT III complex, Shrub, also seems to be involved in Deltex-Krz dependent Notch degradation. Shrub affects Notch trafficking and induces Notch accumulation in MVBs by promoting its polyubiquitylation and antagonizing Deltex, that instead promotes monoubiquitylation in the absence of Shrub [294]. Those opposite ubiquitylation states modulate Notch ligand-independent activation, by regulating how the receptor is trafficked in the endocytic path: polyubiquitylation targets Notch to MVBs for degradation, whereas monoubiquitylation is associated to Notch activation by γ-secretase [294]. Taken together, these results strongly support a preferential activation of Notch receptor during its intracellular trafficking and, more specifically, after its delivery to the (late) endosomal compartment, a trafficking event that is critically controlled by Deltex, at least in invertebrates.

A key aspect of Notch signaling is the need to establish differential signaling between two cell populations, i.e. the signal-receiving cells in which Notch activation can be triggered and the signal-sending cells in which Notch activation is suppressed. In invertebrates, Notch expression at the cell surface of the signal-sending cell is dramatically downregulated in order to inactivate Notch signaling in this cell population. One way of obtaining this effect is to target the Notch receptor to endosomal degradation with a specialized machinery. During the first division of the SOP, a membrane-associated protein called Numb is asymmetrically partitioned in the pIIb cell, which is committed to become the Notch signal-sending cell [295]. Loss of *NUMB* function causes all SOP descendants to differentiate in outer support cells, i.e. in Notch signal-receiving cells. Conversely, ectopic Numb expression during SOP division results in overproduction of neuronal precursors, i.e. of cells with the Notch signal-sending phenotype [295]. The epistaticity between Numb and Notch is further supported by genetic data in which reduction of Notch function can partially suppress the phenotypes resulting from loss of Numb [296]. Experiments in mammalian cells have shown that Numb is an endocytic factor, which binds the α-adaptin subunit of the clathrin-adaptor AP-2 [297, 298] and, together with this adaptor, co-localizes with internalizing receptors in mammalian cells [297]. In the SOP system in flies, Numb asymmetrically segregates AP-2 in the pIIb cell, and mutant isoforms of α-adaptin that no longer bind Numb fail to asymmetrically partition and cause Numb-like defects in SOP division [299]. Direct evidence of Numb function in Notch internalization was recently seen in anti-Notch antibody uptake experiments in the SOP lacking Numb expression [300].

Since Numb can co-exist with Notch in some cell systems without antagonizing its function (as in the lateral inhibition of *Drosophila* neuroectoderm [300]), it is plausible that other factors come into play to force the functional interaction of Numb with Notch receptors. In particular, two proteins have been identified as critical Numb-Notch interactors: (i) the

HECT-domain E3 ubiquitin ligase Itch, which ubiquitylates Notch receptors [301, 302] and many other proteins involved in key signaling pathways in mammalian cells (for a review see [303]) and (ii) the four-pass transmembrane protein Sanpodo (Spdo) [304, 305]. Numb, cooperatively enhances Itch-dependent ubiquitylation of Notch-1 [302]. This action requires direct and simultaneous binding of Numb to the ANK repeats of the Notch receptor and to the WW1-2 domains of Itch [302]. Therefore, Numb acts as an adaptor, facilitating or stabilizing the interaction between Itch and its substrate and, therefore, its catalytic activity. Numb-dependent endosomal sorting of Notch-1 in C2C12 cells critically depends on Itch function, since Numb mutants that do not interact with Itch (or that cannot interact with endocytic proteins) fail to promote Notch-1 degradation. All together, these experiments support a scenario in which Itch-mediated ubiquitylation is used to re-route Notch receptors to the late endosome for degradation and signaling suppression [306].

Another key interactor of Numb is Spdo, which is expressed in flies in both the Notch ligand-bearing and Notch receptor-bearing cells, where it acts differentially: in neuroblast division, Spdo is required for the activation of the Notch receptor in the A cell (a cell with Notch-dependent fate) [305] while in the B (signal-sending) cells it stimulates the endocytic degradation of the Notch receptor, in concert with Numb [305]. Notably, Numb in pIIb (signal-sending) cells of SOP induces the endocytosis of Spdo in early and late (but not recycling) endosomal vesicles. As for Numb internalization, Spdo endocytosis requires α–adaptin both in SOP [307] and in the neuroblast divisions in the flies [298]. As a result of *SPDO* loss-of-function, SOP cell stem cells divide symmetrically into two pIIb (signal-sending) cells, confirming that Spdo is required for Notch activation [300, 307]. In the case of Spdo ectopic overexpression, pIIb cells are generated as a result of Numb/Spdo-induced downregulation of Notch from the PM [307, 308]. Hence, Spdo may either activate or inhibit Notch signaling, depending on the presence or absence of Numb, and both actions are related to endocytosis.

4. Conclusions

In 2013, it will be one hundred years since the first Notch gene was discovered. During this century, fundamental aspects of gene functioning have been uncovered, including the key molecular mechanisms involved in the normal and pathological activation of Notch signaling. What emerged is that endocytosis is the master regulator of Notch activation. This function is exerted by means of a specialized endocytic machinery, which acts differentially in the Notch signal-sending cell compared with the Notch signal-receiving cell.

In Notch signal-sending, genetic, cell biology, structural, and biophysical studies point to a mechanical action of the Notch ligand on its receptor, so that critical proteolytic sites are uncovered for constitutive activation. Although the molecular machinery has not been fully characterized, genetic evidence in vertebrates and invertebrates supports clathrin-mediated endocytosis of ubiquitylated DSLs, as being the key mechanism that exerts the pulling action on the Notch receptor. In some developmental and cell culture systems, trafficking of the Notch ligand by transcytosis is another crucial mechanism which exerts the fundamental action of locating the Notch ligands in PM domains, where the interaction with the Notch receptor occurs with the highest efficiency. Evidence in specialized developmental systems in invertebrates supports a third function of endocytosis in the Notch signal-receiving cell, where DSL trafficking through the recycling endosome may serve the purpose of making Notch ligands competent for interaction with Notch receptors. However, the molecular event that pre-activates the Notch ligand is unknown, and no evidence has been provided yet to support a similar request in mammalian cells.

In comparison with the Notch signal-sending cell, where an endocytosis requirement is well established and many molecular details of its action are known, very little information is available, especially in vertebrates, to help us understand the need for endocytosis in the Notch signal-receiving cell. Genetic and cell-biology studies suggest that Notch signaling preferentially spreads from the endosomal compartment, where the acidic environment favors the γ-secretase release of the Notch active fragment (i.e. the NICD). As in the signal-sending cell, ubiquitylation is requested for this process, and its modulation by a variety of factors either firmly localizes Notch in a membrane trafficking compartment for signal activation, or quickly moves it to lysosomes for signal suppression.

Although we are beginning to see the "the big picture", crucial mechanisms are still missing. Although incomplete, some of the available endocytosis-related information has already entered medical experimentation [309]. A clear example is γ-secretase inhibitors (GIS), whose action is exploited in many current clinical trials for T-ALL, breast carcinoma, colon cancer, medulloblastoma, glioblastoma, osteosarcoma, pancreatic cancer, small-cell lung carcinoma, and melanoma, just to cite some of these studies. Analyses of GIS have also been extended to basically all cell lines and animal models in which a function of Notch for tumor promotion, progression and spreading was not only proved, but merely supposed. However, GIS use in current medical practice is far from established since the molecules that have so far been tested are plagued by significant human toxicity involving gastrointestinal bleeding and immunosuppression, which is attributable to widespread suppression of Notch signaling in many tissues. As discussed throughout this review, Notch actually plays a key role in the homeostasis of a variety of adult tissues, and its suppression thus hampers the functionality of many organs and systems. More unconventional approaches of Notch-related therapy are based on raising inhibiting or activating antibodies that regulate the level of Notch signaling by interfering with the Notch ligand-Notch receptor interaction, and, consequently, by directly or indirectly affecting the endocytic regulation of Notch signaling. Some of these antibodies are already in the initial phases of clinical trials, and they promise to offer better selectivity in targeting specific Notch components, thus minimizing side effects.

Notch-targeting therapies have a wide potential spectrum of application besides cancer, which includes developmental, vascular, cardiac, and other diseases associated with Notch pathway malfunction, or where Notch function could be exploited profitably for their treatment. It is not difficult to envisage a future interest for a highly-specific "endocytic-based" therapeutic approach to Notch dysregulation.

Author details

Elisa Sala, Luca Ruggiero and Giuseppina Di Giacomo
Università Vita-Salute San Raffaele, Milano, Italy

Ottavio Cremona
Università Vita-Salute San Raffaele, Milano, Italy
Istituto Nazionale di Neuroscienze, Italy

Acknowledgement

We thank Michael John for critically editing this review and Lorenzo Cremona for the artwork. Our work is supported by grants to OC from Ministero dell'Istruzione dell'Università e della Ricerca Scientifica (MIUR - PRIN2008 grant), Fondazione Cariplo, Fondazione Telethon (grant GGP10099), and Associazione Italiana per la Ricerca sul Cancro.

5. References

[1] Mercer J, Schelhaas M, Helenius A (2010) Virus entry by endocytosis. Annu. Rev. Biochem. 79: 803-833.
[2] Margadant C, Monsuur HN, Norman JC, Sonnenberg A (2011) Mechanisms of integrin activation and trafficking. Curr. Opin. Cell Biol. 23: 607-614.
[3] Shilo BZ, Schejter ED (2011) Regulation of developmental intercellular signalling by intracellular trafficking. EMBO J. 30: 3516-3526.
[4] McMahon HT, Boucrot E (2011) Molecular mechanism and physiological functions of clathrin-mediated endocytosis. Nat. Rev. Mol. Cell Biol. 12: 517-533.
[5] Haucke V, Neher E, Sigrist SJ (2011) Protein scaffolds in the coupling of synaptic exocytosis and endocytosis. Nat. Rev. Neurosci. 12: 127-138.
[6] Jiang X, Chen ZJ (2012) The role of ubiquitylation in immune defence and pathogen evasion. Nat. Rev. Immunol. 12: 35-48.
[7] Fu W, Jiang Q, Zhang C (2011) Novel functions of endocytic player clathrin in mitosis. Cell Res. 21: 1655-1661.
[8] Dittman J, Ryan TA (2009) Molecular circuitry of endocytosis at nerve terminals. Annu. Rev. Cell Dev. Biol. 25: 133-160.
[9] Sigismund S, Confalonieri S, Ciliberto A, Polo S, Scita G, Di Fiore PP (2012) Endocytosis and signaling: cell logistics shape the eukaryotic cell plan. Physiol. Rev. 92: 273-366.
[10] Sorkin A, von Zastrow M (2009) Endocytosis and signalling: intertwining molecular networks. Nat. Rev. Mol. Cell Biol. 10: 609-622.
[11] Haglund K, Dikic I (2012) The role of ubiquitylation in receptor endocytosis and endosomal sorting. J. Cell Sci. 125: 265-275.
[12] Abella JV, Park M (2009) Breakdown of endocytosis in the oncogenic activation of receptor tyrosine kinases. Am J Physiol Endocrinol Metab. 296: E973-984.
[13] Bache KG, Slagsvold T, Stenmark H (2004) Defective downregulation of receptor tyrosine kinases in cancer. EMBO J. 23: 2707-2712.

[14] Haglund K, Rusten TE, Stenmark H (2007) Aberrant receptor signaling and trafficking as mechanisms in oncogenesis. Crit. Rev. Oncog. 13: 39-74.

[15] Baass PC, Di Guglielmo GM, Authier F, Posner BI, Bergeron JJ (1995) Compartmentalized signal transduction by receptor tyrosine kinases. Trends Cell Biol. 5: 465-470.

[16] Mosesson Y, Mills GB, Yarden Y (2008) Derailed endocytosis: an emerging feature of cancer. Nat. Rev. Cancer. 8: 835-850.

[17] Pitulescu ME, Adams RH (2010) Eph/ephrin molecules--a hub for signaling and endocytosis. Genes Dev. 24: 2480-2492.

[18] Haugh JM, Meyer T (2002) Active EGF receptors have limited access to PtdIns(4,5)P(2) in endosomes: implications for phospholipase C and PI 3-kinase signaling. J. Cell Sci. 115: 303-310.

[19] Howe CL, Mobley WC (2004) Signaling endosome hypothesis: A cellular mechanism for long distance communication. J. Neurobiol. 58: 207-216.

[20] Hupalowska A, Miaczynska M (2012) The new faces of endocytosis in signaling. Traffic. 13: 9-18.

[21] Wegner CS, Rodahl LM, Stenmark H (2011) ESCRT proteins and cell signalling. Traffic. 12: 1291-1297.

[22] Gonnord P, Blouin CM, Lamaze C (2012) Membrane trafficking and signaling: Two sides of the same coin. Semin. Cell Dev. Biol. 23: 154-164.

[23] Disanza A, Frittoli E, Palamidessi A, Scita G (2009) Endocytosis and spatial restriction of cell signaling. Mol. Oncol. 3: 280-296.

[24] Winckler B, Mellman I (2010) Trafficking guidance receptors. Cold Spring Harb Perspect Biol. 2: a001826.

[25] Wilde A, Beattie EC, Lem L, Riethof DA, Liu SH, Mobley WC, Soriano P, Brodsky FM (1999) EGF receptor signaling stimulates SRC kinase phosphorylation of clathrin, influencing clathrin redistribution and EGF uptake. Cell. 96: 677-687.

[26] Ferguson S, Raimondi A, Paradise S, Shen H, Mesaki K, Ferguson A, Destaing O, Ko G, Takasaki J, Cremona O, E OT, De Camilli P (2009) Coordinated actions of actin and BAR proteins upstream of dynamin at endocytic clathrin-coated pits. Dev. Cell. 17: 811-822.

[27] Romer W, Pontani LL, Sorre B, Rentero C, Berland L, Chambon V, Lamaze C, Bassereau P, Sykes C, Gaus K, Johannes L (2010) Actin dynamics drive membrane reorganization and scission in clathrin-independent endocytosis. Cell. 140: 540-553.

[28] Shupliakov O, Bloom O, Gustafsson JS, Kjaerulff O, Low P, Tomilin N, Pieribone VA, Greengard P, Brodin L (2002) Impaired recycling of synaptic vesicles after acute perturbation of the presynaptic actin cytoskeleton. Proc. Natl. Acad. Sci. U. S. A. 99: 14476-14481.

[29] Soldati T, Schliwa M (2006) Powering membrane traffic in endocytosis and recycling. Nat. Rev. Mol. Cell Biol. 7: 897-908.

[30] Puthenveedu MA, Lauffer B, Temkin P, Vistein R, Carlton P, Thorn K, Taunton J, Weiner OD, Parton RG, von Zastrow M (2010) Sequence-dependent sorting of recycling proteins by actin-stabilized endosomal microdomains. Cell. 143: 761-773.

[31] Doherty GJ, McMahon HT (2009) Mechanisms of endocytosis. Annu. Rev. Biochem. 78: 857-902.
[32] Sandvig K, Pust S, Skotland T, van Deurs B (2011) Clathrin-independent endocytosis: mechanisms and function. Curr. Opin. Cell Biol. 23: 413-420.
[33] Grant BD, Donaldson JG (2009) Pathways and mechanisms of endocytic recycling. Nat. Rev. Mol. Cell Biol. 10: 597-608.
[34] Traub LM (2009) Tickets to ride: selecting cargo for clathrin-regulated internalization. Nat. Rev. Mol. Cell Biol. 10: 583-596.
[35] Cheng Y, Boll W, Kirchhausen T, Harrison SC, Walz T (2007) Cryo-electron tomography of clathrin-coated vesicles: structural implications for coat assembly. J. Mol. Biol. 365: 892-899.
[36] Kozlov MM, McMahon HT, Chernomordik LV (2010) Protein-driven membrane stresses in fusion and fission. Trends Biochem. Sci. 35: 699-706.
[37] Ferguson SM, Raimondi A, Paradise S, Shen H, Mesaki K, Ferguson A, Destaing O, Ko G, Takasaki J, Cremona O, E OT, De Camilli P (2009) Coordinated actions of actin and BAR proteins upstream of dynamin at endocytic clathrin-coated pits. Dev. Cell. 17: 811-822.
[38] Boulant S, Kural C, Zeeh JC, Ubelmann F, Kirchhausen T (2011) Actin dynamics counteract membrane tension during clathrin-mediated endocytosis. Nat. Cell Biol. 13: 1124-1131.
[39] Ferguson SM, De Camilli P (2012) Dynamin, a membrane-remodelling GTPase. Nat. Rev. Mol. Cell Biol. 13: 75-88.
[40] Schmid SL, Frolov VA (2011) Dynamin: functional design of a membrane fission catalyst. Annu. Rev. Cell Dev. Biol. 27: 79-105.
[41] Campelo F, Malhotra V (2012) Membrane Fission: The Biogenesis of Transport Carriers. Annu. Rev. Biochem. DOI: 10.1146/annurev-biochem-051710-094912.
[42] Schlossman DM, Schmid SL, Braell WA, Rothman JE (1984) An enzyme that removes clathrin coats: purification of an uncoating ATPase. J. Cell Biol. 99: 723-733.
[43] Ungewickell E, Ungewickell H, Holstein SE, Lindner R, Prasad K, Barouch W, Martin B, Greene LE, Eisenberg E (1995) Role of auxilin in uncoating clathrin-coated vesicles. Nature. 378: 632-635.
[44] Harrison SC, Kirchhausen T (2010) Structural biology: Conservation in vesicle coats. Nature. 466: 1048-1049.
[45] Ungewickell EJ, Hinrichsen L (2007) Endocytosis: clathrin-mediated membrane budding. Curr. Opin. Cell Biol. 19: 417-425.
[46] Cremona O, Di Paolo G, Wenk MR, Luthi A, Kim WT, Takei K, Daniell L, Nemoto Y, Shears SB, Flavell RA, McCormick DA, De Camilli P (1999) Essential role of phosphoinositide metabolism in synaptic vesicle recycling. Cell. 99: 179-188.
[47] Kim WT, Chang S, Daniell L, Cremona O, Di Paolo G, De Camilli P (2002) Delayed reentry of recycling vesicles into the fusion-competent synaptic vesicle pool in synaptojanin 1 knockout mice. Proc. Natl. Acad. Sci. U. S. A. 99: 17143-17148.
[48] Howes MT, Mayor S, Parton RG (2010) Molecules, mechanisms, and cellular roles of clathrin-independent endocytosis. Curr. Opin. Cell Biol. 22: 519-527.

[49] Hansen CG, Nichols BJ (2009) Molecular mechanisms of clathrin-independent endocytosis. J. Cell Sci. 122: 1713-1721.

[50] Mayor S, Pagano RE (2007) Pathways of clathrin-independent endocytosis. Nat. Rev. Mol. Cell Biol. 8: 603-612.

[51] Rodal SK, Skretting G, Garred O, Vilhardt F, van Deurs B, Sandvig K (1999) Extraction of cholesterol with methyl-beta-cyclodextrin perturbs formation of clathrin-coated endocytic vesicles. Mol. Biol. Cell. 10: 961-974.

[52] Lingwood D, Simons K (2010) Lipid rafts as a membrane-organizing principle. Science. 327: 46-50.

[53] Husnjak K, Dikic I (2012) Ubiquitin-Binding Proteins: Decoders of Ubiquitin-Mediated cellular functions. Annu. Rev. Biochem. DOI: 10.1146/annurev-biochem-051810-094654.

[54] Rao Y, Haucke V (2011) Membrane shaping by the Bin/amphiphysin/Rvs (BAR) domain protein superfamily. Cell. Mol. Life Sci. 68: 3983-3993.

[55] Canton I, Battaglia G (2012) Endocytosis at the nanoscale. Chem. Soc. Rev. 41: 2718-2739.

[56] Dexter JS (1914) The analysis of a case of continuous variation in Drosophila by a study of its linkage relations. Am. Nat. 48: 712–758.

[57] Welshons WJ (1958) The analysis of a pseudoallelic recessive lethal system at the notch locus of Drosophila melanogaster. Cold Spring Harb. Symp. Quant. Biol. 23: 171-176.

[58] Poulson DF (1937) Chromosomal Deficiencies and the Embryonic Development of Drosophila Melanogaster. Proc. Natl. Acad. Sci. U. S. A. 23: 133-137.

[59] Lehmann R, Jimenez F, Dietrich U, Campos-Ortega JA (1983) On the phenotype and development of mutants of early neurogenesis in Drosophila melanogaster. Dev. Genes Evol. 192: 62-74.

[60] Artavanis-Tsakonas S, Rand MD, Lake RJ (1999) Notch signaling: cell fate control and signal integration in development. Science. 284: 770-776.

[61] Artavanis-Tsakonas S, Muskavitch MA (2010) Notch: the past, the present, and the future. Curr. Top. Dev. Biol. 92: 1-29.

[62] van Tetering G, van Diest P, Verlaan I, van der Wall E, Kopan R, Vooijs M (2009) Metalloprotease ADAM10 is required for Notch1 site 2 cleavage. J. Biol. Chem. 284: 31018-31027.

[63] Brou C, Logeat F, Gupta N, Bessia C, LeBail O, Doedens JR, Cumano A, Roux P, Black RA, Israel A (2000) A novel proteolytic cleavage involved in Notch signaling: the role of the disintegrin-metalloprotease TACE. Mol. Cell. 5: 207-216.

[64] Peschon JJ, Slack JL, Reddy P, Stocking KL, Sunnarborg SW, Lee DC, Russell WE, Castner BJ, Johnson RS, Fitzner JN, Boyce RW, Nelson N, Kozlosky CJ, Wolfson MF, Rauch CT, Cerretti DP, Paxton RJ, March CJ, Black RA (1998) An essential role for ectodomain shedding in mammalian development. Science. 282: 1281-1284.

[65] Hartmann D, de Strooper B, Serneels L, Craessaerts K, Herreman A, Annaert W, Umans L, Lubke T, Lena Illert A, von Figura K, Saftig P (2002) The disintegrin/metalloprotease ADAM 10 is essential for Notch signalling but not for alpha-secretase activity in fibroblasts. Hum. Mol. Genet. 11: 2615-2624.

[66] Lieber T, Kidd S, Young MW (2002) kuzbanian-mediated cleavage of Drosophila Notch. Genes Dev. 16: 209-221.
[67] Sotillos S, Roch F, Campuzano S (1997) The metalloprotease-disintegrin Kuzbanian participates in Notch activation during growth and patterning of Drosophila imaginal discs. Development. 124: 4769-4779.
[68] Zhang C, Tian L, Chi C, Wu X, Yang X, Han M, Xu T, Zhuang Y, Deng K (2010) Adam10 is essential for early embryonic cardiovascular development. Dev. Dyn. 239: 2594-2602.
[69] De Strooper B, Annaert W, Cupers P, Saftig P, Craessaerts K, Mumm JS, Schroeter EH, Schrijvers V, Wolfe MS, Ray WJ, Goate A, Kopan R (1999) A presenilin-1-dependent gamma-secretase-like protease mediates release of Notch intracellular domain. Nature. 398: 518-522.
[70] Schroeter EH, Kisslinger JA, Kopan R (1998) Notch-1 signalling requires ligand-induced proteolytic release of intracellular domain. Nature. 393: 382-386.
[71] Francis R, McGrath G, Zhang J, Ruddy DA, Sym M, Apfeld J, Nicoll M, Maxwell M, Hai B, Ellis MC, Parks AL, Xu W, Li J, Gurney M, Myers RL, Himes CS, Hiebsch R, Ruble C, Nye JS, Curtis D (2002) aph-1 and pen-2 are required for Notch pathway signaling, gamma-secretase cleavage of betaAPP, and presenilin protein accumulation. Dev. Cell. 3: 85-97.
[72] Morel V, Lecourtois M, Massiani O, Maier D, Preiss A, Schweisguth F (2001) Transcriptional repression by suppressor of hairless involves the binding of a hairless-dCtBP complex in Drosophila. Curr. Biol. 11: 789-792.
[73] Ghai V, Gaudet J (2008) The CSL transcription factor LAG-1 directly represses hlh-6 expression in C. elegans. Dev. Biol. 322: 334-344.
[74] Dou S, Zeng X, Cortes P, Erdjument-Bromage H, Tempst P, Honjo T, Vales LD (1994) The recombination signal sequence-binding protein RBP-2N functions as a transcriptional repressor. Mol. Cell. Biol. 14: 3310-3319.
[75] Andersson ER, Sandberg R, Lendahl U (2011) Notch signaling: simplicity in design, versatility in function. Development. 138: 3593-3612.
[76] Tanigaki K, Honjo T (2010) Two opposing roles of RBP-J in Notch signaling. Curr. Top. Dev. Biol. 92: 231-252.
[77] Kopan R, Ilagan MX (2009) The canonical Notch signaling pathway: unfolding the activation mechanism. Cell. 137: 216-233.
[78] Borggrefe T, Oswald F (2009) The Notch signaling pathway: transcriptional regulation at Notch target genes. Cell. Mol. Life Sci. 66: 1631-1646.
[79] Hansson EM, Lendahl U, Chapman G (2004) Notch signaling in development and disease. Semin. Cancer Biol. 14: 320-328.
[80] Chillakuri CR, Sheppard D, Lea SM, Handford PA (2012) Notch receptor-ligand binding and activation: Insights from molecular studies. Semin. Cell Dev. Biol DOI: 10.1016/j.semcdb.2012.01.009.
[81] van Tetering G, Vooijs M (2011) Proteolytic cleavage of Notch: "HIT and RUN". Curr. Mol. Med. 11: 255-269.
[82] Yochem J, Greenwald I (1989) glp-1 and lin-12, genes implicated in distinct cell-cell interactions in C. elegans, encode similar transmembrane proteins. Cell. 58: 553-563.

[83] Fitzgerald K, Wilkinson HA, Greenwald I (1993) glp-1 can substitute for lin-12 in specifying cell fate decisions in Caenorhabditis elegans. Development. 119: 1019-1027.
[84] Blaumueller CM, Qi H, Zagouras P, Artavanis-Tsakonas S (1997) Intracellular cleavage of Notch leads to a heterodimeric receptor on the plasma membrane. Cell. 90: 281-291.
[85] Logeat F, Bessia C, Brou C, LeBail O, Jarriault S, Seidah NG, Israel A (1998) The Notch1 receptor is cleaved constitutively by a furin-like convertase. Proc. Natl. Acad. Sci. U. S. A. 95: 8108-8112.
[86] Bush G, diSibio G, Miyamoto A, Denault JB, Leduc R, Weinmaster G (2001) Ligand-induced signaling in the absence of furin processing of Notch1. Dev. Biol. 229: 494-502.
[87] Nichols JT, Miyamoto A, Olsen SL, D'Souza B, Yao C, Weinmaster G (2007) DSL ligand endocytosis physically dissociates Notch1 heterodimers before activating proteolysis can occur. J. Cell Biol. 176: 445-458.
[88] Gordon WR, Vardar-Ulu D, L'Heureux S, Ashworth T, Malecki MJ, Sanchez-Irizarry C, McArthur DG, Histen G, Mitchell JL, Aster JC, Blacklow SC (2009) Effects of S1 cleavage on the structure, surface export, and signaling activity of human Notch1 and Notch2. PLoS ONE. 4: e6613.
[89] Kido H, Okumura Y, Takahashi E, Pan HY, Wang S, Yao D, Yao M, Chida J, Yano M (2012) Role of host cellular proteases in the pathogenesis of influenza and influenza-induced multiple organ failure. Biochim. Biophys. Acta. 1824: 186-194.
[90] Day PM, Schiller JT (2009) The role of furin in papillomavirus infection. Future Microbiol. 4: 1255-1262.
[91] Moulard M, Hallenberger S, Garten W, Klenk HD (1999) Processing and routage of HIV glycoproteins by furin to the cell surface. Virus Res. 60: 55-65.
[92] Bolt G, Pedersen IR (1998) The role of subtilisin-like proprotein convertases for cleavage of the measles virus fusion glycoprotein in different cell types. Virology. 252: 387-398.
[93] Harrison SC (2005) Mechanism of membrane fusion by viral envelope proteins. Adv. Virus Res. 64: 231-261.
[94] Sanchez-Irizarry C, Carpenter AC, Weng AP, Pear WS, Aster JC, Blacklow SC (2004) Notch subunit heterodimerization and prevention of ligand-independent proteolytic activation depend, respectively, on a novel domain and the LNR repeats. Mol. Cell. Biol. 24: 9265-9273.
[95] Gordon WR, Roy M, Vardar-Ulu D, Garfinkel M, Mansour MR, Aster JC, Blacklow SC (2009) Structure of the Notch1-negative regulatory region: implications for normal activation and pathogenic signaling in T-ALL. Blood. 113: 4381-4390.
[96] Li K, Li Y, Wu W, Gordon WR, Chang DW, Lu M, Scoggin S, Fu T, Vien L, Histen G, Zheng J, Martin-Hollister R, Duensing T, Singh S, Blacklow SC, Yao Z, Aster JC, Zhou BB (2008) Modulation of Notch signaling by antibodies specific for the extracellular negative regulatory region of NOTCH3. J. Biol. Chem. 283: 8046-8054.
[97] Tiyanont K, Wales TE, Aste-Amezaga M, Aster JC, Engen JR, Blacklow SC (2011) Evidence for increased exposure of the Notch1 metalloprotease cleavage site upon conversion to an activated conformation. Structure. 19: 546-554.
[98] Gordon WR, Vardar-Ulu D, Histen G, Sanchez-Irizarry C, Aster JC, Blacklow SC (2007) Structural basis for autoinhibition of Notch. Nat. Struct. Mol. Biol. 14: 295-300.

[99] Spinner NB, Colliton RP, Crosnier C, Krantz ID, Hadchouel M, Meunier-Rotival M (2001) Jagged1 mutations in alagille syndrome. Hum. Mutat. 17: 18-33.
[100] Ropke A, Kujat A, Graber M, Giannakudis J, Hansmann I (2003) Identification of 36 novel Jagged1 (JAG1) mutations in patients with Alagille syndrome. Hum. Mutat. 21: 100.
[101] Mizuhara E, Nakatani T, Minaki Y, Sakamoto Y, Ono Y, Takai Y (2005) MAGI1 recruits Dll1 to cadherin-based adherens junctions and stabilizes it on the cell surface. J. Biol. Chem. 280: 26499-26507.
[102] Six EM, Ndiaye D, Sauer G, Laabi Y, Athman R, Cumano A, Brou C, Israel A, Logeat F (2004) The notch ligand Delta1 recruits Dlg1 at cell-cell contacts and regulates cell migration. J. Biol. Chem. 279: 55818-55826.
[103] Popovic M, Bella J, Zlatev V, Hodnik V, Anderluh G, Barlow PN, Pintar A, Pongor S (2011) The interaction of Jagged-1 cytoplasmic tail with afadin PDZ domain is local, folding-independent, and tuned by phosphorylation. J. Mol. Recognit. 24: 245-253.
[104] LaVoie MJ, Selkoe DJ (2003) The Notch ligands, Jagged and Delta, are sequentially processed by alpha-secretase and presenilin/gamma-secretase and release signaling fragments. J. Biol. Chem. 278: 34427-34437.
[105] Parr-Sturgess CA, Rushton DJ, Parkin ET (2010) Ectodomain shedding of the Notch ligand Jagged1 is mediated by ADAM17, but is not a lipid-raft-associated event. Biochem. J. 432: 283-294.
[106] Fehon RG, Kooh PJ, Rebay I, Regan CL, Xu T, Muskavitch MA, Artavanis-Tsakonas S (1990) Molecular interactions between the protein products of the neurogenic loci Notch and Delta, two EGF-homologous genes in Drosophila. Cell. 61: 523-534.
[107] Rebay I, Fleming RJ, Fehon RG, Cherbas L, Cherbas P, Artavanis-Tsakonas S (1991) Specific EGF repeats of Notch mediate interactions with Delta and Serrate: implications for Notch as a multifunctional receptor. Cell. 67: 687-699.
[108] Rebay I, Fehon RG, Artavanis-Tsakonas S (1993) Specific truncations of Drosophila Notch define dominant activated and dominant negative forms of the receptor. Cell. 74: 319-329.
[109] Xu A, Lei L, Irvine KD (2005) Regions of Drosophila Notch that contribute to ligand binding and the modulatory influence of Fringe. J. Biol. Chem. 280: 30158-30165.
[110] Hambleton S, Valeyev NV, Muranyi A, Knott V, Werner JM, McMichael AJ, Handford PA, Downing AK (2004) Structural and functional properties of the human notch-1 ligand binding region. Structure. 12: 2173-2183.
[111] Shimizu K, Chiba S, Kumano K, Hosoya N, Takahashi T, Kanda Y, Hamada Y, Yazaki Y, Hirai H (1999) Mouse jagged1 physically interacts with notch2 and other notch receptors. Assessment by quantitative methods. J. Biol. Chem. 274: 32961-32969.
[112] Cordle J, Johnson S, Tay JZ, Roversi P, Wilkin MB, de Madrid BH, Shimizu H, Jensen S, Whiteman P, Jin B, Redfield C, Baron M, Lea SM, Handford PA (2008) A conserved face of the Jagged/Serrate DSL domain is involved in Notch trans-activation and cis-inhibition. Nat. Struct. Mol. Biol. 15: 849-857.

[113] Cordle J, Redfieldz C, Stacey M, van der Merwe PA, Willis AC, Champion BR, Hambleton S, Handford PA (2008) Localization of the delta-like-1-binding site in human Notch-1 and its modulation by calcium affinity. J. Biol. Chem. 283: 11785-11793.

[114] Wang Y, Spellman MW (1998) Purification and characterization of a GDP-fucose:polypeptide fucosyltransferase from Chinese hamster ovary cells. J. Biol. Chem. 273: 8112-8118.

[115] Okajima T, Irvine KD (2002) Regulation of notch signaling by o-linked fucose. Cell. 111: 893-904.

[116] Moloney DJ, Panin VM, Johnston SH, Chen J, Shao L, Wilson R, Wang Y, Stanley P, Irvine KD, Haltiwanger RS, Vogt TF (2000) Fringe is a glycosyltransferase that modifies Notch. Nature. 406: 369-375.

[117] Sasamura T, Sasaki N, Miyashita F, Nakao S, Ishikawa HO, Ito M, Kitagawa M, Harigaya K, Spana E, Bilder D, Perrimon N, Matsuno K (2003) neurotic, a novel maternal neurogenic gene, encodes an O-fucosyltransferase that is essential for Notch-Delta interactions. Development. 130: 4785-4795.

[118] Shi S, Stanley P (2003) Protein O-fucosyltransferase 1 is an essential component of Notch signaling pathways. Proc. Natl. Acad. Sci. U. S. A. 100: 5234-5239.

[119] Yao D, Huang Y, Huang X, Wang W, Yan Q, Wei L, Xin W, Gerson S, Stanley P, Lowe JB, Zhou L (2011) Protein O-fucosyltransferase 1 (Pofut1) regulates lymphoid and myeloid homeostasis through modulation of Notch receptor ligand interactions. Blood. 117: 5652-5662.

[120] Ge C, Stanley P (2008) The O-fucose glycan in the ligand-binding domain of Notch1 regulates embryogenesis and T cell development. Proc. Natl. Acad. Sci. U. S. A. 105: 1539-1544.

[121] Okajima T, Xu A, Lei L, Irvine KD (2005) Chaperone activity of protein O-fucosyltransferase 1 promotes notch receptor folding. Science. 307: 1599-1603.

[122] Becker DJ, Lowe JB (2003) Fucose: biosynthesis and biological function in mammals. Glycobiology. 13: 41R-53R.

[123] Glavic A, Lopez-Varea A, de Celis JF (2011) The balance between GMD and OFUT1 regulates Notch signaling pathway activity by modulating Notch stability. Biol. Res. 44: 25-34.

[124] Stahl M, Uemura K, Ge C, Shi S, Tashima Y, Stanley P (2008) Roles of Pofut1 and O-fucose in mammalian Notch signaling. J. Biol. Chem. 283: 13638-13651.

[125] Vodovar N, Schweisguth F (2008) Functions of O-fucosyltransferase in Notch trafficking and signaling: towards the end of a controversy? J. Biol. 7: 7.

[126] Okajima T, Reddy B, Matsuda T, Irvine KD (2008) Contributions of chaperone and glycosyltransferase activities of O-fucosyltransferase 1 to Notch signaling. BMC Biol. 6: 1.

[127] Lowe JB (2005) Cell biology. Does Notch take the sweet road to success? Science. 307: 1570-1572.

[128] Irvine KD, Wieschaus E (1994) fringe, a Boundary-specific signaling molecule, mediates interactions between dorsal and ventral cells during Drosophila wing development. Cell. 79: 595-606.

[129] Johnston SH, Rauskolb C, Wilson R, Prabhakaran B, Irvine KD, Vogt TF (1997) A family of mammalian Fringe genes implicated in boundary determination and the Notch pathway. Development. 124: 2245-2254.

[130] Tien AC, Rajan A, Bellen HJ (2009) A Notch updated. J. Cell Biol. 184: 621-629.

[131] Rana NA, Haltiwanger RS (2011) Fringe benefits: functional and structural impacts of O-glycosylation on the extracellular domain of Notch receptors. Curr. Opin. Struct. Biol. 21: 583-589.

[132] Rampal R, Li AS, Moloney DJ, Georgiou SA, Luther KB, Nita-Lazar A, Haltiwanger RS (2005) Lunatic fringe, manic fringe, and radical fringe recognize similar specificity determinants in O-fucosylated epidermal growth factor-like repeats. J. Biol. Chem. 280: 42454-42463.

[133] Visan I, Tan JB, Yuan JS, Harper JA, Koch U, Guidos CJ (2006) Regulation of T lymphopoiesis by Notch1 and Lunatic fringe-mediated competition for intrathymic niches. Nat. Immunol. 7: 634-643.

[134] Evrard YA, Lun Y, Aulehla A, Gan L, Johnson RL (1998) lunatic fringe is an essential mediator of somite segmentation and patterning. Nature. 394: 377-381.

[135] Zhang N, Gridley T (1998) Defects in somite formation in lunatic fringe-deficient mice. Nature. 394: 374-377.

[136] Zhang N, Norton CR, Gridley T (2002) Segmentation defects of Notch pathway mutants and absence of a synergistic phenotype in lunatic fringe/radical fringe double mutant mice. Genesis. 33: 21-28.

[137] Moran JL, Shifley ET, Levorse JM, Mani S, Ostmann K, Perez-Balaguer A, Walker DM, Vogt TF, Cole SE (2009) Manic fringe is not required for embryonic development, and fringe family members do not exhibit redundant functions in the axial skeleton, limb, or hindbrain. Dev. Dyn. 238: 1803-1812.

[138] Svensson P, Bergqvist I, Norlin S, Edlund H (2009) MFng is dispensable for mouse pancreas development and function. Mol. Cell. Biol. 29: 2129-2138.

[139] Xu A, Haines N, Dlugosz M, Rana NA, Takeuchi H, Haltiwanger RS, Irvine KD (2007) In vitro reconstitution of the modulation of Drosophila Notch-ligand binding by Fringe. J. Biol. Chem. 282: 35153-35162.

[140] Panin VM, Papayannopoulos V, Wilson R, Irvine KD (1997) Fringe modulates Notch-ligand interactions. Nature. 387: 908-912.

[141] Fleming RJ, Gu Y, Hukriede NA (1997) Serrate-mediated activation of Notch is specifically blocked by the product of the gene fringe in the dorsal compartment of the Drosophila wing imaginal disc. Development. 124: 2973-2981.

[142] Acar M, Jafar-Nejad H, Takeuchi H, Rajan A, Ibrani D, Rana NA, Pan H, Haltiwanger RS, Bellen HJ (2008) Rumi is a CAP10 domain glycosyltransferase that modifies Notch and is required for Notch signaling. Cell. 132: 247-258.

[143] Fernandez-Valdivia R, Takeuchi H, Samarghandi A, Lopez M, Leonardi J, Haltiwanger RS, Jafar-Nejad H (2011) Regulation of mammalian Notch signaling and embryonic development by the protein O-glucosyltransferase Rumi. Development. 138: 1925-1934.

[144] Irvine KD (2008) A notch sweeter. Cell. 132: 177-179.

[145] Kosaka T, Ikeda K (1983) Possible temperature-dependent blockage of synaptic vesicle recycling induced by a single gene mutation in Drosophila. J. Neurobiol. 14: 207-225.
[146] Poodry CA (1990) shibire, a neurogenic mutant of Drosophila. Dev. Biol. 138: 464-472.
[147] Ramaswami M, Rao S, van der Bliek A, Kelly RB, Krishnan KS (1993) Genetic studies on dynamin function in Drosophila. J. Neurogenet. 9: 73-87.
[148] Bejsovec A, Wieschaus E (1995) Signaling activities of the Drosophila wingless gene are separately mutable and appear to be transduced at the cell surface. Genetics. 139: 309-320.
[149] Seto ES, Bellen HJ (2006) Internalization is required for proper Wingless signaling in Drosophila melanogaster. J. Cell Biol. 173: 95-106.
[150] Rives AF, Rochlin KM, Wehrli M, Schwartz SL, DiNardo S (2006) Endocytic trafficking of Wingless and its receptors, Arrow and DFrizzled-2, in the Drosophila wing. Dev. Biol. 293: 268-283.
[151] Marois E, Mahmoud A, Eaton S (2006) The endocytic pathway and formation of the Wingless morphogen gradient. Development. 133: 307-317.
[152] Fischer JA, Eun SH, Doolan BT (2006) Endocytosis, endosome trafficking, and the regulation of Drosophila development. Annu. Rev. Cell Dev. Biol. 22: 181-206.
[153] Seugnet L, Simpson P, Haenlin M (1997) Requirement for dynamin during Notch signaling in Drosophila neurogenesis. Dev. Biol. 192: 585-598.
[154] Kooh PJ, Fehon RG, Muskavitch MA (1993) Implications of dynamic patterns of Delta and Notch expression for cellular interactions during Drosophila development. Development. 117: 493-507.
[155] Huppert SS, Jacobsen TL, Muskavitch MA (1997) Feedback regulation is central to Delta-Notch signalling required for Drosophila wing vein morphogenesis. Development. 124: 3283-3291.
[156] Parks AL, Huppert SS, Muskavitch MA (1997) The dynamics of neurogenic signalling underlying bristle development in Drosophila melanogaster. Mech. Dev. 63: 61-74.
[157] Klueg KM, Muskavitch MA (1999) Ligand-receptor interactions and trans-endocytosis of Delta, Serrate and Notch: members of the Notch signalling pathway in Drosophila. J. Cell Sci. 112:3289-3297.
[158] Klueg KM, Parody TR, Muskavitch MA (1998) Complex proteolytic processing acts on Delta, a transmembrane ligand for Notch, during Drosophila development. Mol. Biol. Cell. 9: 1709-1723.
[159] Parks AL, Turner FR, Muskavitch MA (1995) Relationships between complex Delta expression and the specification of retinal cell fates during Drosophila eye development. Mech. Dev. 50: 201-216.
[160] Kramer H, Phistry M (1996) Mutations in the Drosophila hook gene inhibit endocytosis of the boss transmembrane ligand into multivesicular bodies. J. Cell Biol. 133: 1205-1215.
[161] Fehon RG, Johansen K, Rebay I, Artavanis-Tsakonas S (1991) Complex cellular and subcellular regulation of notch expression during embryonic and imaginal development of Drosophila: implications for notch function. J. Cell Biol. 113: 657-669.

[162] Le Borgne R, Remaud S, Hamel S, Schweisguth F (2005) Two distinct E3 ubiquitin ligases have complementary functions in the regulation of delta and serrate signaling in Drosophila. PLoS Biol. 3: e96.

[163] Le Borgne R, Schweisguth F (2003) Unequal segregation of Neuralized biases Notch activation during asymmetric cell division. Dev. Cell. 5: 139-148.

[164] Itoh M, Kim CH, Palardy G, Oda T, Jiang YJ, Maust D, Yeo SY, Lorick K, Wright GJ, Ariza-McNaughton L, Weissman AM, Lewis J, Chandrasekharappa SC, Chitnis AB (2003) Mind bomb is a ubiquitin ligase that is essential for efficient activation of Notch signaling by Delta. Dev. Cell. 4: 67-82.

[165] Parks AL, Klueg KM, Stout JR, Muskavitch MA (2000) Ligand endocytosis drives receptor dissociation and activation in the Notch pathway. Development. 127: 1373-1385.

[166] Hicks C, Ladi E, Lindsell C, Hsieh JJ, Hayward SD, Collazo A, Weinmaster G (2002) A secreted Delta1-Fc fusion protein functions both as an activator and inhibitor of Notch1 signaling. J. Neurosci. Res. 68: 655-667.

[167] Ladi E, Nichols JT, Ge W, Miyamoto A, Yao C, Yang LT, Boulter J, Sun YE, Kintner C, Weinmaster G (2005) The divergent DSL ligand Dll3 does not activate Notch signaling but cell autonomously attenuates signaling induced by other DSL ligands. J. Cell Biol. 170: 983-992.

[168] Chitnis A, Henrique D, Lewis J, Ish-Horowicz D, Kintner C (1995) Primary neurogenesis in Xenopus embryos regulated by a homologue of the Drosophila neurogenic gene Delta. Nature. 375: 761-766.

[169] Jen WC, Wettstein D, Turner D, Chitnis A, Kintner C (1997) The Notch ligand, X-Delta-2, mediates segmentation of the paraxial mesoderm in Xenopus embryos. Development. 124: 1169-1178.

[170] Sun X, Artavanis-Tsakonas S (1996) The intracellular deletions of Delta and Serrate define dominant negative forms of the Drosophila Notch ligands. Development. 122: 2465-2474.

[171] Wang W, Struhl G (2004) Drosophila Epsin mediates a select endocytic pathway that DSL ligands must enter to activate Notch. Development. 131: 5367-5380.

[172] Chen WJ, Goldstein JL, Brown MS (1990) NPXY, a sequence often found in cytoplasmic tails, is required for coated pit-mediated internalization of the low density lipoprotein receptor. J. Biol. Chem. 265: 3116-3123.

[173] Jafar-Nejad H, Andrews HK, Acar M, Bayat V, Wirtz-Peitz F, Mehta SQ, Knoblich JA, Bellen HJ (2005) Sec15, a component of the exocyst, promotes notch signaling during the asymmetric division of Drosophila sensory organ precursors. Dev. Cell. 9: 351-363.

[174] Emery G, Hutterer A, Berdnik D, Mayer B, Wirtz-Peitz F, Gaitan MG, Knoblich JA (2005) Asymmetric Rab 11 endosomes regulate delta recycling and specify cell fate in the Drosophila nervous system. Cell. 122: 763-773.

[175] Heuss SF, Ndiaye-Lobry D, Six EM, Israel A, Logeat F (2008) The intracellular region of Notch ligands Dll1 and Dll3 regulates their trafficking and signaling activity. Proc. Natl. Acad. Sci. U. S. A. 105: 11212-11217.

[176] Rajan A, Tien AC, Haueter CM, Schulze KL, Bellen HJ (2009) The Arp2/3 complex and WASp are required for apical trafficking of Delta into microvilli during cell fate specification of sensory organ precursors. Nat. Cell Biol. 11: 815-824.
[177] Gho M, Bellaiche Y, Schweisguth F (1999) Revisiting the Drosophila microchaete lineage: a novel intrinsically asymmetric cell division generates a glial cell. Development. 126: 3573-3584.
[178] Reddy GV, Rodrigues V (1999) A glial cell arises from an additional division within the mechanosensory lineage during development of the microchaete on the Drosophila notum. Development. 126: 4617-4622.
[179] Zeng C, Younger-Shepherd S, Jan LY, Jan YN (1998) Delta and Serrate are redundant Notch ligands required for asymmetric cell divisions within the Drosophila sensory organ lineage. Genes Dev. 12: 1086-1091.
[180] Benhra N, Vignaux F, Dussert A, Schweisguth F, Le Borgne R (2010) Neuralized promotes basal to apical transcytosis of delta in epithelial cells. Mol. Biol. Cell. 21: 2078-2086.
[181] Firat-Karalar EN, Welch MD (2011) New mechanisms and functions of actin nucleation. Curr. Opin. Cell Biol. 23: 4-13.
[182] Rottner K, Hanisch J, Campellone KG (2010) WASH, WHAMM and JMY: regulation of Arp2/3 complex and beyond. Trends Cell Biol. 20: 650-661.
[183] Kurisu S, Takenawa T (2010) WASP and WAVE family proteins: friends or foes in cancer invasion? Cancer Sci. 101: 2093-2104.
[184] Schejter ED (2009) Delta traffic takes a sh-Arp turn. Nat. Cell Biol. 11: 791-793.
[185] Kaksonen M, Toret CP, Drubin DG (2006) Harnessing actin dynamics for clathrin-mediated endocytosis. Nat. Rev. Mol. Cell Biol. 7: 404-414.
[186] Anitei M, Hoflack B (2012) Bridging membrane and cytoskeleton dynamics in the secretory and endocytic pathways. Nat. Cell Biol. 14: 11-19.
[187] Conibear E (2010) Converging views of endocytosis in yeast and mammals. Curr. Opin. Cell Biol. 22: 513-518.
[188] Galletta BJ, Mooren OL, Cooper JA (2010) Actin dynamics and endocytosis in yeast and mammals. Curr. Opin. Biotechnol. 21: 604-610.
[189] Banks SM, Cho B, Eun SH, Lee JH, Windler SL, Xie X, Bilder D, Fischer JA (2011) The functions of auxilin and Rab11 in Drosophila suggest that the fundamental role of ligand endocytosis in notch signaling cells is not recycling. PLoS ONE. 6: e18259.
[190] Windler SL, Bilder D (2010) Endocytic internalization routes required for delta/notch signaling. Curr. Biol. 20: 538-543.
[191] Benoit M, Gabriel D, Gerisch G, Gaub HE (2000) Discrete interactions in cell adhesion measured by single-molecule force spectroscopy. Nat. Cell Biol. 2: 313-317.
[192] Ahimou F, Mok LP, Bardot B, Wesley C (2004) The adhesion force of Notch with Delta and the rate of Notch signaling. J. Cell Biol. 167: 1217-1229.
[193] Hoppe PE, Greenspan RJ (1990) The Notch locus of Drosophila is required in epidermal cells for epidermal development. Development. 109: 875-885.

[194] Baker NE, Yu SY (1998) The R8-photoreceptor equivalence group in Drosophila: fate choice precedes regulated Delta transcription and is independent of Notch gene dose. Mech. Dev. 74: 3-14.
[195] Okajima T, Xu A, Irvine KD (2003) Modulation of notch-ligand binding by protein O-fucosyltransferase 1 and fringe. J. Biol. Chem. 278: 42340-42345.
[196] Musse AA, Meloty-Kapella L, Weinmaster G (2012) Notch ligand endocytosis: Mechanistic basis of signaling activity. Semin. Cell Dev. Biol DOI: 10.1016/j.semcdb.2012.01.011.
[197] Struhl G, Adachi A (1998) Nuclear access and action of notch in vivo. Cell. 93: 649-660.
[198] Kopan R, Schroeter EH, Weintraub H, Nye JS (1996) Signal transduction by activated mNotch: importance of proteolytic processing and its regulation by the extracellular domain. Proc. Natl. Acad. Sci. U. S. A. 93: 1683-1688.
[199] Lieber T, Kidd S, Alcamo E, Corbin V, Young MW (1993) Antineurogenic phenotypes induced by truncated Notch proteins indicate a role in signal transduction and may point to a novel function for Notch in nuclei. Genes Dev. 7: 1949-1965.
[200] Aste-Amezaga M, Zhang N, Lineberger JE, Arnold BA, Toner TJ, Gu M, Huang L, Vitelli S, Vo KT, Haytko P, Zhao JZ, Baleydier F, L'Heureux S, Wang H, Gordon WR, Thoryk E, Andrawes MB, Tiyanont K, Stegmaier K, Roti G, Ross KN, Franlin LL, Wang F, Chastain M, Bett AJ, Audoly LP, Aster JC, Blacklow SC, Huber HE (2010) Characterization of Notch1 antibodies that inhibit signaling of both normal and mutated Notch1 receptors. PLoS ONE. 5: e9094.
[201] Berry LW, Westlund B, Schedl T (1997) Germ-line tumor formation caused by activation of glp-1, a Caenorhabditis elegans member of the Notch family of receptors. Development. 124: 925-936.
[202] Greenwald I, Seydoux G (1990) Analysis of gain-of-function mutations of the lin-12 gene of Caenorhabditis elegans. Nature. 346: 197-199.
[203] Weng AP, Ferrando AA, Lee W, Morris JPt, Silverman LB, Sanchez-Irizarry C, Blacklow SC, Look AT, Aster JC (2004) Activating mutations of NOTCH1 in human T cell acute lymphoblastic leukemia. Science. 306: 269-271.
[204] Wu Y, Cain-Hom C, Choy L, Hagenbeek TJ, de Leon GP, Chen Y, Finkle D, Venook R, Wu X, Ridgway J, Schahin-Reed D, Dow GJ, Shelton A, Stawicki S, Watts RJ, Zhang J, Choy R, Howard P, Kadyk L, Yan M, Zha J, Callahan CA, Hymowitz SG, Siebel CW (2010) Therapeutic antibody targeting of individual Notch receptors. Nature. 464: 1052-1057 .
[205] Wales TE, Engen JR (2006) Hydrogen exchange mass spectrometry for the analysis of protein dynamics. Mass Spectrom. Rev. 25: 158-170.
[206] Raphael K (2011) Deuterium reveals the dynamics of notch activation. Structure. 19: 443-444.
[207] Rand MD, Grimm LM, Artavanis-Tsakonas S, Patriub V, Blacklow SC, Sklar J, Aster JC (2000) Calcium depletion dissociates and activates heterodimeric notch receptors. Mol. Cell. Biol. 20: 1825-1835.

[208] Chen J, Zolkiewska A (2011) Force-induced unfolding simulations of the human Notch1 negative regulatory region: possible roles of the heterodimerization domain in mechanosensing. PLoS ONE. 6: e22837.
[209] Weinmaster G, Fischer JA (2011) Notch ligand ubiquitylation: what is it good for? Dev. Cell. 21: 134-144.
[210] Le Bras S, Loyer N, Le Borgne R (2011) The multiple facets of ubiquitination in the regulation of notch signaling pathway. Traffic. 12: 149-161.
[211] Baron M, Aslam H, Flasza M, Fostier M, Higgs JE, Mazaleyrat SL, Wilkin MB (2002) Multiple levels of Notch signal regulation (review). Mol. Membr. Biol. 19: 27-38.
[212] Macgurn JA, Hsu PC, Emr SD (2012) Ubiquitin and Membrane Protein Turnover: From Cradle to Grave. Annu. Rev. Biochem.
[213] Emmerich CH, Schmukle AC, Walczak H (2011) The emerging role of linear ubiquitination in cell signaling. Sci. Signal. 4: re5.
[214] Schaefer A, Nethe M, Hordijk PL (2012) Ubiquitin links to cytoskeletal dynamics, cell adhesion and migration. Biochem. J. 442: 13-25.
[215] Piper RC, Lehner PJ (2011) Endosomal transport via ubiquitination. Trends Cell Biol. 21: 647-655.
[216] Ikeda F, Dikic I (2008) Atypical ubiquitin chains: new molecular signals. 'Protein Modifications: Beyond the Usual Suspects' review series. EMBO Rep. 9: 536-542.
[217] Dikic I, Wakatsuki S, Walters KJ (2009) Ubiquitin-binding domains - from structures to functions. Nat. Rev. Mol. Cell Biol. 10: 659-671.
[218] Maupin-Furlow J (2012) Proteasomes and protein conjugation across domains of life. Nat. Rev. Microbiol. 10: 100-111.
[219] Mocciaro A, Rape M (2012) Emerging regulatory mechanisms in ubiquitin-dependent cell cycle control. J. Cell Sci. 125: 255-263.
[220] Vucic D, Dixit VM, Wertz IE (2011) Ubiquitylation in apoptosis: a post-translational modification at the edge of life and death. Nat. Rev. Mol. Cell Biol. 12: 439-452.
[221] Voutsadakis IA (2012) Ubiquitination and the Ubiquitin-Proteasome System as regulators of transcription and transcription factorsin epithelial mesenchymal transition of cancer. Tumour Biol DOI: 10.1007/s13277-012-0355-x.
[222] Wright DE, Wang CY, Kao CF (2012) Histone ubiquitylation and chromatin dynamics. Front. Biosci. 17: 1051-1078.
[223] Hammond-Martel I, Yu H, Affar el B (2012) Roles of ubiquitin signaling in transcription regulation. Cell. Signal. 24: 410-421.
[224] Benyair R, Ron E, Lederkremer GZ (2011) Protein quality control, retention, and degradation at the endoplasmic reticulum. Int Rev Cell Mol Biol. 292: 197-280.
[225] Burrows JF, Johnston JA (2012) Regulation of cellular responses by deubiquitinating enzymes: an update. Front. Biosci. 17: 1184-1200.
[226] Todi SV, Paulson HL (2011) Balancing act: deubiquitinating enzymes in the nervous system. Trends Neurosci DOI:10.1016/j.tins.2011.05.004.
[227] Ramakrishna S, Suresh B, Baek KH (2011) The role of deubiquitinating enzymes in apoptosis. Cell. Mol. Life Sci. 68: 15-26.

[228] Shabek N, Ciechanover A (2010) Degradation of ubiquitin: the fate of the cellular reaper. Cell Cycle. 9: 523-530.
[229] Komander D, Clague MJ, Urbe S (2009) Breaking the chains: structure and function of the deubiquitinases. Nat. Rev. Mol. Cell Biol. 10: 550-563.
[230] Reyes-Turcu FE, Ventii KH, Wilkinson KD (2009) Regulation and cellular roles of ubiquitin-specific deubiquitinating enzymes. Annu. Rev. Biochem. 78: 363-397.
[231] Lai EC, Deblandre GA, Kintner C, Rubin GM (2001) Drosophila neuralized is a ubiquitin ligase that promotes the internalization and degradation of delta. Dev. Cell. 1: 783-794.
[232] Yeh E, Dermer M, Commisso C, Zhou L, McGlade CJ, Boulianne GL (2001) Neuralized functions as an E3 ubiquitin ligase during Drosophila development. Curr. Biol. 11: 1675-1679.
[233] Deblandre GA, Lai EC, Kintner C (2001) Xenopus neuralized is a ubiquitin ligase that interacts with XDelta1 and regulates Notch signaling. Dev. Cell. 1: 795-806.
[234] Pavlopoulos E, Pitsouli C, Klueg KM, Muskavitch MA, Moschonas NK, Delidakis C (2001) neuralized Encodes a peripheral membrane protein involved in delta signaling and endocytosis. Dev. Cell. 1: 807-816.
[235] De Renzis S, Yu J, Zinzen R, Wieschaus E (2006) Dorsal-ventral pattern of Delta trafficking is established by a Snail-Tom-Neuralized pathway. Dev. Cell. 10: 257-264.
[236] Parks AL, Stout JR, Shepard SB, Klueg KM, Dos Santos AA, Parody TR, Vaskova M, Muskavitch MA (2006) Structure-function analysis of delta trafficking, receptor binding and signaling in Drosophila. Genetics. 174: 1947-1961.
[237] Glittenberg M, Pitsouli C, Garvey C, Delidakis C, Bray S (2006) Role of conserved intracellular motifs in Serrate signalling, cis-inhibition and endocytosis. EMBO J. 25: 4697-4706.
[238] Fontana JR, Posakony JW (2009) Both inhibition and activation of Notch signaling rely on a conserved Neuralized-binding motif in Bearded proteins and the Notch ligand Delta. Dev. Biol. 333: 373-385.
[239] Commisso C, Boulianne GL (2007) The NHR1 domain of Neuralized binds Delta and mediates Delta trafficking and Notch signaling. Mol. Biol. Cell. 18: 1-13.
[240] Koo BK, Yoon MJ, Yoon KJ, Im SK, Kim YY, Kim CH, Suh PG, Jan YN, Kong YY (2007) An obligatory role of mind bomb-1 in notch signaling of Mammalian development. PLoS ONE. 2: e1221.
[241] Ruan Y, Tecott L, Jiang MM, Jan LY, Jan YN (2001) Ethanol hypersensitivity and olfactory discrimination defect in mice lacking a homolog of Drosophila neuralized. Proc. Natl. Acad. Sci. U. S. A. 98: 9907-9912.
[242] Vollrath B, Pudney J, Asa S, Leder P, Fitzgerald K (2001) Isolation of a murine homologue of the Drosophila neuralized gene, a gene required for axonemal integrity in spermatozoa and terminal maturation of the mammary gland. Mol. Cell. Biol. 21: 7481-7494.
[243] Chen W, Casey Corliss D (2004) Three modules of zebrafish Mind bomb work cooperatively to promote Delta ubiquitination and endocytosis. Dev. Biol. 267: 361-373.

[244] Pitsouli C, Delidakis C (2005) The interplay between DSL proteins and ubiquitin ligases in Notch signaling. Development. 132: 4041-4050.
[245] Wang W, Struhl G (2005) Distinct roles for Mind bomb, Neuralized and Epsin in mediating DSL endocytosis and signaling in Drosophila. Development. 132: 2883-2894.
[246] Carrasco-Rando M, Ruiz-Gomez M (2008) Mind bomb 2, a founder myoblast-specific protein, regulates myoblast fusion and muscle stability. Development. 135: 849-857.
[247] Hansson EM, Lanner F, Das D, Mutvei A, Marklund U, Ericson J, Farnebo F, Stumm G, Stenmark H, Andersson ER, Lendahl U (2010) Control of Notch-ligand endocytosis by ligand-receptor interaction. J. Cell Sci. 123: 2931-2942.
[248] Yamamoto M, Morita R, Mizoguchi T, Matsuo H, Isoda M, Ishitani T, Chitnis AB, Matsumoto K, Crump JG, Hozumi K, Yonemura S, Kawakami K, Itoh M (2010) Mib-Jag1-Notch signalling regulates patterning and structural roles of the notochord by controlling cell-fate decisions. Development. 137: 2527-2537.
[249] Wendland B, Steece KE, Emr SD (1999) Yeast epsins contain an essential N-terminal ENTH domain, bind clathrin and are required for endocytosis. EMBO J. 18: 4383-4393.
[250] Cadavid AL, Ginzel A, Fischer JA (2000) The function of the Drosophila fat facets deubiquitinating enzyme in limiting photoreceptor cell number is intimately associated with endocytosis. Development. 127: 1727-1736.
[251] Chen H, Fre S, Slepnev VI, Capua MR, Takei K, Butler MH, Di Fiore PP, De Camilli P (1998) Epsin is an EH-domain-binding protein implicated in clathrin-mediated endocytosis. Nature. 394: 793-797.
[252] Rosenthal JA, Chen H, Slepnev VI, Pellegrini L, Salcini AE, Di Fiore PP, De Camilli P (1999) The epsins define a family of proteins that interact with components of the clathrin coat and contain a new protein module. J. Biol. Chem. 274: 33959-33965.
[253] Spradling KD, McDaniel AE, Lohi J, Pilcher BK (2001) Epsin 3 is a novel extracellular matrix-induced transcript specific to wounded epithelia. J. Biol. Chem. 276: 29257-29267.
[254] Ko G, Paradise S, Chen H, Graham M, Vecchi M, Bianchi F, Cremona O, Di Fiore PP, De Camilli P (2010) Selective high-level expression of epsin 3 in gastric parietal cells, where it is localized at endocytic sites of apical canaliculi. Proc. Natl. Acad. Sci. U. S. A. 107: 21511-21516.
[255] Itoh T, Koshiba S, Kigawa T, Kikuchi A, Yokoyama S, Takenawa T (2001) Role of the ENTH Domain in Phosphatidylinositol-4,5-Bisphosphate Binding and Endocytosis. Science. 291: 1047-1051.
[256] Ford MG, Mills IG, Peter BJ, Vallis Y, Praefcke GJ, Evans PR, McMahon HT (2002) Curvature of clathrin-coated pits driven by epsin. Nature. 419: 361-366.
[257] Chen H, Slepnev VI, Di Fiore PP, De Camilli P (1999) The interaction of epsin and Eps15 with the clathrin adaptor AP-2 is inhibited by mitotic phosphorylation and enhanced by stimulation-dependent dephosphorylation in nerve terminals. J. Biol. Chem. 274: 3257-3260.
[258] Polo S, Sigismund S, Faretta M, Guidi M, Capua MR, Bossi G, Chen H, De Camilli P, Di Fiore PP (2002) A single motif responsible for ubiquitin recognition and monoubiquitination in endocytic proteins. Nature. 416: 451-455.

[259] Chang Q, Balice-Gordon RJ (2000) Highwire, rpm-1, and futsch: balancing synaptic growth and stability. Neuron. 26: 287-290.

[260] Bao H, Reist NE, Zhang B (2008) The Drosophila epsin 1 is required for ubiquitin-dependent synaptic growth and function but not for synaptic vesicle recycling. Traffic . 9: 2190-2205.

[261] DiAntonio A, Haghighi AP, Portman SL, Lee JD, Amaranto AM, Goodman CS (2001) Ubiquitination-dependent mechanisms regulate synaptic growth and function. Nature. 412: 449-452.

[262] Chen H, Polo S, Di Fiore PP, De Camilli PV (2003) Rapid Ca2+-dependent decrease of protein ubiquitination at synapses. Proc. Natl. Acad. Sci. U. S. A. 100: 14908-14913.

[263] Stang E, Blystad FD, Kazazic M, Bertelsen V, Brodahl T, Raiborg C, Stenmark H, Madshus IH (2004) Cbl-dependent ubiquitination is required for progression of EGF receptors into clathrin-coated pits. Mol. Biol. Cell. 15: 3591-3604.

[264] Sigismund S, Woelk T, Puri C, Maspero E, Tacchetti C, Transidico P, Di Fiore PP, Polo S (2005) Clathrin-independent endocytosis of ubiquitinated cargos. Proc. Natl. Acad. Sci. U. S. A. 102: 2760-2765.

[265] Hawryluk MJ, Keyel PA, Mishra SK, Watkins SC, Heuser JE, Traub LM (2006) Epsin 1 is a polyubiquitin-selective clathrin-associated sorting protein. Traffic. 7: 262-281.

[266] Zoncu R, Perera RM, Sebastian R, Nakatsu F, Chen H, Balla T, Ayala G, Toomre D, De Camilli PV (2007) Loss of endocytic clathrin-coated pits upon acute depletion of phosphatidylinositol 4,5-bisphosphate. Proc. Natl. Acad. Sci. U. S. A. 104: 3793-3798.

[267] Kazazic M, Bertelsen V, Pedersen KW, Vuong TT, Grandal MV, Rodland MS, Traub LM, Stang E, Madshus IH (2009) Epsin 1 is involved in recruitment of ubiquitinated EGF receptors into clathrin-coated pits. Traffic. 10: 235-245.

[268] Kweon DH, Shin YK, Shin JY, Lee JH, Lee JB, Seo JH, Kim YS (2006) Membrane topology of helix 0 of the Epsin N-terminal homology domain. Mol. Cells. 21: 428-435.

[269] Aguilar RC, Longhi SA, Shaw JD, Yeh LY, Kim S, Schon A, Freire E, Hsu A, McCormick WK, Watson HA, Wendland B (2006) Epsin N-terminal homology domains perform an essential function regulating Cdc42 through binding Cdc42 GTPase-activating proteins. Proc. Natl. Acad. Sci. U. S. A. 103: 4116-4121.

[270] Brady RJ, Damer CK, Heuser JE, O'Halloran TJ (2010) Regulation of Hip1r by epsin controls the temporal and spatial coupling of actin filaments to clathrin-coated pits. J. Cell Sci. 123: 3652-3661.

[271] Tian X, Hansen D, Schedl T, Skeath JB (2004) Epsin potentiates Notch pathway activity in Drosophila and C. elegans. Development. 131: 5807-5815.

[272] Overstreet E, Fitch E, Fischer JA (2004) Fat facets and Liquid facets promote Delta endocytosis and Delta signaling in the signaling cells. Development. 131: 5355-5366.

[273] Chen H, Ko G, Zatti A, Di Giacomo G, Liu L, Raiteri E, Perucco E, Collesi C, Min W, Zeiss C, De Camilli P, Cremona O (2009) Embryonic arrest at midgestation and disruption of Notch signaling produced by the absence of both epsin 1 and epsin 2 in mice. Proc. Natl. Acad. Sci. U. S. A. 106: 13838-13843.

[274] Boucrot E, Pick A, Camdere G, Liska N, Evergren E, McMahon HT, Kozlov MM (2012) Membrane Fission Is Promoted by Insertion of Amphipathic Helices and Is Restricted by Crescent BAR Domains. Cell. 149: 124-136.

[275] Sorensen EB, Conner SD (2010) gamma-secretase-dependent cleavage initiates notch signaling from the plasma membrane. Traffic. 11: 1234-1245.

[276] Struhl G, Adachi A (2000) Requirements for presenilin-dependent cleavage of notch and other transmembrane proteins. Mol. Cell. 6: 625-636.

[277] Tagami S, Okochi M, Yanagida K, Ikuta A, Fukumori A, Matsumoto N, Ishizuka-Katsura Y, Nakayama T, Itoh N, Jiang J, Nishitomi K, Kamino K, Morihara T, Hashimoto R, Tanaka T, Kudo T, Chiba S, Takeda M (2008) Regulation of Notch signaling by dynamic changes in the precision of S3 cleavage of Notch-1. Mol. Cell. Biol. 28: 165-176.

[278] Sakata T, Sakaguchi H, Tsuda L, Higashitani A, Aigaki T, Matsuno K, Hayashi S (2004) Drosophila Nedd4 regulates endocytosis of notch and suppresses its ligand-independent activation. Curr. Biol. 14: 2228-2236.

[279] Vaccari T, Lu H, Kanwar R, Fortini ME, Bilder D (2008) Endosomal entry regulates Notch receptor activation in Drosophila melanogaster. J. Cell Biol. 180: 755-762.

[280] Spang A (2009) On the fate of early endosomes. Biol. Chem. 390: 753-759.

[281] Vaccari T, Duchi S, Cortese K, Tacchetti C, Bilder D (2010) The vacuolar ATPase is required for physiological as well as pathological activation of the Notch receptor. Development. 137: 1825-1832.

[282] Yan Y, Denef N, Schupbach T (2009) The vacuolar proton pump, V-ATPase, is required for notch signaling and endosomal trafficking in Drosophila. Dev. Cell. 17: 387-402.

[283] Pasternak SH, Bagshaw RD, Guiral M, Zhang S, Ackerley CA, Pak BJ, Callahan JW, Mahuran DJ (2003) Presenilin-1, nicastrin, amyloid precursor protein, and gamma-secretase activity are co-localized in the lysosomal membrane. J. Biol. Chem. 278: 26687-26694.

[284] Baron M (2012) Endocytic routes to Notch activation. Semin. Cell Dev. Biol. DOI: 10.1016/j.semcdb.2012.01.008.

[285] Yamada K, Fuwa TJ, Ayukawa T, Tanaka T, Nakamura A, Wilkin MB, Baron M, Matsuno K (2011) Roles of Drosophila deltex in Notch receptor endocytic trafficking and activation. Genes Cells. 16: 261-272.

[286] Hori K, Fostier M, Ito M, Fuwa TJ, Go MJ, Okano H, Baron M, Matsuno K (2004) Drosophila deltex mediates suppressor of Hairless-independent and late-endosomal activation of Notch signaling. Development. 131: 5527-5537.

[287] Wilkin M, Tongngok P, Gensch N, Clemence S, Motoki M, Yamada K, Hori K, Taniguchi-Kanai M, Franklin E, Matsuno K, Baron M (2008) Drosophila HOPS and AP-3 complex genes are required for a Deltex-regulated activation of notch in the endosomal trafficking pathway. Dev. Cell. 15: 762-772.

[288] Peden AA, Oorschot V, Hesser BA, Austin CD, Scheller RH, Klumperman J (2004) Localization of the AP-3 adaptor complex defines a novel endosomal exit site for lysosomal membrane proteins. J. Cell Biol. 164: 1065-1076.

[289] Rink J, Ghigo E, Kalaidzidis Y, Zerial M (2005) Rab conversion as a mechanism of progression from early to late endosomes. Cell. 122: 735-749.
[290] Luttrell LM, Lefkowitz RJ (2002) The role of beta-arrestins in the termination and transduction of G-protein-coupled receptor signals. J. Cell Sci. 115: 455-465.
[291] Chen W, Kirkbride KC, How T, Nelson CD, Mo J, Frederick JP, Wang XF, Lefkowitz RJ, Blobe GC (2003) Beta-arrestin 2 mediates endocytosis of type III TGF-beta receptor and down-regulation of its signaling. Science. 301: 1394-1397.
[292] Chen W, ten Berge D, Brown J, Ahn S, Hu LA, Miller WE, Caron MG, Barak LS, Nusse R, Lefkowitz RJ (2003) Dishevelled 2 recruits beta-arrestin 2 to mediate Wnt5A-stimulated endocytosis of Frizzled 4. Science. 301: 1391-1394.
[293] Mukherjee A, Veraksa A, Bauer A, Rosse C, Camonis J, Artavanis-Tsakonas S (2005) Regulation of Notch signalling by non-visual beta-arrestin. Nat. Cell Biol. 7: 1191-1201.
[294] Hori K, Sen A, Kirchhausen T, Artavanis-Tsakonas S (2011) Synergy between the ESCRT-III complex and Deltex defines a ligand-independent Notch signal. J. Cell Biol. 195: 1005-1015.
[295] Rhyu MS, Jan LY, Jan YN (1994) Asymmetric distribution of numb protein during division of the sensory organ precursor cell confers distinct fates to daughter cells. Cell. 76: 477-491.
[296] Guo M, Jan LY, Jan YN (1996) Control of daughter cell fates during asymmetric division: interaction of Numb and Notch. Neuron. 17: 27-41.
[297] Santolini E, Puri C, Salcini AE, Gagliani MC, Pelicci PG, Tacchetti C, Di Fiore PP (2000) Numb is an endocytic protein. J. Cell Biol. 151: 1345-1352.
[298] Song Y, Lu B (2012) Notch Signaling Modulator Numb's Interaction with alpha-Adaptin Regulates Endocytosis of Notch Pathway Components and Cell Fate Determination of Neural Stem Cells. J. Biol. Chem DOI: 10.1074/jbc.M112.360719..
[299] Berdnik D, Torok T, Gonzalez-Gaitan M, Knoblich JA (2002) The endocytic protein alpha-Adaptin is required for numb-mediated asymmetric cell division in Drosophila. Dev. Cell. 3: 221-231.
[300] Couturier L, Vodovar N, Schweisguth F (2012) Endocytosis by Numb breaks Notch symmetry at cytokinesis. Nat. Cell Biol. 14: 131-139 DOI: 10.1074/jbc.M112.360719..
[301] Qiu L, Joazeiro C, Fang N, Wang HY, Elly C, Altman Y, Fang D, Hunter T, Liu YC (2000) Recognition and ubiquitination of Notch by Itch, a hect-type E3 ubiquitin ligase. J. Biol. Chem. 275: 35734-35737.
[302] McGill MA, McGlade CJ (2003) Mammalian numb proteins promote Notch1 receptor ubiquitination and degradation of the Notch1 intracellular domain. J. Biol. Chem. 278: 23196-23203.
[303] Bernassola F, Karin M, Ciechanover A, Melino G (2008) The HECT family of E3 ubiquitin ligases: multiple players in cancer development. Cancer Cell. 14: 10-21.
[304] Dye CA, Lee JK, Atkinson RC, Brewster R, Han PL, Bellen HJ (1998) The Drosophila sanpodo gene controls sibling cell fate and encodes a tropomodulin homolog, an actin/tropomyosin-associated protein. Development. 125: 1845-1856.

[305] O'Connor-Giles KM, Skeath JB (2003) Numb inhibits membrane localization of Sanpodo, a four-pass transmembrane protein, to promote asymmetric divisions in Drosophila. Dev. Cell. 5: 231-243.

[306] McGill MA, Dho SE, Weinmaster G, McGlade CJ (2009) Numb regulates post-endocytic trafficking and degradation of Notch1. J. Biol. Chem. 284: 26427-26438.

[307] Hutterer A, Knoblich JA (2005) Numb and alpha-Adaptin regulate Sanpodo endocytosis to specify cell fate in Drosophila external sensory organs. EMBO Rep. 6: 836-842.

[308] Babaoglan AB, O'Connor-Giles KM, Mistry H, Schickedanz A, Wilson BA, Skeath JB (2009) Sanpodo: a context-dependent activator and inhibitor of Notch signaling during asymmetric divisions. Development. 136: 4089-4098.

[309] Groth C, Fortini ME (2012) Therapeutic approaches to modulating Notch signaling: Current challenges and future prospects. Semin. Cell Dev. Biol. DOI: 10.1016/ j.semcdb.2012.01.016.

Chapter 13

Mutual Regulation of Receptor-Mediated Cell Signalling and Endocytosis: EGF Receptor System as an Example

Zhixiang Wang

Additional information is available at the end of the chapter

1. Introduction

Epidermal growth factor (EGF) receptor (EGFR), also known as HER1 or ErbB1, is the prototypal member of the superfamily of receptors with intrinsic tyrosine kinase activity and is widely expressed in many cell types including epithelial and mesenchymal lineages [1, 2]. The other three members of the ErbB receptor family include Her2/ErbB2/neu [3, 4], Her3/ErbB3 [5] and Her4/ErbB4 [6] (Fig. 1A). EGFR is a 170 kDa membrane glycoprotein composed of three domains. The heavily glycosylated 622-amino acid extracellular domain containing two cysteine rich regions is responsible for ligand binding. The transmembrane domain is a single 23-amino acid α-helical transmembrane peptide. The 542-residue intracellular cytoplasmic domain contains a 250-amino acid conserved protein tyrosine kinase core followed by a 229-amino acid C-terminal tail with regulatory tyrosine residues (Fig. 1B) [7]. Eleven ligands have been identified for ErbB receptors. These ligands can be classified into three groups based on their ability to bind to different ErbB receptors. The first group of ligands includes EGF, transforming growth factor-α, amphiregulin and epgen, which specifically binds to EGFR. The second group of ligands includes betacellulin, heparin-binding EGF and epiregulin, which binds to both EGFR and ErbB4. The third group of ligands includes neuregulin/heregulin, which binds to ErbB3 and ErbB4 [8, 9] (Fig. 1A).

The EGFR family of receptor tyrosine kinases lies at the head of a complex signal transduction cascade that modulates cell proliferation, survival, adhesion, migration and differentiation [1, 2, 10]. While growth factor-induced EGFR signalling is essential for many normal morphogenic processes and is involved in numerous additional cellular responses, the aberrant activity of the members of this receptor family has been shown to play a key role in the development and growth of tumour cells [10-12]. The ErbB receptors were first implicated in cancer when the avian erythroblastosis tumor virus was found to encode an aberrant form of EGFR. Now, EGFR has been implicated in many cancers including squamous cell head and neck cancer, colorectal cancer, non-small cell lung cancer, gastric cancer, pancreatic cancer, breast cancer, ovarian cancer, renal cancer, glimas prostatic cancer and cervical cancer [13]. The dysregulation of ErbB receptor signalling in cancer can occur by various mechanisms, including overexpression due to gene amplification, autocrine ligand production, heterodimerization, deficiency in endocytosis, and gene mutations that increase receptor transcription, translation, protein stability and kinase activity[14, 15].

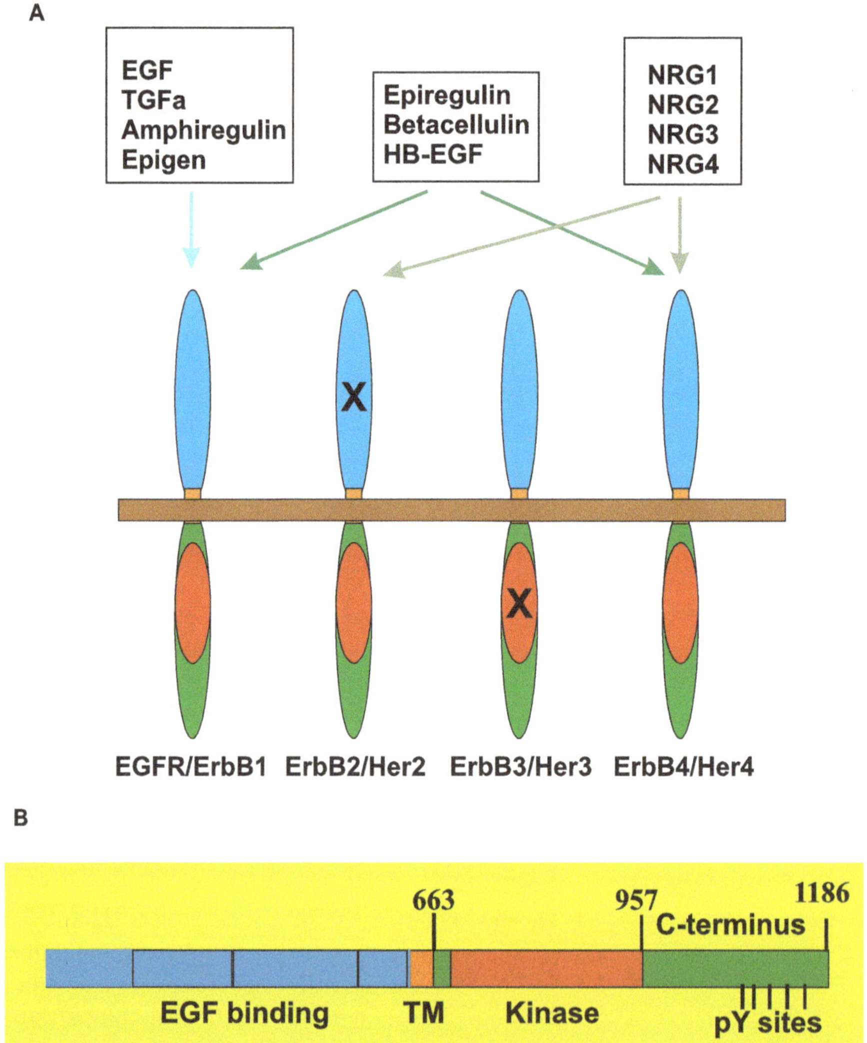

Figure 1. ErbB receptor family. (A) ErbB receptors and their ligands. ErbB family receptors are composed of four members: EGFR/ErbB1/Her1, ErbB2/Her2/neu, ErbB3/Her3 and ErbB4/Her4. Eleven ligands are identified for ErbB family receptors. (B) Linear structure of EGFR.

The binding of EGF at the cell surface induces dimerization of EGFR, which results in the activation of EGFR tyrosine kinase and receptor trans-autophosphorylation [16, 17]. EGFR activation stimulates various signaling pathways, leading to cell mitogenesis and survival [9, 10]. EGFR is overexpressed or hyper activated in many epithelial tumors and plays important roles in cancer development and progression [12]. The activated receptors are rapidly internalized into endosomes and eventually degraded in lysosomes [18]. Initially, the endocytosis of ligand-activated receptors was considered a mechanism to attenuate signaling. Recently, more evidence suggests that the internalized receptors may maintain their ability to generate cell signaling in endosomes [19-23]. Thus, the alteration of EGFR endocytosis may result in abnormal cell signaling, leading to cancer. On the other hand, EGFR endocytosis is firmly regulated by signal recognition and various signaling proteins at every step.

2. EGFR-mediated cell signaling

EGFR plays important roles in initiating cell signaling to produce specific effects on cell growth and development [9, 10]. EGFR is activated through the homodimerization or heterodimerization with other ErbBs such as ErbB2 and ErbB3 in response to ligand stimulation (Fig. 2)[2]. The dimerization of EGFR at the plasma membrane induces the activation of the EGFR tyrosine kinase and trans-autophosphorylation. The sites of tyrosine phosphorylation in the activated EGFR form signaling complexes with many signaling proteins, including Grb2, Shc, phospholipase C-γ1 (PLC-γ1), the p85α subunit of PI3K (p85), p120 rasGAP, Src, Stats, and Cbl [2, 24-26] [2]. The formation of the receptor-signaling protein complexes then initiates the activation of various signaling pathways (Fig. 3A)[9-11, 27-29].

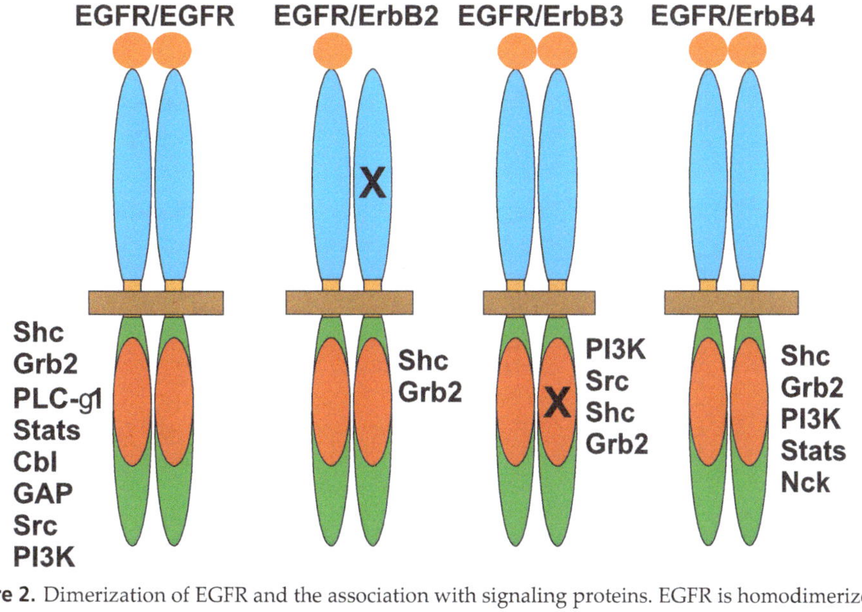

Figure 2. Dimerization of EGFR and the association with signaling proteins. EGFR is homodimerized or heterodimerized with other ErbB proteins in response to ligand.

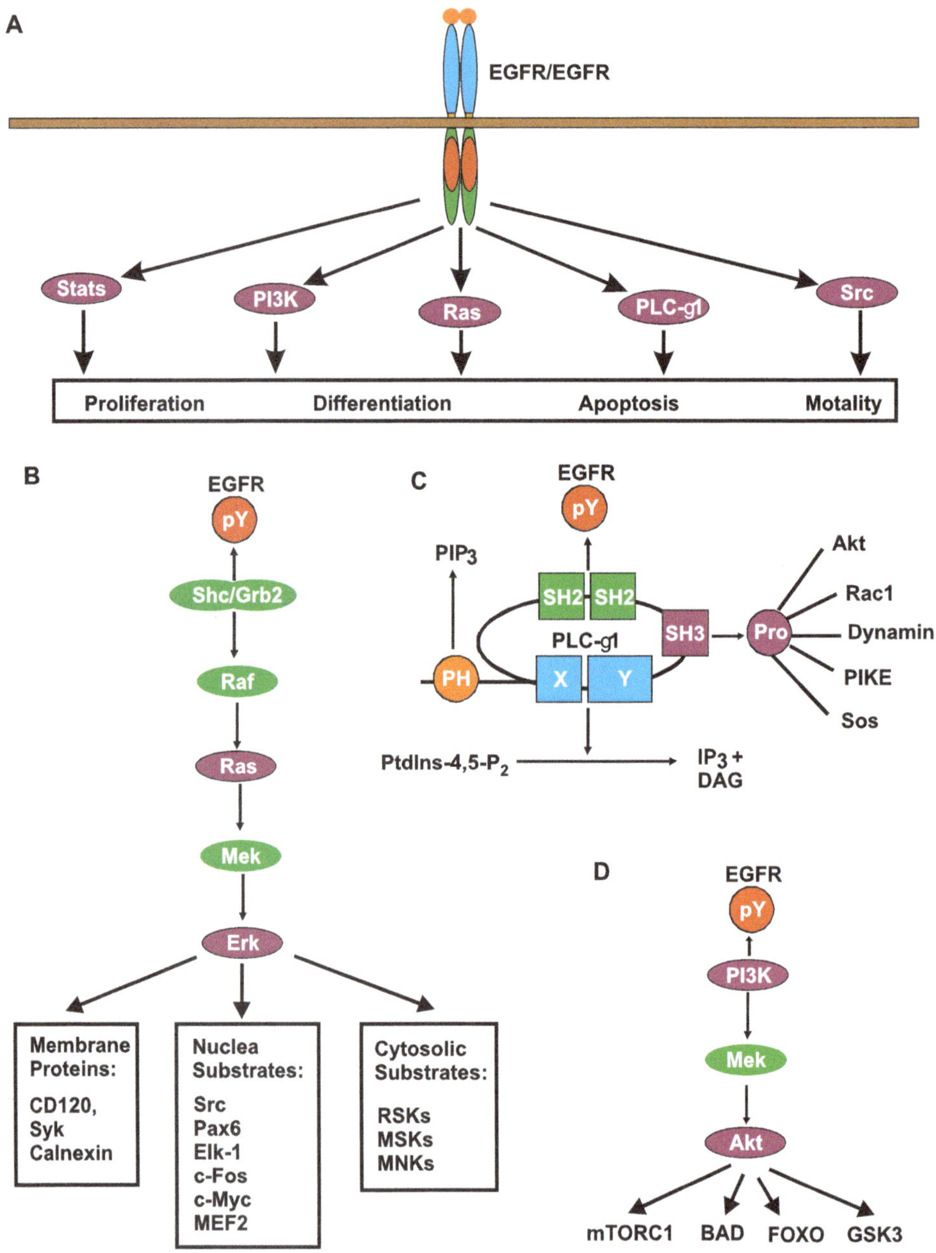

Figure 3. Signaling pathways activated by EGFR. (A) Binding of EGF to EGFR at the plasma membrane initiate the activation of various signaling pathways. The well-defined pathways include Ras-Erk pathway, PI3K-Akt pathway, PLC-γ1 pathway, Stat pathway and Src pathway. (B)The signaling cascade of Ras-Erk pathway. (C) The signaling cascade of PLC-γ1 pathway. (D) The signaling cascade of PI3K-Akt pathway.

The activated EGFR interacts with Shc and Grb2 through multiple phosphorylated tyrosine (pY) residues localized at the C-terminus, which results in the recruitment of Sos to the plasma membrane to activate Ras. Activated Ras mediates Raf activation, which then phosphorylates and activates MEK. Activated MEK then phosphorylates and activates ERK. Activated ERK phosphorylates Rsk, which in turn translocates into the nucleus to activate transcription factors such as c-fos and SRF. Activated ERK may also translocate into the nucleus to activate transcription factors such as Elk1 and c-fos, which is critical in controlling cell mitogenesis (Fig. 3B) [2, 24, 30-34].

Activated EGFR also interacts with PLC-γ1 with multiple pY residues at the C-terminal regulatory domain, which results in the phosphorylation of PLC-γ1 and an increase in its enzymatic activity [35-37]. Active PLC-γ1 hydrolyzes phosphatidylinositol 4, 5-bisphosphate (PtdIns(4,5)P_2) to form the second messengers inositol 1, 4, 5-triphosphate ($InsP_3$) and diacylglycerol (DAG). $InsP_3$ and DAG mobilize intracellular calcium and activate protein kinase C (PKC), respectively. Recent studies have shown that PLC-γ1 is involved in broad cell signaling. Interestingly, most recently identified interactions between PLC-γ1 and its binding proteins are mediated by its SH3 domain. EGF stimulates the interaction between PLC-γ1 and PLD_2, which is mediated by the PLC-γ1 SH3 domain [38]. PLC-γ1 binds directly to Akt in response to EGF through its SH3 domain [39]. The PLC-γ1 SH3 domain acts as a guanine nucleotide exchange factor (GEF) for PIKE [40], dynamin [41] and Rac1 [42]. The activated PLC-γ1 regulates cell mitogenesis and migration (Fig. 3C)[39, 42-44].

Activated EGFR also activates PI3K either through its direct interaction with the p85α subunit or through the activated Ras [45, 46]. Activated PI3K then catalyzes the production of the second messenger phosphatidylinositol-3,4,5-trisphosphate (PIP_3) by phosphorylating phosphatidylinositol-4,5-bisphosphate (PIP_2). A direct antagonist of PI3K is the phosphatase and tensin homologue deleted on chromosome 10 (PTEN). PTEN dephosphorylates PIP_3 into PIP_2 to reverse the activity of PI3K and therefore function as an important negative controlling element of incoming signals. PIP_3 transduces activating signals by binding to pleckstrin homology (PH) domains of proteins to recruit them to the cell membrane. One centrally important downstream mediator of the PI3K signalling cascade is the serine threonine (Thr) kinase Akt. Akt is recruited to the plasma membrane by its SH3 domain interaction with PIP3, which exposes Akt Thr 308 for phosphorylation by 3-phosphoinositide-dependent kinase 1 (PDK-1), which is already located at the membrane. The rapamycin complex 2 (mTORC2) phosphorylates Ser 473 in the C-terminus, which leads to full Akt activation. Activated Akt then mediates signals promoting cellular growth and survival and suppresses pro-apoptotic signals. Akt phosphorylates several intracellular proteins, including forkhead box O transcription factors (FoxO), the BCL2-associated agonist of cell death (BAD), and the glycogen synthase kinase 3 (GSK3), to promote cell cycle entry and cell survival. The proteins TSC1 (Hamartin) and TSC2 (Tuberin) form a complex that inhibits the activity of the small G-protein ras homologue enriched in the brain (Rheb), which is necessary for mTORC1 activation. The Akt-mediated phosphorylation of TSC2 releases Rheb from its inhibited state. Rheb then accumulates in a GTP-bound state and can directly activate mTORC1, which phosphorylates the p70S6 kinase (S6K1) and the eukaryotic translation initiation factor 4E binding protein 1 (4EBP1), leading to increased protein translation (Fig. 3D), which protects the cell from undergoing apoptosis [45, 47, 48].

Activated EGFR also activate Stats directly by binding to and phosphorylating Stats, or indirectly by activating c-Src. Activation can occur via cytokine signaling (IL-6), growth factor receptor signaling (EGFR), or non-receptor tyrosine kinase signaling (Src). JAK is not required when Stats bind directly to EGFR for activation, but JAK provides maximal activation of Stats phosphorylated by EGFR-activated Src. Grb2 and SOCS can inhibit Stat-mediated EGFR signaling respectively, by either binding to the STAT activation site on EGFR or by binding to JAK to suppress Src activation of Stats. Once activated, Stats dimerize and translocate to the nucleus where they activate the transcription of genes involved in proliferation, differentiation, and survival [49].

Importantly, Src kinases, which have been reported to be activated in many cancers with high EGFR levels, have been shown to potentiate EGFR signaling [50-52]. The c-Src potentiation of EGFR has been demonstrated to be associated with the c-Src-dependent phosphorylation of EGFR and the complex formation between c-Src and EGFR [50, 51]. In addition to focal adhesion kinase (FAK), which is involved in the regulation of adhesion and migration, PI3K and Stat3 are also substrates for c-Src [53]. Although the Src kinase has been linked with the development and progression of cancer for many years, we still do not completely understand its role in cancer [54]. Src is a member of a ten-gene family (FYN, YES, BLK,FRK, FGR, HCK, LCK LYN, and SRMS) of non-RTKs that play a fundamental role in the regulation of cell proliferation, migration, adhesion, and tumor angiogenesis [55, 56]. Src signaling is cross-connected with many signaling pathways, such as the PI3K and Stat pathway [55, 56]. Even though tyrosine kinase activity of Src is independent of RTK signaling, it may interact with RTKs such as EGFR. As such, Src-EGFR interaction may enhance EGFR signaling, and on the other hand it may be involved in resistance to EGFR-targeted therapy [54, 57].

3. EGFR-mediated endocytosis

3.1. Clathrin-dependent and clathrin-independent endocytic pathways

The concept of receptor-mediated endocytosis was formulated in 1974 to explain how the sequential cell surface binding, internalization, and intracellular degradation of plasma low density lipoprotein (LDL) regulates cellular cholesterol metabolism [58]. Receptor-mediated endocytosis is a multiple step event [58]. In general, receptor-mediated endocytosis consists of two stages: internalization and intracellular trafficking. Endocytic pathways are generally classified as either clathrin-dependent or clathrin-independent. Much work has focused on clathrin-mediated endocytosis (CME). In this process, cargo proteins are recruited into developing clathrin-coated pits (CCPs), and subsequently form clathrin-coated vesicles (CCVs) [59]. Several proteins or protein complexes, including clathrin, adaptin AP-2, dynamin and Eps15, participate in the CME of all receptors. Adaptin AP-2 is a cytoplasmic protein complex that interacts with the cytoplasmic tails of various receptors. These interactions are thought to account for the ability of cells to selectively direct receptors to CCVs. Clathrin is ideally designed to form a scaffold which, when attached to the membrane, causes the membrane to deform into a budding vesicle. Clathrin presumably binds to the membrane by interacting with membrane-bound AP-2. Dynamin has been identified as a major player in the endocytic pathway and is essential for the scission of coated vesicles. Eps15 is an essential component of the early endocytic pathway [59-61].

Although CME is certainly an extremely important endocytic mechanism, accounting for a large proportion of endocytic events, an ever expanding array of cargos has been shown to undergo non-clathrin-mediated endocytosis (NCE) [62]. Many NCE pathways have been reported, including caveolar-type endocytosis, CLIC/GEEC-type endocytosis, the putative flotillin-associated endocytic structures, phagocytosis, macropinocytosis, dorsal ruffles (or waves), and entosis [62, 63]. Caveolar-type endocytosis is the best studied NCE.

3.2. Endocytic and sorting signals

The targeting of transmembrane proteins to different compartments of the endocytic pathways is largely dependent upon sorting signals contained within the cytoplasmic domains of the proteins [64-66]. Most of these sorting signals are short, linear sequences of amino acid residues. These signals can be classified to two groups. One group of signals is referred to as tyrosine-based sorting signals and the other group of signals is known as dileucine-based signals. All of these signals are recognized by components of protein coats peripherally associated with the cytosolic face of membranes [66].

Tyrosine-based signals constitute a family of degenerate motifs minimally defined by the presence of a critical tyrosine residue [66]. Most tyrosine-based signals conform to the consensus motifs YXXΦ (Y is tyrosine, X is any amino acid and Φ is an amino acid with a bulky hydrophobic side chain) [67] or NPXY (N is asparagine and P is proline) [68-71]. It was shown by several groups that the substitution of tyrosine residues in the cytosolic domains of various endocytic receptors devoid of NPXY motifs impaired internalization [72-77]. NPXY signals have been shown to mediate only the rapid internalization of a subset of type I integral membrane proteins, and not mediate other intracellular sorting events. The interaction between NPXY motif and endocytic protein is less understood. However, several proteins, including clathrin, AP-2, and Dab2, have been proposed to function as recognition proteins for NPXY signals. There are several NPXY motifs located in the EGFR C-terminus [78]. Systematic mutational analyses led to the identification of another tyrosine-based motif, YXXΦ, as the major determinant of endocytosis of the mannose 6-phosphate as well as many other transmembrane proteins [67, 78, 79]. In mammalian cells, virtually all YXXΦ signals mediate rapid internalization from the cell surface. Some YXXΦ signals can additionally mediate lysosomal targeting [64-66]. Recent evidence suggests that the μ2 subunit of AP2 directly interacts with YXXΦ to mediate rapid internalization [80-83].

Di-leucine-based sorting signals have been implicated in various sorting process [78]. Two classes of di-leucine-based sorting motifs have been distinguished. [DE]XXXL[LI] signals play critical roles in the sorting of many type I, type II and multispanning transmembrane proteins. The [DE]XXXL[LI] signals in mammalian proteins mediate rapid internalization and target the proteins to endosomal-lysosomal compartments, suggesting that they can be recognized both at the plasma membrane and at intracellular locations. [DE]XXXL[LI] signals are recognized by the adaptor protein (AP) complexes. DXXLL signals are present in several transmembrane receptors and other proteins that cycle between the TGN and endosomes. DXXLL signals are recognized by another family of adaptors known as GGAs.

Ubiquitination of cytosolic lysine residues constitutes another important signal for sorting transmembrane receptors at various stages of the endosomal-lysosomal system. Ubiquitin is a globular protein consisting of 76 amino acids that is able to covalently conjugate to other proteins [84]. Ubiquitin is covalently conjugated to proteins by forming a bond between the carboxy-terminal glycine of ubiquitin and the ε-NH_2 group of a lysine residue in the substrate protein. Alternatively, ubiquitin can be conjugated to the α-NH_2 group of the N-terminal amino acid of the substrate [85, 86]. Conjugated ubiquitin is recognized by UIM, UBA, or UBC domains present within many components of the internalization and lysosomal targeting machinery. It has been shown that EGFR is ubiquitinated in response to EGF, which plays an important role in EGFR degradation in lysosomes. The presence of these various type of sorting signals within the transmembrane receptors and their interaction with the signal recognition proteins ensures the dynamic but accurate distribution of transmembrane proteins to different compartments along the endocytic pathways.

3.3. Endocytosis of EGFR

The first comprehensive study of EGF endocytosis, in which many of the key concepts of internalization and lysosomal degradation of EGF have been established, was published by Carpenter and Cohen [87]. The binding of EGF results in the clustering and internalization of EGFR. The accumulation of EGF and EGFR can be detected in the early endosome after 1-5 min of incubation with EGF at 37°C. EGF and EGFR accumulate in late endosomes after 10-20 min at 37°C. A substantial number of EGFR can be detected in organelles with typical biochemical and morphological features of mature lysosomes only after 40-60 min of continuous internalization at 37°C [16, 88]. Intracellular trafficking of receptors involves a series of membrane budding and fusion events [89]. Endosome fusion is regulated by specific cytosolic and membrane-associated protein factors, including a group of Ras-like small guanosine triphosphatases (GTPases) called Rabs [90-92]. Four classes of endocytic organelles are typically distinguished based largely on their relative kinetics of labeling by endocytic tracers: early endosomes (EEs), late endosomes (LEs), recycling vesicles (RVs), and lysosomes [65]. The precise relationship among these structures has yet to be determined, and in fact may never be known because of the great plasticity and dynamics of the system.

The internalization of constitutively internalized receptors is largely mediated by sorting signals such as YXXΦ and NPXY. However, for the receptors that are internalized in response to ligand binding, there is likely some means of switching their sorting signals on and off [93]. Given that ligand binding is essential for the rapid internalization of EGFR, the events induced by the ligand binding likely contribute to the regulation of ligand-induced EGFR internalization. These events include receptor dimerization, activation of intrinsic tyrosine kinase activity, autophosphorylation and association with various binding proteins.

The initial results are very controversial regarding the role of EGFR kinase activity in EGFR internalization. Data from some research groups suggest that kinase-dead EGFR is deficient in EGF-induced internalization [94, 95]; however, data from other research groups suggest that kinase-dead EGFR is internalized normally like wild type EGFR, but is quickly recycled back to the plasma membrane [96, 97]. Since the mid 1990s, most studies suggest that EGFR kinase activity is required for EGF-induced EGFR internalization. It was reported that EGFR kinase activation is required for the recruitment of EGFR into coated pits [98]. The EGFR activation of c-Src tyrosine kinase has been implicated in the regulation of the clathrin-dependent endocytosis of EGFR through the ability to phosphorylate clathrin heavy chain [99]. The EGFR activation of Eps15 has been shown to be required for the internalization of EGFR [100]. The inhibition of EGFR kinase activity by AG1478 and PD158780 was shown to block EGFR internalization [101, 102].

While most studies indicate that EGFR kinase activity is essential for EGF-induced EGFR internalization [94, 95, 98-102], it was shown recently that EGFR kinase activity is not required for EGF-induced EGFR internalization [103-107]. Inhibition of EGFR kinase activation by the specific inhibitor AG1478 and PD158780 in BT20, MDCK, Cos7 and Hela cells did not block EGFR internalization. When transiently expressed in 293T cells or stably expressed in CHO cells, a kinase-dead EGFR (EGFR K721A), was internalized following EGF stimulation in a similar pattern to wild type EGFR, which indicates that EGFR kinase activation is not required for EGFR internalization [103-107].

If kinase activity is not necessary for EGF-induced EGFR endocytosis, an EGF-induced event before or independent of EGF-induced EGFR kinase activation must be responsible for mediating EGF-induced EGFR endocytosis. The only significant event induced by EGF before the activation of EGFR kinase is the dimerization of EGFR. It is well established that receptor dimerization is critical for EGFR kinase activation [108]. In fact, it is generally believed that the only function of receptor dimerization is to allow the activation of EGFR kinase and the trans-autophosphoylation of the two receptors. However, it was shown recently that EGFR dimerization is necessary and sufficient to stimulate EGFR internalization, independent of EGFR kinase activation [105]. EGF-induced EGFR dimerization in the absence of kinase activation is sufficient to stimulate EGFR internalization. Non-ligand-induced dimerization of EGFR without kinase activation is sufficient to stimulate EGFR internalization. Moreover, the inhibition of EGF-induced EGFR dimerization by deleting the receptor dimerization loop abolishes EGF-induced EGFR internalization [105]. It has also been reported that the crosslinking of two EGFR with antibody stimulates the endocytosis of EGFR without activating EGFR kinase [109]. How dimerization may mediate EGFR endocytosis independent of its role in EGFR kinase activation is not known. Several possibilities have been suggested. It is possible that EGF-induced EGFR dimerization causes necessary conformational changes of the receptor to expose the cryptic internalization codes. Alternatively, the internalization regulating proteins essential for EGFR internalization may have a dimeric nature and can only bind to dimerized EGFR [105].

Many studies have also focused on the role of EGFR C-terminus in EGFR internalization (Fig. 4). The EGFR mutants truncated from the C-terminus to residue 991 [110] or to residue 973 [111] are internalized inefficiently and the mutant truncated at residue 958 is not internalized [110]. Simultaneous point mutation of the five-tyrosine residues (Y992, Y1068, Y 1086, Y1148 and Y1173) to phenylalanines significantly reduces EGFR internalization [112]. EGFR is co-immunoprecipitated with adaptin AP-2 [88]. The binding between EGFR and AP-2 is mediated by EGFR amino acid residues 970-991, especially Y974 [113, 114]. This interaction accelerates EGFR internalization when EGFR is expressed at high levels, but is not required for EGFR internalization when EGFR is expressed at low levels [83, 113, 114]. A 15-amino acid domain (residues 943-957) was found to be essential for binding sorting nexin-1 (SNX1) which is involved in targeting EGFR to lysosome [115], but not EGFR internalization. It was shown that the EGFR C-terminal sequences from 992 to 1044 are essential for mediating EGF-induced EGFR internalization with or without the inhibition of EGFR kinase activation [105]. It was further shown that EGFR residues 1005-1017, especially the di-leucine ^{1010}LL^{1011} is required for EGF induced rapid internalization of full length EGFR and the role of ^{1010}LL^{1011} in EGFR internalization is independent of EGFR kinase activation [106]. The identification of ^{1010}LL^{1011} as essential for EGFR internalization is very interesting. EGFR di-leucine motif ^{1010}LL^{1011} proceeded with TSRTP, which is different from the two classes of di-leucine-based sorting signals described above. Two di-leucine motifs including ^{679}LL^{680} and ^{1010}LL^{1011} have been implicated in EGFR sorting. It was reported that ^{679}LL^{680} is required for the efficient transport of EGFR to lysosomes and for the retention of EGFR in endosomes [116, 117]. It was also shown that ^{1010}LL^{1011} is critical in the tyrosine phosphorylation of β2 subunit of clathrin adaptor complex AP-2 and is required for EGFR degradation [118]. The only data suggesting a possible role of ^{1010}LL^{1011} in EGFR endocytosis is that it regulates the slow endocytosis of a mutant EGFR truncated at amino acid 1022 [119]. However, other data from the same group showed that ^{1010}LL^{1011} is not involved in the endocytosis of full length EGFR [118, 119].

The role of various EGFR binding proteins in EGFR endocytosis has also been extensively studied. Some proteins that bind to pY sites of EGFR have also been implicated in EGFR endocytosis. These proteins including Grb2, Eps15, PLD, Cbl, Rin1, and Src [41, 99, 100, 119-123]. Grb2 regulates EGFR endocytosis, possibly through its SH3 domain interaction with dynamin [120]. Knocking-down Grb2 with siRNA also blocks EGFR endocytosis [119, 124]. EGF receptor endocytosis is dependent upon PLD and the PLD1 regulators, protein kinase C alpha and RalA [125]. Tyrosine phosphorylation of Eps15 is necessary for the internalization of EGFR [100]. Eps15 functions as a scaffolding adaptor protein and is involved in both secretion and endocytosis. Eps15 has been shown to bind to AP-1 and AP-2 complexes, to inositol lipids, and to several other proteins involved in the regulation of intracellular trafficking [126]. Phosphorylation of clathrin heavy chain by Src facilitates EGFR endocytosis [99]. Rin1 binds to EGFR and regulates EGFR endocytosis through its SH2 domain [123]. Although it is generally agreed that Cbl acts to negatively regulate EGFR activity by promoting the intracellular trafficking and degradation of EGFR, it is still disputed whether Cbl binding or Cbl-mediated ubiquitination is altogether required for ligand-induced EGFR endocytosis [122, 127]. Some recent data indicate that Cbl-mediated ubiquitin of EGFR is not required for EGFR endocytosis [127, 128]. While PI3K is required for β-PDGFR endocytosis and down-regulation [129-131], PI3K activity is not required for EGFR endocytosis [132].

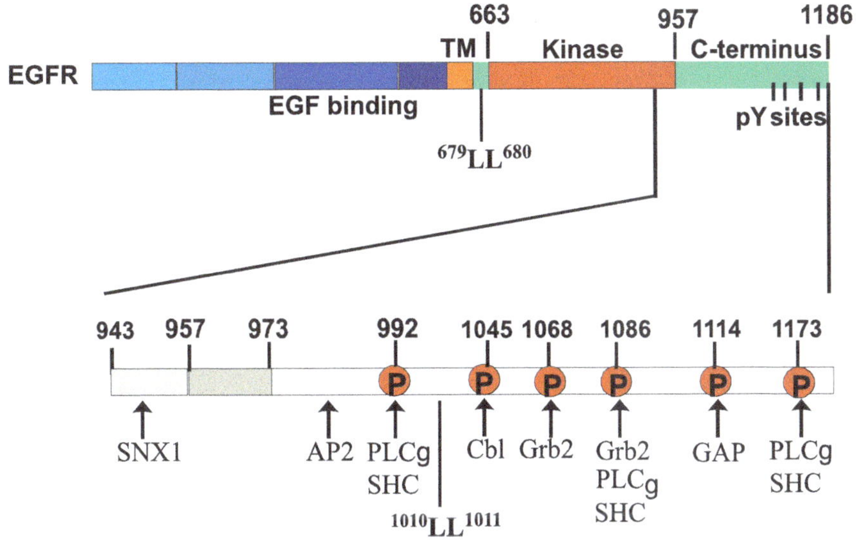

Figure 4. Internalization and sorting signals within EGFR intracellular domain.

Strong evidence suggests that CME is the major pathway of EGFR endocytosis. EGF and EGFR are found concentrated in CCP and CCV. EGFR endocytosed with a rate similar to those of other receptors that are internalized by CME, such as LDL and transferrin (Tfn). Knockdown of clathrin heavy chain or dynamin by RNA interference (RNAi) inhibits EGFR endocytosis [18]. Although CME is certainly an extremely important endocytic mechanism, accounting for a large proportion of endocytic events, an ever expanding array of cargos has been show to undergo endocytosis in clathrin-independent manner [62, 133]. Clathrin-independent endocytosis itself has been further dissected into seemingly distinct pathways, based on the reliance of these pathways on certain proteins and lipids, their differential drug sensitivities and their abilities to internalize particular cargos [134]. Many NCE pathways have been reported including caveolar-type endocytosis, CLIC/GEEC-type endocytosis, the putative flotillin-associated endocytic structures, phagocytosis, macropinocytosis, dorsal ruffles (or waves), and entosis [62, 63]. New evidence suggests that EGF-induced EGFR endocytosis may also be mediated by NCE. NCE of EGFR via dorsal waves was observed in several types of cells [135]. This pathway required the activity of the EGFR kinase, PI3K and dynamin [135]. The NCE of EGFR involving cholesterol-rich lipid rafts and/or caveolar has also been reported [136]. This cholesterol-dependent endocytosis was observed under conditions of high EGFR occupancy by EGF in Hela cells. All of the reported NCE pathways are significantly slower than CME, although they are faster than constitutive receptor endocytosis [18].

4. Regulation of EGFR signalling by endocytosis

Endocytosis is essential for cells to perceive extracellular signals and transduce them in a temporally and spatially controlled fashion, directly influencing not only the duration and intensity of the signaling output, but also their correct location. It is well established that the endocytosis of EGFR from the plasma membrane to lysosomes results in the degradation of the receptor, which can attenuate receptor signaling and may even be conceived of as a tumor suppressor pathway [19]. On the other hand, accumulated evidence suggests that the internalized EGF-EGFR complex may maintain its ability to generate cell signaling from endosomes [19-23].

4.1. Downregulation of EGFR by endocytosis

Endocytosis has been recognized as the most significant pathway to downregulate EGFR by removing the receptor from the cell surface for degradation in lysosomes [137]. This downregulation of EGFR is a complicated and tightly regulated process. During this process the EGFR-containing internalized vesicles mature into multivesicular bodies (MVBs), which then fuse with lysosomes to allow degradation of their content. This was first shown by Cohen and his colleagues, who observed that ferritin-conjugated EGF was rapidly internalized upon binding to EGFR and trafficked to MVBs within 15 minutes exposure of cells to ligand [87, 138]. The impaired endocytic downregulation of signaling receptors is frequently associated with cancer, since it can lead to increased and uncontrolled receptor signaling [139].

The role of endocytosis in the downregulation of EGFR signalling is best illustrated by the findings that the inhibition of EGFR endocytosis frequently leads to cancer. The best characterized EGFR mutant with impaired endocytosis is EGFRvIII. EGFRvIII has been implicated in many types of tumors [140-145]. EGFRvIII is a mutant EGFR with the deletion of amino acid residues 6–273 in the extracellular domain of EGFR. This results a truncated 145 kDa receptor with a non-functional ligand binding pocket and no dimerization arm. In spite of not binding any ligands, the receptor is constitutively active [146], and is able to activate Ras-Erk1/2 and PI 3-kinase-Akt pathways [142, 147]. In concordance with this, EGFRvIII was shown to transform fibroblasts and to enhance the proliferation and/or tumorigenicity of cells both in vivo and in vitro [142, 148-152]. The constitutive activity may be important for tumorigenicity, but impaired downregulation certainly enhances the effect. Two recent reports show that EGFRvIII is not degraded in cells with endogenous levels of Cbl, instead, internalized EGFRvIII is recycled back to the plasma membrane [153].

It is generally accepted that ErbB2 avoids efficient endocytic downregulation [154-158], which contributes to its important role in the development of various cancers [2, 10]. As ErbB2 and the EGFR-ErbB2 heterodimers are impaired in EGF-induced endocytosis [158], EGFR-mediated cell signaling are significantly sustained in the cells with overexpressed

ErbB2 due to the formation of EGFR-ErbB2 heterodimers. EGFR signaling can also be sustained if the molecular machinery normally involved in receptor downregulation does not function optimally. Indeed, several mutations of such proteins have been found in tumors, including Cbl, TSG101 (an ESCRT-I subunit), and VPS25 (an ESCRT-II subunit) (recently reviewed in [159]. In conclusion, endocytic impairment may be a returning theme of oncogenic EGFR mutants.

The role of endocytosis in the downregulation of EGFR signalling is also frequently and successfully explored as a therapy for cancer. Since the lack of endocytic downregulation is an emerging theme in ErbB cancer biology, it is evident that the stimulation of ErbB endocytosis and lysosomal degradation is an attractive means to inhibit tumor growth. Polyvalent antibodies have been developed to stimulate EGFR and other ErbB endocytosis by crosslinking the receptors together [156, 160]. One good example is trastuzumab (Herceptin). Trastuzumab is a humanized recombinant mAb that binds to the extracellular domain of ErbB2 protein [161, 162]. Currently, it is the only ErbB2-targetted therapy approved by FDA for metastatic breast cancer treatment [163]. Although several recent studies suggest that Trastuzumab does not induce the endocytosis of ErbB2 to a significant degree [154, 156, 157], the dominating opinion has been that Trastuzumab causes endocytic downregulation of ErbB2 [2, 164-166].

Cetuximab is an antibody targeting EGFR that is currently used in treatment of colorectal cancer and head and neck cancer [167]. Several studies have shown that Cetuximab induces the internalization of EGFR [109, 168]. Cetuximab-induced EGFR internalization is independent of receptor tyrosine kinase activity, and it is both slower and less efficient in terms of receptor downregulation than ligand-induced endocytosis [109]. At present, the knowledge of mechanisms underlying antibody-mediated endocytic downregulation is relatively sparse. A useful observation is that extensive antibody-based crosslinking of ErbB receptors is far more efficient at inducing ErbB endocytic downregulation than single antibodies are [156, 160]. Crosslinking can either be done using antibodies that form multivalent aggregates via secondary antibodies or gold particles (Hommelgaard2004}, or by a more clinically relevant approach using combinations of monoclonal antibodies against distinct epitopes in an ErbB receptor [160]. Thus, whereas the administration of Trastuzumab alone did not induce significant ErbB2 endocytosis, the combination of Trastuzumab with another monoclonal antibody to ErbB2 was very efficient at downregulating ErbB2. In addition, the combination of two antibodies was much more efficient at inhibiting tumor growth in a mouse model compared to Trastuzumab administered alone [160, 166].

Although the endocytic downregulation of EGFR has been mostly attributed to clathrin-dependent endocytosis [18], other endocytic pathways have also been proposed during recent years, especially following stimulation with high concentrations of EGF [136]. The concentration of EGF varies greatly throughout the human body. The EGF concentration in most tissue fluid is about 1–2 ng/ml, but it is much higher, up to 100 ng/ml or more, in tubular duct lumens of kidney, salivary glands, and the mammary gland [87, 169]. Normally, EGFR is not reached by the high luminal concentrations of EGF in these systems, since the receptor is present at the basolateral site of the epithelial cells. However, during wound healing or malignant transformation, the tight junctions disappear and allow the high concentrations of EGF to access the receptor [170]. Very high EGF concentration can also be found in solid tumors [171]. It was reported that at high concentrations of EGF (20 ng/ml) the receptor became ubiquitinated and was to a high degree internalized by caveolae [136]. Incubation of epithelial cells with 30 ng/ml of EGF for 5–20 min resulted in an eight to tenfold increase in the number of plasma membrane caveolae due to EGF-induced tyrosine phosphorylation of caveolin-1 [172]. Moreover, live cell imaging revealed increased dynamics of green fluorescent protein (GFP)-tagged caveolin upon stimulation of cells with 30 ng/ml EGF. Thus, some studies suggest a role of caveolae in EGFR endocytosis. More interesting, it was further revealed by Sigismund et al that EGFRs internalized via CME are not targeted for degradation, but instead are recycled to the cell surface. By contrast, clathrin-independent internalization preferentially commits the receptor to degradation [173]. A prior study has shown that TGFβ receptor is internalized by two distinct endocytic pathways, clathrin-mediated endocytosis leading to TGFβ receptor signaling and lipid-raft-mediated endocytosis leading to the degradation of TGFβ receptor [174].

4.2. Signalling endosomes

The concept of EGFR signalling from endosomes or "signalling endosomes" has been gradually developed. Early evidence to support signalling from endosomes was reported in middle to late 1980s. These researches showed that internalized EGFR is autophosphorylated and catalytically active [175-177]. Various signaling molecules that regulate Ras activity, including Grb2, SHC, Sos and GAP, are co-internalized with EGFR into endosomes and remain associated with the receptor in endosomes [20, 178-181]. Afterwards, more results confirmed the interaction between EGFR and various signaling proteins in endosomes [182-186].

The major evidence supporting endosomal EGFR signalling came from endocytosis inhibition experiments. Since the mid 1990s, researchers have developed many ways to inhibit EGFR endocytosis and then examine the effects on cell signalling. These experiments have yielded mixed results regarding what signalling pathways activated by endosomal EGFR and the physiological relevance of EGFR signaling from endosomes. The inhibition of EGFR endocytosis by a dominant-negative mutant dynamin enhances the activation of PLC-γ1 and cell proliferation, but decreases ERK activation [187]. In a study of EGFR transactivation by G-protein coupled receptors, it was found that the inhibition of EGFR endocytosis by either mutant dynamin or β-arrestin abolished ERK activation [188, 189]. The inhibition of EGFR endocytosis by phospholipase D also blocks EGF stimulated-ERK activation [125]. However, none of these researches provided a mechanism to explain why activated EGFR at the plasma membrane is unable to activate ERK. On the other hand, other research showed that the inhibition of EGFR internalization enhances ERK activation [190, 191]. EGFR efficiently activates mitogen-activated protein kinase in HeLa cells and Hep2 cells, which is conditionally defective in clathrin-dependent endocytosis by overexpressing dominant negative dynamin [191]. Sprouty2 attenuates EGFR ubiquitination and endocytosis, and consequently enhances Ras/ERK signalling [190]. Initially, in the few cases where biological end points were measured, inhibition of endocytosis did not result in the attenuation of biological effects [187, 192]. These results argue against a physiological relevance of endosome-originated signals [193].

The controversy over endosomal signaling and its physiological relevance is in part due to the limitation of current approaches. For example, while it has made significant contribution and remains a powerful tool to study endosomal signaling, this endocytosis-inhibition approach has its limitations. While the inhibition of EGFR endocytosis eliminates endosomal signaling, the retention of EGFR at the cell surface also enhances signaling from the plasma membrane. Thus, it is difficult to determine whether the observed effects are due to the lack of endosomal signaling or due to prolonged plasma membrane signaling. Blocking EGFR endocytosis by mutant dynamin or β-arrestin affects all endocytic events mediated by these factors. Thus, it is difficult to determine whether the observed effects are due to the inhibition of EGFR endosomal signaling or due to a broad inhibition of endocytosis. Moreover, this approach is not suitable for studying the dynamics of endosomal signaling. None of these approaches offered mean to get activated receptors inside a cell without initial activation at the cell surface [194]

In the early 2000s, a novel system was established to allow the specific activation of endosome-associated EGFR without the initial activation at the plasma membrane and without disrupting the overall endocytosis pathway. To specifically activate endosomal EGFR, cells were treated with EGF in the presence of AG1478, a specific EGFR tyrosine kinase inhibitor, and monensin that blocks the recycling of EGFR. This treatment led to the internalization of inactive EGF-EGFR complex into endosomes. The endosome-associated EGFR was then activated by removing AG1478 and monensin. No surface EGFR phosphorylation was detected [103, 104]. The specific activation of endosome-associated EGFR was also achieved without using monensin[103, 104]. In this system EGFR follows the same endocytic pathway as the control: EGF receptor is first internalized into Rab5-positive endosomes and eventually traffics to lysosomes for degradation. The only difference is that the EGF receptor is not activated during its internalization from the plasma membrane to endosomes and stops at endosomes until being activated. Thus, this system not only allows the generation of specific endosomal signaling of EGFR, but also under a condition very similar to the endosomal signaling of EGFR following its activation at the plasma membrane. By using this system, it was shown that 1) endosomes can serve as a nucleation site for the formation of signaling complexes, 2) endosomal EGFR signaling is sufficient to activate the major signaling pathways leading to cell proliferation and survival, and 3) endosomal EGFR signaling is sufficient to suppress apoptosis induced by serum-withdrawal [103] and to stimulate cell proliferation [195].

In most cases, the endosomal EGFR signaling is the continuation of EGFR signaling at the plasma membrane, serving to maintain EGFR signaling and provide spatial-temporal regulation of EGFR signaling. However, in some cases, specific and novel signaling may be initiated only from endosomes as these signaling events require factors to be brought together by endocytosis. While specific signaling complexes can be assembled through their recruitment to the early endosomal resident protein RAB5, there are no convincing examples that specific and novel signaling is initiated from endosomes in the context of EGFR signaling. However, it is well illustrated in TGFβ signaling that specific and novel signaling may be initiated only from endosomes. TGFβ receptors (TGFβR) become phosphorylated at Ser residues and are internalized by endocytosis following ligand binding. Once localized into endosomes, TGFβR can bind to SMAD anchor for receptor activation (SARA). The protein complex induced phosphorylation of the transcription factors SMAD1 or SMAD2 by their Ser/Thr kinase receptors. Upon phosphorylation, SMADs are released into the cytoplasm, bind to a cofactor (SMAD4), enter the nucleus, and promote gene transcription [137, 174].

Together, it is clear that EGFR signals from both the plasma membrane and the endosomes, and that the signals from both locations are able to activate major signaling pathways, stimulate cell proliferation, and promote cell survival. However, following EGF stimulation, activated EGF receptors only stay at the plasma membrane briefly (5-10 min), but stay in the endosome much longer (1 h) (Fig. 5) [88]. This argues for a more physiologically important role for endosomal signaling. The plasma membrane EGFR signaling are usually exaggerated by studies with the inhibition EGFR endocytosis, as activated EGFR that stay at the plasma membrane are artificially over extended.

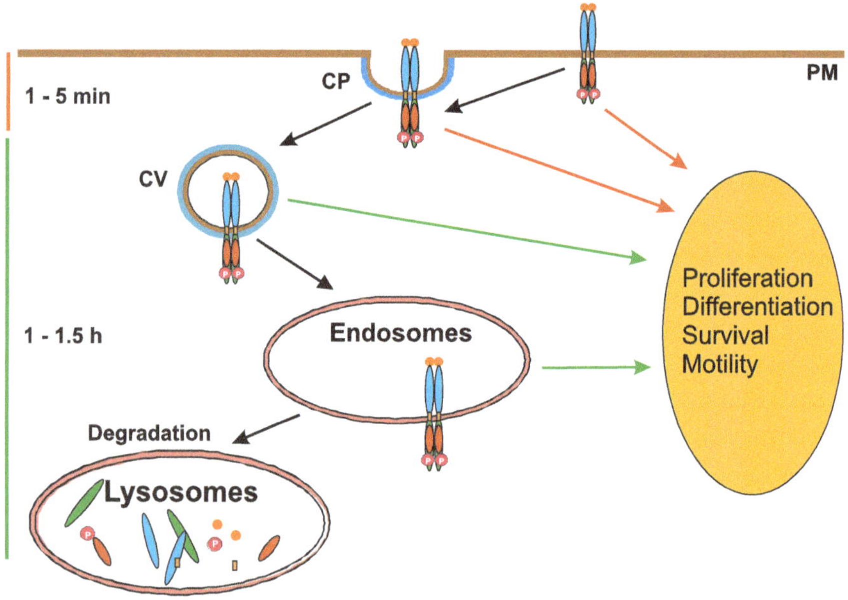

Figure 5. A model to describe the EGFR signaling along its endocytic route. Following its activation at the plasma membrane EGFR continues to signal along its endocytic route till its degradation in lysosome. EGFR signaling from both the plasma membrane and the intracellular endocytic compartment regulates major signaling pathways leading to cell proliferation and survival. The Activated EGFR may stay 1-5 minutes at the plasma membrane and 1 - 1.5 h along the endocytic compartments. PM: plasma membrane. CP: coated pit. CV: coated vesicle.

5. Remaining questions, perspectives and future research

In spite of intensive research and significant progress, many important issues remain unsolved regarding EGFR endocytosis and its regulation of cell signaling. One issue is clathrin-mediated endocytosis vs non-clathrin-mediated endocytosis. Significant evidence supportting the presence of non-clathrin mediated endocytosis and EGFR also internalized through non-clathrin mediated endocytic pathways. However, the mechanisms dictating which endocytic pathways EGFR follows under various conditions are far from clear and the functions of these different endocytic pathways are not clear either. While a few recent researchers showed that EGFR undergoes non-clathrin mediated endocytosis at high EGF concentrations and leads to EGFR degradation [137], the extensive data that support the role of chathrin-mediate endocytosis in EGFR internalization and degradation in lysosomes in the past several decades are mostly obtained at high EGF concentrations. A recent study showed that during cell mitosis, EGFR follows non-clathrin mediated endocytic pathway under both low and high EGF concentrations [107], which suggest that EGF concentration is, at least, not the only factor dictating the entry of EGFR into different endocytic pathways. It is also difficult to explain why cells choose the much slower non-clathrin mediated endocytosis to degrade EGFR, because it provides the heavily phosphorylated EGFR too much time to signal before degradation.

Another issue is the role of EGFR kinase activity in EGFR endocytosis. Both of the opposing claims that EGF-induced EGFR endocytosis is dependent on EGFR kinase activity and that it is independent of EGFR kinase activity are supported by many data. It is difficult to reconcile the differences in the literature. However, a recent piece of research may shed some light. It was recently reported that EGF-induced EGFR endocytosis is independent of EGFR kinase activity during interphase, but is dependent on EGFR kinase activity during mitosis [107]. During mitosis, EGF-induced EGFR endocytosis is slower and independent of clathrin [107]. As previous research never distinguished the cells at interphase from cells at mitosis and at any given time there is a portion of cells at mitosis, the reported results are always a combination of kinase-independent and kinase-dependent endocytosis. Depending on the cell type and experimental conditions, the data may vary significantly. Moreover, EGFR may also undergo both kinase-dependent and kinase independent endocytosis during interphase depending on the cell type and experimental conditions. It has been shown that antibody-induced EGFR endocytosis is independent of EGFR kinase activity [109]. Future research is needed to elucidate the molecular mechanisms underlying kinase-independent and kinase-dependent endocytic pathways.

Last, but not least, the function and significance of EGFR signaling from endosome vs EGFR signaling from the plasma membrane provides room for further research. It is clear that EGFR signals from both the plasma membrane and the endosomes, and that the signals from both locations are able to activate major signaling pathways, stimulate cell proliferation, and promote cell survival. However, the extent of the difference between these two signals is unclear. So far, the results have come from either the comparison between endosomal signaling and standard EGFR signaling, or the comparison between plasma membrane signaling and standard EGFR signaling. A direct comparison between endosomal EGFR signaling and the plasma membrane EGFR signaling is needed to define the functional difference and their physiological significance of these two signals. The spatio-temporal dynamics of EGFR signaling in controlling cell function has become a new focus of current research. EGFR signaling along the endocytic route from the plasma membrane to endosomes allows a vigorous regulation of spatio-temporal dynamics of EGFR signaling.

Author details

Zhixiang Wang
Department of Medical Genetics, Signal Transduction Research Group,
Faculty of Medicine and Dentistry, University of Alberta, Edmonton, AB, Canada

6. References

[1] Wells A. EGF receptor. International Journal of Biochemistry & Cell Biology 1999 Jun;31(6):637-43.

[2] Yarden Y, Sliwkowski. Untangling the ErbB signalling network. Nature Reviews 2001;2:127-37.

[3] Bargmann CI, Hung MC, Weinberg RA. Multiple independent activations of the neu oncogene by a point mutation altering the transmembrane domain of p185. Cell 1986;45(5):649-57.

[4] Yamamoto T, Ikawa S, Akiyama T, Nomura N, Miyajima N, Saito T, et al. Similarity of protein encoded by the human c-erb-B-2 gene to epidermal growth factor receptor. Nature 1986;319(6050):230-4.

[5] Kraus MH, Issing W, Miki T, Popescu NC, Aaronson SA. Isolation and characterization of ERBB3, a third member of the ERBB/epidermal growth factor receptor family: evidence for overexpression in a subset of human mammary tumors. Proc Natl Acad Sci U S A 1989;86(23):9193-7.

[6] Plowman GD, Culouscou JM, Whitney GS, Green JM, Carlton GW, Foy L, et al. Ligand-specific activation of HER4/p180erbB4, a fourth member of the epidermal growth factor receptor family. Proc Natl Acad Sci U S A 1993;90(5):1746-50.

[7] Coussens L, Yang-Feng TL, Liao YC, Chen E, Gray A, McGrath J, et al. Tyrosine kinase receptor with extensive homology to EGF receptor shares chromosomal location with neu oncogene. Science 1985;230(4730):1132-9.

[8] Hynes NE, Lane HA. ERBB receptors and cancer: the complexity of targeted inhibitors. Nat Rev Cancer 2005 May;5(5):341-54.

[9] Mitsudomi T, Yatabe Y. Epidermal growth factor receptor in relation to tumor development: EGFR gene and cancer. FEBS J 2010 Jan;277(2):301-8.

[10] Citri A, Yarden Y. EGF-ERBB signalling: towards the systems level. Nat Rev Mol Cell Biol 2006 Jul;7(7):505-16.

[11] Burgess AW. EGFR family: structure physiology signalling and therapeutic targets. Growth Factors 2008 Oct;26(5):263-74.

[12] Hynes NE, MacDonald G. ErbB receptors and signaling pathways in cancer. Curr Opin Cell Biol 2009 Apr;21(2):177-84.
[13] Rivera F, Vega-Villegas ME, Lopez-Brea MF, Marquez R. Current situation of panitumumab, matuzumab, nimotuzumab and zalutumumab. Acta Oncologica 2008;47(1):9-19.
[14] de Bono JS, Rowinsky EK. The ErbB receptor family: a therapeutic target for cancer. Trends in Molecular Medicine 2002;8(4):S19-S26.
[15] Zandi R, Larsen AB, Andersen P, Stockhausen MT, Poulsen HS. Mechanisms for oncogenic activation of the epidermal growth factor receptor. Cell Signal 2007 Oct;19(10):2013-23.
[16] Carpenter G. Receptors for epidermal growth factor and other polypeptide mitogens. Annu Rev Biochem 1987;56:881-914:881-914.
[17] Schlessinger J, Ullrich A. Growth factor signaling by receptor tyrosine kinases. Neuron 1992;9(3):383-91.
[18] Sorkin A, Goh LK. Endocytosis and intracellular trafficking of ErbBs. Exp Cell Res 2009 Feb 15;315(4):683-96.
[19] Di Fiore PP, Gill GN. Endocytosis and mitogenic signaling. Curr Opin Cell Biol 1999 Aug;11(4):483-8.
[20] Clague MJ, Urbe S. The interface of receptor trafficking and signalling. J Cell Sci 2001 Sep;114(Pt 17):3075-81.
[21] Grandal MV, Madshus IH. Epidermal growth factor receptor and cancer: control of oncogenic signalling by endocytosis. J Cell Mol Med 2008 Sep;12(5A):1527-34.
[22] Murphy JE, Padilla BE, Hasdemir B, Cottrell GS, Bunnett NW. Endosomes: a legitimate platform for the signaling train. Proc Natl Acad Sci U S A 2009 Oct 20;106(42):17615-22.
[23] Sadowski L, Pilecka I, Miaczynska M. Signaling from endosomes: location makes a difference. Exp Cell Res 2009 May 15;315(9):1601-9.
[24] Pawson T. Protein modules and signalling networks. Nature 1995;373(6515):573-80.
[25] Pawson T. New impressions of Src and Hck [news; comment]. Nature 1997;385(6617):582-3, 585.
[26] Olayioye MA, Beuvink I, Horsch K, Daly JM, Hynes NE. ErbB receptor-induced activation of stat transcription factors is mediated by Src tyrosine kinases. J Biol Chem 1999 Jun 11;274(24):17209-18.
[27] Pawson T. Specificity in signal transduction: from phosphotyrosine-SH2 domain interactions to complex cellular systems. Cell 2004 Jan 23;116(2):191-203.
[28] Marmor MD, Skaria KB, Yarden Y. Signal transduction and oncogenesis by ErbB/HER receptors. Int J Radiat Oncol Biol Phys 2004 Mar 1;58(3):903-13.
[29] Lemmon MA, Schlessinger J. Cell signaling by receptor tyrosine kinases. Cell 2010 Jun 25;141(7):1117-34.
[30] Pearson G, Robinson F, Gibson TB, Xu BE, Karandikar M, Berman K, et al. Mitogen-activated protein (MAP) kinase pathways: Regulation and physiological functions. Endocrine Reviews 2001 Apr;22(2):153-83.
[31] Avruch J. MAP kinase pathways: The first twenty years. Biochimica et Biophysica Acta-Molecular Cell Research 2007 Aug;1773(8):1150-60.

[32] Marshall CJ. MAP kinase kinase kinase, MAP kinase kinase and MAP kinase. Curr Opin Genet Dev 1994;4(1):82-9.
[33] Marshall CJ. Cell signalling. Raf gets it together. Nature 1996 Sep 12;383(6596):127-8.
[34] Roux PP, Blenis J. ERK and p38 MAPK-activated protein kinases: a family of protein kinases with diverse biological functions. Microbiol Mol Biol Rev 2004 Jun;68(2):320-44.
[35] Kim HK, Kim JW, Zilberstein A, Margolis B, Kim JG, Schlessinger J, et al. PDGF stimulation of inositol phospholipid hydrolysis requires PLC-gamma 1 phosphorylation on tyrosine residues 783 and 1254. Cell 1991;65(3):435-41.
[36] Ronnstrand L, Mori S, Arridsson AK, Eriksson A, Wernstedt C, Hellman U, et al. Identification of two C-terminal autophosphorylation sites in the PDGF beta-receptor: involvement in the interaction with phospholipase C- gamma. EMBO J 1992;11(11):3911-9.
[37] Rotin D, Margolis B, Mohammadi M, Daly RJ, Daum G, Li N, et al. SH2 domains prevent tyrosine dephosphorylation of the EGF receptor: identification of Tyr992 as the high-affinity binding site for SH2 domains of phospholipase C gamma. EMBO J 1992;11(2):559-67.
[38] Jang IH, Lee S, Park JB, Kim JH, Lee CS, Hur EM, et al. The direct interaction of phospholipase C-gamma 1 with phospholipase D2 is important for epidermal growth factor signaling. J Biol Chem 2003 May 16;278(20):18184-90.
[39] Wang Y, Wu J, Wang Z. Akt Binds to and Phosphorylates Phospholipase C-{gamma}1 in Response to Epidermal Growth Factor. Mol Biol Cell 2006 May;17(5):2267-77.
[40] Ye K, Aghdasi B, Luo HR, Moriarity JL, Wu FY, Hong JJ, et al. Phospholipase C gamma 1 is a physiological guanine nucleotide exchange factor for the nuclear GTPase PIKE. Nature 2002 Jan 31;415(6871):541-4.
[41] Choi JH, Park JB, Bae SS, Yun S, Kim HS, Hong WP, et al. Phospholipase C-gamma1 is a guanine nucleotide exchange factor for dynamin-1 and enhances dynamin-1-dependent epidermal growth factor receptor endocytosis. J Cell Sci 2004 Aug 1;117(Pt 17):3785-95.
[42] Li S, Wang Q, Wang Y, Chen X, Wang Z. PLC-gamma1 and Rac1 coregulate EGF-induced cytoskeleton remodeling and cell migration. Mol Endocrinol 2009 Jun;23(6):901-13.
[43] Wells A, Kassis J, Solava J, Turner T, Lauffenburger DA. Growth factor-induced cell motility in tumor invasion. Acta Oncol 2002;41(2):124-30.
[44] Wang Z, Gluck S, Zhang L, Moran MF. Requirement for phospholipase C-gamma1 enzymatic activity in growth factor-induced mitogenesis. Mol Cell Biol 1998 Jan;18(1):590-7.
[45] Downward J. Mechanisms and consequences of activation of protein kinase B/Akt. Curr Opin Cell Biol 1998 Apr;10(2):262-7.
[46] Castellano E, Downward J. RAS Interaction with PI3K: More Than Just Another Effector Pathway. Genes Cancer 2011 Mar;2(3):261-74.
[47] Burgering BM, Coffer PJ. Protein kinase B (c-Akt) in phosphatidylinositol-3-OH kinase signal transduction [see comments]. Nature 1995 Aug 17;376(6541):599-602.

[48] Okano J, Gaslightwala I, Birnbaum MJ, Rustgi AK, Nakagawa H. Akt/protein kinase B isoforms are differentially regulated by epidermal growth factor stimulation. J Biol Chem 2000 Oct 6;275(40):30934-42.
[49] Quesnelle KM, Boehm AL, Grandis JR. STAT-mediated EGFR signaling in cancer. J Cell Biochem 2007 Oct 1;102(2):311-9.
[50] Maa MC, Leu TH, McCarley DJ, Schatzman RC, Parsons SJ. Potentiation of epidermal growth factor receptor-mediated oncogenesis by c-Src: implications for the etiology of multiple human cancers. Proc Natl Acad Sci U S A 1995 Jul 18;92(15):6981-5.
[51] Tice DA, Biscardi JS, Nickles AL, Parsons SJ. Mechanism of biological synergy between cellular Src and epidermal growth factor receptor. Proc Natl Acad Sci U S A 1999 Feb 16;96(4):1415-20.
[52] Egloff AM, Grandis JR. Targeting epidermal growth factor receptor and SRC pathways in head and neck cancer. Semin Oncol 2008 Jun;35(3):286-97.
[53] Scaltriti M, Baselga J. The epidermal growth factor receptor pathway: a model for targeted therapy. Clin Cancer Res 2006 Sep 15;12(18):5268-72.
[54] Lurje G, Lenz HJ. EGFR signaling and drug discovery. Oncology 2009;77(6):400-10.
[55] Summy JM, Gallick GE. Treatment for advanced tumors: SRC reclaims center stage. Clin Cancer Res 2006 Mar 1;12(5):1398-401.
[56] Yeatman TJ. A renaissance for SRC. Nat Rev Cancer 2004 Jun;4(6):470-80.
[57] Leu TH, Maa MC. Functional implication of the interaction between EGF receptor and c-Src. Front Biosci 2003 Jan 1;8:s28-s38.
[58] Goldstein JL, Brown MS, Anderson RG, Russell DW, Schneider WJ. Receptor-mediated endocytosis: concepts emerging from the LDL receptor system. Annu Rev Cell Biol 1985;1:1-39:1-39.
[59] Schmid EM, McMahon HT. Integrating molecular and network biology to decode endocytosis. Nature 2007 Aug 23;448(7156):883-8.
[60] Kelly BT, McCoy AJ, Spate K, Miller SE, Evans PR, Honing S, et al. A structural explanation for the binding of endocytic dileucine motifs by the AP2 complex. Nature 2008 Dec 18;456(7224):976-9.
[61] Pryor PR, Jackson L, Gray SR, Edeling MA, Thompson A, Sanderson CM, et al. Molecular basis for the sorting of the SNARE VAMP7 into endocytic clathrin-coated vesicles by the ArfGAP Hrb. Cell 2008 Sep 5;134(5):817-27.
[62] Doherty GJ, McMahon HT. Mechanisms of endocytosis. Annu Rev Biochem 2009;78:857-902.
[63] Hansen CG, Nichols BJ. Molecular mechanisms of clathrin-independent endocytosis. J Cell Sci 2009 Jun 1;122(Pt 11):1713-21.
[64] Kirchhausen T, Bonifacino JS, Riezman H. Linking cargo to vesicle formation: receptor tail interactions with coat proteins. Curr Opin Cell Biol 1997 Aug;9(4):488-95.
[65] Mellman I. Endocytosis and molecular sorting. Annu Rev Cell Dev Biol 1996;12:575-625:575-625.
[66] Bonifacino JS, Dell'Angelica EC. Molecular bases for the recognition of tyrosine-based sorting signals. J Cell Biol 1999 May 31;145(5):923-6.

[67] Canfield WM, Johnson KF, Ye RD, Gregory W, Kornfeld S. Localization of the signal for rapid internalization of the bovine cation-independent mannose 6-phosphate/insulin-like growth factor-II receptor to amino acids 24-29 of the cytoplasmic tail. J Biol Chem 1991 Mar 25;266(9):5682-8.
[68] Chen WJ, Goldstein JL, Brown MS. NPXY, a sequence often found in cytoplasmic tails, is required for coated pit-mediated internalization of the low density lipoprotein receptor. J Biol Chem 1990 Feb 25;265(6):3116-23.
[69] Paccaud JP, Reith W, Johansson B, Magnusson KE, Mach B, Carpentier JL. Clathrin-coated pit-mediated receptor internalization. Role of internalization signals and receptor mobility. J Biol Chem 1993 Nov 5;268(31):23191-6.
[70] Hsu D, Knudson PE, Zapf A, Rolband GC, Olefsky JM. NPXY motif in the insulin-like growth factor-I receptor is required for efficient ligand-mediated receptor internalization and biological signaling. Endocrinology 1994 Feb;134(2):744-50.
[71] Vignoud L, Albiges-Rizo C, Frachet P, Block MR. NPXY motifs control the recruitment of the alpha5beta1 integrin in focal adhesions independently of the association of talin with the beta1 chain. J Cell Sci 1997 Jun;110(Pt 12):1421-30.
[72] Lazarovits J, Roth M. A single amino acid change in the cytoplasmic domain allows the influenza virus hemagglutinin to be endocytosed through coated pits. Cell 1988 Jun 3;53(5):743-52.
[73] Lobel P, Fujimoto K, Ye RD, Griffiths G, Kornfeld S. Mutations in the cytoplasmic domain of the 275 kd mannose 6-phosphate receptor differentially alter lysosomal enzyme sorting and endocytosis. Cell 1989 Jun 2;57(5):787-96.
[74] McGraw TE, Maxfield FR. Human transferrin receptor internalization is partially dependent upon an aromatic amino acid on the cytoplasmic domain. Cell Regul 1990 Mar;1(4):369-77.
[75] Collawn JF, Stangel M, Kuhn LA, Esekogwu V, Jing SQ, Trowbridge IS, et al. Transferrin receptor internalization sequence YXRF implicates a tight turn as the structural recognition motif for endocytosis. Cell 1990 Nov 30;63(5):1061-72.
[76] Alvarez E, Girones N, Davis RJ. A point mutation in the cytoplasmic domain of the transferrin receptor inhibits endocytosis. Biochem J 1990 Apr 1;267(1):31-5.
[77] Breitfeld PP, Casanova JE, McKinnon WC, Mostov KE. Deletions in the cytoplasmic domain of the polymeric immunoglobulin receptor differentially affect endocytotic rate and postendocytotic traffic. J Biol Chem 1990 Aug 15;265(23):13750-7.
[78] Bonifacino JS, Traub LM. Signals for sorting of transmembrane proteins to endosomes and lysosomes. Annu Rev Biochem 2003;72:395-447. Epub;%2003 Mar 06.:395-447.
[79] Jadot M, Canfield WM, Gregory W, Kornfeld S. Characterization of the signal for rapid internalization of the bovine mannose 6-phosphate/insulin-like growth factor-II receptor. J Biol Chem 1992 Jun 5;267(16):11069-77.
[80] Ohno H, Stewart J, Fournier MC, Bosshart H, Rhee I, Miyatake S, et al. Interaction of tyrosine-based sorting signals with clathrin-associated proteins. Science 1995 Sep 29;269(5232):1872-5.

[81] Ohno H, Fournier MC, Poy G, Bonifacino JS. Structural determinants of interaction of tyrosine-based sorting signals with the adaptor medium chains. J Biol Chem 1996 Nov 15;271(46):29009-15.
[82] Ohno H, Aguilar RC, Yeh D, Taura D, Saito T, Bonifacino JS. The medium subunits of adaptor complexes recognize distinct but overlapping sets of tyrosine-based sorting signals. J Biol Chem 1998 Oct 2;273(40):25915-21.
[83] Nesterov A, Carter RE, Sorkina T, Gill GN, Sorkin A. Inhibition of the receptor-binding function of clathrin adaptor protein AP-2 by dominant-negative mutant mu2 subunit and its effects on endocytosis. EMBO J 1999 May 4;18(9):2489-99.
[84] Hershko A, Ciechanover A. The ubiquitin system. Annu Rev Biochem 1998;67:425-79.
[85] Ben-Saadon R, Fajerman I, Ziv T, Hellman U, Schwartz AL, Ciechanover A. The tumor suppressor protein p16(INK4a) and the human papillomavirus oncoprotein-58 E7 are naturally occurring lysine-less proteins that are degraded by the ubiquitin system. Direct evidence for ubiquitination at the N-terminal residue. J Biol Chem 2004 Oct 1;279(40):41414-21.
[86] Ciechanover A, Ben-Saadon R. N-terminal ubiquitination: more protein substrates join in. Trends Cell Biol 2004 Mar;14(3):103-6.
[87] Carpenter G, King L, Jr., Cohen S. Rapid enhancement of protein phosphorylation in A-431 cell membrane preparations by epidermal growth factor. J Biol Chem 1979;254(11):4884-91.
[88] Sorkin A, Carpenter G. Interaction of activated EGF receptors with coated pit adaptins. Science 1993;261(5121):612-5.
[89] Rothman JE. Mechanisms of intracellular protein transport. Nature 1994;372(6501):55-63.
[90] Valencia A, Chardin P, Wittinghofer A, Sander C. The ras protein family: evolutionary tree and role of conserved amino acids. Biochemistry 1991;30(19):4637-48.
[91] Clark J, Moore L, Krasinskas A, Way J, Battey J, Tamkun J, et al. Selective amplification of additional members of the ADP- ribosylation factor (ARF) family: cloning of additional human and Drosophila ARF- like genes. Proc Natl Acad Sci U S A 1993;90(19):8952-6.
[92] Tsuchiya M, Price SR, Tsai SC, Moss J, Vaughan M. Molecular identification of ADP-ribosylation factor mRNAs and their expression in mammalian cells. J Biol Chem 1991 Feb 15;266(5):2772-7.
[93] Trowbridge IS, Collawn JF, Hopkins CR. Signal-dependent membrane protein trafficking in the endocytic pathway. Annu Rev Cell Biol 1993;9:129-61:129-61.
[94] Chen WS, Lazar CS, Poenie M, Tsien RY, Gill GN, Rosenfeld MG. Requirement for intrinsic protein tyrosine kinase in the immediate and late actions of the EGF receptor. Nature 1987;328(6133):820-3.
[95] Glenney JR, Jr., Chen WS, Lazar CS, Walton GM, Zokas LM, Rosenfeld MG, et al. Ligand-induced endocytosis of the EGF receptor is blocked by mutational inactivation and by microinjection of anti- phosphotyrosine antibodies. Cell 1988;52(5):675-84.

[96] Felder S, Miller K, Moehren G, Ullrich A, Schlessinger J, Hopkins CR. Kinase activity controls the sorting of the epidermal growth factor receptor within the multivesicular body. Cell 1990;61(4):623-34.
[97] Felder S, LaVin J, Ullrich A, Schlessinger J. Kinetics of binding, endocytosis, and recycling of EGF receptor mutants. J Cell Biol 1992;117(1):203-12.
[98] Lamaze C, Schmid SL. Recruitment of epidermal growth factor receptors into coated pits requires their activated tyrosine kinase. J Cell Biol 1995 Apr;129(1):47-54.
[99] Wilde A, Beattie EC, Lem L, Riethof DA, Liu SH, Mobley WC, et al. EGF receptor signaling stimulates SRC kinase phosphorylation of clathrin, influencing clathrin redistribution and EGF uptake. Cell 1999 Mar 5;96(5):677-87.
[100] Confalonieri S, Salcini AE, Puri C, Tacchetti C, Di Fiore PP. Tyrosine phosphorylation of Eps15 is required for ligand-regulated, but not constitutive, endocytosis. J Cell Biol 2000 Aug 21;150(4):905-12.
[101] Sorkina T, Huang F, Beguinot L, Sorkin A. Effect of tyrosine kinase inhibitors on clathrin-coated pit recruitment and internalization of epidermal growth factor receptor. J Biol Chem 2002 Jul 26;277(30):27433-41.
[102] Schmidt MH, Furnari FB, Cavenee WK, Bogler O. Epidermal growth factor receptor signaling intensity determines intracellular protein interactions, ubiquitination, and internalization. Proc Natl Acad Sci U S A 2003 May 27;100(11):6505-10.
[103] Wang Y, Pennock S, Chen X, Wang Z. Endosomal signaling of epidermal growth factor receptor stimulates signal transduction pathways leading to cell survival. Mol Cell Biol 2002 Oct;22(20):7279-90.
[104] Wang Y, Pennock S, Chen X, Wang Z. Internalization of inactive EGF receptor into endosomes and the subsequent activation of endosome-associated EGF receptors. Epidermal growth factor. Sci STKE 2002 Dec 3;2002(161):L17.
[105] Wang Q, Villeneuve G, Wang Z. Control of epidermal growth factor receptor endocytosis by receptor dimerization, rather than receptor kinase activation. EMBO Rep 2005 Oct;6(10):942-8.
[106] Wang Q, Zhu F, Wang Z. Identification of EGF Receptor C-terminal Sequences 1005-1017 and Di-leucine Motif $^{1010}LL^{1011}$ as Essential in EGF Receptor Endocytosis. Exp Cell Res 2007;313:3349-63.
[107] Liu L, Shi H, Chen X, Wang Z. Regulation of EGF-stimulated EGF receptor endocytosis during M phase. Traffic 2011 Feb;12(2):201-17.
[108] Schlessinger J. Ligand-induced, receptor-mediated dimerization and activation of EGF receptor. Cell 2002 Sep;%20;110(6):669-72.
[109] Jaramillo ML, Leon Z, Grothe S, Paul-Roc B, Abulrob A, O'Connor MM. Effect of the anti-receptor ligand-blocking 225 monoclonal antibody on EGF receptor endocytosis and sorting. Exp Cell Res 2006 Sep 10;312(15):2778-90.
[110] Chang CP, Lazar CS, Walsh BJ, Komuro M, Collawn JF, Kuhn LA, et al. Ligand-induced internalization of the epidermal growth factor receptor is mediated by multiple endocytic codes analogous to the tyrosine motif found in constitutively internalized receptors. J Biol Chem 1993;268(26):19312-20.

[111] Decker SJ, Alexander C, Habib T. Epidermal growth factor (EGF)-stimulated tyrosine phosphorylation and EGF receptor degradation in cells expressing EGF receptors truncated at residue 973. J Biol Chem 1992;267(2):1104-8.
[112] Sorkin A, Helin K, Waters CM, Carpenter G, Beguinot L. Multiple autophosphorylation sites of the epidermal growth factor receptor are essential for receptor kinase activity and internalization. Contrasting significance of tyrosine 992 in the native and truncated receptors. J Biol Chem 1992;267(12):8672-8.
[113] Nesterov A, Wiley HS, Gill GN. Ligand-induced endocytosis of epidermal growth factor receptors that are defective in binding adaptor proteins. Proc Natl Acad Sci U S A 1995;92(19):8719-23.
[114] Sorkin A, Mazzotti M, Sorkina T, Scotto L, Beguinot L. Epidermal growth factor receptor interaction with clathrin adaptors is mediated by the Tyr974-containing internalization motif. J Biol Chem 1996 Jun 7;271(23):13377-84.
[115] Kurten RC, Cadena DL, Gill GN. Enhanced degradation of EGF receptors by a sorting nexin, SNX1. Science 1996;272(5264):1008-10.
[116] Kil SJ, Hobert M, Carlin C. A leucine-based determinant in the epidermal growth factor receptor juxtamembrane domain is required for the efficient transport of ligand-receptor complexes to lysosomes. J Biol Chem 1999 Jan 29;274(5):3141-50.
[117] Tsacoumangos A, Kil SJ, Ma L, Sonnichsen FD, Carlin C. A novel dileucine lysosomal-sorting-signal mediates intracellular EGF-receptor retention independently of protein ubiquitylation. J Cell Sci 2005 Sep 1;118(Pt 17):3959-71.
[118] Huang F, Jiang X, Sorkin A. Tyrosine phosphorylation of the beta2 subunit of clathrin adaptor complex AP-2 reveals the role of a di-leucine motif in the epidermal growth factor receptor trafficking. J Biol Chem 2003 Oct 31;278(44):43411-7.
[119] Jiang X, Huang F, Marusyk A, Sorkin A. Grb2 regulates internalization of EGF receptors through clathrin-coated pits. Mol Biol Cell 2003 Mar;14(3):858-70.
[120] Wang Z, Moran MF. Requirement for the adapter protein GRB2 in EGF receptor endocytosis. Science 1996;272(5270):1935-9.
[121] Johannessen LE, Pedersen NM, Pedersen KW, Madshus IH, Stang E. Activation of the epidermal growth factor (EGF) receptor induces formation of EGF receptor- and Grb2-containing clathrin-coated pits. Mol Cell Biol 2006 Jan;26(2):389-401.
[122] Huang F, Sorkin A. Growth factor receptor binding protein 2-mediated recruitment of the RING domain of Cbl to the epidermal growth factor receptor is essential and sufficient to support receptor endocytosis. Mol Biol Cell 2005 Mar;16(3):1268-81.
[123] Barbieri MA, Kong C, Chen PI, Horazdovsky BF, Stahl PD. The SRC homology 2 domain of Rin1 mediates its binding to the epidermal growth factor receptor and regulates receptor endocytosis. J Biol Chem 2003 Aug 22;278(34):32027-36.
[124] Huang F, Khvorova A, Marshall W, Sorkin A. Analysis of clathrin-mediated endocytosis of epidermal growth factor receptor by RNA interference. J Biol Chem 2004 Apr 16;279(16):16657-61.
[125] Shen Y, Xu L, Foster DA. Role for phospholipase D in receptor-mediated endocytosis. Mol Cell Biol 2001 Jan;21(2):595-602.

[126] van Bergen en Henegouwen PM. Eps15: a multifunctional adaptor protein regulating intracellular trafficking. Cell Commun Signal 2009;7:24.

[127] Huang F, Goh LK, Sorkin A. EGF receptor ubiquitination is not necessary for its internalization. Proc Natl Acad Sci U S A 2007 Oct 23;104(43):16904-9.

[128] Pennock S, Wang Z. A tale of two Cbls: interplay of c-Cbl and Cbl-b in epidermal growth factor receptor downregulation. Mol Cell Biol 2008 May;28(9):3020-37.

[129] Joly M, Kazlauskas A, Fay FS, Corvera S. Disruption of PDGF receptor trafficking by mutation of its PI-3 kinase binding sites. Science 1994;263(5147):684-7.

[130] Joly M, Kazlauskas A, Corvera S. Phosphatidylinositol 3-kinase activity is required at a postendocytic step in platelet-derived growth factor receptor trafficking. J Biol Chem 1995;270(22):13225-30.

[131] Shpetner H, Joly M, Hartley D, Corvera S. Potential sites of PI-3 kinase function in the endocytic pathway revealed by the PI-3 kinase inhibitor, wortmannin. J Cell Biol 1996;132(4):595-605.

[132] Chen X, Wang Z. Regulation of intracellular trafficking of the EGF receptor by Rab5 in the absence of phosphatidylinositol 3-kinase activity. EMBO rep 2001 Jan ;2 (1):68 -74 2001;2(1):68-74.

[133] Kirkham M, Parton RG. Clathrin-independent endocytosis: new insights into caveolae and non-caveolar lipid raft carriers. Biochim Biophys Acta 2005 Dec 30;1746(3):349-63.

[134] Mayor S, Pagano RE. Pathways of clathrin-independent endocytosis. Nat Rev Mol Cell Biol 2007 Aug;8(8):603-12.

[135] Orth JD, Krueger EW, Weller SG, McNiven MA. A novel endocytic mechanism of epidermal growth factor receptor sequestration and internalization. Cancer Res 2006 Apr 1;66(7):3603-10.

[136] Sigismund S, Woelk T, Puri C, Maspero E, Tacchetti C, Transidico P, et al. Clathrin-independent endocytosis of ubiquitinated cargos. Proc Natl Acad Sci U S A 2005 Feb 22;102(8):2760-5.

[137] Dobrowolski R, De Robertis EM. Endocytic control of growth factor signalling: multivesicular bodies as signalling organelles. Nat Rev Mol Cell Biol 2012 Jan;13(1):53-60.

[138] McKanna JA, Haigler HT, Cohen S. Hormone receptor topology and dynamics: morphological analysis using ferritin-labeled epidermal growth factor. Proc Natl Acad Sci U S A 1979 Nov;76(11):5689-93.

[139] Roepstorff K, Grovdal L, Grandal M, Lerdrup M, van DB. Endocytic downregulation of ErbB receptors: mechanisms and relevance in cancer. Histochem Cell Biol 2008 May;129(5):563-78.

[140] Garcia dP, I, Adams GP, Sundareshan P, Wong AJ, Testa JR, Bigner DD, et al. Expression of mutated epidermal growth factor receptor by non-small cell lung carcinomas. Cancer Res 1993 Jul 15;53(14):3217-20.

[141] Ge H, Gong X, Tang CK. Evidence of high incidence of EGFRvIII expression and coexpression with EGFR in human invasive breast cancer by laser capture microdissection and immunohistochemical analysis. Int J Cancer 2002 Mar 20;98(3):357-61.

[142] Moscatello DK, Holgado-Madruga M, Emlet DR, Montgomery RB, Wong AJ. Constitutive activation of phosphatidylinositol 3-kinase by a naturally occurring mutant epidermal growth factor receptor. J Biol Chem 1998 Jan 2;273(1):200-6.
[143] Okamoto I, Kenyon LC, Emlet DR, Mori T, Sasaki J, Hirosako S, et al. Expression of constitutively activated EGFRvIII in non-small cell lung cancer. Cancer Sci 2003 Jan;94(1):50-6.
[144] Olapade-Olaopa EO, Moscatello DK, MacKay EH, Horsburgh T, Sandhu DP, Terry TR, et al. Evidence for the differential expression of a variant EGF receptor protein in human prostate cancer. Br J Cancer 2000 Jan;82(1):186-94.
[145] Sugawa N, Yamamoto K, Ueda S, Morita N, Kita M, Nishino H, et al. Function of aberrant EGFR in malignant gliomas. Brain Tumor Pathol 1998;15(1):53-7.
[146] Kuan CT, Wikstrand CJ, Bigner DD. EGF mutant receptor vIII as a molecular target in cancer therapy. Endocr Relat Cancer 2001 Jun;8(2):83-96.
[147] Antonyak MA, Moscatello DK, Wong AJ. Constitutive activation of c-Jun N-terminal kinase by a mutant epidermal growth factor receptor. J Biol Chem 1998 Jan 30;273(5):2817-22.
[148] Batra SK, Castelino-Prabhu S, Wikstrand CJ, Zhu X, Humphrey PA, Friedman HS, et al. Epidermal growth factor ligand-independent, unregulated, cell-transforming potential of a naturally occurring human mutant EGFRvIII gene. Cell Growth Differ 1995 Oct;6(10):1251-9.
[149] Nagane M, Coufal F, Lin H, Bogler O, Cavenee WK, Huang HJ. A common mutant epidermal growth factor receptor confers enhanced tumorigenicity on human glioblastoma cells by increasing proliferation and reducing apoptosis. Cancer Res 1996 Nov 1;56(21):5079-86.
[150] Nishikawa R, Ji XD, Harmon RC, Lazar CS, Gill GN, Cavenee WK, et al. A mutant epidermal growth factor receptor common in human glioma confers enhanced tumorigenicity. Proc Natl Acad Sci U S A 1994 Aug 2;91(16):7727-31.
[151] Pedersen MW, Pedersen N, Ottesen LH, Poulsen HS. Differential response to gefitinib of cells expressing normal EGFR and the mutant EGFRvIII. Br J Cancer 2005 Oct 17;93(8):915-23.
[152] Tang CK, Gong XQ, Moscatello DK, Wong AJ, Lippman ME. Epidermal growth factor receptor vIII enhances tumorigenicity in human breast cancer. Cancer Res 2000 Jun 1;60(11):3081-7.
[153] Han W, Zhang T, Yu H, Foulke JG, Tang CK. Hypophosphorylation of residue Y1045 leads to defective downregulation of EGFRvIII. Cancer Biol Ther 2006 Oct;5(10):1361-8.
[154] Austin CD, De Maziere AM, Pisacane PI, van Dijk SM, Eigenbrot C, Sliwkowski MX, et al. Endocytosis and sorting of ErbB2 and the site of action of cancer therapeutics trastuzumab and geldanamycin. Mol Biol Cell 2004 Dec;15(12):5268-82.
[155] Baulida J, Kraus MH, Alimandi M, Di Fiore PP, Carpenter G. All ErbB receptors other than the epidermal growth factor receptor are endocytosis impaired. J Biol Chem 1996;271(9):5251-7.
[156] Hommelgaard AM, Lerdrup M, van DB. Association with membrane protrusions makes ErbB2 an internalization-resistant receptor. Mol Biol Cell 2004 Apr;15(4):1557-67.

[157] Longva KE, Pedersen NM, Haslekas C, Stang E, Madshus IH. Herceptin-induced inhibition of ErbB2 signaling involves reduced phosphorylation of Akt but not endocytic down-regulation of ErbB2. Int J Cancer 2005 Sep 1;116(3):359-67.

[158] Wang Z, Zhang L, Yeung TK, Chen X. Endocytosis deficiency of epidermal growth factor (EGF) receptor-ErbB2 heterodimers in response to EGF stimulation. Mol Biol Cell 1999 May;10(5):1621-36.

[159] Haglund K, Rusten TE, Stenmark H. Aberrant receptor signaling and trafficking as mechanisms in oncogenesis. Crit Rev Oncog 2007 Aug;13(1):39-74.

[160] Friedman LM, Rinon A, Schechter B, Lyass L, Lavi S, Bacus SS, et al. Synergistic down-regulation of receptor tyrosine kinases by combinations of mAbs: implications for cancer immunotherapy. Proc Natl Acad Sci U S A 2005 Feb 8;102(6):1915-20.

[161] Fendly BM, Kotts C, Vetterlein D, Lewis GD, Winget M, Carver ME, et al. The extracellular domain of HER2/neu is a potential immunogen for active specific immunotherapy of breast cancer. J Biol Response Mod 1990 Oct;9(5):449-55.

[162] Lewis GD, Figari I, Fendly B, Wong WL, Carter P, Gorman C, et al. Differential responses of human tumor cell lines to anti-p185HER2 monoclonal antibodies. Cancer Immunol Immunother 1993 Sep;37(4):255-63.

[163] Hudis CA. Current status and future directions in breast cancer therapy. Clin Breast Cancer 2003 Nov;4 Suppl 2:S70-S75.

[164] Menard S, Pupa SM, Campiglio M, Tagliabue E. Biologic and therapeutic role of HER2 in cancer. Oncogene 2003 Sep 29;22(42):6570-8.

[165] Rubin I, Yarden Y. The basic biology of HER2. Ann Oncol 2001;12 Suppl 1:S3-S8.

[166] Ben-Kasus T, Schechter B, Lavi S, Yarden Y, Sela M. Persistent elimination of ErbB-2/HER2-overexpressing tumors using combinations of monoclonal antibodies: relevance of receptor endocytosis. Proc Natl Acad Sci U S A 2009 Mar 3;106(9):3294-9.

[167] Astsaturov I, Cohen RB, Harari P. EGFR-targeting monoclonal antibodies in head and neck cancer. Curr Cancer Drug Targets 2007 Nov;7(7):650-65.

[168] Sunada H, Magun BE, Mendelsohn J, MacLeod CL. Monoclonal antibody against epidermal growth factor receptor is internalized without stimulating receptor phosphorylation. Proc Natl Acad Sci U S A 1986 Jun;83(11):3825-9.

[169] Hayashi T, Sakamoto S. Radioimmunoassay of human epidermal growth factor--hEGF levels in human body fluids. J Pharmacobiodyn 1988 Mar;11(3):146-51.

[170] Mullin JM. Epithelial barriers, compartmentation, and cancer. Sci STKE 2004 Jan 20;2004(216):e2.

[171] Salomon DS, Normanno N, Ciardiello F, Brandt R, Shoyab M, Todaro GJ. The role of amphiregulin in breast cancer. Breast Cancer Res Treat 1995;33(2):103-14.

[172] Orlichenko L, Huang B, Krueger E, McNiven MA. Epithelial growth factor-induced phosphorylation of caveolin 1 at tyrosine 14 stimulates caveolae formation in epithelial cells. J Biol Chem 2006 Feb 24;281(8):4570-9.

[173] Sigismund S, Argenzio E, Tosoni D, Cavallaro E, Polo S, Di Fiore PP. Clathrin-mediated internalization is essential for sustained EGFR signaling but dispensable for degradation. Dev Cell 2008 Aug;15(2):209-19.

[174] Di Guglielmo GM, Le RC, Goodfellow AF, Wrana JL. Distinct endocytic pathways regulate TGF-beta receptor signalling and turnover. Nat Cell Biol 2003 May;5(5):410-21.
[175] Cohen S, Fava RA. Internalization of functional epidermal growth factor: receptor/kinase complexes in A-431 cells. J Biol Chem 1985;260(22):12351-8.
[176] Kay DG, Lai WH, Uchihashi M, Khan MN, Posner BI, Bergeron JJ. Epidermal growth factor receptor kinase translocation and activation in vivo. J Biol Chem 1986;261(18):8473-80.
[177] Lai WH, Cameron PH, Doherty JJ2, Posner BI, Bergeron JJ. Ligand-mediated autophosphorylation activity of the epidermal growth factor receptor during internalization. J Cell Biol 1989;109(6 Pt 1):2751-60.
[178] Di Guglielmo GM, Baass PC, Ou WJ, Posner BI, Bergeron JJ. Compartmentalization of SHC, GRB2 and mSOS, and hyperphosphorylation of Raf-1 by EGF but not insulin in liver parenchyma. EMBO J 1994;13(18):4269-77.
[179] Wang Z, Tung PS, Moran MF. Association of p120 ras GAP with endocytic components and colocalization with epidermal growth factor (EGF) receptor in response to EGF stimulation. Cell Growth Differ 1996;7(1):123-33.
[180] Fukazawa T, Miyake S, Band V, Band H. Tyrosine phosphorylation of Cbl upon epidermal growth factor (EGF) stimulation and its association with EGF receptor and downstream signaling proteins. J Biol Chem 1996;271(24):14554-9.
[181] Levkowitz G, Waterman H, Zamir E, Kam Z, Oved S, Langdon WY, et al. c-Cbl/Sli-1 regulates endocytic sorting and ubiquitination of the epidermal growth factor receptor. Genes Dev 1998 Dec 1;12(23):3663-74.
[182] Haugh JM, Huang AC, Wiley HS, Wells A, Lauffenburger DA. Internalized epidermal growth factor receptors participate in the activation of p21(ras) in fibroblasts. J Biol Chem 1999 Nov 26;274(48):34350-60.
[183] Kuruvilla R, Ye H, Ginty DD. Spatially and functionally distinct roles of the PI3-K effector pathway during NGF signaling in sympathetic neurons. Neuron 2000 Sep;27(3):499-512.
[184] Burke P, Schooler K, Wiley HS. Regulation of epidermal growth factor receptor signaling by endocytosis and intracellular trafficking. Mol Biol Cell 2001 Jun;12(6):1897-910.
[185] Slepnev VI, De Camilli P. Accessory factors in clathrin-dependent synaptic vesicle endocytosis. Nat Rev Neurosci 2000 Dec;1(3):161-72.
[186] Miller WE, Lefkowitz RJ. Expanding roles for beta-arrestins as scaffolds and adapters in GPCR signaling and trafficking. Curr Opin Cell Biol 2001 Apr;13(2):139-45.
[187] Vieira AV, Lamaze C, Schmid SL. Control of EGF receptor signaling by clathrin-mediated endocytosis. Science 1996;274(5295):2086-9.
[188] Ahn S, Maudsley S, Luttrell LM, Lefkowitz RJ, Daaka Y. Src-mediated tyrosine phosphorylation of dynamin is required for beta2-adrenergic receptor internalization and mitogen-activated protein kinase signaling. J Biol Chem 1999 Jan 15;274(3):1185-8.
[189] Maudsley S, Pierce KL, Zamah AM, Miller WE, Ahn S, Daaka Y, et al. The beta(2)-adrenergic receptor mediates extracellular signal-regulated kinase activation via

assembly of a multi-receptor complex with the epidermal growth factor receptor. J Biol Chem 2000 Mar 31 ;275 (13):9572 -80 2000;275(13):9572-80.

[190] Wong ES, Fong CW, Lim J, Yusoff P, Low BC, Langdon WY, et al. Sprouty2 attenuates epidermal growth factor receptor ubiquitylation and endocytosis, and consequently enhances Ras/ERK signalling. EMBO J 2002 Sep 16;21(18):4796-808.

[191] Johannessen LE, Ringerike T, Molnes J, Madshus IH. Epidermal growth factor receptor efficiently activates mitogen-activated protein kinase in HeLa cells and Hep2 cells conditionally defective in clathrin-dependent endocytosis. Exp Cell Res 2000 Oct 10;260(1):136-45.

[192] Ceresa BP, Kao AW, Santeler SR, Pessin JE. Inhibition of clathrin-mediated endocytosis selectively attenuates specific insulin receptor signal transduction pathways. Mol Cell Biol 1998 Jul;18(7):3862-70.

[193] Di Fiore PP, De Camilli P. Endocytosis and signaling. an inseparable partnership. Cell 2001 Jul 13;106(1):1-4.

[194] Wiley HS, Burke PM. Regulation of receptor tyrosine kinase signaling by endocytic trafficking. Traffic 2001 Jan;2(1):12-8.

[195] Pennock S, Wang Z. Stimulation of cell proliferation by endosomal epidermal growth factor receptor as revealed through two distinct phases of signaling. Mol Cell Biol 2003 Aug;23(16):5803-15.

Chapter 14

Hyaluronan Endocytosis: Mechanisms of Uptake and Biological Functions

Ronny Racine and Mark E. Mummert

Additional information is available at the end of the chapter

1. Introduction

Hyaluronan (HA) is a non-sulfated linear glycosaminoglycan composed of multiple copies of the disaccharide unit of D-glucuronic acid (GlcA) and N-acetyl-D-glucosamine (GlcNAc); [β-1,4-GlcA-β -1,3-GlcNAc-]$_n$ where n is the number of repeating disaccharide subunits. HA is synthesized by the HA synthase family of enzymes. Three HA synthases, termed HAS1 through HAS3, have been identified in humans and in mice. These enzymes differ from each other in their catalytic activities (HAS3 > HAS2 > HAS1) as well as in the sizes of their final products. HAS1 and HAS2 polymerize long stretches of GlcA-GlcNAc disaccharide chains, whereas HAS3 polymerizes relatively short stretches (<300 kDa). Biosynthesis of HA is regulated by exogenous stimuli. For example, HA synthesis in fibroblasts is upregulated by phorbol esters, tumor growth factor alpha, and platelet derived growth factor, whereas HA synthesis in keratinocytes is upregulated by retinoic acid, epidermal growth factor, and tumor growth factor alpha and suppressed by corticosteroids [1-4]. HA is unique among extracellular matrix components in that it is not synthesized within the cell and transported to the surface via vesicles. Hyaluronan synthase is an integral membrane protein on the surface of cells. It links together UDP-α-N-acetyl-D-glucosamine and UDP-α-D-glucuronate to spin out long strands of HA. Because, unlike other extracellular matrix carbohydrates, HA is spooled out from the cell surface, it can achieve molecular weights ranging from five thousand Daltons to twenty million Daltons.

Although HA was originally considered to be an inert filling material in the extracellular matrix and intercellular spaces, this simple carbohydrate is now known to have a number of functions in several different biological processes including development, cancer biology, wound healing and the immune response.

2. HA endocytic pathway

A scheme for the endocytosis of high molecular weight HA (HMW-HA) and its catabolism to HA oligomers have been suggested previously [5]. Figure 1 illustrates the current model for uptake of HMW-HA and its processing to bioactive oligomeric fragments.

HMW-HA in the extracellular matrix can be degraded to fragments that are 50-100 saccharides in length by the HA digesting enzyme, hyaluronidase 2 (Hyal2) [6]. Hyal2 is expressed in the lysosome and also as a GPI-linked cell surface protein [7]. This raises the interesting question of how cell surface Hyal2 retains catalytic activity for HA digestion given its requirement for an acid environment to function. Previous investigations have shown that binding of HA to the HA-receptor, CD44, leads to the interaction of CD44 with the NHE1 Na+/H+ exchanger. In turn, the NHE1 Na+-H+ exchanger creates an acidic environment facilitating Hyal2 activity [8]. The HA saccharides generated by Hyal2 can then be endocytosed by one of multiple pathways. To date, receptor mediated endocytosis of HA and macropinocytosis of bulk phase HA have been reported. Receptor mediated endocytosis can occur via lipid rafts or by the clathrin coated pit pathway. Receptors for HA endocytosis may be recycled to the cell surface or turned-over. HA saccharides are further digested to HA oligosaccharides by hyaluronidase 1 (Hyal1) in the endosome. The HA oligosaccharides could then potentially be degraded into its GlcA and GlcNAc building blocks by the concerted activities of β-D-glucuronidase and β-N-acetyl-D-hexosaminidase [9] or HA oligosaccharides could be exocytosed. The exocytosed HA fragments could have myriad biological functions which will be detailed below.

Although HA endocytosis and subsequent degradation may be important for generation of bioactive HA fragments, the HA endocytic pathway is also essential for HA homeostasis. In a 70 kilogram human there is approximately 15 grams of total HA [10]. Up to 50% of the total HA in the body is expressed in the skin [11]. HA is turned over at a rate of approximately 5 grams per day [10]. In the skin, HA has a metabolic half-life <1.5 days [11]. HA is turned-over locally in tissues while systemic HA is cleared mostly in the liver and to a lesser extent the kidneys and spleen [12,13]. HA in the tissue extracellular matrix is thought to be partially degraded and then enters the lymph nodes via the draining lymphatics. Specific HA receptors in the lymphatics will be discussed below.

3. Mechanisms of hyaluronan endocytosis

3.1. Receptor mediated endocytosis

Uptake of hyaluronan by various receptors has been widely studied, and several key receptors have been identified. Some receptors, such as CD44 and LYVE-1, serve dual purposes in that they not only facilitate the endocytosis of HA, but also trigger signaling events that generate cell specific responses to HA binding. ICAM-1 was initially believed to serve as a metabolic receptor for HA only but is now suspected to have cell signaling roles in response to HA binding [14,15].

3.1.1. CD44

CD44 is a cell surface glycoprotein that serves as the endocytic receptor for HA in keratinocytes, chondrocytes, and breast cancer cells [16-18]. It is important to stress that the binding of HA to CD44 and the uptake of HA by CD44 mediated endocytosis are two separate events that often do not take place at the same time [19]. Internalization of HA through CD44 mediated endocytosis has been shown to require acylation of the CD44 cytoplasmic tail [20]. CD44 associates with lipid rafts for internalization as determined by gradient ultracentrifugation. Palmitoylation of CD44 on two cysteine residues, Cys^{286} in the transmembrane domain and Cys^{295} in the cytoplasmic domain, was found to be essential for lipid raft association, but not for HA binding. These acylation reactions could be cell type specific, which may explain why CD44 does not endocytose HA in all CD44 expressing cell types, such as B16-F10 melanoma cells [19]. There is also evidence that CD44 interacts directly with endocytosis proteins such as coatomer protein complexes [21]. Previous investigations have suggested that endocytosed CD44 can be recycled to the cell surface provided it is not ubiquitinated after endocytosis [22]. Recent studies in fibroblasts have shown that clathrin-independent carriers (CLIC) form an endocytic sorting system at the leading edge of migrating cells. Adhesion molecules, including CD44, are recycled in the CLIC pathway. CD44 and other CLIC cargo are concentrated within flotillin-1 and cholesterol enriched microdomains. Actin and GRAF-1 form the initial carriers within 15 seconds. Next, Rab11 and Rab5 / EEA-1 complexes allow bulk membrane flow to early endosomes and plasma membrane recycling [23]. It is tempting to speculate that the CLIC pathway in CD44 recycling is also involved in HA endocytosis.

3.1.2. RHAMM

The **R**eceptor for **H**yaluronic **A**cid **M**ediated **M**otility (RHAMM) was discovered originally as a soluble protein that altered the migration of cells and could bind HA [24]. RHAMM has no cytoplasmic or transmembrane domain and has no signaling domains, but it has been implicated in ERK1/2 signaling through a complex with CD44 upon HA binding [25]. RHAMM is also found in the cytoplasm where it associates with the mitotic spindle apparatus, which is responsible for establishing cell polarity and distribution of chromosomes during mitosis [25]. RHAMM can be transported out of the cytoplasm to the cell surface. In terms of RHAMM endocytosis, very little is known. By contrast, a number of studies have shown a role for RHAMM in tumor progression and the differentiation of osteoblasts [26,27]. As it is known that CD44 and RHAMM can associate with each other and that RHAMM binds HA, it is possible that RHAMM/CD44/HA complexes can be endocytosed or leads to signaling. There is evidence that RHAMM and CD44 co-signal through the ERK1/2 pathway to increase basal motility in breast cancer cells and increase fibroblast migration and differentiation during wound repair [28,29].

3.1.3. LYVE-1

Lymphatic vessel endothelial-1 (LYVE-1) is expressed on the surface of lymphatic endothelial cells. Interestingly, LYVE-1 has a glycosylation domain on its extracellular domain that renders it inactive. Cleavage of this glycosylated region allows LYVE-1 to bind HA [30]. The LYVE-1 binds HA which is then bound by leukocyte CD44 in order to facilitate their adhesion and entry into the lymphatics [31]. Not much is known about the mechanisms of LYVE-1 endocytosis of HA, but it is thought to occur similarly to CD44 mediated uptake of HA; i.e., LYVE-1 associates with lipid rafts before endocytosis [31]. LYVE-1 may be responsible for the transport of HA to the luminal side of the lymphatics. It has previously been shown that LYVE-1 binds to HA on the lymphatic endothelial cells, endocytosis of the complex occurs, and then the vesicles are released on the lumen side of the lymphatics allowing for release of HA which can possibly modulate immune responses or mediate removal of HA from the lymph for clearance [32]. Importantly, the lymph nodes are the first sites of clearance for total body HA turn-over. In fact, about 85% of total body HA is cleared by the lymph nodes. The remaining HA is largely turned over in the liver [33].

3.1.4. HARE

Hyaluronan Receptor for Endocytosis (HARE) is expressed on sinusoid hepatocytes, the venous sinuses of the red pulp in spleen and the medullary sinuses in lymph nodes where it is important for the turnover of systemic HA [34]. Indeed, blocking HARE results in an inhibition of HA clearance in the liver [35]. HARE also plays a role in chondroitin sulfate proteoglycan endocytosis [36]. Previous investigations infer that HARE endocytosis occurs by the clathrin coated pit pathway and it appears that HARE is recycled to the cell surface [37]. Binding of HARE to HA was observed using ligand blotting and immunohistochemistry which shows that the HARE HA binding event occurs prior to HA internalization [38]. Four putative AP-2 / clathrin mediated endocytosis signaling domains have been identified in the HARE cytoplasmic domain: YSYFRI^{2485}, FQHF^{2495}, NPLY^{2519}, and DPF^{2534} (315-HARE numbering). Deletion analyses of the signaling domains showed that three signal sequences (YSYFRI, FQHF, and NPLY) provide redundancy to mediate coated pit targeting and endocytosis of HARE. Importantly, the coated pit targeting domains did not impact binding of the HARE ectodomain to HA showing that HA binding to HARE and HARE mediated endocytosis of HA are separate events [39].

It is noted that HARE and LYVE-1 are both expressed in the lymph nodes. Interestingly, HARE and LYVE-1 show different and non-overlapping distributions [33]. How and if LYVE-1 and HARE coordinate HA turn-over remains to be determined.

3.1.5. ICAM-1

Intercellular adhesion molecule 1 (ICAM-1) is perhaps the least studied of the receptors for HA. It appears not to facilitate HA endocytosis but rather, it is a receptor for HA that has signaling capacity. Preliminary data suggests that ICAM-1 functions as a signaling molecule when HA binds to it. When HA is added to the macrophage cell line U937, it induces Akt phosphorylation which activates the nuclear factor-kappa B pathway, inducing interleukin-6 production, but blocking of ICAM-1 with an antibody stops this from occurring [14,15].

3.2. Non-receptor mediated endocytosis

3.2.1. Macropinocytosis of hyaluronan

It has been demonstrated that HA uptake can occur without the aid of endocytic receptors. In [19], it was determined that B16-F10 melanoma cells endocytosed HA through macropinocytosis. The cells were observed to display membrane ruffling and localization of HA within vesicles as well as strong co-localization of HA with fluorescently labeled dextran, a macropinocytosis tracer. The uptake was also inhibited by amiloride, an inhibitor of macropinocytosis. Interestingly, the B16-F10 melanoma cells expressed surface CD44. Moreover, CD44 mediated adhesion of the melanoma cells to HA-coated plates. On the other hand, removal of CD44 from the B16-F10 melanoma cells by proteolytic cleavage failed to impact HA uptake. These results showed that CD44 did not play a significant role in the endocytosis of soluble HA. As mentioned in the previous section, the uptake of HA by CD44 requires CD44 to be in a specific state. It is possible, through splice variants of CD44, that these modifications were not present, and that there are distinct pathways in which HA can be taken up by cells. The ability of HA to induce macropinocytosis in B16-F10 melanoma cells is most likely due to its ability to non-specifically interact with the cell surface. HA has been shown to interact with and cause rearrangement of the cell surface [40]. [41] also showed that blocking of CD44 failed to inhibit the uptake of HA by 10T1/2 fibroblasts.

4. Potential biological functions of endocytosed HA

As shown in Figure 1, HA oligomers are produced by degradation of HA following its endocytosis. On the other hand, most studies on the biological activities of HA have been performed by adding exogenous HA oligomers to cell cultures and then determining their effect on biological activity. Presumably, HMW-HA co-polymers are endocytosed and digested and the resulting low molecular weight-HA (LMW-HA) fragments exocytosed under physiological conditions. Indeed, the hyaluronidase inhibitor apigenin resulted in accumulation of HA in pre-lysosomal endosomes in rat keratinocytes in vitro [18]. On the other hand, direct evidence for exocytosis of HA fragments are currently lacking. It is important to note that in some cases, the hyaluronidases may be secreted into the extracellular matrix or in the case of Hyal-2, expressed on the cell surface. HA degrading activity can also be exocytosed [42]. Obiviously, secreted hyaluronidase and cell surface Hyal-2 could circumvent the requirement for endocytosis.

4.1. Cell activation via HA

A number of investigators have reported that LMW-HA can induce molecular pathways culminating in gene expression. [43] has shown that oligomeric HA stimulated various transcription factors in chondrocytes including Sp1 and NF-κB. The same group showed that HA oligosaccharides induce expression of matrix metalloproteinase 13 by p38 MAPK and transcriptional activation of NF-κB [44]. It has also been shown that HA oligosaccharides

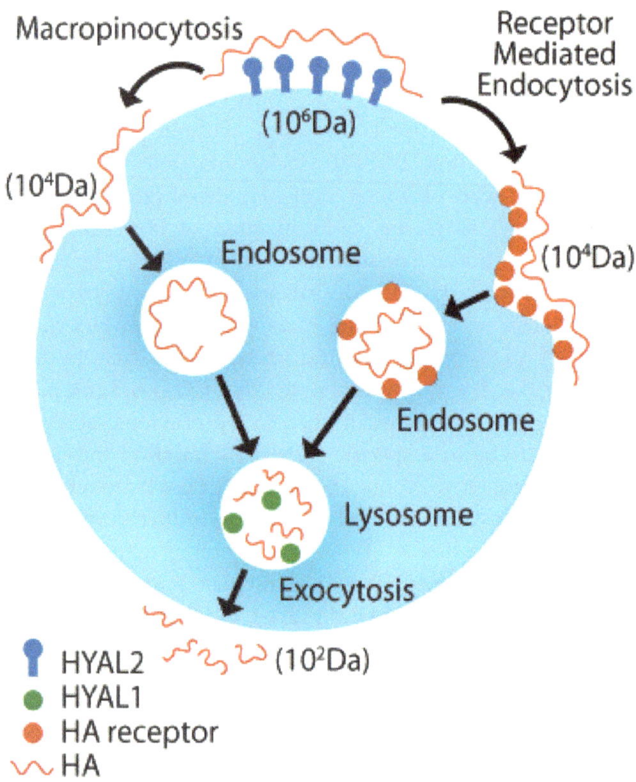

HMW-HA (~10^6Da) is first degraded by hyaluronidase 2 (HYAL2) into smaller 10^4Da sized fragments before it is taken up by a cell. The cell can either utilize surface HA receptors for receptor mediated endocytosis or macropinocytosis. Once internalized the HA is degraded by hyaluronidase 1 (HYAL1) into small 10^2Da fragments and then exocytosed.

Figure 1. Schematic overview of HA endocytosis and processing.

inhibit the expression of runt-related gene 2 (Runx2) in chondrocytes. Runx2 is a transcription factor for chondrocyte differentiation in hypertrophic chondrocytes [45,46]. Thus, HA oligosaccharides may impact the differentiation of chondrocytes during endochondrol ossification. [46] showed that binding of HA fragments to the HA receptor CD44 induces the Nanog-Stat-3 signaling pathway culminating in expression of the multidrug resistance gene, MDR-1, in breast and ovarian tumor cells. Expression of MDR-1 in the tumor cells conferred resistance to chemotherapeutic drugs doxorubicin and paclitaxel. Finally, it has been shown that LMW-HA upregulated CD44 expression and increased the expression levels of PKCδ and PKCϵ [47].

4.2. Cancer cell invasion and metastasis

Controversy has surrounded the role of the hyaluronidase enzymes in tumor biology with initial reports suggesting that Hyal1 was a tumor suppressor [48]. Early positional cloning studies identified the Hyal1 locus on 3p21.3 with LuCa1 (Lung Cancer 1). Because LuCa1 was either deleted or there was a loss in heterozygosity in LuCa1 in most lung cancers, it was hypothesized that Hyal1 behaved as a tumor suppressor. However, subsequent studies showed that Hyal1 was not the relevant tumor suppressor at the examined locus [49]. Nonetheless, the suppressive activity of hyaluronidses was confirmed by investigations showing that administration of high doses (300 U) of PH20 to mice with human breast tumor xenografts showed a significant reduction in tumor growth [50]. The over-expression of Hyal1 in a rat colon carcinoma line also inhibited tumor growth further suggesting that Hyal1 is a tumor suppressor [51].

Groundbreaking work by [52] showed that Hyal1 promoted tumor growth, invasion and angiogenesis. On the other hand, overproduction of Hyal1 to high levels (100 mU / 10^6 cells) inhibited tumor growth. These results suggested that high concentrations of Hyal1 may result in tumor inhibition while at lower levels Hyal1 leads to tumor progression.

Previous studies have shown that anti-sense Hyal1 stably expressed in bladder and prostate cancer cells induced down regulation of cdc25c, cyclin B1, cdk1 and cdk1 kinase activity [52,53]. Expression of Hyal1 in oral carcinoma cells resulted in a dramatic increase in cells in S-phase and a decrease in the number of cells in the G0-G1 phase [54]. To date, the mechanism whereby Hyal1 promotes tumor cell growth remains unknown. On the other hand, treatment of mouse fibroblasts with PH20 leads to phosphorylation of JNK-1 and -2 as well as p42 / p44 ERK [55]. Importantly, ERK plays a role in the G1-S transition [56].

The activities of the hyaluronidase enzymes also appear to have roles in the biology of malignant melanoma (MM) tumors. [57] found that aberrant expression of PH20 by MM cells was correlated with their induction of angiogenesis in a mouse model. Histological studies of MM showed that tumor associated HA expression was correlated with patient survival with low HA levels showing poor prognosis [58]. Because the activities of HA synthesis (by the HAS enzymes) and HA degradation (by the hyaluronidase enzymes) is highly regulated and may be interconnected, these results might suggest that an imbalance in HA metabolism in MM tumors may lead to cancer progression in humans.

Hyaluronidase activity has also been evaluated as a biomarker in bladder cancer. Tumor associated Hyal1 is released into the urine of bladder cancer patients [59]. Urinary hyaluronidase activity was elevated in patients with intermediate and high grade bladder cancer as compared with patients with: a) low grade bladder cancer, b) patients with a history of bladder cancer, c) normal individuals, and d) patients with benign urologic conditions [60]. These findings underscore the potential utility of the hyaluronidases to serve as biomarkers for cancer grading.

4.3. Wound healing

Tissue contraction during wound healing is achieved by myofibroblasts. Fibroblasts differentiate into myofibroblasts which line up at the edges of the wound and adhere to each other with desmosomes. They then use their actin networks to contract the ECM around the wound and shrink the size of the wounded area. Fibroblasts in the area around the wound then secrete collagen to stabilize the contraction. The trigger for contraction is HA production, and HA plays a key role in regulating this process by directing fibroblast migration and proliferation at wound sites [61].

HA's role in wound healing is not clearly understood, but it is known that in a fetal state, as compared to an adult state, the extracellular matrix is composed of primarily HA. This abundance of HA rather than collagen reduces scar formation. CD44 mediated endocytosis of HA fragments also aids fibroblasts in migration to wound sites [62].

5. Applications of hyaluronan endocytosis

5.1. Drug delivery

HA may allow the targeted delivery of chemotherapeutic reagents to tumor cells via CD44. HA have a number of functional groups for "decoration" including a carboxylate on the glucuronic acid, the N-acetylglucosamine hydroxyl and the reducing end. Thus, a wide array of different chemotherapeutic reagents can be chemically conjugated to HA. A review of the chemical derivatization of HA and the potential applications of HA to disease treatment is beyond the scope of this chapter but has been recently reviewed [63,64]. In brief, previous investigations in vitro showed that a taxol-HA bioconjugate was cytotoxic to a panel of tumor cell lines (breast, colon and ovarian) but not human fibroblasts [65]. Although there are relatively few in vivo studies that have evaluated the efficacy of HA bioconjugates, previous investigations with a paclitaxol-HA bioconjugate have shown that it inhibits tumor growth of RT-112/84 human transitional cell carcinomas in mice and increases the survival of mice that had been inoculated with the NMP-1 or SK-OV-3ip human ovarian carcinoma lines [66,67]. Similarly, a butyrate-HA bioconjugate was found to inhibit tumor growth and reduced lung metastasis in mice inoculated with LL3 murine lung carcinoma cells [68]. Thus, a number of reports reinforce the concept that HA may be useful as a drug carrier / ligand targeting delivery agent. Other investigators conjugated doxorubicin, a chemotherapeutic agent, to HA and administered it topically to B16-F10 melanoma tumors in mice [69]. They found that doxorubicin-conjugated HA selectively targeted the tumor cells and reduced tumor growth. These findings open the door for future work with drug delivery to tumor cells using HA.

HA may also be a useful carrier of carboranes for boron neutron capture therapy for tumors. Previous investigations have shown that a water soluble HA-polycarborane derivative was taken up and showed toxicity to a number of tumor cell lines in vitro [70].

5.2. Imaging

HA may also be derivatized for imaging tumor cells. An activatable HA molecular probe, called FRET-HA, designed to detect hyaluronidase activity was recently reported [71]. Briefly, the HA co-polymer was chemically labeled with donor and acceptor fluorescent probes. Energy transfer from the donor to acceptor probe resulted in quenching of the acceptor (i.e., fluorescence resonance energy transfer or FRET). Because energy transfer is distance dependent, an increase in the distance between the donor and acceptor probes, for example due to HA degradation, results in de-quenching of the donor. The rate of donor fluorescence change was used to determine the enzyme kinetics of bovine testes hyaluronidase with high precision and accuracy. FRET-HA has also been used to detect hyaluronidase activity in B16-F10 melanoma cells in vitro (unpublished results). The potential utility for FRET-HA to detect increased hyaluronidase activities in vivo is currently under investigation.

Recently, HA has been coated onto superparamagnetic iron oxide nanoparticles [72]. These HA-coated nanoparticles were endocytosed by cancer cells allowing their magnetic resonance imaging in vitro.

6. Conclusion

HA is a glycosaminoglycan with diverse biological functions. The molecular weight of HA is important for dictating its biological functions with HA fragments inducing distinct responses from high molecular weight co-polymers. Endocytosis of HA via receptor dependent and independent pathways is likely required for digestion of HA to biologically active fragments. In addition, endocytosis of HA may be exploited for the uptake of chemotherapeutic drugs for cancer treatment or imaging probes for the detection of metastatic tumors. Future directions include, 1) better understanding of the endocytic mechanism for HA metabolism, 2) better understanding of HA receptor signaling and interactions and 3) the development of second generation HA scaffolds for delivery and medical imaging.

Author details

Ronny Racine

Department of Molecular Biology and Immunology, University of North Texas Health Science Center, Fort Worth, TX, USA

Mark E. Mummert*

Department of Psychiatry and Behavioral Health, University of North Texas Health Science Center, Fort Worth, TX, USA

Acknowledgement

We thank Mrs. Jessica Hall for her helpful suggestions and critiques. This work was supported by NIH grant RO1 AR48840.

7. References

[1] Ogawa M, Hirano H, Tsubaki H, Kodama H, Tanaka T. The role of cytokines in cervical ripening: correlations between the concentrations of cytokines and hyaluronic acid in

* Corresponding Author

cervical mucus and the induction of hyaluronic acid production by inflammatory cytokines by human cervical fibroblasts Am J Obstet Gynecol 1998 Jul;179(1):105-110.
[2] Tammi R, Ripellino JA, Margolis RU, Maibach HI, Tammi M. Hyaluronate accumulation in human epidermis treated with retinoic acid in skin organ culture J Invest Dermatol 1989 Mar;92(3):326-332.
[3] Agren UM, Tammi M, Tammi R. Hydrocortisone regulation of hyaluronan metabolism in human skin organ culture J Cell Physiol 1995 Aug;164(2):240-248.
[4] Lamberg SI, Yuspa SH, Hascall VC. Synthesis of hyaluronic acid is decreased and synthesis of proteoglycans is increased when cultured mouse epidermal cells differentiate J Invest Dermatol 1986 Jun;86(6):659-667.
[5] Stern R. Devising a pathway for hyaluronan catabolism: are we there yet? Glycobiology 2003 Dec;13(12):105R-115R.
[6] Stern R. Hyaluronidases in cancer biology. Semin Cancer Biol 2008 Aug;18(4):275-280.
[7] Rai SK, Duh FM, Vigdorovich V, Danilkovitch-Miagkova A, Lerman MI, Miller AD. Candidate tumor suppressor HYAL2 is a glycosylphosphatidylinositol (GPI)-anchored cell-surface receptor for jaagsiekte sheep retrovirus, the envelope protein of which mediates oncogenic transformation. Proc Natl Acad Sci U S A 2001 Apr 10;98(8):4443-4448.
[8] Bourguignon LY, Singleton PA, Diedrich F, Stern R, Gilad E. CD44 interaction with Na+-H+ exchanger (NHE1) creates acidic microenvironments leading to hyaluronidase-2 and cathepsin B activation and breast tumor cell invasion. J Biol Chem 2004 Jun 25;279(26):26991-27007.
[9] Roden L, Campbell P, Fraser JR, Laurent TC, Pertoft H, Thompson JN. Enzymic pathways of hyaluronan catabolism. Ciba Found Symp 1989;143:60-76; discussion 76-86, 281-5.
[10] Laurent T, Fraser J. Catabolism of Hyaluronan. In: Henriksen J, editor. Degradation of Bioactive Substances: Physiology and Pathophysiology Boca Raton, FL: CRC Press; 1991. p. 249-265.
[11] Tammi R, Saamanen AM, Maibach HI, Tammi M. Degradation of newly synthesized high molecular mass hyaluronan in the epidermal and dermal compartments of human skin in organ culture. J Invest Dermatol 1991 Jul;97(1):126-130.
[12] Fraser JR, Appelgren LE, Laurent TC. Tissue uptake of circulating hyaluronic acid. A whole body autoradiographic study. Cell Tissue Res 1983;233(2):285-293.
[13] Fraser JR, Laurent TC, Pertoft H, Baxter E. Plasma clearance, tissue distribution and metabolism of hyaluronic acid injected intravenously in the rabbit. Biochem J 1981 Nov 15;200(2):415-424.
[14] Yasuda T. Hyaluronan inhibits Akt, leading to nuclear factor-kappaB down-regulation in lipopolysaccharide-stimulated U937 macrophages. J Pharmacol Sci 2011;115(4):509-515.
[15] Yasuda T. Activation of p38 mitogen-activated protein kinase is inhibited by hyaluronan via intercellular adhesion molecule-1 in articular chondrocytes stimulated with type II collagen peptide. J Pharmacol Sci 2012;118(1):25-32.

[16] Culty M, Shizari M, Thompson EW, Underhill CB. Binding and degradation of hyaluronan by human breast cancer cell lines expressing different forms of CD44: correlation with invasive potential. J Cell Physiol 1994 Aug;160(2):275-286.
[17] Hua Q, Knudson CB, Knudson W. Internalization of hyaluronan by chondrocytes occurs via receptor-mediated endocytosis. J Cell Sci 1993 Sep;106 (Pt 1)(Pt 1):365-375.
[18] Tammi R, Rilla K, Pienimaki JP, MacCallum DK, Hogg M, Luukkonen M, et al. Hyaluronan enters keratinocytes by a novel endocytic route for catabolism. J Biol Chem 2001 Sep 14;276(37):35111-35122.
[19] Greyner HJ, Wiraszka T, Zhang LS, Petroll WM, Mummert ME. Inducible macropinocytosis of hyaluronan in B16-F10 melanoma cells. Matrix Biol 2010 Jul;29(6):503-510.
[20] Thankamony SP, Knudson W. Acylation of CD44 and its association with lipid rafts are required for receptor and hyaluronan endocytosis J Biol Chem 2006 Nov 10;281(45):34601-34609.
[21] Skandalis SS, Kozlova I, Engstrom U, Hellman U, Heldin P. Proteomic identification of CD44 interacting proteins. IUBMB Life 2010 Nov;62(11):833-840.
[22] Eyster CA, Cole NB, Petersen S, Viswanathan K, Fruh K, Donaldson JG. MARCH ubiquitin ligases alter the itinerary of clathrin-independent cargo from recycling to degradation. Mol Biol Cell 2011 Sep;22(17):3218-3230.
[23] Howes MT, Kirkham M, Riches J, Cortese K, Walser PJ, Simpson F, et al. Clathrin-independent carriers form a high capacity endocytic sorting system at the leading edge of migrating cells. J Cell Biol 2010 Aug 23;190(4):675-691.
[24] Turley EA. Purification of a hyaluronate-binding protein fraction that modifies cell social behavior Biochem Biophys Res Commun 1982 Oct 15;108(3):1016-1024.
[25] Maxwell CA, Keats JJ, Crainie M, Sun X, Yen T, Shibuya E, et al. RHAMM is a centrosomal protein that interacts with dynein and maintains spindle pole stability Mol Biol Cell 2003 Jun;14(6):2262-2276.
[26] Li H, Guo L, Li JW, Liu N, Qi R, Liu J. Expression of hyaluronan receptors CD44 and RHAMM in stomach cancers: relevance with tumor progression. Int J Oncol 2000 Nov;17(5):927-932.
[27] Hatano H, Shigeishi H, Kudo Y, Higashikawa K, Tobiume K, Takata T, et al. Overexpression of receptor for hyaluronan-mediated motility (RHAMM) in MC3T3-E1 cells induces proliferation and differentiation through phosphorylation of ERK1/2. J Bone Miner Metab 2011 Sep 27.
[28] Hamilton SR, Fard SF, Paiwand FF, Tolg C, Veiseh M, Wang C, et al. The hyaluronan receptors CD44 and Rhamm (CD168) form complexes with ERK1,2 that sustain high basal motility in breast cancer cells. J Biol Chem 2007 Jun 1;282(22):16667-16680.
[29] Tolg C, Hamilton SR, Nakrieko KA, Kooshesh F, Walton P, McCarthy JB, et al. Rhamm-/- fibroblasts are defective in CD44-mediated ERK1,2 motogenic signaling, leading to defective skin wound repair. J Cell Biol 2006 Dec 18;175(6):1017-1028.
[30] Nightingale TD, Frayne ME, Clasper S, Banerji S, Jackson DG. A mechanism of sialylation functionally silences the hyaluronan receptor LYVE-1 in lymphatic endothelium J Biol Chem 2009 Feb 6;284(6):3935-3945.

[31] Jackson DG. Immunological functions of hyaluronan and its receptors in the lymphatics Immunol Rev 2009 Jul;230(1):216-231.
[32] Jackson DG, Prevo R, Clasper S, Banerji S. LYVE-1, the lymphatic system and tumor lymphangiogenesis. Trends Immunol 2001 Jun;22(6):317-321.
[33] Weigel PH, Yik JH. Glycans as endocytosis signals: the cases of the asialoglycoprotein and hyaluronan/chondroitin sulfate receptors. Biochim Biophys Acta 2002 Sep 19;1572(2-3):341-363.
[34] Harris EN, Kyosseva SV, Weigel JA, Weigel PH. Expression, processing, and glycosaminoglycan binding activity of the recombinant human 315-kDa hyaluronic acid receptor for endocytosis (HARE) J Biol Chem 2007 Feb 2;282(5):2785-2797.
[35] Weigel JA, Raymond RC, McGary C, Singh A, Weigel PH. A blocking antibody to the hyaluronan receptor for endocytosis (HARE) inhibits hyaluronan clearance by perfused liver J Biol Chem 2003 Mar 14;278(11):9808-9812.
[36] Smedsrod B, Malmgren M, Ericsson J, Laurent TC. Morphological studies on endocytosis of chondroitin sulphate proteoglycan by rat liver endothelial cells. Cell Tissue Res 1988 Jul;253(1):39-45.
[37] McGary CT, Raja RH, Weigel PH. Endocytosis of hyaluronic acid by rat liver endothelial cells. Evidence for receptor recycling. Biochem J 1989 Feb 1;257(3):875-884.
[38] Harris EN, Weigel JA, Weigel PH. Endocytic function, glycosaminoglycan specificity, and antibody sensitivity of the recombinant human 190-kDa hyaluronan receptor for endocytosis (HARE) J Biol Chem 2004 Aug 27;279(35):36201-36209.
[39] Pandey MS, Harris EN, Weigel JA, Weigel PH. The cytoplasmic domain of the hyaluronan receptor for endocytosis (HARE) contains multiple endocytic motifs targeting coated pit-mediated internalization. J Biol Chem 2008 Aug 1;283(31):21453-21461.
[40] Kultti A, Rilla K, Tiihonen R, Spicer AP, Tammi RH, Tammi MI. Hyaluronan synthesis induces microvillus-like cell surface protrusions. J Biol Chem 2006 Jun 9;281(23):15821-15828.
[41] Collis L, Hall C, Lange L, Ziebell M, Prestwich R, Turley EA. Rapid hyaluronan uptake is associated with enhanced motility: implications for an intracellular mode of action. FEBS Lett 1998 Dec 4;440(3):444-449.
[42] Bastow E, Last K, Golub S, Stow J, Stanley A, Fosang A. Evidence for lysosomal exocytosis and release of aggrecan-degrading hydrolases from hypertrophic chondrocytes, *in vitro* and *in vivo*. Biology Open 2012 February 15, 2012.
[43] Ohno S, Im HJ, Knudson CB, Knudson W. Hyaluronan oligosaccharide-induced activation of transcription factors in bovine articular chondrocytes. Arthritis Rheum 2005 Mar;52(3):800-809.
[44] Ohno S, Im HJ, Knudson CB, Knudson W. Hyaluronan oligosaccharides induce matrix metalloproteinase 13 via transcriptional activation of NFkappaB and p38 MAP kinase in articular chondrocytes. J Biol Chem 2006 Jun 30;281(26):17952-17960.
[45] Tanne Y, Tanimoto K, Tanaka N, Ueki M, Lin YY, Ohkuma S, et al. Expression and activity of Runx2 mediated by hyaluronan during chondrocyte differentiation. Arch Oral Biol 2008 May;53(5):478-487.

[46] Bourguignon LY, Peyrollier K, Xia W, Gilad E. Hyaluronan-CD44 interaction activates stem cell marker Nanog, Stat-3-mediated MDR1 gene expression, and ankyrin-regulated multidrug efflux in breast and ovarian tumor cells. J Biol Chem 2008 Jun 20;283(25):17635-17651.
[47] Campo GM, Avenoso A, Campo S, D'Ascola A, Traina P, Calatroni A. Differential effect of molecular size HA in mouse chondrocytes stimulated with PMA. Biochim Biophys Acta 2009 Oct;1790(10):1353-1367.
[48] Csoka AB, Frost GI, Heng HH, Scherer SW, Mohapatra G, Stern R. The hyaluronidase gene HYAL1 maps to chromosome 3p21.2-p21.3 in human and 9F1-F2 in mouse, a conserved candidate tumor suppressor locus. Genomics 1998 Feb 15;48(1):63-70.
[49] Zabarovsky ER, Lerman MI, Minna JD. Tumor suppressor genes on chromosome 3p involved in the pathogenesis of lung and other cancers. Oncogene 2002 Oct 7;21(45):6915-6935.
[50] Shuster S, Frost GI, Csoka AB, Formby B, Stern R. Hyaluronidase reduces human breast cancer xenografts in SCID mice. Int J Cancer 2002 Nov 10;102(2):192-197.
[51] Jacobson A, Rahmanian M, Rubin K, Heldin P. Expression of hyaluronan synthase 2 or hyaluronidase 1 differentially affect the growth rate of transplantable colon carcinoma cell tumors. Int J Cancer 2002 Nov 20;102(3):212-219.
[52] Lokeshwar VB, Cerwinka WH, Lokeshwar BL. HYAL1 hyaluronidase: a molecular determinant of bladder tumor growth and invasion. Cancer Res 2005 Mar 15;65(6):2243-2250.
[53] Lokeshwar VB, Cerwinka WH, Isoyama T, Lokeshwar BL. HYAL1 hyaluronidase in prostate cancer: a tumor promoter and suppressor. Cancer Res 2005 Sep 1;65(17):7782-7789.
[54] Lin G, Stern R. Plasma hyaluronidase (Hyal-1) promotes tumor cell cycling. Cancer Lett 2001 Feb 10;163(1):95-101.
[55] Chang NS. Hyaluronidase activation of c-Jun N-terminal kinase is necessary for protection of L929 fibrosarcoma cells from staurosporine-mediated cell death. Biochem Biophys Res Commun 2001 May 4;283(2):278-286.
[56] Liu X, Yan S, Zhou T, Terada Y, Erikson RL. The MAP kinase pathway is required for entry into mitosis and cell survival. Oncogene 2004 Jan 22;23(3):763-776.
[57] Liu D, Pearlman E, Diaconu E, Guo K, Mori H, Haqqi T, et al. Expression of hyaluronidase by tumor cells induces angiogenesis in vivo. Proc Natl Acad Sci U S A 1996 Jul 23;93(15):7832-7837.
[58] Karjalainen JM, Tammi RH, Tammi MI, Eskelinen MJ, Agren UM, Parkkinen JJ, et al. Reduced level of CD44 and hyaluronan associated with unfavorable prognosis in clinical stage I cutaneous melanoma. Am J Pathol 2000 Sep;157(3):957-965.
[59] Hautmann SH, Lokeshwar VB, Schroeder GL, Civantos F, Duncan RC, Gnann R, et al. Elevated tissue expression of hyaluronic acid and hyaluronidase validates the HA-HAase urine test for bladder cancer. J Urol 2001 Jun;165(6 Pt 1):2068-2074.
[60] Pham HT, Block NL, Lokeshwar VB. Tumor-derived hyaluronidase: a diagnostic urine marker for high-grade bladder cancer. Cancer Res 1997 Feb 15;57(4):778-783.

[61] Slavkovsky R, Kohlerova R, Jiroutova A, Hajzlerova M, Sobotka L, Cermakova E, et al. Effects of hyaluronan and iodine on wound contraction and granulation tissue formation in rat skin wounds. Clin Exp Dermatol 2010 Jun;35(4):373-379.
[62] Maharjan AS, Pilling D, Gomer RH. High and low molecular weight hyaluronic acid differentially regulate human fibrocyte differentiation PLoS One 2011;6(10):e26078.
[63] Yadav AK, Mishra P, Agrawal GP. An insight on hyaluronic acid in drug targeting and drug delivery. J Drug Target 2008 Feb;16(2):91-107.
[64] Gaffney J, Matou-Nasri S, Grau-Olivares M, Slevin M. Therapeutic applications of hyaluronan. Mol Biosyst 2010 Mar;6(3):437-443.
[65] Luo Y, Prestwich GD. Synthesis and selective cytotoxicity of a hyaluronic acid-antitumor bioconjugate. Bioconjug Chem 1999 Sep-Oct;10(5):755-763.
[66] Rosato A, Banzato A, De Luca G, Renier D, Bettella F, Pagano C, et al. HYTAD1-p20: a new paclitaxel-hyaluronic acid hydrosoluble bioconjugate for treatment of superficial bladder cancer. Urol Oncol 2006 May-Jun;24(3):207-215.
[67] Auzenne E, Ghosh SC, Khodadadian M, Rivera B, Farquhar D, Price RE, et al. Hyaluronic acid-paclitaxel: antitumor efficacy against CD44(+) human ovarian carcinoma xenografts. Neoplasia 2007 Jun;9(6):479-486.
[68] Coradini D, Pellizzaro C, Abolafio G, Bosco M, Scarlata I, Cantoni S, et al. Hyaluronic-acid butyric esters as promising antineoplastic agents in human lung carcinoma: a preclinical study. Invest New Drugs 2004 Aug;22(3):207-217.
[69] Zhang LS, Greyner HJ, Mummert ME, Petroll WM. Development of a hyaluronan bioconjugate for the topical treatment of melanoma. J Dermatol Sci 2009 Jul;55(1):56-59.
[70] Di Meo C, Panza L, Campo F, Capitani D, Mannina L, Banzato A, et al. Novel types of carborane-carrier hyaluronan derivatives via "click chemistry". Macromol Biosci 2008 Jul 7;8(7):670-681.
[71] Zhang LS, Mummert ME. Development of a fluorescent substrate to measure hyaluronidase activity. Anal Biochem 2008 Aug 1;379(1):80-85.
[72] El-Dakdouki MH, Zhu DC, El-Boubbou K, Kamat M, Chen J, Li W, et al. Development of Multifunctional Hyaluronan-Coated Nanoparticles for Imaging and Drug Delivery to Cancer Cells. Biomacromolecules 2012, Apr 9; 13(4):1144-1151.

Chapter 15

Identification of Ubiquitin System Factors in Growth Hormone Receptor Transport

Johan A. Slotman, Peter van Kerkhof,
Gerco Hassink, Hendrik J. Kuiken and Ger J. Strous

Additional information is available at the end of the chapter

1. Introduction

Proteins can be covalently modified by ubiquitin. These modifications are the result of a coordinated enzymatic process and regulate many cellular networks and processes (Schwartz & Ciechanover, 2009, Weissman et al., 2011). First, ubiquitin is activated by an ubiquitin activating enzyme (E1), then the activated ubiquitin is transferred to the active cysteine of an ubiquitin conjugating enzyme (E2), which interacts with a substrate recognizing enzyme, the ubiquitin ligating enzyme (E3). This enzyme positions ubiquitin towards the substrate, after which ubiquitin is covalently linked via its C-terminal glycine residue to the ε-amino group of a lysine residue. The human genome contains two E1, 35 E2, and more than 1000 E3 enzymes. The E3s are subdivided in three types: Hect, Ring and U-box domain containing. Ubiquitin has seven internal lysine residues that can be modified with ubiquitin, resulting into chains of multiple ubiquitin moieties. Depending on the internal lysine that is modified, different types of ubiquitin chains can be synthesized. The most abundant ubiquitin chains are linked via lysine-48 (K48), lysine-63 (K63), or combinations, resulting in mixed chains (Goto et al., 2010). Additionally, it was recently discovered that linear ubiquitin chains are important in NFκB activation. These chains are formed by LUBAC ("linear ubiquitin chain-assembly complex"), by conjugating ubiquitin moieties head-to-tail (Iwai K & Tokunaga, 2009, Rahighi et al., 2009). Substrates can be modified by one ubiquitin moiety (mono-ubiquitylation), by one ubiquitin moiety on more than one lysine in the same substrate (multiple mono-ubiquitylation) or by chains of ubiquitin (poly-ubiquitylation). Analogous to phosphorylation, ubiquitin can be removed from substrates. A special class of proteases, the deubiquitylating enzymes (DUBs) of which there are approximately 100 genes in the human genome, can specifically remove ubiquitin moieties (reviewed in Soboleva & Baker, 2004). Modification by ubiquitin can have several outcomes for a substrate protein. Besides degradation by the proteasome, ubiquitin modification can regulate a.o. DNA repair as well as sorting of transmembrane proteins, which is the topic of this chapter. Clathrin mediated endocytosis, a well characterized mode of endocytosis of membrane and cargo molecules, involves the recognition of cargo, assembly of the coat and the pinching off of the invagination (Marsh & McMahon, 1999). The cargo is acquired by adapter proteins such as AP-2 or by AP-2 binding proteins such as Epsin1 or Eps15, after which the clathrin lattice is formed and the membrane is curved by BAR domain-containing proteins (Kirchhausen, 1999). The curved domain is pinched off by dynamin and forms a cargo-containing coated vesicle. Subsequently, this vesicle is uncoated and fused with early endosomes (van der Bliek et al., 1993). From the early endosomes the cargo can either be transported toward the lysosome, ultimately resulting in degradation, or recycled back to the plasma membrane. These decisions are mediated by three endosomal sorting complexes, ESCRT-I, -II and -III (Jovic et al., 2010).

The involvement of ubiquitin in membrane traffic was first described in yeast, where ubiquitylation of both the α-factor receptor and carboxypeptidase S by the HECT E3 ligase Rsp5 is essential for endocytosis as well as for their correct trafficking into the multi vesicular bodies (MVB) (Hicke & Riezman, 1996). In mammalian cells, the role of ubiquitin was first studied for the growth hormone (GH) receptor and the sodium channel ENaC (Staub et al., 1996, Strous et al., 1996). ENaC is ubiquitylated by the HECT E3 Nedd4-2 and is subsequently recognized by the ubiquitin binding domain containing clathrin adapter Epsin1. After endocytosis, deubiquitylation of ENaC by the DUBs UHC-L3 and USP2 determines its fate at the multivesicular bodies. ENaC that remains ubiquitylated is recognized by the ESCRT complexes and degraded in the lysosome, whereas deubiquitylated ENaC is recycled back toward the plasma membrane (Butterworth & Johnson, 2008). A second DUB involved in ENaC down regulation is USP10, which can deubiquitylate sorting nexin 3, resulting in both decreased ENaC endocytosis and increased recycling. USP10 is also implied in endocytosis and recycling of the cystic fibrosis transmembrane conductance regulator CFTR (Bomberger et al., 2010, Boulkroun et al., 2008).

The role of ubiquitylation in endocytosis of receptor tyrosine kinases has been extensively studied for the tyrosine kinase receptor, the epidermal growth factor (EGF) receptor. The RING E3 ligase c-Cbl and the E2 enzyme Ube2D1-4 are involved in ubiquitylation of the EGF receptor. Its clathrin-dependent endocytosis is regulated via binding to the clathrin adapter AP-2, either directly or via the ubiquitin binding endocytosis adapter Eps15 or Epsin1. Ubiquitin plays also an important role in sorting of the EGF receptor from the endosomal system toward the lysosome. This receptor is actively sorted away from the recycling endosome by binding to Hrs, which in turn binds the flat clathrin coat on the endosome (Madshus & Stang, 2009). Hrs binds to the ESCRT-I component Tsg101 that delivers the EGF receptor to the ESCRT machinery. Incorporation of the EGF receptor into intraluminal vesicles depends on the DUBs AMSH and USP8 that are associated with the ESCRT-III machinery (Row et al., 2007). However, other studies have shown that USP8 can also act , earlier at the level of ESCRT-I, by preventing entry into multivesicular bodies promoting recycling. The interaction of USP8 to both ESCRT-I and ESCRT-III indicates a complex role for this DUB in sorting of endosomal cargo (Berlin et al., 2010).

For the cytokine receptors interferon-α receptor chain 1 (IFNAR1), the prolactin, the erythropoietin and GH receptor, βTrCP, the substrate recognizing subunit of the E3 ligase complex skip-culling-F-box (SCF), is essential for endocytosis and degradation (Kumar et al., 2004, Li et al., 2004, Meyer et al., 2007, van Kerkhof et al., 2007). Clearly, cytokine receptors appear particularly well regulated by the ubiquitin system. To identify additional regulators of ubiquitylation events involved in trafficking of cytokine receptors we used our model, the GH receptor.

The GH receptor plays an important role in growth and metabolic pathways. In contrast to the EGF receptor, the GH receptor is constitutively endocytosed and degraded in lysosomes. Using a temperature-sensitive E1 mutant, we previously showed that endocytosis is clathrin-mediated and depends on an intact ubiquitin system (Strous et al., 1996). This activity is mediated via the ubiquitin-dependent endocytosis (UbE) motif, a 12 amino acid stretch in the cytosolic tail of the GH receptor (Govers et al., 1999). The ring E3 ligase SCF(βTrCP) binds to the UbE motif and this binding is required for proper endocytosis (van Kerkhof et al., 2007). Recently, it was shown that, in addition to endocytosis, βTrCP is involved in sorting the GH receptor from multivesicular bodies towards the lysosome (van Kerkhof et al., 2011). In the absence of βTrCP, transport of the GH-GH receptor complex is halted at the multivesicular bodies and routed back to the plasma membrane. Interestingly, ubiquitylation of the receptor itself is not required for proper sorting, but it does require binding of this E3 ligase. These findings suggest that the GH receptor travels from the cell surface to the lysosome using βTrCP as a cargo-specific adapter that ubiquitylates (an ancillary factor of) the machinery instead of the cargo itself. In this study we used small interfering ribonucleic acid (siRNA) in a cell-based assay to find additional regulators of ubiquitylation involved in the fate of GH receptors expressed at the cell surface. We screened a library of siRNAs targeting DUBs and other ubiquitylation factors for their involvement in GH receptor sorting towards the lysosomes.

2. Materials and methods

2.1. Antibodies, chemicals, and cells

The DUB siRNA library (Table 1) was obtained from Dharmacon, Thermo Scientific. Anti GH receptor (B) was previously described (van Kerkhof et al., 2000). Anti-actin was obtained from ICN. Lipofectamine2000 was obtained from InVitrogen. Butyrate was obtained from Sigma. NHis6-GH receptor expressing HepG2 cells ware grown in MEM (InVitrogen), supplemented with 10% FCS, 100 units/ml penicillin and 0.1 mg/ml streptavidin and geneticin. GH receptor U2OS cells were generated and propagated as described in van Kerkhof et al., 2011.

2.2. SiRNA transfection and screening

Mixtures of 0.07 μl lipofectamine2000 and siRNAs in a total volume of 10 μl Optimem with a final concentration of 48 nM siRNA per well (348 well plates) were incubated for 30 min at room temperature. Four thousand Nhis6-GH receptor expressing HepG2 cells were added to the mixture and allowed to propagate. After 48 h, 10 mM butyrate was added to increase the expression of GH receptor. After 72 h, the cells were incubated at 37°C with 5 ng/μl Cy3-GH for 30 min, and fixed with 3% paraformaldehyde for 4 h at room temperature. After three 5-min washes with PBS, cells were incubated with 50 μl DAPI, 0.5 μg/ml for 10 min and washed with PBS. Automated image acquisition was performed by the BD Pathway 855 system (BD Bioscience). DAPI was used for focusing and 4 images per well were acquired using a 20x objective.

Gene Symbol	Accession Number	Gene Symbol	Accession Number	Gene Symbol	Accession Number	Gene Symbol	Accession Number
ATXN3L	XM045705	UBE1C	NM003968	UBL4	NM014235	USP33	NM015017
BAP1	NM004656	UBE1DC1	NM024818	UBL5	NM024292	USP34	XM291018
COPS5	NM006837	UBE1L	NM003335	UBR1	NM174916	USP35	XM290527
CXORF53	NM024332	UBE2A	NM003336	UBTD1	NM024954	USP36	NM025090
CYLD	NM015247	UBE2B	NM003337	UCHL1	NM004181	USP37	NM020935
DUB1A	XM377830	UBE2C	NM007019	UCHL3	NM006002	USP38	NM032557
DUB3	NM201402	UBE2D1	NM003338	UCHL5	NM015984	USP39	NM006590
FBXO7	NM012179	UBE2D2	NM003339	UEV3	NM018314	USP4	NM003363
FBXO8	NM012180	UBE2D3	NM003340	UFD1L	NM005659	USP40	NM018218
FLJ14981	NM032868	UBE2E1	NM003341	USP1	NM003368	USP41	XM036729
JOSD1	NM014876	UBE2E2	NM152653	USP10	NM005153	USP42	XM166526
LOC391622	NM212553	UBE2E3	NM006357	USP11	NM004651	USP43	XM371015
MJD	NM004993	UBE2G1	NM003342	USP12	NM182488	USP44	NM032147
MYSM1	XM055481	UBE2G2	NM003343	USP13	NM003940	USP45	XM371838
OTUB1	NM017670	UBE2H	NM003344	USP14	NM005151	USP46	NM022832
OTUB2	NM023112	UBE2I	NM003345	USP15	NM006313	USP47	NM017944
OTUD1	XM166659	UBE2J1	NM016021	USP16	NM006447	USP48	NM033236
OTUD4	NM017493	UBE2J2	NM058167	USP18	NM017414	USP5	NM003481
OTUD5	NM017602	UBE2L3	NM003347	USP19	XM496642	USP50	NM203494
OTUD6B	NM016023	UBE2L6	NM004223	USP2	NM004205	USP51	NM201286
OTUD7	NM130901	UBE2M	NM003969	USP20	NM006676	USP52	NM014871
PARP11	NM020367	UBE2N	NM003348	USP21	NM012475	USP53	XM052597
PRPF8	NM006445	UBE2NL	XM372257	USP22	XM042698	USP54	NM152586
PSMD14	NM005805	UBE2Q	NM017582	USP24	XM165973	USP6	NM004505
SBBI54	NM138334	UBE2R2	NM017811	USP25	NM013396	USP7	NM003470
SENP2	NM021627	UBE2S	NM014501	USP26	NM031907	USP8	NM005154
SHFM3P1	AF174606	UBE2V2	NM003350	USP28	NM020886	USP9X	NM004652
STAMBP	NM006463	UBE3A	NM000462	USP29	NM020903	USP9Y	NM004654
STAMBPL1	NM020799	UBE3B	NM130466	USP3	NM006537	VCPIP1	NM025054
TNFAIP3	NM006290	UBE4A	NM004788	USP30	NM032663	YOD1	NM018566
TRFP	NM004275	UBE4B	NM006048	USP31	NM020718	ZA20D1	NM020205
UBE1	NM003334	UBL3	NM007106	USP32	NM032582	ZRANB1	NM017580

Table 1. DUB siRNA library. [2]Underlined genes are ubiquitin-related proteins without DUB activity

2.3. Western blotting

GH receptor expressing U2OS cells were transfected with the siRNAs to silence the indicated genes. Cells were lysed in 1% Triton X-100, 10 μg/ml leupeptin, 10 μg/ml aprotinin and 1 mM PMSF in PBS for 20 min and centrifuged at 13,000 x g for 5 min at 4°C. The supernatant was boiled in Laemmli sample buffer, and the proteins were separated on SDS-PAGE and transferred to Immobilon-FL polyvinylidenedifluoride membrane (Millipore). Blots were immunostained with primary antibodies followed by Alexa Fluor 680 or IRDye 800 conjugated anti-mouse or anti-rabbit antibodies. An Odyssey system was used for detection (Li-Cor Biosciences, Lincoln NE).

3. Results

3.1. High throughput screening

GH receptor transport from the plasma membrane toward the lysosome is a complex process that requires, in addition to general endocytosis factors, specific regulatory proteins. As the ubiquitin system was implied in this process (Govers et al., 1997), we set up an assay to identify novel modifiers of regulatory ubiquitilation involved in GH uptake.

Since the liver is an important target of GH, we used the human hepatocellular carcinoma cell line HepG2 for this assay. These epithelial cells are non-tumorigenic and synthesize a variety of liver-specific proteins, such as the asialoglycoprotein receptor and several plasma proteins (albumin, transferrin and the acute phase proteins fibrinogen, β2-macroglobulin, α1-antitrypsin, and plasminogen). Although HepG2 cells are GH sensitive, they have insufficient GH receptors to probe the effect of gene silencing. Therefore, we constructed a HepG2 cell line that expresses 100-200,000 rabbit GH receptors per cell. To identify genes involved in GH receptor endocytosis and degradation, we transfected the GH receptor expressing HepG2 cells with siRNAs as indicated. After 72 h, the cells were allowed to take up GH conjugated with Cy3 (Cy3-GH) for 30 min at 37°C, after which they were immediately fixed and assessed for Cy3-GH uptake (Fig. 1).

We screened a commercial siRNA library targeting 84 DUBs and 44 other ubiquitylation-related enzymes (Table 1) for regulators of GH uptake and receptor sorting (Dharmacon). This arrayed library consists of siRNA pools constituted of 4 unique siRNA sequences, targeting a specific gene (Dharmacon). The screen was performed in a 384 well plate format and an automated spinning-disc confocal microscope was used for unbiased image acquisition, 4 images per well, of Cy3-GH and DAPI signal. The experiment itself was performed in duplicate yielding eight data points per gene. Hits were called when at least 3 out of 4 wells showed a similar phenotype. The assessment of hits was done manually in a double blind fashion by two researchers, independently. The hits were ordered in three categories: Category I, Cy3-GH enrichment on the plasma membrane, category II, Cy3-GH enrichment on both the plasma membrane and intracellular, and category III, intracellular enrichment only. Clathrin heavy chain and Tsg101 depletion, involved in clathrin-mediated endocytosis and in sorting at multivesicular bodies, respectively, were used as controls. Fig.

1 (first row) shows that clathrin depletion resulted in accumulation of Cy3-GH at the plasma membrane, whereas Tsg101 depletion resulted in Cy3-GH accumulation in both internal vesicles and on the plasma membrane. These data show that our screen setup has sufficient sensitivity to detect accumulation of GH receptor. Using the same method we identified 13 genes: 7 in category I, 3 in category II and 3 in category III (Table 2). The E1 enzyme (UBE1) was identified as a category I hit, in concordance with its previously described role in GH receptor endocytosis (Strous et al., 1997)

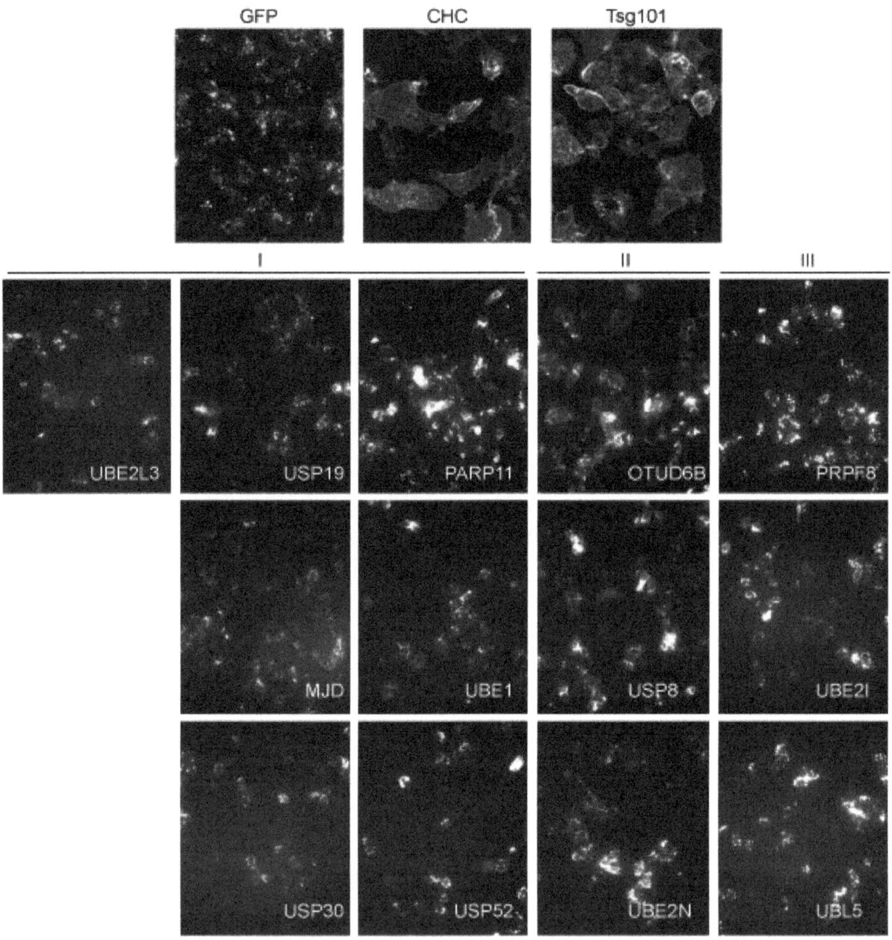

Figure 1. Representative images of observed phenotypes in the screen. Cells were transfected with smartpool siRNA, containing 4 individual siRNAs directed against a specific target. After 72 h, cells were incubated with Cy3-GH at 37°C for 30 min and fixed. Four images of the samples were automatically acquired, and analysed by eye. All the targets that were considered hits are shown. The image is a representative of 2 experiments. The images are grouped according to three pheno-typical categories: I: Cy3-GH enrichment on the plasma membrane, II: Cy3-GH enrichment on both the plasma membrane and intracellular, and III: intracellular enrichment only.

Gene	Protein	Function	Cat
USP19	Ubiquitin carboxyl-terminal hydrolase 19	DUB functioning in ER folding	I
MJD	Ataxin-3	DUB functions in IGF-1 signalling and longevity in *C. Elegans*	I
USP30	Ubiquitin specific protease 30	DUB located at outer mitochondrial membrane, function unknown	I
OTUD6B	OTU domain containing 6B	Member of the otubain DUB family, function unknown	II
PARP11	poly (ADP-ribose) polymerase family, member 11	Poly (ADP-ribose) polymerase family member, mRNA processing	I
PRPF8	pre-mRNA processing factor 8 (PRPF8 homolog, yeast)	mRNA processing factor functions in the spliceosome	III
UBE1	Ubiquitin activating enzyme E1	Human ubiquitin-activating enzyme E1	I
USP52	Ubiquitin specific peptidase 52	DUB, function unknown	I
USP8	ubiquitin specific peptidase 8	DUB, involved in EGF receptor sorting at multivesicular bodies	II
UBE2I	ubiquitin-conjugating enzyme E2I (UBC9 homolog, yeast)	Human homolog of UBC9, activating enzyme for ubiquitin like protein SUMO	III
UBE2L3	ubiquitin-conjugating enzyme E2L 3	Ubiquitin conjugating enzyme, cell cycle, can function together with Triad1	I
UBE2N	ubiquitin-conjugating enzyme E2N (UBC13 homolog, yeast)	Ubiquitin conjugating enzyme able to catalyse K-63 linked ubiquitin chains	II
UBL5	ubiquitin-like protein 5 (BEACON homolog, human)	Ubiquitin like protein, differentially expressed between lean and obese	III

Table 2. Screening results: Genes with a phenotype

3.2. Validation of hits

To validate candidate genes, we tested 8 siRNA pools with the strongest phenotypes in the first screen (USP19, Ubc9, Usp52, Usp8, Ubc13, Ubl5, PRPF8, and OTUD6B) in a second screen using a different cell type. For this purpose, we transfected osteosarcoma U2OS cells with rabbit GH receptor and selected a clone that expressed a limited number of exogenous GH receptors (Fig. 2A). In our experience, U2OS cells have a limited capacity to endocytose cargo via the clathrin-mediated pathway, a feature that renders the endocytosis assay particularly sensitive. In addition to the siRNA pools we transfected the individual siRNAs that constituted the pools (data not shown). In all cases at least 2 out of 4 siRNAs showed a similar phenotype as the pool, decreasing the likelihood of off-target effects. These 8 hits were examined in more detail.

3.2.1. Usp19

Usp19 depletion showed an overall increase in Cy3-GH signal as well as some cell surface labelling (Fig 2A). Usp19 is an ER-localized membrane protein (Hassink et al., 2009), which was originally identified as a DUB induced in atrophying muscle (Combaret et al., 2005). It stabilizes ER localized proteins like CFTR and TCRα (Hassink et al., 2009) as well as cytosolic proteins like Hif1α, Siah1 and 2, c-IAP1 and 2, and the cyclin-dependent kinase inhibitor regulating ubiquitin ligases KPC1 and 2 (Altun et al., 2012, Lu et al., 2009, Mei et al., 2011). Interestingly, the catalytic activity of USP19 does not seem to be required for the stabilization of its substrates, but for stabilization of itself, facilitated via self-association (Altun et al., Mei et al., 2011). It also modulates transcription of major myofibrillar proteins (Sundaram et al., 2009), probably in a similar way as the earlier mentioned proteins. Furthermore, Usp19 is upregulated in smoke-induced muscle atrophy conditions in mice (Liu et al., 2011).

A

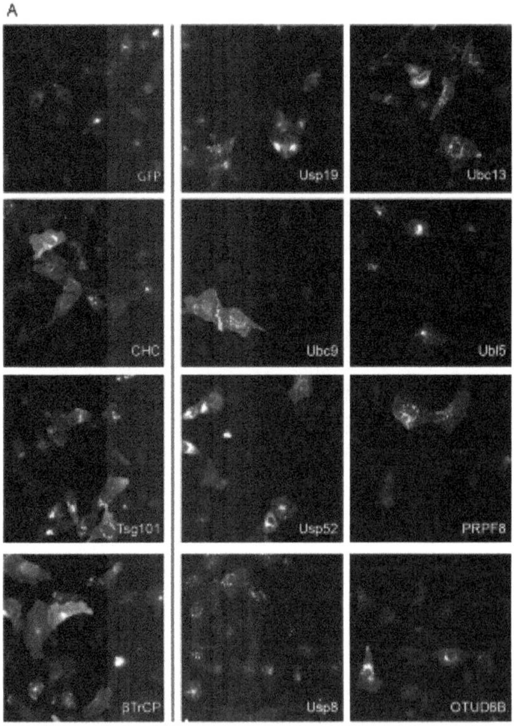

B

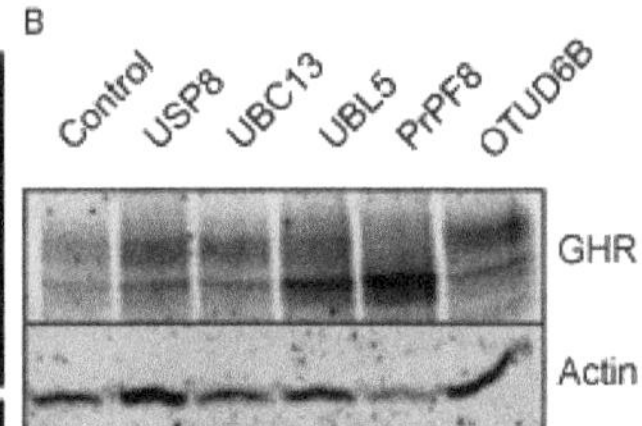

	Ratio mature/immature
Control	1,09
USP8	1,26
UBC13	1,27
UBL5	0,73
PrPF8	0,42
OTUD6B	1,36

Figure 2. Validation of hits obtained by the initial screen. GH receptor expressing U2OS cells were transfected with siRNAs as indicated. A) After 72 h, the cells were allowed to take up Cy3-GH for 30 min at 37°C and immediately fixed. Cy3-GH uptake was assessed by automated confocal microscopy; images are representative of 4 experiments. B) After 72 h, cells were lysed and proteins were separated by SDS-PAGE followed by Western blot analysis for GH receptor (GHR) using actin as a loading control. The upper GHR band represents the mature GHR, while the lower band is the immature (ER) form. The ratio mature/immature for the GHR species was taken as a measure for the turnover of the GHR related to the steady state situation in control cells.

Since the predominant splice variant is a type IV membrane protein, USP19 might act as an ER resident protein with no direct role in GH receptor regulation at the endosomal level. Further characterization was beyond the scope of this study. Alternative roles in the cytosol will undoubtedly become clear in the future.

3.2.2. *Usp8*

Usp8, also known as UBPY, is a deubiquitylating enzyme that functions in protein sorting at multivesicular bodies. Depletion of Usp8 showed a perinuclear accumulation of Cy3-GH punctae (Fig. 2A), reminiscent of the phenotype resulting from Tsg101 depletion. To characterize the effects of gene silencing on the GH receptor degradation further we performed Western blot analysis for the GH receptor. This approach reveals additional details related to the role of USP8 in GH receptor trafficking. At steady state, Western blot analyses of the GH receptor show an approximately 1:1 ratio of the amount of 'immature' (high mannose oligosaccharides-containing) receptor, mainly located in the rough ER, and the amount of the 'mature', complex-glycosylated GH receptor, mainly present at the plasma membrane. Previously, we determined the half-life of both GH receptor species to be 50 min at 37°C. The half-life of the high-mannose GH receptor is defined as the time it takes for the GH receptor to exit the ER and arrive at the Golgi complex to receive the 'complex' sugar moieties, while the half-life of the 'mature' GH receptor is defined as the time required for its transport from plasma membrane to the lysosomes. As Usp8 depletion caused a relative increase in mature over immature GH receptor levels, the results suggest that Usp8 promotes GH receptor degradation.

Usp8 can bind to ESCRT-III components and its depletion leads to enlarged endosomes and diminished incorporation of EGF receptors into the intraluminal vesicles (ILV) of the multivesicular bodies (Bowers et al., 2006). In addition, it has been reported that Usp8 can interact with the ESCRT-0 component STAM via its SH3 domain and that its depletion results in accelerated degradation of EGF receptor in the lysosome, strongly suggesting that Usp8-dependent deubiquitylation of EGF receptor prevents EGF receptor from being recognized by the ESCRT machinery, a recognition depending on ESCRT-0 component Hrs (Berlin et al., 2010, Rao et al., 2011). Furthermore, the ability of Usp8 to bind both ESCRT-I and ESCRT-III suggests a dual function for Usp8 in receptor sorting. Indeed, at the plasma membrane Usp8 can inhibit degradation of receptors by removing the ubiquitin tag that directs them to the multivesicular bodies, while at the multivesicular bodies Usp8 can promote degradation of receptors by allowing incorporation of receptors into the ILVs (Wright et al., 2011). We demonstrate here that depletion of Usp8 increases the level of intracellular GH receptor indicative of a defect in sorting towards the lysosome, most likely at the level of ESCRT-III. The latter is supported by previous data showing that the transport of GH receptor from multivesicular bodies to the lysosome does not depend on ESCRT-0 components. Apparently, the GH receptor enters the ESCRT complexes at the level of ESCRT-I (van Kerkhof et al., 2011).

3.2.3. *Ubl5*

Ubl5 depletion leads to an increased accumulation of Cy3-GH in endosomes. The Western blot analyses showed a complex pattern wherein both the immature GH receptor band and the mature GH receptor were increased compared to control cells. Ubl5 is a small ubiquitin like modifier that, in contrast to other ubiquitin like molecules, does not have a C-terminal glycine residue. A remarkable finding implicated a mutation in non-coding regions of Ubl5 in metabolic syndrome-related phenotypes (Bozaoglu et al., 2006). In a recent study, Mishra and co-workers show that its yeast homologue Hub1 alters the splice specificity for certain non-canonical 5' splice sites by non-covalent attachment to elements of the spliceosomes (Mishra et al., 2011). Combining the above renders it plausible that expression levels of Ubl5, altered by mutations in the non-coding region, attenuate the expression of proteins ultimately affecting metabolism. Since the GH receptor is tightly involved in regulation of metabolic pathways, the involvement of Ubl5 in GH receptor trafficking toward the lysosome is interesting. Either the GH receptor itself or GH receptor-regulating factors might be affected by Ubl5. Preliminary data from ^{125}I-GH uptake experiments suggest that depletion of Ubl5 does not alter the internalization and degradation kinetics of the receptor. Furthermore, a direct interaction between the GH receptor and Ubl5 could not be identified (data not shown). Since steady state levels of GH receptor as measured with Western blot analyses did increase as a result of Ubl5 silencing (Fig. 2B), the data suggest that Ubl5 either influences transcription or synthesis of the GH receptor. To fully understand the role of Ubl5 more investigation is required.

3.2.4. *PrPF8*

Pre-mRNA-processing-splicing factor 8 (PrPF8) is a large nuclear protein that functions in the U2, U12 and trans-spliceosome, and has an active role in processing of pre-mRNA (reviewed in Grainger & Beggs, 2005). PrPF8 knockdown showed an intracellular accumulation of Cy3-GH. In addition, the Western blot analysis showed an increase in the immature form of GH receptor and a complex band pattern of slower migrating bands. This might indicate that multiple pathways are affected. Because of PrpF8's role in the spliceosome, it is likely that depletion of this protein can influence correct splicing of many factors including the GH receptor and proteins of the endocytosis machinery. We did not yet elucidate whether PrPF8 is directly or indirectly involved in GH receptor endocytosis. For example, previously, we have shown that alternative splicing of βTrCP isoforms may affect their localization, which may have consequences for GH receptor endocytosis (Putters et al., 2011b). Because the GH receptor mRNA was artificially transcribed from a plasmid driven by a CMV promoter (pcDNA3) it is possible that this might indirectly lead to alterations in sorting of GH receptor at the plasma membrane or at endosomes by exceeding the sorting capacity of these systems. Alternatively, absence of PrPF8 may affect the balance of folding factors in the ER, resulting in ER overload of GH receptors that cannot leave the ER, e.g. due to defective dimerization (van den Eijnden et al., 2006).

3.2.5. OTUD6B

Ovarian tumour (OTU) domain containing 6B (OTUD6B) is a member of the OTU-domain Ubal-binding protein (otubain) domain containing protein family that constitutes a specific subset of deubiquitylating enzymes. The function of OTUD6B is not known. Substrates of two other OTU domain containing proteins, OTUB1 and OTUB2, are GRAIL (Soares et al., 2004), oestrogen receptor (Stanisic et al., 2009), p97 (Ernst et al., 2009), RhoA (Edelmann et al.), Traf3 and Traf6 (Li et al., 2010). They have been implicated in non-canonical DNA-damage response (Nakada et al., 2010). Furthermore, OTUB1 shows substrate specificity towards K48 linked ubiquitin chains (Wang et al., 2009). We have identified OTUD6B as a novel factor in GH receptor endocytosis. Its depletion leads to a phenotype where mature GH receptor accumulates, mainly in the endosomes but also at the plasma membrane. Fig. 2B clearly shows a decreased degradation with little effect on its biosynthesis. Analogous to Tsg101 depletion, this might point to a role in multivesicular body function. Further investigations are necessary to determine whether OTUB6D acts specific on the GH receptor or has a general role in endosomal sorting.

3.2.6. Ubc13 and associated E3s

The presence of UBE2N, encoding for the E2 enzyme Ubc13, in our hit list was striking. Previously, after our finding that ubiquitylation is required for GH receptor endocytosis and degradation, we identified SCF(βTrCP) as an E3 that specifically binds to the UbE motif of the GH receptor and conjugates K48 poly-ubiquitin chains to the GH receptor (Putters et al., 2011a, van Kerkhof et al., 2007). With the identification of Ubc13 it became plausible that also K63-linked ubiquitin chains might be involved in GH receptor endocytosis, as Ubc13 together with the pseudo E2s UEV1A and MMS2 can synthesize K63 linked ubiquitin chains (Deng et al., 2000).

Gene	Protein	Function	Reference
ARIH2*	Triad1	Cell cycle, cell growth	(Marteijn et al., 2009)
CHFR*	CHFR	Cell cycle	(Bothos et al., 2003, Loring et al., 2008)
PARK2*	Parkin	Neuroprotection, cell death	(Lo Bianco et al., 2004)
RNF8*	RNF8	DNA repair	(Plans et al., 2006)
Rad5	Rad5	DNA repair	(Torres-Ramos et al., 2002)
Rad18	PCNA	DNA repair	(Ulrich & Jentsch, 2000)
STUB1*	CHIP	Heat shock system	(Zhang et al., 2005)
TRAF2	Traf2	TNF signalling	(Habelhah et al., 2004)
TRAF6	Traf6	TNF signalling	(Wooff et al., 2004)

Table 3. E3 ligases known to interact with Ubc13

Furthermore, these K63-linked ubiquitin chains have been implied in endocytosis in yeast (Lauwers et al., 2009) and in mammalian cells (Kamsteeg et al., 2006). To investigate whether K63-linked ubiquitylation is indeed involved in GH receptor endocytosis, we selected nine E3 ubiquitin ligases that are known to interact with Ubc13 (listed in Table 3) (Bothos et al., 2003, Deng et al., 2000, Loring et al., 2008, Marteijn et al., 2009, Plans et al., 2006, Sun et al., 2004, Torres-Ramos et al., 2002, Zhang et al., 2005). As gene silencing of Traf2 and Traf6 induced cell death in previous experiments (data not shown), and rad5 and rad18 are mainly involved in DNA repair, we excluded them from further investigation and performed gene silencing experiments for the 5 remaining E3 ligases using validated siRNAs (Table 3, asterisk).

We transfected GH receptor-expressing U2OS cells with these siRNAs and propagated them for 3 days prior to performing the assay as described above. Cells, treated with siRNAs for ARIH2 and STUB1, encoding Triad1 and CHIP, respectively, showed a strong phenotype (Fig. 3). Not only was the label intensity much higher, silencing of both genes also induced accumulation of Cy3-GH signal on the plasma membrane. Recently, we published details about the role of the ubiquitin ligase CHIP in GH receptor endocytosis involving a direct interaction with specific amino acid sequences in the cytosolic tail of the receptor. The same study shows that CHIP acts in collaboration with UBC13, implicating K63-specific ubiquitylation in the GH receptor endocytosis (Slotman et al., 2012). This is in accordance with Fig. 2.

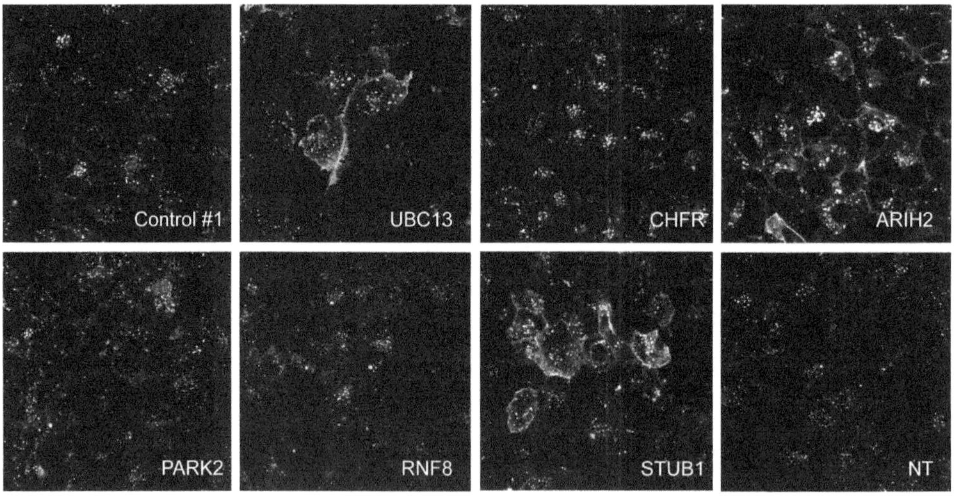

Figure 3. Mini-screen for E3s involved in Ubc13 mediated ubiquitylation. GH receptor-expressing U2OS cells were transfected with validated siRNAs targeting 5 E3 ligases that can function together with the E2 Ubc13. After 72 h, the cells were allowed to take up Cy3-GH at 37°C for 30 min after which they were fixed. Cy3-GH uptake was assessed by confocal microscopy. Shown cells are representative for all cells in the samples of 2 independent experiments.

Triad1-depleted cells showed larger internal vesicles with an altered pattern as compared to wild type. Triad1 is an E3 ligase that has 2 RING domains and, consequently, has the ability to bind two ubiquitin conjugases. Previously, Marteijn and co-workers identified both Ubc13 and UbcH7 as E2s involved in Triad1 action (Marteijn et al., 2009). As seen in Fig. 1 and Table 2, UBCH7 (UBE2L3) also tested positive in our initial screen, strengthening the involvement of Triad1 in the sorting of GH receptors in endosomes (Fig. 1, Table 2)

In conclusion, we identified three novel genes in GH-induced endocytosis via the GH receptor, two of which (UBC13 and STUB1) demonstrated identical phenotypes, while silencing of ARIH2 resulted in a different pattern. All three enzymes are involved in K63-linked ubiquitin chain formation. Further studies are being performed to determine whether ARIH2 acts specifically on the GH receptor or whether it has a broader function in endocytosis and/or lysosomal degradation (Hassink et al., 2012).

4. Discussion

We performed a siRNA screen in which we assayed 84 DUBs and 44 other enzymes of the ubiquitin and related systems. In the first screen we identified 13 genes that showed an altered uptake or degradation pattern of Cy3-GH. As expected, the UBE1 gene, encoding the E1 enzyme, one of two ubiquitin activating enzymes in mammalian cells, was among the 13 hits. It is known that by disrupting the enzymatic activity of the E1 enzyme GH receptor endocytosis is blocked (Strous et al., 1996). The observation that E1 silencing results in a similar phenotype shows that the screen setup is valid and able to pick up factors of the ubiquitin system that are involved in (GH receptor) endocytosis. Furthermore, the large number of hits in this screen clearly shows that GH receptor trafficking by ubiquitin is a complex and well regulated process.

Among the proteins that did not appear in our screen as a hit, but would be expected to show a phenotype, is the activating enzyme (E1) of the ubiquitin like modifier ISG15. ISG15ylation of Ubc13 disrupts its function (Zou et al., 2005) and depletion of this E1 would lead to less ISG15ylation of Ubc13 that might result in an increased endocytosis of GH receptor. Unfortunately, our approach lacks the sensitivity to register accelerated endocytosis. An assay that would pick up such an effect must be suited to measure endocytosis in a quantitative way rather than qualitatively. Another factor that we expected to find was UBE2M, which functions as the E2 for neddylation of cullins, the modification by the ubiquitin like protein Nedd8. Cullin neddylation activates SCF(βTrCP), which is essential for GH receptor ubiquitylation and degradation (van Kerkhof et al., 2007). Recently, the ubiquitin conjugase UBE2F was identified as a Nedd8 conjugating enzyme (Huang et al., 2009). Therefore, UBE2M and UBE2F might have a redundant function, possibly explaining the absence of UBE2M among our hits. Of note, UBE2F was not targeted by our siRNA library and was therefore not tested in our assay.

The screen identified two proteins involved in ubiquitin-like mechanisms, UBE2I, also known as UBC9, the E2 for SUMOylation, and UBL5, an ubiquitin like protein. This is the first time that these two pathways are implicated in GH receptor trafficking. Many similarities exist between the ubiquitin-based and ubiquitin-like mechanisms with many examples of interplay. In GH receptor trafficking SUMOylation might block ubiquitylation and *vice versa* (Denuc & Marfany, 2010). The action of UBL5 is more mysterious, since it cannot be conjugated to substrates due to a lack of a C-terminal glycine residue. Hypothetically, ubiquitin binding domains could be blocked by the structurally very similar UBL5 and in this a way UBL5 may regulate ubiquitylation events (McNally et al., 2003). A correlation between obesity and UBL5 mRNA levels was found in P. *obesus* (Walder et al., 2002) but this finding is still debated in humans (Bozaoglu et al., 2006, McNally et al., 2003). Since GH receptor signalling is highly involved in growth and obesity (Erman et al., 2010, Gao et al., 2010), altering the sorting of GH receptor could be involved. Hence, GH receptor sorting defects, UBL5, and growth and metabolic diseases may be linked.

We identified two novel ubiquitin ligases that are involved in GH receptor trafficking, CHIP and Triad1; thus, together with SCF(βTrCP), at least three E3s have a regulatory role in GH receptor degradation. This indicates that both K48 and K63-linked ubiquitin chain formation are involved. In addition, we identified six DUBs involved in GH receptor endocytosis and homeostasis: USP19, MJD, USP30, OTUD6B, UPS52 and USP8. These DUBs might antagonize the action of the ligases mentioned above, providing a high level regulation. Together, these proteins are part of a complex system that regulates ubiquitylation of many substrates. In part by spatial en temporal regulation, ubiquitylation and deubiquitylation events modulate the sorting of GH receptor, and thereby respond quickly to different forms of stress. Additionally, the number of distinct E2, E3, and DUB enzymes involved in ubiquitylation provides specificity in regulating different classes of receptors in disparate ways within the same endocytic system.

These findings underscore the complexity of the ubiquitylation system and endocytosis process, and, at the same time, offer interesting opportunities to discover drugs that specifically target the GH receptor. As this receptor is implicated both in anabolism, in longevity and insulin sensitivity (Bartke, 2012), drugs that can either up- or down regulate its activity are highly relevant.

5. Conclusion

The ubiquitin proteasome system plays an essential role in trafficking of the growth hormone receptor from the plasma membrane to the lysosomes, where degradation takes place. When ubiquitylation is disabled, the GH receptor accumulates at the plasma membrane and its degradation is inhibited. As endosomal sorting is a highly regulated process that depends on a variety of ubiquitylation events, we set up a cell-based, high content siRNA screen targeting 128 genes of the ubiquitylation system. In this study we report the identification of 13 modifiers of regulatory ubiquitylation events that are involved in trafficking and degradation of the GH receptor. As the GH receptor is a key regulator of metabolism, this study highlights the complexity of pathways that underlie its regulation. These findings may guide the development of specific drugs that either up- or down-regulate GH-based signal transduction. As GH signalling is implicated in longevity, insulin-sensitivity, and cancer, studies focussed at factors that regulate ubiquitylation and GH receptor levels are highly relevant.

Author details

Johan A. Slotman, Peter van Kerkhof, Gerco Hassink and Ger J. Strous
Department of Cell Biology, University Medical Center Utrecht and
Institute of Biomembranes, Utrecht, The Netherlands

Hendrik J. Kuiken
Netherlands Cancer Institute, Division of Molecular Carcinogenesis, Amsterdam, The Netherlands

Acknowledgement

This research was supported by the European Network of Excellence, Rubicon "Role of ubiquitin and ubiquitin-like modifiers in cellular regulation" (Grant LSHG-CT-2005-018683), the Marie Curie network, "UbiRegulators", (Grant MRTN-CT-2006-034555), and the Netherlands Proteomics Centre, "Proteomic analysis of ubiquitylation in membrane trafficking, NPC3.1 to JAS. We thank all other members of the GH receptor group for suggestions and helpful discussions.

6. References

Altun M, Zhao B, Velasco K, Liu H, Hassink G, Paschke J, Pereira T & Lindsten K (2012). Ubiquitin-specific protease 19 (USP19) regulates hypoxia-inducible factor 1alpha (HIF-1alpha) during hypoxia. *J Biol Chem,* 287:1962-1969.

Bartke A (2012). Pleiotropic effects of growth hormone signaling in aging. *Trends Endocrinol Metab,* 22:437-442.

Berlin I, Schwartz H & Nash PD (2010). Regulation of epidermal growth factor receptor ubiquitination and trafficking by the USP8.STAM complex. *J Biol Chem,* 285:34909-34921.

Bomberger JM, Barnaby RL & Stanton BA (2010). The deubiquitinating enzyme USP10 regulates the endocytic recycling of CFTR in airway epithelial cells. *Channels (Austin),* 4:150-154.

Bothos J, Summers MK, Venere M, Scolnick DM & Halazonetis TD (2003). The Chfr mitotic checkpoint protein functions with Ubc13-Mms2 to form Lys63-linked polyubiquitin chains. *Oncogene,* 22:7101-7107.

Boulkroun S, Ruffieux-Daidie D, Vitagliano JJ, Poirot O, Charles RP, Lagnaz D, Firsov D, Kellenberger S & Staub O (2008). Vasopressin-inducible ubiquitin-specific protease 10 increases ENaC cell surface expression by deubiquitylating and stabilizing sorting nexin 3. *Am J Physiol Renal Physiol,* 295:F889-900.

Bowers K, Piper SC, Edeling MA, Gray SR, Owen DJ, Lehner PJ & Luzio JP (2006). Degradation of Endocytosed Epidermal Growth Factor and Virally Ubiquitinated Major Histocompatibility Complex Class I Is Independent of Mammalian ESCRTII. *J Biol Chem,* 281:5094-5105.

Bozaoglu K, Curran JE, Elliott KS, Walder KR, Dyer TD, Rainwater DL, VandeBerg JL, Comuzzie AG, Collier GR, Zimmet P, MacCluer JW, Jowett JB & Blangero J (2006). Association of genetic variation within UBL5 with phenotypes of metabolic syndrome. *Hum Biol,* 78:147-159.

Butterworth MB & Johnson JP (2008). USP10: the nexus between nexin and vasopressin. *Am J Physiol Renal Physiol,* 295:F888.

Combaret L, Adegoke OA, Bedard N, Baracos V, Attaix D & Wing SS (2005). USP19 is a ubiquitin-specific protease regulated in rat skeletal muscle during catabolic states. *Am J Physiol Endocrinol Metab,* 288:E693-700.

Deng L, Wang C, Spencer E, Yang L, Braun A, You J, Slaughter C, Pickart C & Chen ZJ (2000). Activation of the IkappaB kinase complex by TRAF6 requires a dimeric ubiquitin-conjugating enzyme complex and a unique polyubiquitin chain. *Cell,* 103:351-361.

Denuc A & Marfany G (2010). SUMO and ubiquitin paths converge. *Biochem Soc Trans,* 38:34-39.

Edelmann MJ, Kramer HB, Altun M & Kessler BM (2010). Post-translational modification of the deubiquitinating enzyme otubain 1 modulates active RhoA levels and susceptibility to Yersinia invasion. *Febs J,* 277:2515-2530.

Erman A, Veilleux A, Tchernof A & Goodyer CG (2010). Human growth hormone receptor (GHR) expression in obesity: I. GHR mRNA expression in omental and subcutaneous adipose tissues of obese women. *Int J Obes (Lond),* 35:1511-1519.

Ernst R, Mueller B, Ploegh HL & Schlieker C (2009). The otubain YOD1 is a deubiquitinating enzyme that associates with p97 to facilitate protein dislocation from the ER. *Mol Cell,* 36:28-38.

Gao L, Zheng Z, Cao L, Shen S, Yang Y, Zhao Z, Zhi D, Cheng R, Pei Z, Yongfu Y & Luo F (2010). The growth hormone receptor (GHR) exon 3 polymorphism and its correlation with metabolic profiles in obese Chinese children. *Pediatr Diabetes,* 12:429-434.

Goto E, Yamanaka Y, Ishikawa A, Aoki-Kawasumi M, Mito-Yoshida M, Ohmura-Hoshino M, Matsuki Y, Kajikawa M, Hirano H & Ishido S (2010). Contribution of lysine 11-linked ubiquitination to MIR2-mediated major histocompatibility complex class I internalization. *J Biol Chem,* 285:35311-35319.

Govers R, ten Broeke T, van Kerkhof P, Schwartz AL & Strous GJ (1999). Identification of a novel ubiquitin conjugation motif, required for ligand-induced internalization of the growth hormone receptor. *Embo J,* 18:28-36.

Govers R, van Kerkhof P, Schwartz AL & Strous GJ (1997). Linkage of the ubiquitin-conjugating system and the endocytic pathway in ligand-induced internalization of the growth hormone receptor. *Embo J,* 16:4851-4858.

Grainger RJ & Beggs JD (2005). Prp8 protein: at the heart of the spliceosome. *RNA,* 11:533-557.

Habelhah H, Takahashi S, Cho SG, Kadoya T, Watanabe T & Ronai Z (2004). Ubiquitination and translocation of TRAF2 is required for activation of JNK but not of p38 or NF-kappaB. *Embo J*, 23:322-332.

Hassink GC, Zhao B, Sompallae R, Altun M, Gastaldello S, Zinin NV, Masucci MG & Lindsten K (2009). The ER-resident ubiquitin-specific protease 19 participates in the UPR and rescues ERAD substrates. *EMBO Rep*, 10:755-761.

Hassink GC, Slotman J, Oorschot V, van der Reijden BA, Monteferrario D, Noordermeer SM, van Kerkhof P, Klumperman J & Strous GJ (2012). Identification of the ubiquitin ligase Triad1 as regulator of endosomal transport. *Biol Open.*, on line doi: 10.1242/bio.2012778.

Hicke L & Riezman H (1996). Ubiquitination of a yeast plasma membrane receptor signals its ligand-stimulated endocytosis. *Cell*, 84:277-287.

Huang DT, Ayrault O, Hunt HW, Taherbhoy AM, Duda DM, Scott DC, Borg LA, Neale G, Murray PJ, Roussel MF & Schulman BA (2009). E2-RING expansion of the NEDD8 cascade confers specificity to cullin modification. *Mol Cell*, 33:483-495.

Iwai K & Tokunaga F (2009). Linear polyubiquitination: a new regulator of NF-kappaB activation. *EMBO Rep*, 10:706-713.

Jovic M, Sharma M, Rahajeng J & Caplan S (2010). The early endosome: a busy sorting station for proteins at the crossroads. *Histol Histopathol*, 25:99-112.

Kamsteeg EJ, Hendriks G, Boone M, Konings IB, Oorschot V, van der Sluijs P, Klumperman J & Deen PM (2006). Short-chain ubiquitination mediates the regulated endocytosis of the aquaporin-2 water channel. *Proc Natl Acad Sci U S A*, 103:18344-18349.

Kirchhausen T (1999). Adaptors for clathrin-mediated traffic. *Annu Rev Cell Devel Biol*, 15:705-732.

Kumar KGS, Krolewski JJ & Fuchs SY (2004). Phosphorylation and specific ubiquitin acceptor are required for ubiquitination and degradation of the IFNAR1 subunit of type I Interferon receptor. *J. Biol. Chem.*, 279:46614-46620.

Lauwers E, Jacob C & Andre B (2009). K63-linked ubiquitin chains as a specific signal for protein sorting into the MVB pathway. *J Cell Biol*, 185:493-502.

Li S, Zheng H, Mao AP, Zhong B, Li Y, Liu Y, Gao Y, Ran Y, Tien P & Shu HB (2010). Regulation of virus-triggered signaling by OTUB1- and OTUB2-mediated deubiquitination of TRAF3 and TRAF6. *J Biol Chem*, 285:4291-4297.

Li Y, Suresh Kumar KG, Tang W, Spiegelman VS & Fuchs SY (2004). Negative Regulation of Prolactin Receptor Stability and Signaling Mediated by SCFβTrCP E3 Ubiquitin Ligase. *Mol. Cell. Biol.*, 24:4038-4048.

Liu Q, Xu WG, Luo Y, Han FF, Yao XH, Yang TY, Zhang Y, Pi WF & Guo XJ (2011). Cigarette smoke-induced skeletal muscle atrophy is associated with up-regulation of USP-19 via p38 and ERK MAPKs. *J Cell Biochem*, 112:2307-2316.

Lo Bianco C, Schneider BL, Bauer M, Sajadi A, Brice A, Iwatsubo T & Aebischer P (2004). Lentiviral vector delivery of parkin prevents dopaminergic degeneration in an alpha-synuclein rat model of Parkinson's disease. *Proc Natl Acad Sci U S A*, 101:17510-17515.

Loring GL, Christensen KC, Gerber SA & Brenner C (2008). Yeast Chfr homologs retard cell cycle at G1 and G2/M via Ubc4 and Ubc13/Mms2-dependent ubiquitination. *Cell Cycle*, 7:96-105.

Lu Y, Adegoke OA, Nepveu A, Nakayama KI, Bedard N, Cheng D, Peng J & Wing SS (2009). USP19 deubiquitinating enzyme supports cell proliferation by stabilizing KPC1, a ubiquitin ligase for p27Kip1. *Mol Cell Biol,* 29:547-558.

Madshus IH & Stang E (2009). Internalization and intracellular sorting of the EGF receptor: a model for understanding the mechanisms of receptor trafficking. *J Cell Sci,* 122:3433-3439.

Marsh M & McMahon HT (1999). The structural era of endocytosis. *Science,* 285:215-220.

Marteijn JA, van der Meer LT, Smit JJ, Noordermeer SM, Wissink W, Jansen P, Swarts HG, Hibbert RG, de Witte T, Sixma TK, Jansen JH & van der Reijden BA (2009). The ubiquitin ligase Triad1 inhibits myelopoiesis through UbcH7 and Ubc13 interacting domains. *Leukemia,* 23:1480-1489.

Martin S, Nishimune A, Mellor JR & Henley JM (2007). SUMOylation regulates kainate-receptor-mediated synaptic transmission. *Nature,* 447:321-325.

McNally T, Huang Q, Janis RS, Liu Z, Olejniczak ET & Reilly RM (2003). Structural analysis of UBL5, a novel ubiquitin-like modifier. *Protein Sci,* 12:1562-1566.

Mei Y, Hahn AA, Hu S & Yang X (2011). The USP19 deubiquitinase regulates the stability of c-IAP1 and c-IAP2. *J Biol Chem,* 286:35380-35387.

Meyer L, Deau B, Forejtnikova H, Dumenil D, Margottin-Goguet F, Lacombe C, Mayeux P & Verdier F (2007). β-TrCP mediates ubiquitination and degradation of the erythropoietin receptor and controls cell proliferation. *Blood,* 109:5215-5222.

Mishra SK, Ammon T, Popowicz GM, Krajewski M, Nagel RJ, Ares M, Jr., Holak TA & Jentsch S (2011). Role of the ubiquitin-like protein Hub1 in splice-site usage and alternative splicing. *Nature,* 474:173-178.

Nakada S, Tai I, Panier S, Al-Hakim A, Iemura S, Juang YC, O'Donnell L, Kumakubo A, Munro M, Sicheri F, Gingras AC, Natsume T, Suda T & Durocher D (2010). Non-canonical inhibition of DNA damage-dependent ubiquitination by OTUB1. *Nature,* 466:941-946.

Plans V, Scheper J, Soler M, Loukili N, Okano Y & Thomson TM (2006). The RING finger protein RNF8 recruits UBC13 for lysine 63-based self polyubiquitylation. *J Cell Biochem,* 97:572-582.

Putters J, da Silva Almeida AC, van Kerkhof P, van Rossum AG, Gracanin A & Strous GJ (2011a). Jak2 is a negative regulator of ubiquitin-dependent endocytosis of the growth hormone receptor. *PLoS ONE,* 6:e14676.

Putters J, Slotman JA, Gerlach JP & Strous GJ (2011b). Specificity, location and function of betaTrCP isoforms and their splice variants. *Cell Signal,* 23:641-647.

Rahighi S, Ikeda F, Kawasaki M, Akutsu M, Suzuki N, Kato R, Kensche T, Uejima T, Bloor S, Komander D, Randow F, Wakatsuki S & Dikic I (2009). Specific recognition of linear ubiquitin chains by NEMO is important for NF-kappaB activation. *Cell,* 136:1098-1109.

Rao Y, Ruckert C, Saenger W & Haucke V (2011). The early steps of endocytosis: From cargo selection to membrane deformation. *Eur J Cell Biol,*

Row PE, Liu H, Hayes S, Welchman R, Charalabous P, Hofmann K, Clague MJ, Sanderson CM & Urbe S (2007). The MIT domain of UBPY constitutes a CHMP binding and endosomal localization signal required for efficient epidermal growth factor receptor degradation. *J Biol Chem,* 282:30929-30937.

Schwartz AL & Ciechanover A (2009). Targeting proteins for destruction by the ubiquitin system: implications for human pathobiology. *Annu Rev Pharmacol Toxicol,* 49:73-96.

Slotman JA, da Silva Almeida AC, Hassink GC, van de Ven RH, van Kerkhof P, Kuiken HJ & Strous GJ (2012). UBC13 and COOH-terminus of HSP70 interacting protein (CHIP) are required for growth hormone receptor endocytosis. *J Biol Chem,* 287:15533-15543.

Soares L, Seroogy C, Skrenta H, Anandasabapathy N, Lovelace P, Chung CD, Engleman E & Fathman CG (2004). Two isoforms of otubain 1 regulate T cell anergy via GRAIL. *Nat Immunol,* 5:45-54.

Soboleva TA & Baker RT (2004). Deubiquitinating enzymes: their functions and substrate specificity. *Curr Protein Pept Sci,* 5:191-200.

Stanisic V, Malovannaya A, Qin J, Lonard DM & O'Malley BW (2009). OTU Domain-containing ubiquitin aldehyde-binding protein 1 (OTUB1) deubiquitinates estrogen receptor (ER) alpha and affects ERalpha transcriptional activity. *J Biol Chem,* 284:16135-16145.

Staub O, Dho S, Henry PC, Correa J, Ishikawa T, McGlade J & Torin D (1996). WW domains of NEDD4 bind to the proline-rich PY motifs in the epithelial Na channel deleted in Liddle's syndrome. *Embo J,* 15,:2371-2380.

Strous GJ, van Kerkhof P, Govers R, Ciechanover A & Schwartz AL (1996). The ubiquitin conjugation system is required for ligand-induced endocytosis and degradation of the growth hormone receptor. *Embo J,* 15:3806-3812.

Strous GJ, van Kerkhof P, Govers R, Rotwein P & Schwartz AL (1997). Growth hormone-induced signal tranduction depends on an intact ubiquitin system. *J Biol Chem,* 272:40-43.

Sun L, Deng L, Ea CK, Xia ZP & Chen ZJ (2004). The TRAF6 ubiquitin ligase and TAK1 kinase mediate IKK activation by BCL10 and MALT1 in T lymphocytes. *Mol Cell,* 14:289-301.

Sundaram P, Pang Z, Miao M, Yu L & Wing SS (2009). USP19-deubiquitinating enzyme regulates levels of major myofibrillar proteins in L6 muscle cells. *Am J Physiol Endocrinol Metab,* 297:E1283-1290.

Torres-Ramos CA, Prakash S & Prakash L (2002). Requirement of RAD5 and MMS2 for postreplication repair of UV-damaged DNA in Saccharomyces cerevisiae. *Mol Cell Biol,* 22:2419-2426.

Ulrich HD & Jentsch S (2000). Two RING finger proteins mediate cooperation between ubiquitin-conjugating enzymes in DNA repair. *Embo J,* 19:3388-3397.

van den Eijnden MJ, Lahaye LL & Strous GJ (2006). Disulfide bonds determine growth hormone receptor folding, dimerisation and ligand binding. *J Cell Sci,* 119:3078-3086.

van der Bliek AM, Redelmeier TE, Damke H, Tisdale EJ, Meyerowitz EM & Schmid SL (1993). Mutations in human dynamin block an intermediate stage in coated vesicle formation. *J Cell Biol,* 122:553-563.

van Kerkhof P, Govers R, Alves dos Santos CM & Strous GJ (2000). Endocytosis and degradation of the growth hormone receptor are proteasome-dependent. *J Biol Chem,* 275:1575-1580.

van Kerkhof P, Putters J & Strous GJ (2007). The ubiquitin ligase SCF(betaTrCP) regulates the degradation of the growth hormone receptor. *J Biol Chem,* 282:20475-20483.

van Kerkhof P, Westgeest M, Hassink G & Strous GJ (2011). SCF(TrCP) acts in endosomal sorting of the GH receptor. *Exp Cell Res,* 317:1071-1082.

Walder K, Ziv E, Kalman R, Whitecross K, Shafrir E, Zimmet P & Collier GR (2002). Elevated hypothalamic beacon gene expression in Psammomys obesus prone to develop obesity and type 2 diabetes. *Int J Obes Relat Metab Disord,* 26:605-609.

Wang T, Yin L, Cooper EM, Lai MY, Dickey S, Pickart CM, Fushman D, Wilkinson KD, Cohen RE & Wolberger C (2009). Evidence for bidentate substrate binding as the basis for the K48 linkage specificity of otubain 1. *J Mol Biol,* 386:1011-1023.

Weissman AM, Shabek N & Ciechanover A (2011). The predator becomes the prey: regulating the ubiquitin system by ubiquitylation and degradation. *Nat Rev Mol Cell Biol,* 12:605-620.

Wooff J, Pastushok L, Hanna M, Fu Y & Xiao W (2004). The TRAF6 RING finger domain mediates physical interaction with Ubc13. *FEBS Lett,* 566:229-233.

Wright MH, Berlin I & Nash PD (2011). Regulation of endocytic sorting by ESCRT-DUB-mediated deubiquitination. *Cell Biochem Biophys,* 60:39-46.

Wyatt D, Malik R, Vesecky AC & Marchese A (2010). Small ubiquitin-like modifier modification of arrestin-3 regulates receptor trafficking. *J Biol Chem,* 286:3884-3893.

Zhang M, Windheim M, Roe SM, Peggie M, Cohen P, Prodromou C & Pearl LH (2005). Chaperoned ubiquitylation-crystal structures of the CHIP U box E3 ubiquitin ligase and a CHIP-Ubc13-Uev1a complex. *Mol Cell,* 20:525-538.

Zou W, Papov V, Malakhova O, Kim KI, Dao C, Li J & Zhang DE (2005). ISG15 modification of ubiquitin E2 Ubc13 disrupts its ability to form thioester bond with ubiquitin. *Biochem Biophys Res Commun,* 336:61-68.

Section 6

Drug Delivery

Chapter 16

Endosomal Escape Pathways for Non-Viral Nucleic Acid Delivery Systems

Wanling Liang and Jenny K. W. Lam

Additional information is available at the end of the chapter

1. Introduction

Nucleic acids, including plasmid DNA (pDNA) and small interfering RNA (siRNA), are potential therapeutic macromolecules that have been widely explored for the treatment or prevention of various human diseases in the last three decades. pDNA encoding a therapeutic gene sequence can be introduced into the nuclei of the target cells to express functional proteins through transcription and translation in order to produce therapeutic effects. The therapeutic scope of pDNA includes a vast number of diseases, such as cancers (El-Aneed 2004; Yamamoto and Curiel 2005; Johnson, Morgan et al. 2009), genetic disorders (Gaspar, Parsley et al. 2004; Aiuti, Cattaneo et al. 2009; Griesenbach and Alton 2009), infections (Yu, Poeschla et al. 1994; Hashiba, Suzuki et al. 2001; Cull, Bartlett et al. 2002) and cardiovascular diseases (Stewart, Hilton et al. 2006; Vinge, Raake et al. 2008; Henry and Satran 2012). To date, over 1600 gene therapy clinical trials have been initiated (http://www.abedia.com/wiley/phases.php; Edelstein, Abedi et al. 2007). Apart from pDNA, siRNA has recently been intensively studied as a new therapeutic agent. RNA interference (RNAi) was discovered by Fire and colleagues based on the study of *C. elegans* (Fire, Xu et al. 1998). According to their observation, double-stranded RNA (dsRNA) mediates potent and specific silencing of homologous genes. Elbashir *et al.* demonstrated that the sequence-specific mediator of RNAi is 21-23-nucleotide siRNA produced from the cleavage of longer dsRNA by ribonuclease III (Elbashir, Lendeckel et al. 2001). Since then the mechanism of RNAi has been revealed and reviewed in many publications (Bernstein, Caudy et al. 2001; Hammond, Boettcher et al. 2001; Ketting, Fischer et al. 2001; Hannon and Rossi 2004; Mello and Conte 2004). Briefly, siRNA interacts with the RNA-induced silencing complex (RISC) located in the cell cytoplasm and subsequently induces cleavage of mRNA with complementary sequences, thereby inhibiting the translation of a specific protein. Soon after RNAi was discovered to work in mammalian cells (Sui, Soohoo et al. 2002), it quickly emerged as a new therapeutic strategy to suppress the expression of disease-causing gene.

So far, the therapeutic potential of siRNA has been demonstrated successfully both *in vitro* and *in vivo* (Shim and Kwon 2010) while a number of siRNA-based therapy clinical trials have been initiated, including therapeutics directed against inherited skin disorder (Leachman, Hickerson et al. 2009), solid tumor (Davis, Zuckerman et al. 2010), respiratory syncytial virus (RSV) infection (DeVincenzo, Cehelsky et al. 2008; DeVincenzo, Lambkin-Williams et al. 2010) and age-related macular degeneration (AMD) (Kaiser, Symons et al. 2010). Until now however, there are no pDNA or siRNA-based therapeutic products approved by the FDA; the lack of an efficient and safe delivery system being the major hurdle to limit the clinical application of nucleic acids.

1.1. Nucleic acid delivery

In terms of delivery, therapeutic nucleic acids must be transported to their target sites (nucleus for pDNA or RISC in the cytoplasm for siRNA) before producing their biological effects. A delivery system must overcome a series of extracellular and intracellular barriers (Sanders, Rudolph et al. 2009). Nucleic acids are susceptible to endogenous nuclease degradation in the serum and the half-life of unprotected nucleic acids is approximately 10 minutes in mouse whole blood (Kawabata, Takakura et al. 1995). In addition, nucleic acids are negatively charged, hydrophilic macromolecules and are incapable of crossing the plasma membrane unassisted (Khalil, Kogure et al. 2006; Lam, Liang et al. 2012). To achieve successful transfection, an effective nucleic acid delivery system must be able to perform several functions: (i) bind or condense nucleic acids into nanoparticles, (ii) protect nucleic acids from enzymatic degradation, (iii) promote cellular uptake, (iv) release nucleic acids into the cytoplasm, (v) promote nuclear entry (for pDNA delivery) (Bally, Harvie et al. 1999). The use of a carrier system such as cationic polymers (Laga, Carlisle et al. 2012), lipids/ liposomes (Ewert, Zidovska et al. 2010) or peptides (Hassane, Saleh et al. 2010) is the most commonly investigated delivery method for clinical purposes. It was soon found that the transfection efficiency of nucleic acid delivery systems is correlated to not only the level of cellular uptake but also with their ability to escape from endosomal compartments (El Ouahabi, Thiry et al. 1997). Some nucleic acid delivery systems successfully attain high cellular uptake, but fail to achieve good transfection, partly due to their deficiency of endosomolytic property (Medina-Kauwe, Xie et al. 2005). Therefore additional measures must be adopted to promote endosomal escape of the nucleic acid delivery system.

1.2. Intracellular delivery

Viral vectors are known for their high efficiency in transferring nucleic acids into host cells as they have evolved sophisticated endosomal releasing mechanisms which take advantage of the acidic environment inside the endosomes (Cho, Kim et al. 2003). However, the clinical application of viral vectors is limited because of the strong immunogenicity and potential fatal adverse effects (Baum, Düllmann et al. 2003; Hacein-Bey-Abina, von Kalle et al. 2003; Raper, Chirmule et al. 2003; Hacein-Bey-Abina, Garrigue et al. 2008). Compared with viral vectors, non-viral vectors offer advantages of relatively low toxicity and immune response.

However, the delivery efficiency of non-viral vectors is generally poor (Pérez-Martínez, Guerra et al. 2011). To enhance transfection efficiency, substantial efforts have been made to elicit effective endosomal escape. Endocytosis is the major route of cellular entry for non-viral nucleic acid delivery (Khalil, Kogure et al. 2006; Pathak, Patnaik et al. 2009). A number of endocytosis pathways are known to be involved in the uptake of non-viral gene delivery systems, including clathrin-mediated endocytosis, caveolae-mediated endocytosis, macropinocytosis and phagocytosis. However the contribution of each pathway in the internalization of non-viral vectors is not clearly understood due to the diversity of carriers (Morille, Passirani et al. 2008).

After non-viral delivery systems enter into the cells via endocytosis they are immediately transported into the endocytic vesicles. Initially, the delivery vectors are trapped in the early endosomes where the pH drops from neutral to around pH 6. Early endosomes may fuse with sorting endosomes in which the internalized content can be recycled back to the membrane and transported out of the cell by exocytosis. More often, the delivery systems are trafficked to late endosomes which are rapidly acidified to pH 5–6 by the action of the membrane-bound ATPase proton-pump. Subsequently, the late endosomes fuse with lysosomes concomitant with a further pH reduction to approximately pH 4.5 and the existence of various degradative enzymes. The low pH of lysosomes facilitates substrate denaturation and aids lysosomal hydrolases, most of which operate optimally in the range of pH 4.5-5.5 (Mellman, Fuchs et al. 1986). Nucleic acids that fail to be released from these acidic vesicles will ultimately be degraded (Pack, Hoffman et al. 2005; Khalil, Kogure et al. 2006). Therefore, the pH reduction and the enzymatic degradation process in endosomes/lysosomes is an efficiency-limiting step for successful nucleic acid delivery (Whitehead, Langer et al. 2009). Ideal vectors should release their contents from these acidic compartments at an early stage to prevent the fate of lysosomal destruction.

1.3. Endosomal escape

Various approaches have been attempted to promote early endosomal escape of non-viral gene delivery vehicles and many hypotheses have been suggested to explain these processes. The proton sponge hypothesis has been proposed for cationic polymers such as polyethylenimine (PEI) (Boussif, Lezoualc'h et al. 1995; Behr 1997) and polyamidoamine (PAMAM) dendrimers (Zhou, Wu et al. 2006). For cationic-lipid based delivery systems, the flip-flop mechanism was proposed for their endosomal escape mechanism (Xu and Szoka Jr 1996). Cell-penetrating peptides (CPPs) represent another category of promising candidates as non-viral nucleic acid carriers, for example TAT (Torchilin 2008), pep analogues (Gros, Deshayes et al. 2006), GALA (Li, Nicol et al. 2004), MPG (Simeoni, Morris et al. 2003), CADY (Konate, Crombez et al. 2010) and LAH4 derivatives (Lam, Liang et al. 2012) etc. Their mechanisms of promoting endosomal release are still controversial; it has been suggested that the opening of transient pores in the lipid bilayer of endosome is involved (Melikov and Chernomordik 2005), alternatively CPPs may undergo conformational changes in response to the acidification inside the endosomes, leading to destabilization of the endosomal membrane bilayer (Kichler, Mason et al. 2006). Last but not least, photochemical internalization (PCI) is a technique that aims to improve endosomal release. A photosensitizer is localized in the endosomal membrane and destabilizes the membrane upon illumination, triggering the release of endosomal content into cytosol (Berg, Kristian Selbo et al. 1999; Endoh and Ohtsuki 2009).

In order to achieve high levels of transfection, different strategies have been employed to protect nucleic acids from degradation inside endosomes and facilitate their early release from acidic compartments into the cytosol. Various endosomal escape mechanisms of non-viral vectors as well as the endosomolytic reagents which can promote endosomal release are introduced in detail here.

2. Strategies of non-viral vectors for endosomal escape

2.1. The 'proton sponge' hypotheses (pH-buffering effect)

There is a long history regarding the application of cationic polymers to mediate nucleic acid transfer into cells. Cationic polymers can form polyplexes with nucleic acids through electrostatic interaction. Polylysine (PLL) was one of the first cationic polymers investigated for nucleic acid delivery although it failed to display desirable transfection efficiency (Pack, Hoffman et al. 2005). Later, it was discovered that polymers that contain protonable residues at physiological pH, like polyamidoamine (PAMAM) dendrimers and lipopolyamines (Remy, Sirlin et al. 1994) successfully achieve high transfection efficiency in contrast to PLL, which does not possess buffering capacity because of the presence of the strongly charged amino groups (Haensler and Szoka Jr 1993). This pH-buffering property was soon shown to be an important feature of cationic polymers that may induce endosomal disruption and prevent nucleic acids from lysosomal degradation.

Polyethylenimine (PEI) is a synthetic cationic polymer with high amine density and various applications (Godbey, Wu et al. 1999). In 1995, Boussif *et al.* investigated the DNA delivery potential of PEI and found that this polymer can effectively transfer luciferase reporter gene into a variety of cell lines and primary cells (Boussif, Lezoualc'h et al. 1995). The remarkable nucleic acid-delivery ability of PEI was attributed to a "proton sponge" effect in which the extensive buffering capacity of PEI serves a dual purpose: (i) to inhibit the activity of lysosomal nuclease and (ii) to change the osmolarity of acidic vesicles resulting in endosomal swelling and rupture.

The 'proton sponge' phenomenon has been observed in certain cationic polymers with a high pH buffering capability over a wide pH range. These polymers usually contain protonatable secondary and/or tertiary amine groups with pKa close to endosomal/lysosomal pH. During the maturation of endosomes, the membrane-bound ATPase proton pumps actively translocate protons from the cytosol into the endosomes, leading to the acidification of endosomal compartments and activation of hydrolytic enzymes. At this stage, polymers with the 'proton sponge' property will become protonated and resist the acidification of endosomes (fig. 1). As a result, more protons will be continuously pumped into the endosomes with the attempt to lower the pH. The proton pumping action is followed by passive entry of chloride ions, increasing ionic concentration and leading to water influx. Eventually the osmotic pressure causes swelling and rupture of endosomes, releasing their contents to the cytosol (Boussif, Lezoualc'h et al. 1995; Behr 1997; Pack, Hoffman et al. 2005). Sonawane *et al.* (Sonawane, Szoka Jr et al. 2003) tested this hypothesis by studying the concentration of chloride ions, pH and the volume of endosomes after internalization of polyplexes composed of pDNA, non-buffering PLL as well as the highly buffering PEI and PAMAM. Significant chloride ion accumulation, volume expansion and membrane lysis were detected in PEI and PAMAM containing endosomes but not the ones with PLL. This finding provides direct support for the proton sponge hypothesis and a rationale for the design of polymer-based nucleic acid delivery vectors.

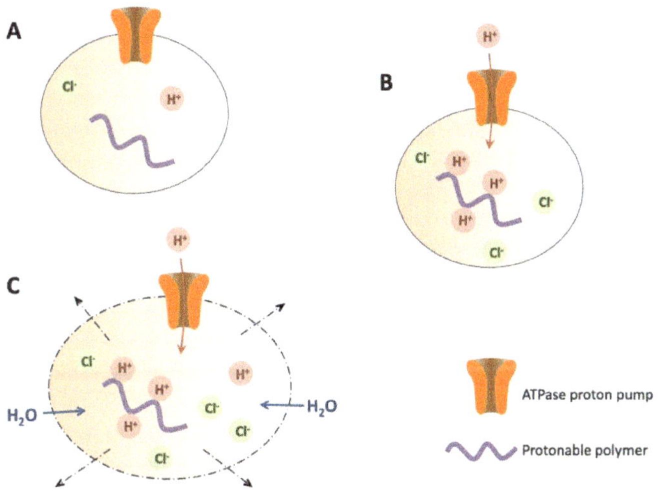

Figure 1. The 'proton sponge' hypotheses (pH-buffering effect). (A) Polyplexes enter cell via endocytosis and are trapped in endosomes. (B) The membrane bound ATPase proton pumps actively translocate protons into endosomes. Polymers become protonated and resist the acidification of endosomes. Hence more protons will be pumped into the endosomes continuously to lower the pH. (C) The proton pumping action is followed by passive chloride ions entry, increasing ionic concentration and hence water influx. High osmotic pressure causes the swelling and rupture of endosomes, releasing their contents to cytosol.

To date, PEI has been demonstrated to transfect nucleic acids successfully into a broad range of cells and tissues both *in vitro* and *in vivo* (Boletta, Benigni *et al.* 1997; Goula, Remy *et al.* 1998; Coll, Chollet *et al.* 1999; Urban-Klein, Werth *et al.* 2004; Merkel, Beyerle *et al.* 2009; Pfeifer, Hasenpusch *et al.* 2011). However, the clinical application of PEI is limited by its substantial toxicity. Chollet *et al.* (Chollet, Favrot *et al.* 2002) injected linear PEI (L-PEI) –pDNA polyplexes into mice intravenously. Signs of toxicity caused by L-PEI were observed 15 minutes after injection of 50 mg of L-PEI/DNA complexes. Increasing the dose up to 100 mg remarkably enhanced the transfection efficiency, but most of the animals suffered from liver necrosis. If the dose was increased to 150 mg, all the treated animals died of shock within the first 30 minutes.

Modified versions of PEI, such as low molecular weight PEI (LMW-PEI) (Kunath, von Harpe et al. 2003) and low branching degree PEI (Fischer, Bieber et al. 1999), have been investigated to reduce the toxicity of the polymers without compromising their pH buffering capacity (Kunath, von Harpe et al. 2003; Arote, Kim et al. 2007).

On the contrary, polymeric carriers that do not possess pH buffering properties, such as chitosan and PLL, fail to achieve satisfactory nucleic acid delivery efficiencies because of their inability to induce endosomal escape even though they are capable of binding to nucleic acids and promoting cellular entry (Wagner and Kloeckner 2006). To enhance transfection efficiency, functional moieties were included into these polymeric systems for improving their buffering capacity. Histidine is one of the commonly employed molecules that was added as functional group of polymers or incorporated into peptide sequences to enhance their pH buffering capacity.

The buffering capacity of histidine is due to the presence of an imidazole ring that has a pKa around 6 and hence can be protonated in a slightly acidic pH (Midoux, Pichon et al. 2009). Midoux *et al.* (Midoux and Monsigny 1999) reported that partial substitution of PLL with histidine residues resulted in transfection 4-5 orders of magnitude higher than the unmodified PLL/pDNA polyplexes. Upon protonation of the imidazole groups, the histidine residues trigger destabilization of polyplexes and fusion with endosomal membrane, leading to the release of polyplexes into cytosol. Chang *et al.* (Chang, Higuchi et al. 2010) introduced histidine to the amino groups of chitosan via disulfide bonds. The result showed that histidine-modified chitosan has a broader pH buffering range, wider distribution in the cytosol as well as a higher transfection level. It is evident that histidine can facilitate pDNA escape from endosomes by increasing the buffering capacity of chitosan inside the acidic compartments. A similar approach was adopted by other researchers to enhance other polycations by introducing imidazole moieties to enhance the pH buffering capacity and ultimately promote endosomal escape. (Yang, Lee et al. 2006; Moreira, Oliveira et al. 2009).

2.2. Flip-flop mechanism

Lipids and liposomes, whether anionic, cationic, neutral and/or pH-sensitive, present another category of non-viral carriers that have been extensively investigated for delivering nucleic acid into mammalian cells. In general, the *in vitro* and *in vivo* transfection efficiency of non-cationic lipids or liposomes is relatively low when compared with their cationic counterparts (Legendre and Szoka Jr 1992). Cationic lipids or liposomes form lipoplexes with the anionic nucleic acids through electrostatic interactions and the overall charge of the lipoplexes are usually positive so that they can easily associate with the negatively charged cell surface to promote cellular entry (Felgner, Gadek et al. 1987).

The mechanism of how lipoplexes gain entry into the cells is controversial. According to literature, there are at least two routes of cellular uptake: (i) direct fusion with the plasma membrane; and (ii) endocytosis (Pedroso de Lima, Simões et al. 2001). Physicochemical properties of lipoplexes such as particle size distribution, lipid composition and charge ratio may also influence their uptake route. In order to gain a better insight of the uptake mechanism of lipid-based system, Wrobel and Collins studied the interaction between cationic liposomes and model anionic membrane as well as cultured mammalian cells. The results indicated that cell surface binding alone is insufficient for cationic liposomes to gain entry into cells via membrane fusion in the absence of endocytosis (Wrobel and Collins 1995). Zhou *et al.* investigated the DNA transfection of cationic liposomes containing lipopolylysine and found that only 2% of the treated cells were transfected and the uptake was mediated by membrane fusion (Zhou and Huang 1994). No correlation between fusion of the lipoplexes with the plasma membrane and the transfection level in monocytic THP-1 cells was observed (Pires, Simões et al. 1999). Nevertheless, some investigators believed that membrane fusion is significant in the internalization of lipoplexes and the contribution of this uptake pathway cannot be completely ruled out.

To find out the intracellular fate of the lipoplexes following endocytosis, an electron microscopic study was carried out by Zabner *et al.* (Zabner, Fasbender et al. 1995). It was observed that after cellular uptake, lipoplexes were delivered to perinuclear vesicular compartments which fuse with early endosomes. It was also noticed that the dissociation of nucleic acids from lipoplexes and their escape from endosomes are crucial barriers for successful transfection. To elucidate this endosomal escape mechanism of cationic liposomes, Zelphatl *et al.* identified the biomolecules that are responsible for dissociating of nucleic acids from the lipoplexes and releasing them into cytosol. Anionic liposomes with similar components with the cytoplasmic-facing monolayer of plasma membrane were used as an endosomal model. It was found that the anionic liposomes were able to trigger a rapid release of nucleic acid from lipoplexes. On the basis of this result, a 'flip-flop' mechanism (fig. 2) was proposed by the authors to describe how nucleic acids were able to dissociate from the lipoplexes and escape from the endosomes into the cytosol (Zelphati and Szoka 1996; Zelphati and Szoka 1996). Once inside the endosomes, there is an electrostatic interaction between the cationic lipoplexes and the negatively charged lipids (mainly found in the cytoplasmic-facing leaflet) of the endosomal membrane. The anionic lipids of the endosomal membrane laterally diffuse into the lipoplexes and form charge-neutralized ion pairs with the cationic lipids of the lipoplexes. As a result, the nucleic acids are displaced from the lipoplexes, allowing the nucleic acids to be released into the cytoplasm (Zhou and Huang 1994; Xu and Szoka Jr 1996; Zelphati and Szoka 1996).

Neutral lipids such as the phosphatidylethanolamine (DOPE) are widely used as helper lipids in combination with cationic liposomes. It is well established that inclusion of DOPE in lipoplexes may significantly improve their transfection activity, whereas replacement of DOPE with other neutral phospholipid dioleoylphosphatidylcholine (DOPC) fails to accomplish the helper function. The role of DOPE as helper lipid is attributed to its endosomolytic activity. Zhou and Huang used transmission electron microscopy to study intracellular trafficking of cationic liposomes and found that DOPE-containing lipoplexes can destabilize the endosomal membrane (Zhou and Huang 1994) whereas the DOPC-containing lipoplexes did not show the same effect. A study carried out by Farhood *et al.* indicated that a substantial amount of DOPE needs to be incorporated into cationic liposomes in order to achieve endosomal membrane destabilization (Farhood, Serbina et al. 1995). Both DOPE and DOPC share very similar acyl chain composition (Hui, Langner et al. 1996) and the major difference between the two phospholipids is their headgroups. The former has an ethanolamine head group whereas the latter contains a choline head group. The cone-shape ethanolamine head group of DOPE displays a high tendency to form inverted hexagonal phase especially at acidic pH while the choline head group of DOPC does not. Zuhorn *et al.* hypothesized that the formation of hexagonal phase of DOPE containing lipoplexes plays a prominent role both in dissociation of nucleic acids from lipoplexes and in efficient destabilizing endosomal membrane (Zuhorn, Bakowsky et al. 2005).

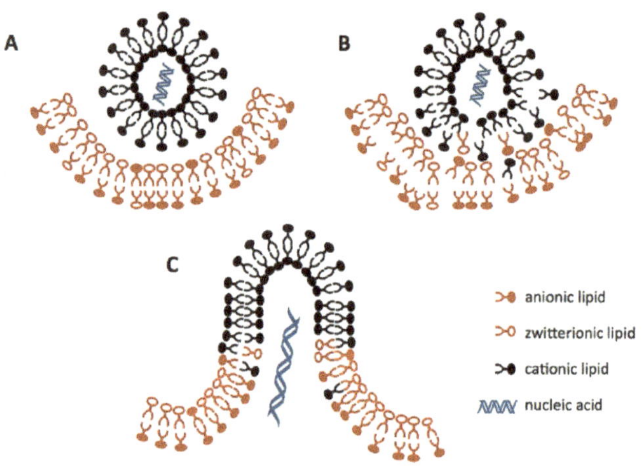

Figure 2. Flip-Flop Mechanism. (A) Lipoplexes are endocytosed and become entrapped inside the early endosomes. (B) There is an electrostatic interaction between the cationic lipoplexes and the anionic lipids which are present in the cytoplasmic-facing side of endosomal membrane. The anionic lipids of the endosomal membrane laterally diffuse into the lipoplexes and form charge-neutralized ion pairs with the cationic lipids of the lipoplexes. (C) The nucleic acids are displaced from the lipoplexes, allowing the nucleic acids entry into the cytoplasm.

Apart from helper lipids, other approaches have been investigated to potentiate endosomal lysis and the release of nucleic acid from lipoplexes to cytosol through the 'flip-flop' mechanism. Simoes *et al.* (Simoes, Slepushkin et al. 1999) described the use of the pH-sensitive endosome-disruptive peptide, GALA, together with lipoplexes (more endosomolytic peptides will be discussed later). A significant improvement in transfection was observed in several cell types and the enhancement was blocked by bafilomycin A1, a specific inhibitor of the vacuolar ATPase proton pump that inhibits the acidification of the endosomes. Since low pH is a triggering factor of the membrane disruption propensity of GALA, the authors speculated that the membrane destabilizing activity of GALA involves structural changes of the peptide which induce the dissociation of nucleic acids from lipoplexes inside the endosomes. As a result, the dissociated cationic lipids become available to interact with the anionic lipids of the endosomal membrane more readily, leading to the release of nucleic acids into the cytosol via the 'flip-flop' mechanism.

2.3. Endosomal membrane fusion or destabilization mechanism

Cell-penetration peptides (CPPs) have attracted tremendous attention as non-viral nucleic acid delivery vectors in recent years. Typically cationic and/or amphipathic in nature, CPPs are short sequences of amino acids, usually 10-30 residues, claimed to have ability to cross the plasma membrane of living cells. They can facilitate the transportation of various hydrophilic macromolecules including proteins, peptides and nucleic acids into cells without the disruption of plasma membrane (Richard, Melikov et al. 2003). CPPs were originally derived from viruses, with TAT peptide being the first CPP identified and was derived from the transcription activating factor of human immunodeficiency virus 1 (HIV-l) (Brooks, Lebleu et al. 2005). Many different sequences of CPPs were soon discovered and synthetic analogues were also rapidly developed. Until now there are a number of CPPs that were documented for nucleic acid delivery.

CPPs either form complexes with nucleic acids, through electrostatic interaction, or can be incorporated into polymeric and lipidic delivery system. In general, they can be categorized into two main classes (Futaki 2006; Patel, Zaro et al. 2007): (i) Cationic peptides that usually contain arginine and lysine residues, e.g., TAT peptide, penetratin (Derossi, Joliot et al. 1994; Derossi, Calvet et al. 1996; Muratovska and Eccles 2004) and oligoarginines (Futaki 2006); (ii) Amphipathic peptides that consist of both hydrophobic and hydrophilic segments. The amphipathicity of these peptides generates from either the primary structure or the secondary structure. Primary amphipathic peptides are sequentially made up of hydrophobic and hydrophilic residues (Fernández-Carneado, Kogan et al. 2004; Deshayes, Morris et al. 2005), and include e.g., MPG (Simeoni, Morris et al. 2003; Veldhoen, Laufer et al. 2006), pep-1 and its analogues. Secondary amphipathic peptides adopt an amphipathic helical conformation with hydrophilic and hydrophobic regions and include e.g., HA-2 (Wagner, Plank et al. 1992; Plank, Oberhauser et al. 1994), GALA (Li, Nicol et al. 2004), KALA (Wyman, Nicol *et al.* 1997), and LAH4 derivatives (Kichler, Leborgne *et al.* 2003; Kichler, Mason *et al.* 2006; Lam, Liang *et al.* 2012). The sequences and the endosomal escape mechanism of these peptides are summarized in table 1. In the past, it was generally accepted that a non-endosomal pathway or direct penetration was the major route of entry for CPPs (Morris, Chaloin et al. 2000). Early studies indicated that the uptake of CPPs into cells could progress at low temperature. It appeared to be energy-independent and insensitive to endocytosis inhibitors (Vivès, Brodin et al. 1997). However recent studies suggested otherwise, that endocytosis may actually be involved in the internalization of CPPs (Richard, Melikov et al. 2003; Lundin, Johansson et al. 2008). Until now, the uptake mechanism still remains controversial (Futaki 2005).

Nevertheless, a variety of CPPs have been shown to enter cells via an endosomal pathway and induce endosomolytic activity. The majority of these membrane-destabilizing peptides were developed to mimic the endosomal disruptive properties of fusogenic sequences of viral fusion proteins. Taking the haemagglutin subunit HA2 of influenza virus as an example, this protein chain is responsible for facilitating membrane fusion. The C-terminal end is embedded in the viral membrane whereas the N-terminal end contains a fusion peptide with a sequence of hydrophobic amino acids. Once inside the endosomes, HA undergoes conformation change in response to the low pH environment and expose the highly conserved hydrophobic N-terminal region. Subsequently, this triggers the fusion of viral membrane with the endosomal membrane, leading to viral genome leakage to cytosol (Stegmann 2000). Wagner *et al.* introduced HA2 as endosomolytic component in polyplexes containing transferrin/PLL/DNA, resulting in significantly augmentation of the delivery efficiency (Wagner, Plank et al. 1992). In contrast, peptides that are derived from HIV-1 fusion protein gp41 usually adopt a pH-independent membrane fusion capacity and are capable of fusing with both plasma membranes and endosomes at neutral pH (Fischer, Krausz et al. 2001).

Since the α-helical component of the HA2 appears to play a crucial role in endosomal membrane destabilization (Oehlke, Scheller et al. 1998), a series of pH-sensitive amphipathic α-helical structural motifs were designed and their structure–activity relationship were investigated. GALA is a synthetic peptide with 30 amino acid residues designed to interact with lipid bilayers at low pH. It contains a histidine and a tryptophan residue, as well as a glutamic acid-alanine-leucine-alanine (EALA) repeat. When the pH of the surrounding environment drops from 7.0 to 5.0, GALA experiences a conformational change from a random coil to an amphipathic α-helix, leading to disruption of model lipid membranes and therefore the release of entrapped aqueous content. The membrane-destabilizing character of this pH sensitive peptide in an acidic environment raises the possibility of enhancing the delivery of nucleic acid by facilitating endosomal escape (Li, Nicol et al. 2004). Haensler and Szoka Jr *et al* reported that when GALA is covalently attached to the dendrimer via a disulfide linkage, an increase of gene expression by 2-3 orders of magnitude was observed (Haensler and Szoka Jr 1993). Simoes's study revealed that by incorporating GALA with transferrin containing lipoplexes, there was a significant increase in luciferase gene expression in COS-7 cells (Simoes, Slepushkin et al. 1999). Similarly, introducing GALA peptide into the multifunctional envelope-type nano device (MEND) can lead to improvement of nucleic acid transfer by facilitating nanoparticle endosomal escape (Sasaki, Kogure et al. 2008).

The negatively charged GALA cannot bind with nucleic acid through electrostatic interaction, GALA can only be added as an additional functional component to polyplexes or lipoplexes. Newer peptides were soon developed to combine both nucleic acid binding and membrane destabilizing properties in order to produce a simpler delivery system. KALA is a modified version of GALA by partially replacing glutamic acid with lysine. It is one of the first generation peptides that is designed to bind nucleic acids as well as destabilize the endosomal membranes. Interestingly, the membrane destabilization mechanism of KALA is substantially different from that of GALA although they share similar amino acid sequence. KALA adopts α-helix conformation in a wide pH range and undergoes a pH-dependent conformational change from amphipathic α-helical to a mixture of α-helix and random coil as the pH is lowered (Wyman, Nicol et al. 1997). Apart from GALA and KALA, other amphipathic peptides that attain endosomal escape ability that is related to their α-helical structure include the Hel series peptides (Niidome, Takaji et al.

Peptide	Primary Sequence	Endosomal Escape Mechanism	Reference
TAT	GRKKRRQRRRPPQ	Unclear, endosomal escape is inefficient	(Vives 2003; Brooks, Lebleu *et al.* 2005; Lee, Johnson *et al.* 2011)
Penetratin	RQIKIWFQNRRMKWKK	Unclear, endosomal escape is inefficient	(Muratovska and Eccles 2004)
EBI	LIRLWSHLIHIWFQNRRLKW KKK	Membrane destabilization	(Lundberg, El-Andaloussi *et al.* 2007)
MPG	GALFLGFLGAAGSTMGAWS QPKKKRKV	Bypass endosomes through non-endosomal uptake	(Morris, Vidal *et al.* 1997; Simeoni, Morris *et al.* 2003)
HGP	LLGRRGWEVLKYWWNLLQ YWSQELC	Membrane destabilization, possibly pore formation	(Kwon, Bergen *et al.* 2008)
Pep-2	KETWFETWFTEWSQPKKKR KV	Bypass endosomes through non-endosomal uptake	(Morris, Depollier *et al.* 2001; Gros, Deshayes *et al.* 2006; Deshayes, Morris *et al.* 2008)
HA-2 derived fusogenic peptide	GLFGAIAGFIEGGWTGMIDG WYG	Membrane fusion and destabilization	(Wagner, Plank *et al.* 1992; Plank, Oberhauser *et al.* 1994; Navarro-Quiroga, Antonio González-Barrios *et al.* 2002)
H5WYG	GLFHAIAHFIHGGWHGLIHG WYG	Membrane destabilization	(Midoux, Kichler *et al.* 1998; Pichon, Gonçalves et al. 2001)
INF-7	GLFEAIEGFIENGWEG MIDGWYG	Membrane fusion and destabilization	(Plank, Oberhauser et al. 1994; van Rossenberg, Sliedregt-Bol *et al.* 2002; Funhoff, van Nostrum *et al.* 2005)

Peptide	Primary Sequence	Endosomal Escape Mechanism	Reference
E5 & E5CA	GLFGAIAGFIEGGWTGMIDG GLFEAIAEFIEGGWEGLIEGCA	Membrane fusion and destabilization	(Midoux, Mendes *et al.* 1993; Pichon, Freulon *et al.* 1997; Vliegenthart, Knollen *et al.* 1999; Klink, Chao *et al.* 2001)
JTS-1	GLFEALLELLESLWELLLEA	Membrane destabilization	(Gottschalk, Sparrow *et al.* 1996; Fominaya, Gasset *et al.* 2000; van Rossenberg, Sliedregt-Bol *et al.* 2002; Vlasov, Lesina *et al.* 2005)
ppTG1	GLFKALLKLLKSLWKLLLKA	Membrane destabilization	(Rittner, Benavente *et al.* 2002; Numata and Kaplan 2010)
GALA	WEAALAEALAEALAEHLAEALAEALEALAA	Membrane destabilization, Pore formation & flip-flop of membrane lipids	(Parente, Nir *et al.* 1990; Haensler and Szoka Jr 1993; Fattal, Nir *et al.* 1994; Plank, Oberhauser *et al.* 1994; Simoes, Slepushkin *et al.* 1999; Li, Nicol *et al.* 2004)
KALA	WEAKLAKALAKALAKHLAKALAKALKACEA	Membrane destabilization	(Wyman, Nicol *et al.* 1997; Li, Nicol *et al.* 2004; Min, Lee *et al.* 2006)
CADY	GLWRALWRLLRSLWRLLWRA	Bypass endosomes through non-endosomal uptake	(Crombez, Aldrian-Herrada *et al.* 2008)
Peptide 4_6 & analogues	LARLLARLLARLLRALLRALLRAL	Membrane destabilization	(Niidome, Ohmori *et al.* 1997)
HEL peptides & analogues	KLLKLLLKLWKKLLKLLK	Membrane destabilization	(Ohmori, Niidome *et al.* 1998; Niidome, Takaji *et al.* 1999)

Peptide	Primary Sequence	Endosomal Escape Mechanism	Reference
LAH4 & analogues	KKALLALALHHLAHLALHL ALALKKA	Membrane destabilization	(Kichler, Leborgne *et al.* 2003; Kichler, Mason *et al.* 2006; Lam, Liang *et al.* 2012)

Table 1. The sequences and the endosomal escape mechanism of selected CPPs that are being investigated for nucleic acid delivery.

1999), INF 7 (Plank, Oberhauser et al. 1994), HGP (Kwon, Bergen et al. 2008), JTS-1 (Gottschalk, Sparrow et al. 1996), EBI (Lundberg, El-Andaloussi et al. 2007), ppTG1 (Rittner, Benavente et al. 2002) and CADY (Crombez, Aldrian-Herrada et al. 2008; Konate, Crombez et al. 2010).

The LAH4 peptide and its derivatives are another class of peptide that exhibits efficient gene transfer activity (Kichler, Leborgne et al. 2003; Lam, Liang et al. 2012). Peptides of the LAH4 family are synthetic cationic amphipathic peptides containing a variable number of histidine residues and hydrophobic amino acids (mainly alanines and leucines). They were initially designed to investigate the interactions that determine the alignment of membrane-associated proteins (Bechinger 1996; Vogt and Bechinger 1999). *In vitro* transfection experiments indicated that peptides with four to five histidine residues in the central region of the sequence achieved high transfection efficiency that is comparable to PEI. The transfection activity was significantly abolished in the presence of proton pump inhibitor bafilomycin A1, suggesting acidification is important for these peptides to mediate high transfection level. In a model membrane experiment, it was noticed that there was a preferential peptide-lipid interaction between LAH4 derivatives and anionic lipids, leading to the disruption of the lipid acyl chains in the acidic environment (Mason, Martinez et al. 2006). Furthermore, LAH4 peptides experience a pH-dependent conformational change from transmembrane orientation at neutral pH to an in-plane orientation at low pH. The pH at which the conformation transition takes place is crucial and highly affects the transfection efficiency (Bechinger 1996; Vogt and Bechinger 1999; Kichler, Leborgne et al. 2003). Based on these observations, it was proposed that at neutral pH, LAH4 adopts a transmembrane orientation without disrupting the membrane integrity. During acidification of the endosomes, the imidazole groups of histidine residues become protonated and the peptide changes to an in-plane alignment and interact with the anionic lipids in the endosomal membrane. Membrane destabilization occurs, followed by the release of nucleic acid into the cytosol (Kichler, Mason et al. 2006).

2.4. Pore formation

Pore formation is another mechanism to explain the endosomal escape of peptide-based nucleic acid delivery systems. Parente *et al.* (Parente, Nir et al. 1990) investigated content-leaking kinetics of peptide GALA from phospholipid vesicles over a wide range of pH. As previously mentioned, GALA undergoes a pH-dependent conformational change to give a helical structure at acidic environment. It was suggested that leakage from phospholipid vesicles is promptly initiated by low pH ($pH<6$) and is rapidly terminated when the pH is raised to 7.5. The author assumes that GALA becomes incorporated into the vesicle bilayer and aggregates to form a pore with diameter ranges from 5 to 10 Å. Fattal *et al.* further investigated the mechanism of pore-forming peptides GALA in details. They concluded that pore formation is a key event that may result in phospholipid flip-flop of biological membranes, leading to the releasing of contents (Fattal, Nir et al. 1994). Simoes *et al.* (Simoes, Slepushkin et al. 1999) combined GALA with lipoplexes and found significant improvement of transfection in several cell cells. The results indicated that the cellular uptake of lipoplexes is though endocytosis and the endosomal escape play a crucial role in intracellular delivery of lipoplexes. However, the authors believe that the dimension of the pores (5 to 10 Å) formed by this peptide may not be big enough to permit the escape of nucleic acid from endosomes. Multiple mechanisms may be accounted for the enhancement of transfection efficiency, including structural changes of the peptide, facilitation of nucleic acid dissociation from the lipoplexes and the flip-flop mechanism.

2.5. Photochemical internalization (PCI)

Photochemical internalization (PCI) is a light-directed delivery technology that utilizes photosensitizers to facilitate the transport of membrane impermeable macromolecules from endocytic vesicles into cytoplasm. The mechanism of PCI as an endosomal escape enhancer strategy is very straight-forward (Fig.3). Photosensitizers that are employed in the PCI technology are usually amphiphilic compounds which can bind to and localize in the plasma membrane. After being taken up by the cells through endocytosis, the photosensitizers are confined to the endosomal membranes and remain inactive until triggered by light with specific wavelengths matching their absorption spectra (Selbo, Weyergang et al. 2010). Once activated, they induce the formation of highly reactive oxygen species, mainly singlet oxygen, leading to the rupture of endosomes and lysosomes membrane. As a result, macromolecules that are trapped inside the endosomes/lysosomes can be liberated into the cytosol (Berg, Kristian Selbo et al. 1999). Photosensitizers used in clinical application are highly reactive reagents with short range of action (10-20 nm) and short life-time (0.01-0.04 μs), thus restricting the damaging effect to the production site (within the endosomal membrane) without affecting other cellular components (Moan and Berg 1991; Berg, Weyergang et al. 2010). Most of the photosensitizers do not localize to the nucleus of the cells, thereby reducing the possibility of causing any mutagenic effects (Dougherty, Henderson et al. 1998).

PCI was initially investigated for anti-tumour drug delivery. A synergistic effect of combining PCI with chemotherapeutic agents was found. PCI principally targets cellular endocytosis that may affect the distribution of molecules that are taken up by the cells via endosomal pathway. It was later employed as a tool to improve the cellular delivery of a large variety of bioactive macromolecules and nucleic acids including pDNA, siRNA and oligonucleotides (Selbo, Weyergang et al. 2010). Examples of photosensitizers that are used in non-viral nucleic acid delivery including disulfonated meso-tetraphenylporphine ($TPPS_{2a}$) (Prasmickaite, Høgset et al. 2000; Kloeckner, Prasmickaite et al. 2004; Maurice-Duelli, Ndoye et al. 2004; Ndoye, Merlin et al. 2004; Ndoye, Dolivet et al. 2006; Oliveira, Fretz et al. 2007; Boe, Longva et al. 2008; Raemdonck, Naeye et al. 2009; Bøe, Sæbøe-Larssen et al. 2010), disulfonated aluminium phthalocyanine ($AlPcS_{2a}$) (Berg, Prasmickaite et al. 2003; Hellum, Høgset et al. 2003; Ndoye, Dolivet et al. 2006; Yip, Weyergang et al. 2007), Zinc-phthalocyanine (Zn-Pc) dendrimer (Nishiyama, Iriyama et al. 2005; Arnida, Nishiyama et al. 2006) and 5,10,15-tri(4-acetamidophenyl)-20-mono(4-carboxyl-phenyl)porphyrin (TAMCPP) conjugated to G4 PAMAM dendrimer (Shieh, Peng et al. 2008).

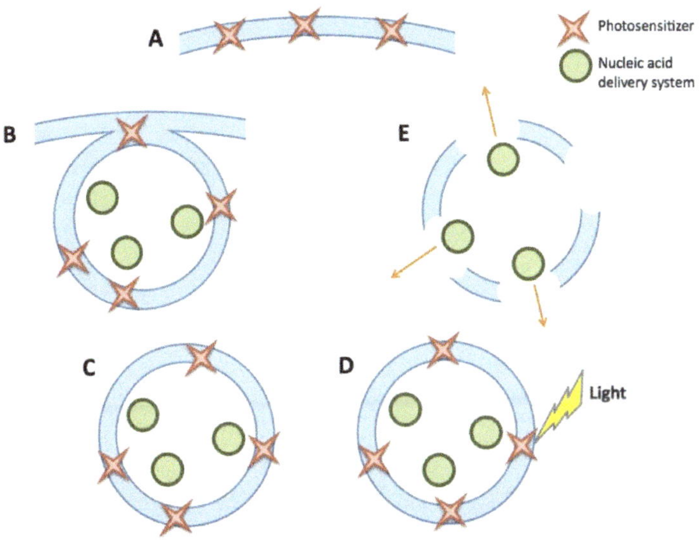

Figure 3. The mechanism of photochemical internalization technology. (A) Photosensitizers bind to and localize in the plasma membrane. (B) Photosensitizers can be taken up by the cells through endocytosis with the delivery systems. (C) Photosensitizers are confined to the endosomal membrane and remain inactive. (D) Photosensitizers are activated by illumination and induce the formation of highly reactive oxygen species, leading to the rupture of endosomes /lysosomes membrane. (E) Molecules that are trapped inside the endosomes/lysosomes can be liberated into the cytosol.

To employ PCI in clinical applications, the penetration of light into the deep tissue is an important issue (Oliveira, Fretz et al. 2007). With the development of fiber optics and laser technology, the control of illumination to sites that are deep inside the human body becomes possible, e.g. gastrointestinal tract, urogenital organs, lungs, brain and pancreas (Dougherty, Henderson et al. 1998; Chatterjee, Fong et al. 2008). PCI mediated therapy can be used in many regions of the body where light delivery can be achieved and where local activation of a drug is desirable. A photosensizer is injected as a single dose prior to light activation. Parameters such as the dose of photosentizers and light, as well as the time interval between administration of photosentiziers and drugs must be carefully optimized. *In vivo* experiments demonstrated that PCI, in combination with a chemotherapeutic agent, has a good effect on cancer treatment (Selbo, Sivam et al. 2001; Berg, Dietze et al. 2005; Hirschberg, Zhang et al. 2009; Norum, Gaustad et al. 2009). In the field of nucleic acid delivery, it has been shown that PCI was able to transfer pDNA and siRNA into cytosol efficiently to enhance their biological effects *in vitro*. Only a number of *in vivo* studies have been carried out to demonstrate the feasibility of PCI technology to enhance the delivery of pDNA to conjunctival tissue (Nishiyama, Iriyama et al. 2005) and tumour tissues (Ndoye, Dolivet et al. 2006), as well as the delivery of siRNA to a tumour site (Oliveira, Hogset et al. 2008). The result was very promising. Perhaps more effort should be invested to further develop this technology for clinical use.

2.6. Other endosomal escape mechanisms

Exogenous additives, such as chloroquine and inactivated adenovirus, have been exploited to promote endosomal escape and enhance the efficiency of nucleic acid delivery. Chloroquine is a weak base that can rapidly penetrate the plasma membrane, accumulate in acidic vesicles and increase the pH of the acidic compartment (Maxfield 1982; Mellman, Fuchs et al. 1986). Preventing endosome acidification may subsequently inhibit hydrolytic enzymes such as proteases and nucleases (Cotten, Längle-Rouault et al. 1990). Chloroquine also causes the swelling and rupture of endosomal vesicle by increasing the osmotic pressure inside the acidic compartment (Khalil, Kogure et al. 2006). Since it can neutralize acidic compartment and induce rupture of endocytic vesicles, adding chloroquine is an alternative measure to improve nucleic acid transfer (Erbacher, Roche et al. 1996).

A number of early studies found that chloroquine is able to enhance DNA transfection in various cell types (Luthman and Magnusson 1983; Cotten, Längle-Rouault *et al.* 1990; Erbacher, Roche *et al.* 1996). Chloroquine promotes escape of polyplexes or lipoplexes from endosome via increasing endosomal pH and hindering endosome fusion with lysosome. To date, chloroquine has been widely used to elucidate the uptake mechanism of non-viral nucleic acid delivery systems (Legendre and Szoka Jr 1992; Simeoni, Morris et al. 2003; Lehto, Abes et al. 2010). However, it is worth noting that chloroquine does not always lead to an improvement of transfection efficiency, depending on the uptake pathway of the delivery systems. Haensler and Szoka reported that transfection of PAMAM was not affected by the presence of chloroquine (Haensler and Szoka Jr 1993). The authors suggested the endosomal lysis activity of PAMAM is strong enough to allow liberation of its content to cytosol with and without the presence of chloroquine. Legendre and Szoka (Legendre and Szoka Jr 1992) described an increase in transfection efficiency after chloroquine was added to DOTMA/DOPE liposomes. Interestingly, chloroquine exerts a negative effect on carcinoma cells transfected by DOPE/CHEMS pH-sensitive liposomes. The authors explained that the membrane destabilization activity of pH-sensitive liposome is attenuated when the acidic environment of endosome is perturbed by chloroquine. Transfection is a complicated process involving multiple steps, such as cellular binding, internalization and nuclear transport (in the case of DNA delivery). Apart from endosomal escape, other factors may also affect the eventual transfection efficiency.

Besides chloroquine, physically coupling chemically inactivated adenovirus particles is another approach for promoting endosomal escape. This method takes the advantage of the endosomolytic activity of adenovirus to facilitate the release of nucleic acids from the endosomes (Curiel, Agarwal et al. 1991; Cotten, Wagner et al. 1992; Wagner, Zatloukal et al. 1992). Curiel *et al.* explored the application of adenovirus as an endosome disruption agent. The inactivated adenoviruses were coupled with transferring-polyplexes. It was observed that pDNA delivery into HeLa cells was improved 1000-fold or more when replication-defective adenovirus particles were present (Curiel, Agarwal et al. 1991).

However, the application of both chloroquine and inactivated adenovirus particles are limited due to safety concern. Although chloroquine is approved by the FDA as an anti-malaria medication, it is found to be toxic to many cell types and can trigger gastrointestinal and nervous adverse effects in high dose (Pack, Putnam et al. 2000). For defective adenovirus particles, the complexity of vector production and potential immunogenicity raised by the virus components make this strategy problematic (Pack, Hoffman et al. 2005). Unless the safety issues can be solved, both methods will remain unsuitable for clinical use.

3. Conclusion

Non-viral vectors are considered to be promising vehicles for delivering therapeutic nucleic acids because of their relatively safe profile and high versatility as compared to their viral counterparts. However, the transfection efficiency of non-viral vectors is less than satisfactory for clinical purpose. The endocytosis pathway is a major route for the cellular entry of non-viral nucleic acid delivery agents. Poor endosomal escape of non-viral systems pose a major challenge for the intracellular delivery of nucleic acids. An ideal nucleic acid delivery system should fulfill several criteria: negligible toxicity, biocompatible and biodegradable, offer protection to nucleic acids from enzymatic degradation, facilitate cellular uptake, promote endosomal escape and release the nucleic acids at site of action. Elucidation of the mechanism of endosomal escape is beneficial in the development of more effective non-viral delivery vectors. However, the uptake and cytoplasmic transportation mechanisms of a variety of non-viral nucleic acid carriers still need to be investigated in more detail. In the future, with the development of cell imaging techniques such as high resolution, spinning disk live cell confocal imaging and the fluorescence correlation spectroscope, the details of intracellular trafficking of non-viral nucleic acid delivery systems will be unveiled. This will guide the future design and development of novel efficient non-viral nucleic acid delivery vectors.

Author details

Wanling Liang and Jenny K.W. Lam
Department of Pharmacology & Pharmacy, Li Ka Shing Faculty of Medicine,
The University of Hong Kong, Hong Kong SAR

Acknowledgement

The authors would like to thank Seed Funding Programme for Basic Research (2009091159006, 201011159060), The University of Hong Kong for financial support. The authors are grateful to Dr A James Mason for critical reading of the manuscript.

4. References

Aiuti, A., F. Cattaneo, et al. (2009). "Gene therapy for immunodeficiency due to adenosine deaminase deficiency." *New England Journal of Medicine* 360(5): 447-458.

Arnida, A., N. Nishiyama, et al. (2006). "Novel ternary polyplex of triblock copolymer, pDNA and anionic dendrimer phthalocyanine for photochemical enhancement of transgene expression." *Journal of Controlled Release* 116(2): e75.

Arote, R., T. H. Kim, et al. (2007). "A biodegradable poly (ester amine) based on polycaprolactone and polyethylenimine as a gene carrier." *Biomaterials* 28(4): 735-744.

Bøe, S., S. Sæbøe-Larssen, et al. (2010). "Light-Induced Gene Expression Using Messenger RNA Molecules." *Oligonucleotides* 20(1): 1-6.

Bally, M. B., P. Harvie, et al. (1999). "Biological barriers to cellular delivery of lipid-based DNA carriers." *Advanced Drug Delivery Reviews* 38(3): 291-315.

Baum, C., J. Düllmann, et al. (2003). "Side effects of retroviral gene transfer into hematopoietic stem cells." *Blood* 101(6): 2099-2113.

Bechinger, B. (1996). "Towards membrane protein design: pH-sensitive topology of histidine-containing polypeptides." *Journal of Molecular Biology* 263(5): 768-775.

Behr, J. P. (1997). "The proton sponge: a trick to enter cells the viruses did not exploit." *Chimia International Journal for Chemistry* 51(1,2): 34-36.

Berg, K., A. Dietze, et al. (2005). "Site-specific drug delivery by photochemical internalization enhances the antitumor effect of bleomycin." *Clinical Cancer Research* 11(23): 8476-8485.

Berg, K., P. Kristian Selbo, et al. (1999). "Photochemical Internalization." *Cancer Research* 59(6): 1180-1183.

Berg, K., A. Weyergang, et al. (2010) "The Potential of Photochemical Internalization (PCI) for the Cytosolic Delivery of Nanomedicines" in Organelle-Specific Pharmaceutical Nanotechnology: John Wiley & Sons, Inc. p. 311-322.

Bernstein, E., A. A. Caudy, et al. (2001). "Role for a bidentate ribonuclease in the initiation step of RNA interference." *Nature* 409(6818): 363-365.

Boe, S., A. S. Longva, et al. (2008). "Evaluation of various polyethylenimine formulations for light-controlled gene silencing using small interfering RNA molecules." *Oligonucleotides* 18(2): 123-132.

Boletta, A., A. Benigni, et al. (1997). "Nonviral gene delivery to the rat kidney with polyethylenimine." *Human Gene Therapy* 8(10): 1243-1251.

Boussif, O., F. Lezoualc'h, et al. (1995). "A versatile vector for gene and oligonucleotide transfer into cells in culture and in vivo: polyethylenimine." *Proceedings of the National Academy of Sciences* 92(16): 7297.

Brooks, H., B. Lebleu, et al. (2005). "Tat peptide-mediated cellular delivery: back to basics." *Advanced Drug Delivery Reviews* 57(4): 559-577.

Chang, K. L., Y. Higuchi, et al. (2010). "Efficient gene transfection by histidine-modified chitosan through enhancement of endosomal escape." *Bioconjugate Chemistry* 21(6): 1087-1095.

Chatterjee, D. K., L. S. Fong, et al. (2008). "Nanoparticles in photodynamic therapy: An emerging paradigm." *Advanced drug delivery reviews* 60(15): 1627-1637.

Cho, Y. W., J. D. Kim, et al. (2003). "Polycation gene delivery systems: escape from endosomes to cytosol." *Journal of Pharmacy and Pharmacology* 55(6): 721-734.

Chollet, P., M. C. Favrot, et al. (2002). "Side-effects of a systemic injection of linear polyethylenimine–DNA complexes." *The Journal of Gene Medicine* 4(1): 84-91.

Coll, J. L., P. Chollet, et al. (1999). "In vivo delivery to tumors of DNA complexed with linear polyethylenimine." *Human Gene Therapy* 10(10): 1659-1666.

Cotten, M., F. Längle-Rouault, et al. (1990). "Transferrin-polycation-mediated introduction of DNA into human leukemic cells: stimulation by agents that affect the survival of transfected DNA or modulate transferrin receptor levels." *Proceedings of the National Academy of Sciences* 87(11): 4033.

Cotten, M., E. Wagner, et al. (1992). "High-efficiency receptor-mediated delivery of small and large (48 kilobase gene constructs using the endosome-disruption activity of defective or chemically inactivated adenovirus particles." *Proceedings of the National Academy of Sciences* 89(13): 6094.

Crombez, L., G. Aldrian-Herrada, et al. (2008). "A new potent secondary amphipathic cell–penetrating peptide for siRNA delivery into mammalian cells." *Molecular Therapy* 17(1): 95-103.

Cull, V. S., E. J. Bartlett, et al. (2002). "Type I interferon gene therapy protects against cytomegalovirus-induced myocarditis." *Immunology* 106(3): 428-437.

Curiel, D. T., S. Agarwal, et al. (1991). "Adenovirus enhancement of transferrin-polylysine-mediated gene delivery." *Proceedings of the National Academy of Sciences* 88(19): 8850.

Davis, M. E., J. E. Zuckerman, et al. (2010). "Evidence of RNAi in humans from systemically administered siRNA via targeted nanoparticles." *Nature* 464(7291): 1067-1070.

Derossi, D., S. Calvet, et al. (1996). "Cell internalization of the third helix of the Antennapedia homeodomain is receptor-independent." *Journal of Biological Chemistry* 271(30): 18188-18193.

Derossi, D., A. H. Joliot, et al. (1994). "The third helix of the Antennapedia homeodomain translocates through biological membranes." *Journal of Biological Chemistry* 269(14): 10444-10450.

Deshayes, S., M. Morris, et al. (2005). "Cell-penetrating peptides: tools for intracellular delivery of therapeutics." *Cellular and Molecular Life Sciences* 62(16): 1839-1849.

Deshayes, S., M. Morris, et al. (2008). "Delivery of proteins and nucleic acids using a non-covalent peptide-based strategy." *Advanced Drug Delivery Reviews* 60(4-5): 537-547.

DeVincenzo, J., J. E. Cehelsky, et al. (2008). "Evaluation of the safety, tolerability and pharmacokinetics of ALN-RSV01, a novel RNAi antiviral therapeutic directed against respiratory syncytial virus (RSV)." *Antiviral Research* 77(3): 225-231.

DeVincenzo, J., R. Lambkin-Williams, et al. (2010). "A randomized, double-blind, placebo-controlled study of an RNAi-based therapy directed against respiratory syncytial virus." *Proceedings of the National Academy of Sciences* 107(19): 8800.

Dougherty, T. J., B. W. Henderson, et al. (1998). "Photodynamic therapy." *Journal of the National Cancer Institute* 90(12): 889-905.

Edelstein, M. L., M. R. Abedi, et al. (2007). "Gene therapy clinical trials worldwide to 2007--an update." *The Journal of Gene Medicine* 9(10): 833-842.

El-Aneed, A. (2004). "An overview of current delivery systems in cancer gene therapy." *Journal of Controlled Release* 94(1): 1-14.

El Ouahabi, A., M. Thiry, et al. (1997). "The role of endosome destabilizing activity in the gene transfer process mediated by cationic lipids." *FEBS letters* 414(2): 187-192.

Elbashir, S. M., W. Lendeckel, et al. (2001). "RNA interference is mediated by 21-and 22-nucleotide RNAs." *Genes & Development* 15(2): 188-200.

Endoh, T. and T. Ohtsuki (2009). "Cellular siRNA delivery using cell-penetrating peptides modified for endosomal escape." *Advanced Drug Delivery Reviews* 61(9): 704-709.

Erbacher, P., A. C. Roche, et al. (1996). "Putative role of chloroquine in gene transfer into a human hepatoma cell line by DNA/lactosylated polylysine complexes." *Experimental Cell Research* 225(1): 186-194.

Ewert, K., A. Zidovska, et al. (2010). "Cationic Liposome–Nucleic Acid Complexes for Gene Delivery and Silencing: Pathways and Mechanisms for Plasmid DNA and siRNA." *Nucleic Acid Transfection*: 191-226.

Farhood, H., N. Serbina et al. (1995). "The role of dioleoyl phosphatidylethanolamine in cationic liposome mediated gene transfer." Biochimica et Biophysica Acta (BBA) - Biomembranes 1235: (2) 289-295 Journal name - italics (Biochimica et Biophysica Acta (BBA) - Biomembranes)

Fattal, E., S. Nir, et al. (1994). "Pore-forming peptides induce rapid phospholipid flip-flop in membranes." *Biochemistry* 33(21): 6721-6731.

Felgner, P. L., T. R. Gadek, et al. (1987). "Lipofection: a highly efficient, lipid-mediated DNA-transfection procedure." *Proceedings of the National Academy of Sciences* 84(21): 7413.

Fernández-Carneado, J., M. J. Kogan, et al. (2004). "Amphipathic peptides and drug delivery." *Peptide Science* 76(2): 196-203.

Fire, A., S. Q. Xu, et al. (1998). "Potent and specific genetic interference by double-stranded RNA in Caenorhabditis elegans." *Nature* 391(6669): 806-811.

Fischer, D., T. Bieber, et al. (1999). "A novel non-viral vector for DNA delivery based on low molecular weight, branched polyethylenimine: effect of molecular weight on transfection efficiency and cytotoxicity." *Pharmaceutical research* 16(8): 1273-1279.

Fischer, P. M., E. Krausz, et al. (2001). "Cellular delivery of impermeable effector molecules in the form of conjugates with peptides capable of mediating membrane translocation." *Bioconjugate chemistry* 12(6): 825-841.

Fominaya, J., M. Gasset, et al. (2000). "An optimized amphiphilic cationic peptide as an efficient non-viral gene delivery vector." *The Journal of Gene Medicine* 2(6): 455-464.

Funhoff, A. M., C. F. van Nostrum, et al. (2005). "Cationic polymethacrylates with covalently linked membrane destabilizing peptides as gene delivery vectors." *Journal of Controlled Release* 101(1): 233-246.

Futaki, S. (2005). "Membrane-permeable arginine-rich peptides and the translocation mechanisms." *Advanced Drug Delivery Reviews* 57(4): 547-558.

Futaki, S. (2006). "Oligoarginine vectors for intracellular delivery: Design and cellular-uptake mechanisms." *Peptide Science* 84(3): 241-249.

Gaspar, H. B., K. L. Parsley, et al. (2004). "Gene therapy of X-linked severe combined immunodeficiency by use of a pseudotyped gammaretroviral vector." *The Lancet* 364(9452): 2181-2187.

Godbey, W., K. K. Wu, et al. (1999). "Poly (ethylenimine) and its role in gene delivery." *Journal of Controlled Release* 60(2-3): 149-160.

Gottschalk, S., J. Sparrow, et al. (1996). "A novel DNA-peptide complex for efficient gene transfer and expression in mammalian cells." *Gene therapy* 3(5): 448.

Goula, D., J. Remy, et al. (1998). "Size, diffusibility and transfection performance of linear PEI/DNA complexes in the mouse central nervous system." *Gene Therapy* 5(5): 712.

Griesenbach, U. and E. W. F. W. Alton (2009). "Cystic fibrosis gene therapy: successes, failures and hopes for the future." *Expert Review of Respiratory Medicine* 3(4): 363-371.

Gros, E., S. Deshayes, et al. (2006). "A non-covalent peptide-based strategy for protein and peptide nucleic acid transduction." *Biochimica et Biophysica Acta (BBA)-Biomembranes* 1758(3): 384-393.

Hacein-Bey-Abina, S., A. Garrigue, et al. (2008). "Insertional oncogenesis in 4 patients after retrovirus-mediated gene therapy of SCID-X1." *The Journal of Clinical Investigation* 118(9): 3132.

Hacein-Bey-Abina, S., C. von Kalle, et al. (2003). "A serious adverse event after successful gene therapy for X-linked severe combined immunodeficiency." *New England Journal of Medicine* 348(3): 255-256.

Haensler, J. and F. C. Szoka Jr (1993). "Polyamidoamine cascade polymers mediate efficient transfection of cells in culture." *Bioconjugate chemistry* 4(5): 372-379.

Hammond, S. M., S. Boettcher, et al. (2001). "Argonaute2, a link between genetic and biochemical analyses of RNAi." *Science* 293(5532): 1146.

Hannon, G. J. and J. J. Rossi (2004). "Unlocking the potential of the human genome with RNA interference." *Nature* 431(7006): 371-378.

Hashiba, T., M. Suzuki, et al. (2001). "Adenovirus-mediated transfer of heme oxygenase-1 cDNA attenuates severe lung injury induced by the influenza virus in mice." *Gene Therapy* 8(19): 1499.

Hassane, F. S., A. F. Saleh, et al. (2010). "Cell penetrating peptides: overview and applications to the delivery of oligonucleotides." *Cellular and Molecular Life Sciences* 67(5): 715-726.

Hellum, M., A. Høgset, et al. (2003). "Photochemically enhanced gene delivery with cationic lipid formulations." *Photochemical & Photobiological Science* 2(4): 407-411.

Henry, T. D. and D. Satran (2012). "Therapeutic angiogenesis." *Coronary Artery Disease*: 67-74.

Hirschberg, H., M. J. Zhang, et al. (2009). "Targeted delivery of bleomycin to the brain using photo-chemical internalization of Clostridium perfringens epsilon prototoxin." *Journal of Neuro-Oncology* 95(3): 317-329.

http://www.abedia.com/wiley/phases.php. date of access: 22/03/2012'

Hui, S. W., M. Langner, et al. (1996). "The role of helper lipids in cationic liposome-mediated gene transfer." *Biophysical Journal* 71(2): 590-599.

Johnson, L. A., R. A. Morgan, et al. (2009). "Gene therapy with human and mouse T-cell receptors mediates cancer regression and targets normal tissues expressing cognate antigen." *Blood* 114(3): 535-546.

Kaiser, P. K., R. C. Symons, et al. (2010). "RNAi-based treatment for neovascular age-related macular degeneration by Sirna-027." *American journal of ophthalmology* 150(1): 33-39

Kawabata, K., Y. Takakura, et al. (1995). "The fate of plasmid DNA after intravenous injection in mice: involvement of scavenger receptors in its hepatic uptake." *Pharmaceutical Research* 12(6): 825-830.

Ketting, R. F., S. E. J. Fischer, et al. (2001). "Dicer functions in RNA interference and in synthesis of small RNA involved in developmental timing in C. elegans." *Genes & Development* 15(20): 2654.

Khalil, I. A., K. Kogure, et al. (2006). "Uptake pathways and subsequent intracellular trafficking in nonviral gene delivery." *Pharmacological Reviews* 58(1): 32-45.

Kichler, A., C. Leborgne, et al. (2003). "Histidine-rich amphipathic peptide antibiotics promote efficient delivery of DNA into mammalian cells." *Proceedings of the National Academy of Sciences* 100(4): 1564.

Kichler, A., A. Mason, et al. (2006). "Cationic amphipathic histidine-rich peptides for gene delivery." *Biochimica et Biophysica Acta (BBA)-Biomembranes* 1758(3): 301-307.

Klink, D. T., S. Chao, et al. (2001). "Nuclear translocation of lactosylated poly-L-lysine/cDNA complex in cystic fibrosis airway epithelial cells." *Molecular Therapy* 3(6): 831-841.

Kloeckner, J., L. Prasmickaite, et al. (2004). "Photochemically enhanced gene delivery of EGF receptor-targeted DNA polyplexes." *Journal of Drug Targeting* 12(4): 205-213.

Konate, K., L. Crombez, et al. (2010). "Insight into the cellular uptake mechanism of a secondary amphipathic cell-penetrating peptide for siRNA delivery." *Biochemistry* 49(16): 3393-3402.

Kunath, K., A. von Harpe, et al. (2003). "Low-molecular-weight polyethylenimine as a non-viral vector for DNA delivery: comparison of physicochemical properties, transfection efficiency and in vivo distribution with high-molecular-weight polyethylenimine." *Journal of Controlled Release* 89(1): 113-125.

Kwon, E. J., J. M. Bergen, et al. (2008). "Application of an HIV gp41-derived peptide for enhanced intracellular trafficking of synthetic gene and siRNA delivery vehicles." *Bioconjugate chemistry* 19(4): 920-927.

Laga, R., R. Carlisle, et al. (2012). "Polymer coatings for delivery of nucleic acid therapeutics." *Journal of Controlled Release* (doi:10.1016/j.jconrel.2012.02.013).

Lam, J. K.-W., W. Liang, et al. (2012). "Pulmonary delivery of therapeutic siRNA." *Advanced Drug Delivery Reviews* 64(1): 1-15.

Lam, J. K. W., W. Liang, et al. (2012). "Effective endogenous gene silencing mediated by pH responsive peptides proceeds via multiple pathways." *Journal of Controlled Release* 158(2): 293-303.

Leachman, S. A., R. P. Hickerson, et al. (2009). "First-in-human mutation-targeted siRNA phase Ib trial of an inherited skin disorder." *Molecular Therapy* 18(2): 442-446.

Lee, Y. J., G. Johnson, et al. (2011). "A HA2-Fusion tag limits the endosomal release of its protein cargo despite causing endosomal lysis." *Biochimica et Biophysica Acta (BBA)-General Subjects* 1810: 752-758.

Legendre, J. Y. and F. C. Szoka Jr (1992). "Delivery of plasmid DNA into mammalian cell lines using pH-sensitive liposomes: comparison with cationic liposomes." *Pharmaceutical Research* 9(10): 1235-1242.

Lehto, T., R. Abes, et al. (2010). "Delivery of nucleic acids with a stearylated (RxR)4 peptide using a non-covalent co-incubation strategy." *Journal of Controlled Release* 141(1): 42-51.

Li, W., F. Nicol, et al. (2004). "GALA: a designed synthetic pH-responsive amphipathic peptide with applications in drug and gene delivery." *Advanced Drug Delivery Reviews* 56(7): 967-985.

Lundberg, P., S. El-Andaloussi, et al. (2007). "Delivery of short interfering RNA using endosomolytic cell-penetrating peptides." *The FASEB Journal* 21(11): 2664-2671.

Lundin, P., H. Johansson, et al. (2008). "Distinct uptake routes of cell-penetrating peptide conjugates." *Bioconjugate Chemistry* 19(12): 2535-2542.

Luthman, H. and G. Magnusson (1983). "High efficiency polyoma DNA transfection of chloroquine treated cells." *Nucleic Acids Research* 11(5): 1295.

Mason, A. J., A. Martinez, et al. (2006). "The antibiotic and DNA-transfecting peptide LAH4 selectively associates with, and disorders, anionic lipids in mixed membranes." *The FASEB Journal* 20(2): 320-322.

Maurice-Duelli, A., A. Ndoye, et al. (2004). "Enhanced cell growth inhibition following PTEN nonviral gene transfer using polyethylenimine and photochemical internalization in endometrial cancer cells." *Technology in Cancer Research & Treatment* 3(5): 459-465.

Maxfield, F. R. (1982). "Weak bases and ionophores rapidly and reversibly raise the pH of endocytic vesicles in cultured mouse fibroblasts." *The Journal of Cell Biology* 95(2): 676-681.

Medina-Kauwe, L., J. Xie, et al. (2005). "Intracellular trafficking of nonviral vectors." *Gene Therapy* 12(24): 1734-1751.

Melikov, K. and L. Chernomordik (2005). "Arginine-rich cell penetrating peptides: from endosomal uptake to nuclear delivery." *Cellular and Molecular Life Sciences* 62(23): 2739-2749.

Mellman, I., R. Fuchs, et al. (1986). "Acidification of the endocytic and exocytic pathways." *Annual Review of Biochemistry* 55(1): 663-700.

Mello, C. C. and D. Conte (2004). "Revealing the world of RNA interference." *Nature* 431(7006): 338-342.

Merkel, O. M., A. Beyerle, et al. (2009). "Nonviral siRNA Delivery to the Lung: Investigation of PEG– PEI Polyplexes and Their In Vivo Performance." *Molecular Pharmaceutics* 6(4): 1246-1260.

Midoux, P., A. Kichler, et al. (1998). "Membrane permeabilization and efficient gene transfer by a peptide containing several histidines." *Bioconjugate Chemistry* 9(2): 260-267.

Midoux, P., C. Mendes, et al. (1993). "Specific gene transfer mediated by lactosylated poly-L-lysine into hepatoma cells." *Nucleic Acids Research* 21(4): 871-878.

Midoux, P. and M. Monsigny (1999). "Efficient gene transfer by histidylated polylysine/pDNA complexes." *Bioconjugate Chemistry* 10(3): 406-411.

Midoux, P., C. Pichon, et al. (2009). "Chemical vectors for gene delivery: a current review on polymers, peptides and lipids containing histidine or imidazole as nucleic acids carriers." *British Journal of Pharmacology* 157(2): 166-178.

Min, S. H., D. C. Lee, et al. (2006). "A composite gene delivery system consisting of polyethylenimine and an amphipathic peptide KALA." *The Journal of Gene Medicine* 8(12): 1425-1434.

Moan, J. and K. Berg (1991). "The photodegradation of porphyrins in cells can be used to estimate the lifetime of singlet oxygen." *Photochemistry and Photobiology* 53(4): 549-553.

Moreira, C., H. Oliveira, et al. (2009). "Improving chitosan-mediated gene transfer by the introduction of intracellular buffering moieties into the chitosan backbone." *Acta Biomaterialia* 5(8): 2995-3006.

Morille, M., C. Passirani, et al. (2008). "Progress in developing cationic vectors for non-viral systemic gene therapy against cancer." *Biomaterials* 29(24-25): 3477-3496.

Morris, M., P. Vidal, et al. (1997). "A new peptide vector for efficient delivery of oligonucleotides into mammalian cells." *Nucleic Acids Research* 25(14): 2730-2736.

Morris, M. C., L. Chaloin, et al. (2000). "Translocating peptides and proteins and their use for gene delivery." *Current Opinion in Biotechnology* 11(5): 461-466.

Morris, M. C., J. Depollier, et al. (2001). "A peptide carrier for the delivery of biologically active proteins into mammalian cells." *Nature Biotechnology* 19(12): 1173-1176.

Muratovska, A. and M. R. Eccles (2004). "Conjugate for efficient delivery of short interfering RNA (siRNA) into mammalian cells." *FEBS letters* 558(1-3): 63-68.

Navarro-Quiroga, I., J. Antonio González-Barrios, et al. (2002). "Improved neurotensin-vector-mediated gene transfer by the coupling of hemagglutinin HA2 fusogenic peptide and Vp1 SV40 nuclear localization signal." *Molecular Brain Research* 105(1-2): 86-97.

Ndoye, A., G. Dolivet, et al. (2006). "Eradication of p53-mutated head and neck squamous cell carcinoma xenografts using nonviral p53 gene therapy and photochemical internalization." *Molecular Therapy* 13(6): 1156-1162.

Ndoye, A., J. L. Merlin, et al. (2004). "Enhanced gene transfer and cell death following p53 gene transfer using photochemical internalisation of glucosylated PEI-DNA complexes." *The Journal of Gene Medicine* 6(8): 884-894.

Niidome, T., N. Ohmori, et al. (1997). "Binding of cationic alpha-helical peptides to plasmid DNA and their gene transfer abilities into cells." *Journal of Biological Chemistry* 272(24): 15307-15312.

Niidome, T., K. Takaji, et al. (1999). "Chain length of cationic alpha-helical peptide sufficient for gene delivery into cells." *Bioconjugate Chemistry* 10(5): 773-780.

Nishiyama, N., A. Iriyama, et al. (2005). "Light-induced gene transfer from packaged DNA enveloped in a dendrimeric photosensitizer." *Nature Materials* 4(12): 934-941.

Norum, O. J., J. V. Gaustad, et al. (2009). "Photochemical internalization of bleomycin is superior to photodynamic therapy due to the therapeutic effect in the tumor periphery." *Photochemistry and Photobiology* 85(3): 740-749.

Numata, K. and D. L. Kaplan (2010). "Silk-based gene carriers with cell membrane destabilizing peptides." *Biomacromolecules* 11: 3189–3195.

Oehlke, J., A. Scheller, et al. (1998). "Cellular uptake of an [alpha]-helical amphipathic model peptide with the potential to deliver polar compounds into the cell interior non-endocytically." *Biochimica et Biophysica Acta (BBA)-Biomembranes* 1414(1-2): 127-139.

Ohmori, N., T. Niidome, et al. (1998). "Importance of Hydrophobic Region in Amphiphilic Structures of α-Helical Peptides for Their Gene Transfer-Ability into Cells." *Biochemical and Biophysical Research Communications* 245(1): 259-265.

Oliveira, S., M. M. Fretz, et al. (2007). "Photochemical internalization enhances silencing of epidermal growth factor receptor through improved endosomal escape of siRNA." *Biochimica et Biophysica Acta (BBA) - Biomembranes* 1768(5): 1211-1217.

Oliveira, S., A. Hogset, et al. (2008). "Delivery of siRNA to the target cell cytoplasm: photochemical internalization facilitates endosomal escape and improves silencing efficiency, in vitro and in vivo." *Current Pharmaceutical Design* 14(34): 3686-3697.

Pérez-Martínez, F., J. Guerra, et al. (2011). "Barriers to Non-Viral Vector-Mediated Gene Delivery in the Nervous System." *Pharmaceutical Research* 28(8): 1843-1858.

Pack, D. W., A. S. Hoffman, et al. (2005). "Design and development of polymers for gene delivery." *Nature Reviews Drug Discovery* 4(7): 581-593.

Pack, D. W., D. Putnam, et al. (2000). "Design of imidazole-containing endosomolytic biopolymers for gene delivery." *Biotechnology and Bioengineering* 67(2): 217-223.

Parente, R. A., S. Nir, et al. (1990). "Mechanism of leakage of phospholipid vesicle contents induced by the peptide GALA." *Biochemistry* 29(37): 8720-8728.

Patel, L. N., J. L. Zaro, et al. (2007). "Cell penetrating peptides: intracellular pathways and pharmaceutical perspectives." *Pharmaceutical Research* 24(11): 1977-1992.

Pathak, A., S. Patnaik, et al. (2009). "Recent trends in non-viral vector-mediated gene delivery." *Biotechnology Journal* 4(11): 1559-1572.

Pedroso de Lima, M. C., S. Simões, et al. (2001). "Cationic lipid-DNA complexes in gene delivery: from biophysics to biological applications." *Advanced Drug Delivery Reviews* 47(2-3): 277-294.

Pfeifer, C., G. Hasenpusch, et al. (2011). "Dry powder aerosols of polyethylenimine (PEI)-based gene vectors mediate efficient gene delivery to the lung." *Journal of Controlled Release* 154(1): 69-76.

Pichon, C., I. Freulon, et al. (1997). "Cytosolic and nuclear delivery of oligonucleotides mediated by an amphiphilic anionic peptide." *Antisense and Nucleic Acid Drug Development* 7(4): 335-343.

Pichon, C., C. Gonçalves, et al. (2001). "Histidine-rich peptides and polymers for nucleic acids delivery." *Advanced Drug Delivery Reviews* 53(1): 75-94.

Pires, P., S. Simões, et al. (1999). "Interaction of cationic liposomes and their DNA complexes with monocytic leukemia cells." *Biochimica et Biophysica Acta (BBA)-Biomembranes* 1418(1): 71-84.

Plank, C., B. Oberhauser, et al. (1994). "The influence of endosome-disruptive peptides on gene transfer using synthetic virus-like gene transfer systems." *Journal of Biological Chemistry* 269(17): 12918-12924.

Prasmickaite, L., A. Høgset, et al. (2000). "Role of endosomes in gene transfection mediated by photochemical internalisation (PCI)." *The Journal of Gene Medicine* 2(6): 477-488.

Raemdonck, K., B. Naeye, et al. (2009). "Biodegradable dextran nanogels for RNA interference: focusing on endosomal escape and intracellular siRNA delivery." *Advanced Functional Materials* 19(9): 1406-1415.

Raper, S. E., N. Chirmule, et al. (2003). "Fatal systemic inflammatory response syndrome in a ornithine transcarbamylase deficient patient following adenoviral gene transfer." *Molecular Genetics and Metabolism* 80(1-2): 148-158.

Remy, J. S., C. Sirlin, et al. (1994). "Gene transfer with a series of lipophilic DNA-binding molecules." *Bioconjugate Chemistry* 5(6): 647-654.

Richard, J. P., K. Melikov, et al. (2003). "Cell-penetrating peptides. A reevaluation of the mechanism of cellular uptake." *Journal of Biological Chemistry* 278(1): 585-590.

Rittner, K., A. Benavente, et al. (2002). "New basic membrane-destabilizing peptides for plasmid-based gene delivery in vitro and in vivo." *Molecular Therapy* 5(2): 104-114.

Sanders, N., C. Rudolph, et al. (2009). "Extracellular barriers in respiratory gene therapy." *Advanced Drug Delivery Reviews* 61(2): 115-127.

Sasaki, K., K. Kogure, et al. (2008). "An artificial virus-like nano carrier system: enhanced endosomal escape of nanoparticles via synergistic action of pH-sensitive fusogenic peptide derivatives." *Analytical and Bioanalytical Chemistry* 391(8): 2717-2727.

Selbo, P. K., G. Sivam, et al. (2001). "In vivo documentation of photochemical internalization, a novel approach to site specific cancer therapy." *International Journal of Cancer* 92(5): 761-766.

Selbo, P. K., A. Weyergang, et al. (2010). "Photochemical internalization provides time- and space-controlled endolysosomal escape of therapeutic molecules." *Journal of Controlled Release* 148(1): 2-12.

Shieh, M. J., C. L. Peng, et al. (2008). "Non-toxic phototriggered gene transfection by PAMAM-porphyrin conjugates." *Journal of Controlled Release* 129(3): 200-206.

Shim, M. S. and Y. J. Kwon (2010). "Efficient and targeted delivery of siRNA in vivo." *FEBS Journal* 277: 4814–4827.

Simeoni, F., M. C. Morris, et al. (2003). "Insight into the mechanism of the peptide-based gene delivery system MPG: implications for delivery of siRNA into mammalian cells." *Nucleic Acids Research* 31(11): 2717-2724.

Simoes, S., V. Slepushkin, et al. (1999). "Mechanisms of gene transfer mediated by lipoplexes associated with targeting ligands or pH-sensitive peptides." *Gene Therapy* 6(11): 1798.

Sonawane, N. D., F. C. Szoka Jr, et al. (2003). "Chloride accumulation and swelling in endosomes enhances DNA transfer by polyamine-DNA polyplexes." *Journal of Biological Chemistry* 278(45): 44826-44831.

Stegmann, T. (2000). "Membrane fusion mechanisms: the influenza hemagglutinin paradigm and its implications for intracellular fusion." *Traffic* 1(8): 598-604.

Stewart, D., J. Hilton, et al. (2006). "Angiogenic gene therapy in patients with nonrevascularizable ischemic heart disease: a phase 2 randomized, controlled trial of AdVEGF121 (AdVEGF121) versus maximum medical treatment." *Gene Therapy* 13(21): 1503-1511.

Sui, G., C. Soohoo, et al. (2002). "A DNA vector-based RNAi technology to suppress gene expression in mammalian cells." *Proceedings of the National Academy of Sciences* 99(8): 5515-5520.

Torchilin, V. P. (2008). "Tat peptide-mediated intracellular delivery of pharmaceutical nanocarriers." *Advanced Drug Delivery Reviews* 60(4): 548-558.

Urban-Klein, B., S. Werth, et al. (2004). "RNAi-mediated gene-targeting through systemic application of polyethylenimine (PEI)-complexed siRNA in vivo." *Gene Therapy* 12(5): 461-466.

van Rossenberg, S. M. W., K. M. Sliedregt-Bol, et al. (2002). "Targeted lysosome disruptive elements for improvement of parenchymal liver cell-specific gene delivery." *Journal of Biological Chemistry* 277(48): 45803-45810.

Veldhoen, S., S. D. Laufer, et al. (2006). "Cellular delivery of small interfering RNA by a non-covalently attached cell-penetrating peptide: quantitative analysis of uptake and biological effect." *Nucleic Acids Research* 34(22): 6561-6573.

Vinge, L. E., P. W. Raake, et al. (2008). "Gene therapy in heart failure." *Circulation Research* 102(12): 1458-1470.

Vivès, E., P. Brodin, et al. (1997). "A truncated HIV-1 Tat protein basic domain rapidly translocates through the plasma membrane and accumulates in the cell nucleus." *Journal of Biological Chemistry* 272(25): 16010-16017.

Vives, E. (2003). "Cellular utake of the Tat peptide: an endocytosis mechanism following ionic interactions." *Journal of Molecular Recognition* 16(5): 265-271.

Vlasov, G., E. Lesina, et al. (2005). "Optimization of transfection properties of DNA-lysine dendrimer complexes." *Russian Journal of Bioorganic Chemistry* 31(2): 153-159.

Vliegenthart, J., W. Knollen, et al. (1999). "Enhanced efficiency of lactosylated poly-L-lysine-mediated gene transfer into cystic fibrosis airway epithelial cells." *American Journal of Respiratory Cell and Molecular Biology* 20(5): 1081-1086.

Vogt, T. C. B. and B. Bechinger (1999). "The interactions of histidine-containing amphipathic helical peptide antibiotics with lipid bilayers." *Journal of Biological Chemistry* 274(41): 29115.

Wagner, E. and J. Kloeckner (2006). "Gene delivery using polymer therapeutics." *Polymer Therapeutics I*: 135-173.

Wagner, E., C. Plank, et al. (1992). "Influenza virus hemagglutinin HA-2 N-terminal fusogenic peptides augment gene transfer by transferrin-polylysine-DNA complexes: toward a synthetic virus-like gene-transfer vehicle." *Proceedings of the National Academy of Sciences* 89(17): 7934-7938.

Wagner, E., K. Zatloukal, et al. (1992). "Coupling of adenovirus to transferrin-polylysine/DNA complexes greatly enhances receptor-mediated gene delivery and expression of transfected genes." *Proceedings of the National Academy of Sciences* 89(13): 6099-6103.

Whitehead, K. A., R. Langer, et al. (2009). "Knocking down barriers: advances in siRNA delivery." *Nature Reviews Drug Discovery* 8(2): 129-138.

Wrobel, I. and D. Collins (1995). "Fusion of cationic liposomes with mammalian cells occurs after endocytosis." *Biochimica et Biophysica Acta (BBA)-Biomembranes* 1235(2): 296-304.

Wyman, T. B., F. Nicol, et al. (1997). "Design, synthesis, and characterization of a cationic peptide that binds to nucleic acids and permeabilizes bilayers." *Biochemistry* 36(10): 3008-3017.

Xu, Y. and F. C. Szoka Jr (1996). "Mechanism of DNA release from cationic liposome/DNA complexes used in cell transfection." *Biochemistry* 35(18): 5616-5623.

Yamamoto, M. and D. T. Curiel (2005). "Cancer gene therapy." *Technology in Cancer Research & Treatment* 4(4): 315-330.

Yang, S. R., H. J. Lee, et al. (2006). "Histidine-conjugated poly (amino acid) derivatives for the novel endosomolytic delivery carrier of doxorubicin." *Journal of Controlled Release* 114(1): 60-68.

Yip, W. L., A. Weyergang, et al. (2007). "Targeted delivery and enhanced cytotoxicity of cetuximab-saporin by photochemical internalization in EGFR-positive cancer cells." *Molecular Pharmaceutics* 4(2): 241-251.

Yu, M., E. Poeschla, et al. (1994). "Progress towards gene therapy for HIV infection." *Gene Therapy* 1(1): 13-26.

Zabner, J., A. J. Fasbender, et al. (1995). "Cellular and molecular barriers to gene transfer by a cationic lipid." *Journal of Biological Chemistry* 270(32): 18997-19007.

Zelphati, O. and F. C. Szoka (1996). "Intracellular distribution and mechanism of delivery of oligonucleotides mediated by cationic lipids." *Pharmaceutical Research* 13(9): 1367-1372.

Zelphati, O. and F. C. Szoka (1996). "Mechanism of oligonucleotide release from cationic liposomes." *Proceedings of the National Academy of Sciences* 93(21): 11493.

Zhou, J., J. Wu, et al. (2006). "PAMAM dendrimers for efficient siRNA delivery and potent gene silencing." *Chemical Communications*(22): 2362-2364.

Zhou, X. and L. Huang (1994). "DNA transfection mediated by cationic liposomes containing lipopolylysine: characterization and mechanism of action." *Biochimica et Biophysica Acta (BBA)-Biomembranes* 1189(2): 195-203.

Zuhorn, I. S., U. Bakowsky, et al. (2005). "Nonbilayer phase of lipoplex–membrane mixture determines endosomal escape of genetic cargo and transfection efficiency." *Molecular Therapy* 11(5): 801-810.

Chapter 17

Endocytosis of Particle Formulations by Macrophages and Its Application to Clinical Treatment

Keiji Hirota and Hiroshi Terada

Additional information is available at the end of the chapter

1. Introduction

Macrophages are known to take up "invaders" such as pathogens and viruses mainly by phagocytosis to protect the host from infections by them. This process of phagocytosis is disadvantageous in general for exhibition of an efficient pharmacological effect of particle formulations containing drugs, because the uptake of particles by macrophages reduces the extracellular drug concentration. Hence, it is important to understand what properties of particles are advantageous or disadvantageous for phagocytic uptake by macrophages. Modification of particles by polyethylene glycol (PEG), which forms a hydrated phase on the surface of particles, enables long-lasting circulation of such particles in the bloodstream by circumventing their uptake by macrophage cells [1].

In contrast, particle formulations that are easily taken up by macrophages would be highly advantageous for macrophage-targeting drug delivery [2]. In this case, a typical example is the treatment of tuberculosis (TB). Namely, *Mycobacterium tuberculosis* (MTB) cells are easily trapped in the phagosomes of alveolar macrophages. However, these cells are not digested by macrophages, because the fusion of the MTB-containing phagosomes with lysosomes, which are indispensable for the digestion of bacteria inside phagosomes, is inhibited. As a result, MTB cells proliferate and accumulate inside macrophages [3]. Hence, the delivery of particles containing antituberculosis agents to alveolar macrophages would be expected to be effective for TB therapy.

As summarized in Figure 1, endocytic uptake including phagocytosis is classified according to the mechanism of vesicle formation as well as the size of particles ingested [4-7]. Phagocytosis is performed by specialized cells such as macrophages, and it plays a role in the clearance of particles having a diameter greater than 0.5 μm. On the other hand, pinocytosis occurs in all cells, including macrophages and cancer cells, for obtaining nutrients and biological mediators. It is noteworthy that macropinocytosis covers a broad range of particle sizes from 100 nm to 5 μm [8-10].

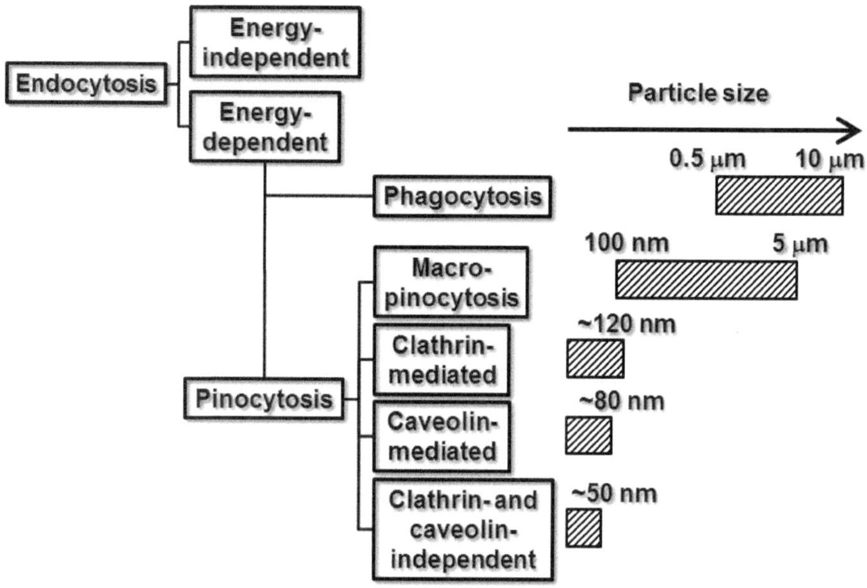

Figure 1. Classification of endocytosis in relation to particle sizes favorable for ingestion.

In this chapter, features of phagocytosis of particles in terms of particle properties, as well as phagocytosis-induced physiological events of macrophages, are described. In addition, the promising aspect of clinical treatment by the utilization of endocytosis-mediated drug action is reviewed.

2. Effect of particle properties on endocytosis

Drug-containing particle formulations are commonly used for delivery of drugs. The particle base is the most important part of a formulation. Poly (lactic-glycolic) acid (PLGA) is one of the candidates for drug-containing particle formulations, because PLGA is biodegradable and biocompatible [11]. Drug release from the particles and its sustainability can be regulated by changing the molecular weight and composition of the lactate and glycolate moieties of PLGA [12].

Phagocytic uptake of particles by macrophage cells proceeds as follows: 1) access of particles to the surface of the macrophage membrane, 2) particle recognition by phagocytic receptors on the macrophage membrane, and 3) dynamic changes in membrane structure (protrusion or invagination). Particle size, shape, and surface properties affect efficient entrapment and subsequent uptake by macrophages.

2.1. Particle size

Particle size is likely the primary factor that governs endocytic uptake of particles. The optimum size of particles for efficient endocytic uptake varies according to the cell type. Macrophage cells are able to ingest large particles having a diameter between 1 μm and 10 μm to eliminate invaders from outside the body [13,14]. The optimal sizes of the particles for the uptake by alveolar macrophages range between 3 μm and 6 μm [15], but those by peritoneal macrophages and peripheral blood mononuclear cells are reportedly from 0.3 μm to 1.1 μm [16-18]. The uptake mechanism of the particles, such as 3-μm particles, was interpreted by the Derjaguin-Landau-Verwey-Overbeek (DLVO) theory [19].

Nanoparticles are advantageous for attacking carcinoma cells, which mainly originate from epithelial cells. The optimal size of the particles for the uptake by carcinoma cells was reported to be around 50 nm, as determined by the use of gold nanoparticles [20]. Besides micrometer-size particles, macrophages also take up nano-size particles [21]. The uptake of nano-size particles mainly proceeds via pinocytosis in such a way that the rate of pinocytosis is dependent on the extracellular particle concentration and time of exposure. For ingestion of liquid phase, this rate for mouse fibroblast L cells is 18.7 μm^3/hr/cell, which value is less than that for mouse resident peritoneal macrophages, 46.5 μm^3/hr/cell [22,23]. Hence, understanding of particle properties other than size is also important for particle delivery.

2.2. Shape

Particle shape is another major factor affecting endocytic uptake by macrophages. The macrophage membrane undergoes structural changes in such a way that the membrane spreads around the particle, starting from the initial contact site between particle and membrane; and the progression of endocytic uptake of particles is dependent on the contact angle between particle and macrophage membrane. For example, an elliptical disk-shaped particle is internalized along its long axis when the particle has attached perpendicularly to the cell membrane, in which case the contact angle is small [24]. In this case, the membrane spreads symmetrically around the particle, engulfing it. In contrast, attachment when the short axis is perpendicular to the membrane increases the angle of contact and the number of contact points with the membrane, which then starts to spread asymmetrically. As a result, the particle is not engulfed. However, when the long axis of particles ranges from 2-3 μm, which corresponds to that of most bacteria, maximum attachment to macrophage cells occurs; and engulfment is successful even though the angle of contact is large [25].

2.3. Surface properties

Most mammalian cells including macrophage cells have negative charges on their surface [26,27]. As the loss of the negative surface charge of the membrane is thought to influence protein localization during endocytosis [28], the surface charge of particles is thought to be also critical for endocytic uptake. In fact, it is reported that changes in cellulose particles by the introduction of extremely negatively charged sulphoethyl residues or of positively charged diethylaminoethyl groups affect the endocytic uptake by mouse peritoneal macrophage cells and that the endocytic uptake is the lowest for particles without having surface charge, as determined in terms of zeta potential [29].

Charge density is also important. The Ohshima theory, based on the analysis of the membrane surface in terms of electrophoretic "softness" and the density of the fixed charge [30], will be effective for understanding of the interaction of particles with cell membranes [31]. Polystyrene particles having electrophoretic softness and a low negative electrical charge density by the introduction of primary amine and carboxyl groups on their surface were reported to be more susceptible to endocytic uptake by rat alveolar macrophage cells than those having more rigid and higher electrical charge density by the introduction of hydroxyl and sulfate groups [13].

2.4. Particle formulations

Particle formulations affect directly the interaction of particles with the endocytic receptors of macrophage cells. The exposure of the phosphatidylserine moiety on the membrane of apoptotic lymphocytes is important for their removal by endocytosis by macrophage cells through recognition via scavenger receptors on the macrophages [32]. Based on this mechanism, liposomes containing phosphatidylserine are more susceptible to uptake by macrophage-like HL-60RG cells than those containing phosphatidylethanolamine or phosphatidic acid [33].

PLGA has been commonly used as a base of particle formulations. Macrophage cells eat PLGA particles more efficiently than polystyrene latex ones. It is noteworthy that phagocytosis of PLGA particles by alveolar macrophage cells stimulates their phagocytic activity in such a way that their uptake increases both the population of phagocytic macrophage cells and the number of particles that have been taken up by individual macrophage [34]. However, the mechanism of interaction of PLGA with macrophage cells is still unknown.

3. Induction of inflammatory responses by endocytosis

Macrophage cells patrol around the tissue where they reside and play a central role in the clearance of invaders. The total surface area of human alveoli is approx. 70 m^2, where 23 billion alveolar macrophage cells reside [35,36]. Namely, a single macrophage cell should monitor invaders in an area of a square with a side length of 55 μm. When macrophage cells encounter invaders, the cells eliminate them by phagocytosis and subsequent digestion with lysosomal enzymes. Simultaneously, macrophage cells generate inflammatory mediators, working as signals to inform the surrounding cells that invaders are coming. Macrophage cells also recognize drug carrier particles as invaders, and then, "undesirable" immune responses such as the production of antibody and inflammatory mediators take place. Hence, silent nature toward macrophage cell functions is required for efficient drug carrier particles.

In the case of endocytosis-mediated drug action (see section 5), high particle uptake by macrophages is favorable. Such efficient uptake will be achieved by up-regulation of endocytic activity, but this action may trigger undesirable immune responses from macrophages. If efficient particle uptake is not associated with the induction of the undesirable responses, the particles could be very useful as drug carriers. As summarized in Table 1, PLGA particles are those having such a desirable silent nature regarding inflammatory responses, although they are yet well phagocytosed by macrophages compared with polystyrene latex (PSL) particles [34,37]. Namely, the PLGA particle behaves like a *"Ninja,"* having stealth and concealment activities. However, it is still unknown how particle formulation and modification are relevant to this silent nature. In this section, we review two distinct pathways involved in signal transduction to generate inflammatory mediators, one using phagocytic receptors and the other, pattern recognition receptors (PRRs).

Responses		**Particles**	
	PLGA	**PSL**	**LPS**
Cell death	-	+	++
TNF-α	-	+	++
NO	-	+	++
IL-10	-	-	+
TGF-β	-	+	-
Phagostimulation	+	-	±

These data are summarized from reports [34,37]. Rat alveolar macrophage cells (NR8383) were exposed to PLGA and PSL particles at a number 10 times greater than the cell number. LPS existing in micellar form in the incubation medium was used at the concentration of 1 μg/mL as a reference. (-), no responses; (+), mild responses; (++), significantly high response than (+).

Table 1. Alveolar macrophage cellular responses induced by particle uptake.

3.1. Phagocytosis-mediated inflammatory response

Phagocytic cells, such as macrophages, monocytes, and polymorphonuclear cells, take up particles and pathogens typically with sizes of more than 0.5 μm to clear them from the body mainly by phagocytosis [4,38]. In the case of macrophages, the ingestion of particles proceeds in such a way that the interaction of the particles with phagocytic receptors causes extension of pseudopods from the plasma membrane to capture the particles, which action is followed by engulfment by these phagocytes [5]. This phagocytic mechanism, called the "zipper" model, requires a reorganization of the actin-based cytoskeleton underlying the region of plasma membrane in contact with the particles and induces signal transduction through the Fcγ receptor (FcγR) and complement-receptor 3 (CR3) [39].

The cross-linking of the FcγR by particles simultaneously initiates a series of signal transduction events mediated by multiple protein tyrosine kinases, phosphoinositide, and free arachidonic acid [40-42]. In the case of monocytes, the cross-linking of FcγR by IgG initiates the release of TNF-α, IL-1β, IL-6, IL-8, and monocyte chemoattractant protein-1 (MCP-1), which molecules are classified as T helper 1 (TH1) cytokines [43-47]. Macrophage cells are also stimulated by the cross-linking of FcγR by IgG; and these cells generate TNF-α through MAPK signal transduction, leading to NF-κB activation [48,49]. In the presence of IL-3 and a high density of IgG, IL-4 and IL-10 of the TH2 inflammatory cytokine family are synthesized in macrophage cells by the cross-linking via IgG [50,51]. Signal transduction initiated from FcγRs is involved in inflammatory immune responses.

In contrast, CR3-mediated phagocytosis by macrophages, which is cooperative with FcγR-mediated phagocytosis [52], seems to be involved in anti-inflammatory responses rather than in inflammatory responses, though various signal transductions mediated by tyrosine kinase are initiated from CR3 [53]. Generation of IL-12 in human monocytes is known to be stimulated by phagocytosis of *Staphylococcus aureus*. However, treatment of the phagocytic monocytes with iC3b, the natural CR3 ligand, down-regulates the generation of IL-12, suggesting that CR3 suppresses the inflammatory response [54]. In addition, ligation of CR3 suppresses the release of TH1 cytokines, such as TNF-α, IL-6 and IL-12, and the TH2 cytokine IL-10 from LPS-stimulated bone marrow-derived mouse dendritic cells [55]. Hence, it is possible that CR3 is associated with the silent nature of the entry of particles, such as PLGA particles, into phagocytic cells.

3.2. Pattern recognition-mediated inflammatory responses

There are macrophage cells in various vertebrates and invertebrates that are capable of recognizing highly conserved pathogenic molecular patterns by receptors called pattern recognition receptors (PRRs) [56,57]. These receptors are classified into two major groups, one involved in endocytic uptake, such as scavenger receptors (SRs), and the other associated with transmission of danger signals independent of endocytosis, such as Toll-like receptors (TLRs) and nucleotide-binding domain leucine-rich repeat-containing receptors (NLRs) [57,58]. It is of importance to understand signal transduction from these receptors for construction of silent drug delivery using particles.

Lipopolysaccharide (LPS) micelles and silica particles are reported to bind with SRs, which binding is followed by their engulfment by macrophages [59-61]. In sepsis patients, SRs play a role in efficient clearance of LPS and attenuation of LPS-induced inflammatory responses [62]. However, signaling pathways initiated from SRs are difficult to identify, because ligands of SRs, such as LPS and polyinosinic acid, simultaneously stimulate TLRs as well. Overexpression of class B SRs in human epithelial carcinoma HeLa cells and human embryonic kidney cells (HEK293) increases the production of the inflammatory mediator IL-8 associated with increased uptake of LPS [63]. This inflammatory response seems to be due to interaction of SRs with LPS. In contrast, CD163, a hemoglobin scavenger receptor, down-regulates TH1 inflammatory responses by initiation of signaling leading to secretion of TH2 cytokines [64]. SRs are associated with TLR-independent signaling pathways and involved in inflammatory responses similarly as FcγR.

TLRs play an important role in innate immunity and recognize various molecules derived from bacteria and viruses [65]. TLR3 and TLR7 express on the endosomal membrane and work as sensors for elimination of unnecessary nucleic acids by induction of inflammatory responses; and these responses should be taken into account for delivery of nucleic acids, such as small interfering RNA (siRNA). Administration of siRNA via liposomes or transgenic reagent Lipofectamine® induce potent immunostimulation generating TH1 inflammatory cytokines from human monocytes and plasmacytoid dendritic cells [66,67]. In addition, immune responses through TLRs' signals cause adoptive immunity, such as the generation of antibody [68]. It is of importance to study the mechanism of signal transduction associated with TLRs for understanding of the generation of undesirable immune responses. To overcome these problems, nanoparticles will be effective, because they are able to escape from endosomes into the cytosol, where these nanoparticles release drugs. Based on this strategy, delivery of siRNA-containing nanoparticles into HeLa cells and human pancreatic carcinoma PanC-1 cells is reported to be successful [69,70].

The role of NLRs should also be taken into consideration in the delivery of particles. One of the most characterized NLRs is NLRP3 (also known as NALP3 or cryopyrin) [71]. It is noteworthy that the inflammasome (NLRP3), which contains procaspase-1, senses lysosomal enzymes in the cytosol, leading to the activation of caspase-1 and that this event is followed by secretion of the inflammatory mediators IL-1β, IL-18, and IL-33 [58,72]. Uptake of micro-particles of silica crystals and aluminum hydroxide causes leakage of the lysosomal enzyme cathepsin B into the cytosol due to destabilization of the lysosomal membrane in human and mouse macrophage cells [73,74]. Possibly, NLRs work as a sensor of danger signals initiated from lysosomal destabilization caused by uptake of such micro-particles. It should be important to know the effect of undegradable particles on the stability of lysosomes for understanding of the onset of cytotoxicity by phagocytosis of particles.

4. Lipid-raft-dependent uptake of particles

When particles are caught by macrophage cells, the macrophage membrane undergoes structural changes after recognition of the particles by endocytic receptors located in the membrane region. This membrane region, referred to as a lipid raft, is enriched in sphingolipids and cholesterol, which serve as a scaffold for the proper functioning of endocytic receptors and various signal transduction pathways [75,76]. In microglia, signaling cascades triggered in response to gangliosides are mediated by recruitment of Src homology 2 domain-containing protein-tyrosine phosphatase 2 (SHP-2) to lipid rafts [77].

Caveolae, a subset of lipid rafts, control various biological events including endocytosis [78] and are associated with the incorporation of pathogens [79]. There is another possibility that inclusion of protein receptors in the rafts is closely associated with phagocytic uptake of particles, because CD36, a class B SR, exists in caveolin-containing lipid rafts in human melanoma cells [80]. A cyclodextrin, MβCD is commonly used as a reagent to disrupt lipid rafts by the extraction of cholesterol [81]. Treatment of mouse macrophage-like J774 cells with MβCD inhibits recruitment of SRs to lipid raft domain [82]. As a result, phagocytic activity toward PSL particles, which are ingested through SRs, decreases. Similarly, recruitment of CD36 to raft domains is necessary for phagocytosis of amyloid β by microglial cells [83]. Lipid rafts are essential for recruitment of phagocytic receptors; and, hence, they are associated with delivery of drug-containing particles by phagocytosis.

An alternative function of lipid rafts is to provide a scaffold for TLRs associated with danger signal transduction. Stimulation of human peripheral blood monocytes with LPS causes clustering of the signaling receptor TLR4 with its accessory protein CD14 [84], and association of this receptor cluster with lipid rafts is thought to be necessary for LPS-induced signal transduction [85]. An increase in membrane fluidity due to ethanol at a concentration of higher than 50 mM inhibits the association of TLR4 with lipid rafts, suppressing LPS-induced TNF-α production in mouse macrophage cells [86]. However, it is interesting to note that treatment with MβCD does not affect LPS-induced gene expression relating to inflammation [87]. This could be because MyD88, the adaptor protein of TLR4, exists in a membrane region other than lipid rafts [88]. In addition, generation of nitric oxide from macrophage cells after disruption of lipid rafts by MβCD is comparable to that of intact macrophage cells, though the MβCD treatment decreases phagocytic activity toward the PSL particles by a half [87]. Further studies on the operation of inflammatory signaling cascades in relation with lipid rafts are needed.

5. Endocytosis-mediated drug action

In the lungs, macrophage cells patrol the air/cell interfaces and play a role in protecting the host from invaders such as pathogens and viruses by phagocytic uptake. However, some pathogens, such as MTB, survive in macrophage cells and proliferate well by using them as incubators after the pathogens have been inhaled into the alveoli by respiration [3]. Owing to this survival strategy, MTB is able to escape from the attack of antitubercular agents, and this is one of the reasons why effective treatment of TB has not been successful till now.

As macrophages phagocytose particle formulations besides bacteria and viruses, utilization of this phagocytosis-mediated transport of these drug formulations into MTB-infected macrophages is expected to be promising for therapy of TB. For this approach, particles containing an antitubercular agent are delivered to the lungs, where alveolar macrophages infected with MTB reside. The macrophages take up the particles, and the antitubercular agent thus phagocytosed in a form of particles attacks the MTB. The effect of PLGA microspheres containing rifampicin (RFP), one of the first-line drugs for TB treatment, on MTB has been well examined to date [89,90]. PLGA microspheres containing RFP (RFP-PLGA) were prepared by various methods such as double-emulsification and spray-drying. The PLGA MS thus prepared deliver an amount of RFP into rat alveolar macrophage NR8383 cells *in vitro* about 20 times greater than that added in the free form in solution [12,37]. Inhalation of PLGA MS containing the antitubercular agent rifabutin increases the drug residence time in the lungs to more than that by intravenous administration in mice due to uptake of the particles by alveolar phagocytic cells [91]. However, the bacterial population in the rat lung is not significantly decreased by pulmonary administration of RFP-PLGA MS, though granuloma formation on the surface of the lung is reduced [92].

To achieve efficient phagocytosis-mediated TB treatment, at least three requirements must be met. Namely, drug-containing particles should be 1) well phagocytosed by alveolar macrophages, 2) exhibit a potent bactericidal effect on MTB inside the macrophages, and 3) should not be toxic to the phagocytes. PLGA particles containing an antitubercular agent well satisfy these three requirements. In addition, homogeneous distribution of drug-containing particles in the target tissue is required to obtain the optimum effect. Understanding of endocytic activities of MTB-infected macrophages toward drug-containing particles *in vivo* is thus important for improving TB therapy.

Another promising aspect of endocytosis-mediated therapy could be the treatment of cancer. One possible way would be the induction of inflammatory mediators, such as NO and TNF-α, in macrophages by immunomodulators such as TLR-ligands, leading to their cytotoxic effects on tumor cells [93-95]. In addition, "re-education" of the healing-type macrophages (M2 macrophages) to the killer-type macrophages (M1 macrophages) by immunomodulators should be effective as well [96]. An increase in the M1 macrophage population could be advantageous for the treatment of tumors. As TLRs are expressed on various cell membranes in the body, endocytosis-mediated delivery of TLR-ligands to macrophage cells should be effective in overcoming malignant neoplasms without the induction of undesirable immune responses.

6. Conclusions

The physiological function of macrophage cells is important in overcoming various diseases, because they rid the body of pathogens by phagocytosis. Hence, phagocytosis-mediated drug delivery is useful for a direct attack against pathogenic bacteria and viruses residing inside macrophage cells. As summarized in Figure 2, the optimum properties of particles targeting macrophage cells are a) "macrophage-philicity," especially toward phagocytic receptors and lipid rafts, b) ability to stimulate actin reorganization, c) a silent nature like a "*Ninja*" with respect to inflammatory responses, and d) ability to allow rapid release of the incorporated drugs. Of these, items "a" and "b" refer to the feasibility of particles for their efficient ingestion; and spherical particles having about 3-μm diameter and surface charges are likely to be favorable for phagocytic uptake. Item "c" is associated with a nontoxic effect on macrophage function; and item "d" is important for exhibition of drug action, in which the release of drugs from the particles modulates the drug action.

In addition, activation of the macrophages of the immune system is advantageous for attack against pathological cells residing close to macrophages, such as tumor cells. It is noteworthy that certain bases, such as PLGA, of the particles themselves modulate the immune functions of macrophages. Development of drug-containing particles, which efficiently attack the pathogens or pathological cells, and which upregulate the immunological function of macrophages, is beneficial to overcome infectious diseases and cancer.

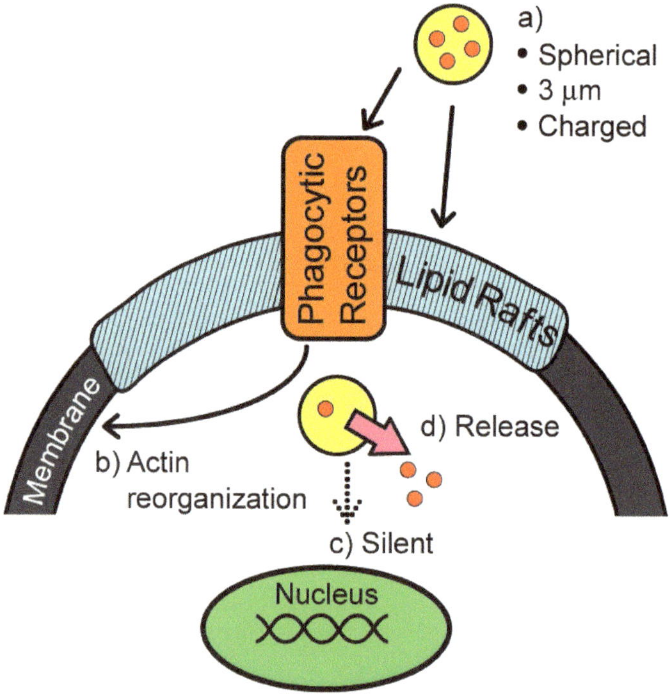

Figure 2. Biochemical events associated with endocytosis-mediated drug delivery via particles.

Author details

Keiji Hirota and Hiroshi Terada
Faculty of Pharmaceutical Sciences, Japan
Center for Drug Delivery Research, Japan
Center for Physical Pharmaceutics, Research Institute for Science and Technology, Tokyo University of Science, Yamazaki, Noda, Chiba, Japan

Acknowledgement

This work was supported, in part, by the program for development of strategic research center in private universities supported by MEXT (Ministry for Education, Culture, Sports, Science, and Technology) of Japan, 2010-2014.

7. References

[1] Gabizon, A.A., Barenholz, Y., and Bialer, M. (1993). Prolongation of the circulation time of doxorubicin encapsulated in liposomes containing a polyethylene glycol-derivatized phospholipid: pharmacokinetic studies in rodents and dogs. *Pharm Res* 10, 703-708.

[2] Gordon, S., and Rabinowitz, S. (1989). Macrophages as targets for drug delivery. *Adv Drug Deliv Rev* 4, 27-47.
[3] Armstrong, J.A., and Hart, P.D. (1975). Phagosome-lysosome interactions in cultured macrophages infected with virulent tubercle bacilli. Reversal of the usual nonfusion pattern and observations on bacterial survival. *J Exp Med* 142, 1-16.
[4] Aderem, A., and Underhill, D.M. (1999). Mechanisms of phagocytosis in macrophages. *Annu Rev Immunol* 17, 593-623.
[5] Conner, S.D., and Schmid, S.L. (2003). Regulated portals of entry into the cell. *Nature* 422, 37-44.
[6] Hansen, C.G., and Nichols, B.J. (2009). Molecular mechanisms of clathrin-independent endocytosis. *J Cell Sci* 122, 1713-1721.
[7] Sahay, G., Alakhova, D.Y., and Kabanov, A.V. (2010). Endocytosis of nanomedicines. *J Control Release* 145, 182-195.
[8] Swanson, J.A., and Watts, C. (1995). Macropinocytosis. *Trends Cell Biol* 5, 424-428.
[9] Tamaru, M., Akita, H., Fujiwara, T., Kajimoto, K., and Harashima, H. (2010). Leptin-derived peptide, a targeting ligand for mouse brain-derived endothelial cells via macropinocytosis. *Biochem Biophys Res Commun* 394, 587-592.
[10] Bhattacharya, S., Roxbury, D., Gong, X., Mukhopadhyay, D., and Jagota, A. (2012). DNA conjugated SWCNTs enter endothelial cells via Rac1 mediated macropinocytosis. *Nano Lett.*
[11] Gilding, D.K., and Reed, A.M. (1979). Biodegradable polymers for use in surgery - Polyglycolic-Poly(Actic Acid) homopolymers and copolymers .1. *Polymer* 20, 1459-1464.
[12] Makino, K., Nakajima, T., Shikamura, M., Ito, F., Ando, S., Kochi, C., Inagawa, H., Soma, G.-I., and Terada, H. (2004). Efficient intracellular delivery of rifampicin to alveolar macrophages using rifampicin-loaded PLGA microspheres: effects of molecular weight and composition of PLGA on release of rifampicin. *Colloid and Surf B: Biointerfaces* 36, 35-42.
[13] Makino, K., Yamamoto, N., Higuchi, K., Harada, N., Ohshima, H., and Terada, H. (2003a). Phagocytic uptake of polystyrene microspheres by alveolar macrophages: effects of the size and surface properties of the microspheres. *Colloid Surf B: Biointerfaces* 27, 33-39.
[14] Hasegawa, T., Hirota, K., Tomoda, K., Ito, F., Inagawa, H., Kochi, C., Soma, G., Makino, K., and Terada, H. (2007). Phagocytic activity of alveolar macrophages toward polystyrene latex microspheres and PLGA microspheres loaded with anti-tuberculosis agent. *Colloid Surf B: Biointerfaces* 60, 221-228.
[15] Hirota, K., Hasegawa, T., Hinata, H., Ito, F., Inagawa, H., Kochi, C., Soma, G., Makino, K., and Terada, H. (2007). Optimum conditions for efficient phagocytosis of rifampicin-loaded PLGA microspheres by alveolar macrophages. *J Control Release* 119, 69-76.
[16] Pratten, M.K., and Lloyd, J.B. (1986). Pinocytosis and phagocytosis: the effect of size of a particulate substrate on its mode of capture by rat peritoneal macrophages cultured in vitro. *Biochim Biophys Acta* 881, 307-313.
[17] Seymour, L., Schacht, E., and Duncan, R. (1991). The effect of size of polystyrene particles on their retention within the rat peritoneal compartment, and on their interaction with rat peritoneal macrophages in vitro. *Cell Biol Int Rep* 15, 277-286.

[18] Kawaguchi, H., Koiwai, N., Ohtsuka, Y., Miyamoto, M., and Sasakawa, S. (1986). Phagocytosis of latex particles by leucocytes. I. Dependence of phagocytosis on the size and surface potential of particles. *Biomaterials* 7, 61-66.
[19] Champion, J.A., Walker, A., and Mitragotri, S. (2008). Role of particle size in phagocytosis of polymeric microspheres. *Pharm Res* 25, 1815-1821.
[20] Chithrani, B.D., Ghazani, A.A., and Chan, W.C. (2006). Determining the size and shape dependence of gold nanoparticle uptake into mammalian cells. *Nano Lett* 6, 662-668.
[21] He, C., Hu, Y., Yin, L., Tang, C., and Yin, C. (2010). Effects of particle size and surface charge on cellular uptake and biodistribution of polymeric nanoparticles. *Biomaterials* 31, 3657-3666.
[22] Steinman, R.M., Silver, J.M., and Cohn, Z.A. (1974). Pinocytosis in fibroblasts. Quantitative studies in vitro. *J Cell Biol* 63, 949-969.
[23] Edelson, P.J., Zwiebel, R., and Cohn, Z.A. (1975). The pinocytic rate of activated macrophages. *J Exp Med* 142, 1150-1164.
[24] Champion, J.A., and Mitragotri, S. (2006). Role of target geometry in phagocytosis. *Proc Natl Acad Sci U S A* 103, 4930-4934.
[25] Doshi, N., and Mitragotri, S. (2010). Macrophages recognize size and shape of their targets. *PLoS One* 5, e10051.
[26] Weiss, L. (1969). The cell periphery. *Int Rev Cytol* 26, 63-105.
[27] Mutsaers, S.E., and Papadimitriou, J.M. (1988). Surface charge of macrophages and their interaction with charged particles. *J Leukoc Biol* 44, 17-26.
[28] Yeung, T., Terebiznik, M., Yu, L., Silvius, J., Abidi, W.M., Philips, M., Levine, T., Kapus, A., and Grinstein, S. (2006). Receptor activation alters inner surface potential during phagocytosis. *Science* 313, 347-351.
[29] Tabata, Y., and Ikada, Y. (1988). Effect of the size and surface charge of polymer microspheres on their phagocytosis by macrophage. *Biomaterials* 9, 356-362.
[30] Ohshima, H. (1994). Electrophoretic Mobility of Soft Particles. *J Colloid Interface Sci* 163, 474-483.
[31] Makino, K., Fukai, F., Hirata, S., and Ohshima, H. (1996). Electro-osmotic studies of endothelial cell surface. *Colloid Surf B: Biointerfaces* 7, 235-238.
[32] Fadok, V.A., Voelker, D.R., Campbell, P.A., Cohen, J.J., Bratton, D.L., and Henson, P.M. (1992). Exposure of phosphatidylserine on the surface of apoptotic lymphocytes triggers specific recognition and removal by macrophages. *J Immunol* 148, 2207-2216.
[33] Makino, K., Tabata, J., Yoshioka, T., Fukuda, M., Ikekita, M., Ohshima, H., and Terada, H. (2003b). Effects of liposomal phosphatidylserine on phagocytic uptake of liposomes by macrophage-like HL-60RG cells. *Colloid Surf B: Biointerfaces* 29, 277-284.
[34] Hirota, K., Hasegawa, T., Nakajima, T., Makino, K., and Terada, H. (2011). Phagostimulatory effect of uptake of PLGA microspheres loaded with rifampicin on alveolar macrophages. *Colloid Surf B: Biointerfaces* 87, 293-298.
[35] Weibel, E.R. (1963). Geometry and dimensions of airways of respiratory zone. In: Morphometry of the human lung. Heidelberg: Springer-Verlag.
[36] Crapo, J.D., Barry, B.E., Gehr, P., Bachofen, M., and Weibel, E.R. (1982). Cell number and cell characteristics of the normal human lung. *Am Rev Respir Dis* 126, 332-337.

[37] Hirota, K., Hasegawa, T., Nakajima, T., Inagawa, H., Kohchi, C., Soma, G., Makino, K., and Terada, H. (2010a). Delivery of rifampicin-PLGA microspheres into alveolar macrophages is promising for treatment of tuberculosis. *J Control Release* 142, 339-346.
[38] Rabinovitch, M. (1995). Professional and non-professional phagocytes: an introduction. *Trends Cell Biol* 5, 85-87.
[39] Swanson, J.A., and Baer, S.C. (1995). Phagocytosis by zippers and triggers. *Trends Cell Biol* 5, 89-93.
[40] Greenberg, S., Chang, P., and Silverstein, S.C. (1993). Tyrosine phosphorylation is required for Fc receptor-mediated phagocytosis in mouse macrophages. *J Exp Med* 177, 529-534.
[41] Cox, D., Tseng, C.C., Bjekic, G., and Greenberg, S. (1999). A requirement for phosphatidylinositol 3-kinase in pseudopod extension. *J Biol Chem* 274, 1240-1247.
[42] Lennartz, M.R., and Brown, E.J. (1991). Arachidonic acid is essential for IgG Fc receptor-mediated phagocytosis by human monocytes. *J Immunol* 147, 621-626.
[43] Debets, J.M., Van Der Linden, C.J., Dieteren, I.E., Leeuwenberg, J.F., and Buurman, W.A. (1988). Fc-receptor cross-linking induces rapid secretion of tumor necrosis factor (cachectin) by human peripheral blood monocytes. *J Immunol* 141, 1197-1201.
[44] Polat, G.L., Laufer, J., Fabian, I., and Passwell, J.H. (1993). Cross-linking of monocyte plasma membrane Fcα, Fcγ or mannose receptors induces TNF production. *Immunology* 80, 287-292.
[45] Krutmann, J., Kirnbauer, R., Kock, A., Schwarz, T., Schopf, E., May, L.T., Sehgal, P.B., and Luger, T.A. (1990). Cross-linking Fc receptors on monocytes triggers IL-6 production. Role in anti-CD3-induced T cell activation. *J Immunol* 145, 1337-1342.
[46] Marsh, C.B., Gadek, J.E., Kindt, G.C., Moore, S.A., and Wewers, M.D. (1995). Monocyte Fcγ receptor cross-linking induces IL-8 production. *J Immunol* 155, 3161-3167.
[47] Marsh, C.B., Wewers, M.D., Tan, L.C., and Rovin, B.H. (1997). Fcγ receptor cross-linking induces peripheral blood mononuclear cell monocyte chemoattractant protein-1 expression: role of lymphocyte FcγRIII. *J Immunol* 158, 1078-1084.
[48] Muroi, M., Muroi, Y., and Suzuki, T. (1994). The binding of immobilized IgG2a to Fcγ2a receptor activates NF-κB via reactive oxygen intermediates and tumor necrosis factor-α[1]. *J Biol Chem* 269, 30561-30568.
[49] Rose, D.M., Winston, B.W., Chan, E.D., Riches, D.W., Gerwins, P., Johnson, G.L., and Henson, P.M. (1997). Fcγ receptor cross-linking activates p42, p38, and JNK/SAPK mitogen-activated protein kinases in murine macrophages: role for $p42^{MAPK}$ in Fcγ receptor-stimulated TNF-α synthesis. *J Immunol* 158, 3433-3438.
[50] Ben-Sasson, S.Z., Le Gros, G., Conrad, D.H., Finkelman, F.D., and Paul, W.E. (1990). Cross-linking Fc receptors stimulate splenic non-B, non-T cells to secrete interleukin 4 and other lymphokines. *Proc Natl Acad Sci U S A* 87, 1421-1425.
[51] Gallo, P., Goncalves, R., and Mosser, D.M. (2010). The influence of IgG density and macrophage Fc (gamma) receptor cross-linking on phagocytosis and IL-10 production. *Immunol Lett* 133, 70-77.
[52] Ehlenberger, A.G., and Nussenzweig, V. (1977). The role of membrane receptors for C3b and C3d in phagocytosis. *J Exp Med* 145, 357-371.

[53] Shi, Y., Tohyama, Y., Kadono, T., He, J., Miah, S.M., Hazama, R., Tanaka, C., Tohyama, K., and Yamamura, H. (2006). Protein-tyrosine kinase Syk is required for pathogen engulfment in complement-mediated phagocytosis. *Blood* 107, 4554-4562.
[54] Marth, T., and Kelsall, B.L. (1997). Regulation of interleukin-12 by complement receptor 3 signaling. *J Exp Med* 185, 1987-1995.
[55] Behrens, E.M., Sriram, U., Shivers, D.K., Gallucci, M., Ma, Z., Finkel, T.H., and Gallucci, S. (2007). Complement receptor 3 ligation of dendritic cells suppresses their stimulatory capacity. *J Immunol* 178, 6268-6279.
[56] Takahashi, K. (2001). Development and differentiation of macrophages and related cells: Historical review and current concepts. *J Clin Exp Hematopathol* 41, 1-33.
[57] Gordon, S. (2002). Pattern recognition receptors: doubling up for the innate immune response. *Cell* 111, 927-930.
[58] Pedra, J.H., Cassel, S.L., and Sutterwala, F.S. (2009). Sensing pathogens and danger signals by the inflammasome. *Curr Opin Immunol* 21, 10-16.
[59] Gough, P.J., and Gordon, S. (2000). The role of scavenger receptors in the innate immune system. *Microbes Infect* 2, 305-311.
[60] Haworth, R., Platt, N., Keshav, S., Hughes, D., Darley, E., Suzuki, H., Kurihara, Y., Kodama, T., and Gordon, S. (1997). The macrophage scavenger receptor type A is expressed by activated macrophages and protects the host against lethal endotoxic shock. *J Exp Med* 186, 1431-1439.
[61] Hamilton, R.F., Jr., Thakur, S.A., and Holian, A. (2008). Silica binding and toxicity in alveolar macrophages. *Free Radic Biol Med* 44, 1246-1258.
[62] Kobayashi, Y., Miyaji, C., Watanabe, H., Umezu, H., Hasegawa, G., Abo, T., Arakawa, M., Kamata, N., Suzuki, H., Kodama, T., *et al.* (2000). Role of macrophage scavenger receptor in endotoxin shock. *J Pathol* 192, 263-272.
[63] Baranova, I.N., Vishnyakova, T.G., Bocharov, A.V., Leelahavanichkul, A., Kurlander, R., Chen, Z., Souza, A.C., Yuen, P.S., Star, R.A., Csako, G., *et al.* (2012). Class B scavenger receptor types I and II and CD36 mediate bacterial recognition and proinflammatory signaling induced by Escherichia coli, lipopolysaccharide, and cytosolic chaperonin 60. *J Immunol* 188, 1371-1380.
[64] Moestrup, S.K., and Moller, H.J. (2004). CD163: a regulated hemoglobin scavenger receptor with a role in the anti-inflammatory response. *Ann Med* 36, 347-354.
[65] Takeda, K., and Akira, S. (2005). Toll-like receptors in innate immunity. *Int Immunol* 17, 1-14.
[66] Judge, A.D., Sood, V., Shaw, J.R., Fang, D., Mcclintock, K., and Maclachlan, I. (2005). Sequence-dependent stimulation of the mammalian innate immune response by synthetic siRNA. *Nat Biotechnol* 23, 457-462.
[67] Hornung, V., Guenthner-Biller, M., Bourquin, C., Ablasser, A., Schlee, M., Uematsu, S., Noronha, A., Manoharan, M., Akira, S., De Fougerolles, A., *et al.* (2005). Sequence-specific potent induction of IFN-alpha by short interfering RNA in plasmacytoid dendritic cells through TLR7. *Nat Med* 11, 263-270.
[68] Steinhagen, F., Kinjo, T., Bode, C., and Klinman, D.M. (2011). TLR-based immune adjuvants. *Vaccine* 29, 3341-3355.

[69] Akita, H., Kogure, K., Moriguchi, R., Nakamura, Y., Higashi, T., Nakamura, T., Serada, S., Fujimoto, M., Naka, T., Futaki, S., *et al.* (2010). Nanoparticles for ex vivo siRNA delivery to dendritic cells for cancer vaccines: programmed endosomal escape and dissociation. *J Control Release* 143, 311-317.
[70] Pittella, F., Zhang, M., Lee, Y., Kim, H.J., Tockary, T., Osada, K., Ishii, T., Miyata, K., Nishiyama, N., and Kataoka, K. (2011). Enhanced endosomal escape of siRNA-incorporating hybrid nanoparticles from calcium phosphate and PEG-block charge-conversional polymer for efficient gene knockdown with negligible cytotoxicity. *Biomaterials* 32, 3106-3114.
[71] Agostini, L., Martinon, F., Burns, K., Mcdermott, M.F., Hawkins, P.N., and Tschopp, J. (2004). NALP3 forms an IL-1beta-processing inflammasome with increased activity in Muckle-Wells autoinflammatory disorder. *Immunity* 20, 319-325.
[72] Cassel, S.L., Joly, S., and Sutterwala, F.S. (2009). The NLRP3 inflammasome: a sensor of immune danger signals. *Semin Immunol* 21, 194-198.
[73] Hornung, V., Bauernfeind, F., Halle, A., Samstad, E.O., Kono, H., Rock, K.L., Fitzgerald, K.A., and Latz, E. (2008). Silica crystals and aluminum salts activate the NALP3 inflammasome through phagosomal destabilization. *Nat Immunol* 9, 847-856.
[74] Halle, A., Hornung, V., Petzold, G.C., Stewart, C.R., Monks, B.G., Reinheckel, T., Fitzgerald, K.A., Latz, E., Moore, K.J., and Golenbock, D.T. (2008). The NALP3 inflammasome is involved in the innate immune response to amyloid-beta. *Nat Immunol* 9, 857-865.
[75] Simons, K., and Ikonen, E. (1997). Functional rafts in cell membranes. *Nature* 387, 569-572.
[76] Simons, K., and Toomre, D. (2000). Lipid rafts and signal transduction. *Nat Rev Mol Cell Biol* 1, 31-39.
[77] Kim, H.Y., Park, S.J., Joe, E.H., and Jou, I. (2006). Raft-mediated Src homology 2 domain-containing protein tyrosine phosphatase 2 (SHP-2) regulation in microglia. *J Biol Chem* 281, 11872-11878.
[78] Hoekstra, D., Maier, O., Van Der Wouden, J.M., Slimane, T.A., and Van, I.S.C. (2003). Membrane dynamics and cell polarity: the role of sphingolipids. *J Lipid Res* 44, 869-877.
[79] Zhang, J., Pekosz, A., and Lamb, R.A. (2000). Influenza virus assembly and lipid raft microdomains: a role for the cytoplasmic tails of the spike glycoproteins. *J Virol* 74, 4634-4644.
[80] Thorne, R.F., Marshall, J.F., Shafren, D.R., Gibson, P.G., Hart, I.R., and Burns, G.F. (2000). The integrins $\alpha_3\beta_1$ and $\alpha_6\beta_1$ physically and functionally associate with CD36 in human melanoma cells - Requirement for the extracellular domain of CD36. *J Biol Chem* 275, 35264-35275.
[81] Rodal, S.K., Skretting, G., Garred, O., Vilhardt, F., Van Deurs, B., and Sandvig, K. (1999). Extraction of cholesterol with methyl-β-cyclodextrin perturbs formation of clathrin-coated endocytic vesicles. *Mol Biol Cell* 10, 961-974.
[82] Nagao, G., Ishii, K., Hirota, K., Makino, K., and Terada, H. (2010). Role of lipid rafts in phagocytic uptake of polystyrene latex microspheres by macrophages. *Anticancer Res.* 30, 3167-3176.
[83] Persaud-Sawin, D.A., Banach, L., and Harry, G.J. (2009). Raft aggregation with specific receptor recruitment is required for microglial phagocytosis of Abeta42. *Glia* 57, 320-335.

[84] Pfeiffer, A., Bottcher, A., Orso, E., Kapinsky, M., Nagy, P., Bodnar, A., Spreitzer, I., Liebisch, G., Drobnik, W., Gempel, K., *et al.* (2001). Lipopolysaccharide and ceramide docking to CD14 provokes ligand-specific receptor clustering in rafts. *Eur J Immunol* 31, 3153-3164.
[85] Kariko, K., Weissman, D., and Welsh, F.A. (2004). Inhibition of toll-like receptor and cytokine signaling--a unifying theme in ischemic tolerance. *J Cereb Blood Flow Metab* 24, 1288-1304.
[86] Dai, Q., and Pruett, S.B. (2006). Ethanol suppresses LPS-induced Toll-like receptor 4 clustering, reorganization of the actin cytoskeleton, and associated TNF-α production. *Alcohol Clin Exp Res* 30, 1436-1444.
[87] Nagao, G., Ishii, K., Hirota, K., Makino, K., and Terada, H. (2011). Role of lipid rafts in innate immunity and phagocytosis of polystyrene latex microspheres. *Colloid Surf B: Biointerfaces* 84, 317-324.
[88] Olsson, S., and Sundler, R. (2006). The role of lipid rafts in LPS-induced signaling in a macrophage cell line. *Mol Immunol* 43, 607-612.
[89] O'hara, P., and Hickey, A.J. (2000). Respirable PLGA microspheres containing rifampicin for the treatment of tuberculosis: manufacture and characterization. *Pharm Res* 17, 955-961.
[90] Sharma, R., Saxena, D., Dwivedi, A.K., and Misra, A. (2001). Inhalable microparticles containing drug combinations to target alveolar macrophages for treatment of pulmonary tuberculosis. *Pharm Res* 18, 1405-1410.
[91] Verma, R.K., Kaur, J., Kumar, K., Yadav, A.B., and Misra, A. (2008). Intracellular time course, pharmacokinetics, and biodistribution of isoniazid and rifabutin following pulmonary delivery of inhalable microparticles to mice. *Antimicrob Agents Chemother* 52, 3195-3201.
[92] Yoshida, A., Matumoto, M., Hshizume, H., Oba, Y., Tomishige, T., Inagawa, H., Kohchi, C., Hino, M., Ito, F., Tomoda, K., *et al.* (2006). Selective delivery of rifampicin incorporated into poly(DL-lactic-co-glycolic) acid microspheres after phagocytotic uptake by alveolar macrophages, and the killing effect against intracellular Mycobacterium bovis Calmette-Guerin. *Microbes Infect* 8, 2484-2491.
[93] Taniguchi, H., Shimada, Y., Sawachi, K., Hirota, K., Inagawa, H., Kohchi, C., Soma, G., Makino, K., and Terada, H. (2010). Lipopolysaccharide-activated alveolar macrophages having cytotoxicity toward lung tumor cells through cell-to-cell binding-dependent mechanism. *Anticancer Res* 30, 3159-3165.
[94] Sawachi, K., Shimada, Y., Taniguchi, H., Hirota, K., Inagawa, H., Kohchi, C., Soma, G., Makino, K., and Terada, H. (2010). Cytotoxic effects of activated alveolar macrophages on lung carcinoma cells via cell-to-cell contact and nitric oxide. *Anticancer Res* 30, 3135-3141.
[95] Hirota, K., Oishi, Y., Taniguchi, H., Sawachi, K., Inagawa, H., Kohchi, C., Soma, G., and Terada, H. (2010b). Antitumor effect of inhalatory lipopolysaccharide and synergetic effect in combination with cyclophosphamide. *Anticancer Res* 30, 3129-3134.
[96] Hagemann, T., Lawrence, T., Mcneish, I., Charles, K.A., Kulbe, H., Thompson, R.G., Robinson, S.C., and Balkwill, F.R. (2008). "Re-educating" tumor-associated macrophages by targeting NF- κB. *J Exp Med* 205, 1261-1268.

Permissions

The contributors of this book come from diverse backgrounds, making this book a truly international effort. This book will bring forth new frontiers with its revolutionizing research information and detailed analysis of the nascent developments around the world.

We would like to thank Brian P. Ceresa, Ph.D., for lending his expertise to make the book truly unique. He has played a crucial role in the development of this book. Without his invaluable contribution this book wouldn't have been possible. He has made vital efforts to compile up to date information on the varied aspects of this subject to make this book a valuable addition to the collection of many professionals and students.

This book was conceptualized with the vision of imparting up-to-date information and advanced data in this field. To ensure the same, a matchless editorial board was set up. Every individual on the board went through rigorous rounds of assessment to prove their worth. After which they invested a large part of their time researching and compiling the most relevant data for our readers. Conferences and sessions were held from time to time between the editorial board and the contributing authors to present the data in the most comprehensible form. The editorial team has worked tirelessly to provide valuable and valid information to help people across the globe.

Every chapter published in this book has been scrutinized by our experts. Their significance has been extensively debated. The topics covered herein carry significant findings which will fuel the growth of the discipline. They may even be implemented as practical applications or may be referred to as a beginning point for another development. Chapters in this book were first published by InTech; hereby published with permission under the Creative Commons Attribution License or equivalent.

The editorial board has been involved in producing this book since its inception. They have spent rigorous hours researching and exploring the diverse topics which have resulted in the successful publishing of this book. They have passed on their knowledge of decades through this book. To expedite this challenging task, the publisher supported the team at every step. A small team of assistant editors was also appointed to further simplify the editing procedure and attain best results for the readers.

Our editorial team has been hand-picked from every corner of the world. Their multi-ethnicity adds dynamic inputs to the discussions which result in innovative outcomes. These outcomes are then further discussed with the researchers and contributors who give their valuable feedback and opinion regarding the same. The feedback is then collaborated with the researches and they are edited in a comprehensive manner to aid the understanding of the subject.

Apart from the editorial board, the designing team has also invested a significant amount of their time in understanding the subject and creating the most relevant covers. They scrutinized every image to scout for the most suitable representation of the subject and create an appropriate cover for the book.

The publishing team has been involved in this book since its early stages. They were actively engaged in every process, be it collecting the data, connecting with the contributors or procuring relevant information. The team has been an ardent support to the editorial, designing and production team. Their endless efforts to recruit the best for this project, has resulted in the accomplishment of this book. They are a veteran in the field of academics and their pool of knowledge is as vast as their experience in printing. Their expertise and guidance has proved useful at every step. Their uncompromising quality standards have made this book an exceptional effort. Their encouragement from time to time has been an inspiration for everyone.

The publisher and the editorial board hope that this book will prove to be a valuable piece of knowledge for researchers, students, practitioners and scholars across the globe.

List of Contributors

Ichirou Karahara
University of Toyama, Japan

L. Andrew Staehelin
University of Colorado, USA

Yoshinobu Mineyuki
University of Hyogo, Japan

Jianjun Sun
Department of Biological Sciences, University of Texas at El Paso, El Paso, TX, USA

Frauke Ackermann, Joshua A. Gregory and Lennart Brodin
Department of Neuroscience, Karolinska Institutet, Stockholm, Sweden

Abbas A. Alibhoy and Hui-Ling Chiang
Department of Cellular and Molecular Physiology,
Penn State University College of Medicine, Hershey, USA

Paola Ramoino
Department of Earth, Environment and Life Sciences (DISTAV), University of Genoa, Genova, Italy

Alberto Diaspro
The Italian Institute of Technology (IIT), Nanophysics Unit, Genova, Italy

Jesse S. Aaron and Jerilyn A. Timlin
Sandia National Laboratories, USA

F. Brent Johnson and Enkhmart Dudleenamjil
Brigham Young University, Provo, Utah, USA

Norica Branza-Nichita, Alina Macovei and Catalin Lazar
Department of Viral Glycoproteins, Institute of Biochemistry Bucharest,
Romanian Academy, Romania

Maude Boisvert and Peter Tijssen
INRS-Institut Armand-Frappier, Institut Pasteur International Network, Université du Québec,
QC, Canada

Thorsten Eierhoff and Winfried Römer
Albert-Ludwigs-University Freiburg,
Institute of Biology II and BIOSS Centre for Biological Signalling Studies, Freiburg, Germany

Bahne Stechmann
Leibniz-Institut für Molekulare Pharmakologie (FMP), Berlin, Germany

Maria C. Touz
Instituto de Investigación Médica Mercedes y Martín Ferreyra, INIMEC - CONICET, National University of Cordoba, Córdoba, Argentina

Elisa Sala, Luca Ruggiero and Giuseppina Di Giacomo
Università Vita-Salute San Raffaele, Milano, Italy

Ottavio Cremona
Università Vita-Salute San Raffaele, Milano, Italy
Istituto Nazionale di Neuroscienze, Italy

Zhixiang Wang
Department of Medical Genetics, Signal Transduction Research Group,
Faculty of Medicine and Dentistry, University of Alberta, Edmonton, AB, Canada

Ronny Racine
Department of Molecular Biology and Immunology, University of North Texas Health Science Center, Fort Worth, TX, USA

Mark E. Mummert
Department of Psychiatry and Behavioral Health, University of North Texas Health Science Center, Fort Worth, TX, USA

Johan A. Slotman, Peter van Kerkhof, Gerco Hassink and Ger J. Strous
Department of Cell Biology, University Medical Center Utrecht and Institute of Biomembranes, Utrecht, The Netherlands

Hendrik J. Kuiken
Netherlands Cancer Institute, Division of Molecular Carcinogenesis, Amsterdam, The Netherlands

Wanling Liang and Jenny K.W. Lam
Department of Pharmacology & Pharmacy, Li Ka Shing Faculty of Medicine, The University of Hong Kong, Hong Kong SAR

Keiji Hirota and Hiroshi Terada
Faculty of Pharmaceutical Sciences, Japan
Center for Drug Delivery Research, Japan
Center for Physical Pharmaceutics, Research Institute for Science and Technology, Tokyo University of Science, Yamazaki, Noda, Chiba, Japan